# ANNUAL REVIEW OF
# NEUROSCIENCE

# ANNUAL REVIEW OF NEUROSCIENCE

## VOLUME 13, 1990

W. MAXWELL COWAN, *Editor*
Howard Hughes Medical Institute

ERIC M. SHOOTER, *Associate Editor*
Stanford University School of Medicine

CHARLES F. STEVENS, *Associate Editor*
Yale University School of Medicine

RICHARD F. THOMPSON, *Associate Editor*
University of Southern California

ANNUAL REVIEWS INC   4139 EL CAMINO WAY   P.O. BOX 10139   PALO ALTO, CALIFORNIA 94303-0897

ANNUAL REVIEWS INC.
Palo Alto, California, USA

*International Standard Serial Number : 0147-006X*
*International Standard Book Number : 0-8243-2413-7*

TYPESET BY AUP TYPESETTERS (GLASGOW) LTD., SCOTLAND
PRINTED AND BOUND IN THE UNITED STATES OF AMERICA

*Annual Review of Neuroscience*
*Volume 13, 1990*

# CONTENTS

# SOME RELATED ARTICLES IN OTHER *ANNUAL REVIEWS*

From the *Annual Review of Biochemistry*, Volume 59 (1990):

How to Succeed in Research Without Being a Genius, O. H. Lowry

Microtubule Motors, R. B. Vallee

Cadherins: A Molecular Family Important in Selective Cell-Cell Adhesion,
M. Takeichi

From the *Annual Review of Cell Biology*, Volume 5 (1989):

Dynein Structure and Function, M. E. Porter

Molecular Genetics of Development Studies in the Transgenic Mouse,
H. Westphal and P. Gruss

From the *Annual Review of Pharmacology and Toxicology*, Volume 30
(1990):

CNS Regulation of Immune Function, R. Ader, D. Felton, and N. Cohen

Excitatory Amino Acids and Neuropsychiatric Disorders, J. W. Olney

Subtypes of Receptors for Serotonin, A. Frazer, B. Wolfe, and S. Maayani

Effects of Opiates on Neuroendocrine Function and Immune Response,
M. J. Kreek

From the *Annual Review of Physiology*, Volume 52 (1990):

Ge: A GTP-Binding Protein Mediating Exocytosis, B. D. Gomperts

Pathways to Regulated Exocytosis in Neurons, P. De Camilli and R. Jahn

Secretory Vesicle-Associated Proteins and Their Role in Exocytosis,
R. D. Burgoyne

Caged Compounds and Striated Muscle Contraction, E. Homsher and
N. C. Millar

Flash Photolysis Studies of Excitation-Contraction Coupling Regulation, and
Contraction in Smooth Muscle, A. P. Somlyo and A. V. Somlyo

Regulation of Cardiac L-Type Calcium Current by Phosphorylation and
G-Proteins, W. Trautwein

G Protein Modulation of Calcium Currents in Neurons, A. Dolphin

G Protein Mediated Regulation of $K^+$ Channels in Heart, G. Szabo and
A. S. Otero

*(Continued)*   vii

Horace Barlow

*Annu. Rev. Neurosci. 1990. 13:1–13*

# CHEMOSENSORY PHYSIOLOGY IN AN AGE OF TRANSITION

*Vincent G. Dethier*

Department of Zoology, University of Massachusetts, Amherst, Massachusetts 01003

In a recent published book entitled *Masks of the Universe*, the cosmologist Edward Harrison reminds us that we as a society create the universe in which we live and that each age fashions its own unique universe. Each of these universes is a mask of The Universe. Similarly, scientific verities are the creations of the society of contemporary scientists who believe as firmly as do cosmologists that their tenets represent reality. This reality changes with time. Each age tends to be but marginally charitable toward the "ignorance" of its predecessors and not especially sensitive to the possibility that, as before, some of today's truths may become tomorrow's heresies. This is not to deny that shards of truth are salvagable from each age to the next; it is to remind us that comprehension at any time is profoundly influenced by contemporary intellectual ambience.

Although historians of science can view from the mountain top of time the totality and comprehensiveness of the scientific endeavor and can discern the slow grandeur of its progress from one period to the next, the proximate eddies and currents that perturb the main stream are perhaps most acutely perceived by those who have lived through transitional periods. During a time of transition one is more deeply aware of the tenuous hold that science has on truth at any one moment, of the fragility as well as power of hypothesis and theory, of the influential role of improbabilities, and of the paradoxical relationship between concept and technology. The last point is particularly relevant in this time of high technology. Although new inventions and developments indubitably stimulate new ideas as well as provide means for solving old questions, the generation of ideas is not inextricably constrained by new technology. At most, realizations may be delayed. As Beidler (1987) has pointed out,

1

technology cannot be substituted for keen insight, philosophical under-standing, and serendipity.

In some small measure the course of biology exemplifies the flux of conceptual cosmologies in general. Biology experienced a rapid transition from one age to another during the first three quarters of the present century. Classical biology was on the wane, molecular biology in the ascendancy, especially in the 1950s. In the 1960s, a new cohesive discipline began to emerge, neuroscience. I consider myself fortunate to have studied and practiced science during this period. It was a time of excitement, expectation, questioning, and discovery. It provided challenges, frus-trations, and expanding vistas of terra incognita.

In presenting this prefatory chapter I offer an account of my area of special interest, the chemical senses, as a vehicle to illustrate some general aspects of a period of transition, to record the importance of time, place, and informal eclectic personal interaction, to describe how ideas may develop, and to note their independence as well as their dependence on technology. Most especially I wish to present the thesis that cherishing a clear goal is essential but that at the same time one profits from peripheral vision. The "surround," to borrow a term from visual psychophysics, modulates and enriches the pursuit of a goal and ultimately places it in its clearest perspective. This essay, then, is not a review of the field. That has been done many times, the most recent being that of Finger & Silver (1987). It is a journey of inquiry through a period when electrophysiology was just coming of age and neuroscience had not yet emerged as a recog-nized discipline.

My early education in science began toward the end of the era of classical biology before any adumbrations of transition were apparent to the bio-logical community as a whole. The starting point for me was a fascination with nature awakened by the beauty and vibrancy of living things, by sensual delight in colors, scents, sounds, and the cycling of the seasons, all enhanced by the inability of adults to answer satisfactorily the childhood "whys." Later, under the stern tutelage of formal science, the "whys" would become "whats" and "hows."

The wonderment that sustained my interest in science throughout child-hood and adolescence was almost obliterated during undergraduate days at Harvard. There, young aspiring biologists absorbed heavy doses of anatomy and taxonomy, preponderantly, of course, by examining pickled, mummified, and skeletonized specimens—or fossils thereof. At the very least, however, one learned the names of what was out there in the plant and animal world and approximately what each looked like externally and internally. Function, on the other hand, remained a mystery and was not related to behavior in any but the most obvious ways. Physiology, particularly the machinery of the human body, flourished in the schools of medicine. Experimental zoology in this country and abroad was not held in high esteem in the first third of the century.

A nodding acquaintance with physiology was acquired en route through two physiology courses. One course offered to our appetites was served up by Cannon, from whom I learned the "wisdom of the body" (1939). He also kept us informed about the controversy between the "electrical" people and the "purveyors of soup" over the nature of transmission at the neuromuscular junction. New discoveries about the role of acetylcholine, esserine, and ATP were only beginning to trickle into the curriculum. The pioneering work of Adrian on electrophysiology in the 1920s had not yet attracted wide attention (Adrian & Forbes 1922, Adrian 1926).

None of this collegial enlightenment satisfied an eccentric interest that I had in the behavior of insects. The closest approach to animal behavior was provided by Welsh's course in invertebrate (mostly marine) physiology. It did not offer any help toward solving questions in the area of my special interest, the obsessive gourmet habits of herbivorous insects. The only explanation of this gastronomic phenomenon extant postulated the existence of a "botanical" sense, a mysterious sixth sense. By analogy with my own eating habits, I suspected that chemical senses were involved. Although we students were constantly warned against the evil of anthropomorphism, I learned over the years that there was considerable heuristic value in posing questions from this point of view.

At Harvard in the 1930s there were two specialists in the field of chemoreception, Parker and Crozier. Neither was currently active in that field, having turned to other matters, but each had written reviews (Parker 1922, Crozier 1934) that summed up knowledge about these least understood of all sensory systems. At this time, knowledge of the chemical senses was restricted to anatomy and histology as constrained by limits of the light microscope and gross anatomy. Only gold and silver impregnation and methylene blue staining were available for revealing the tracery of the nervous system. Generally speaking, investigators were focusing their attention on gross anatomy, histology, thresholds, classification of tastes and odors, relations of sensation to chemical structure, theories of action, psychophysics, flavor, and perfumery. The emphasis throughout lay on vertebrates.

Insofar as insects were concerned, zoologists occupied themselves in free speculation. Despite a great legacy of elegant nineteenth century histology, there was no unanimity of opinion regarding the loci of chemosensory organs, no consensus as to whether there were separate and distinct olfactory and gustatory senses, and no knowledge concerning the identity of the end-organs themselves.

If the problem of insect monophagy was to be solved, the identity and response characteristics of the chemoreceptors had to be revealed. The first and only conceivable approach was to combine ablation, anesthesia, and occlusion of putative end-organs with behavioral observations. By means of these techniques tentative identification of some olfactory organs, among them those of caterpillars, was made during the period 1921–1941 (von Frisch 1921, Dethier 1937, 1941). A new approach was needed.

Although I had acquired a vicarious acquaintance with the pioneering technique of electrophysiology of the 1920s by reading Adrian's *Basis of Sensation* (1928), I had neither the knowledge nor the equipment to apply this new approach to my problem. The instruments that existed were the Einthoven string galvanometer, the Lippmann capillary electrometer, and the Matthews oscillograph. They were available in a few laboratories only, notably Adrian's in Cambridge, Forbes' at Harvard, and Erlanger's and Gasser's at Washington University. Most investigators using this equipment were concerned with trying to understand the nature of the action potential (Gasser & Newcomer 1921, Gasser & Erlanger 1922). Motor nerves and the sensory nerves of mechanoreceptors were the preparations best suited for this research. Only a few attempts to record from chemosensory nerves were made in the early 1930s. Hoagland (1933) detected but could not resolve electrical activity in nerves of the lips and barbels of catfish in response to salt and acid; Zotterman (1935) recorded impulses in the chorda tympani and glossopharyngeal of the rat; Pumphrey (1935) recorded responses to salt and acid placed on the tongue of the frog; Adrian & Ludwig (1938) recorded activity in the olfactory nerves of catfish and carp.

At this time, through chance encounters, I made the acquaintance of Roeder at Tufts College and Prosser at Clark University. Roeder had arrived at Tufts in 1932 and Prosser at Clark in 1934. Roeder's entrée to electrophysiology came as a consequence of his taking a course at Woods Hole, where Prosser was instructing. Prosser was investigating electrical activity in the nerve cord of earthworms and marine invertebrates by means of a Matthews oscillograph and amplifier. Roeder applied this technique together with ablation to studies of copulatory behavior in praying mantids.

At Prosser's invitation I spent part of one summer at Clark, where we attempted to record from chemosensory nerves of caterpillars. Activity from mechanoreceptors was detected, but only physiological silence and instrumental noise followed chemical stimulation. Attempts continued intermittently and unsuccessfully until the advent of World War II.

In the meantime Roeder, a skilled and ingenious tinkerer, was perfecting his instrumentation and techniques for recording from the central nerve cord of the cockroach. Shortly after the war, he and I made another attempt to record from chemoreceptors. Having observed that the very long ovipositors of some parasitic wasps were sensitive to chemical stimulation, I thought that the correspondingly long sensory nerve might lend itself to recording. Again we failed to detect any action potentials.

At this point behavior seemed to offer a more promising approach to the problems of chemoreception. The turning point came as the result of two lucky decisions, namely, that the caterpillar was not the most cooperative animal and that gustation was more tractable than olfaction. The choice of the blowfly was serendipitous. In 1922 Minnich had observed that flies and butterflies extended their proboscises when the legs were touched with sugar. Four years later he reported in one short sentence that touching a single long curved hair on the proboscis of the blowfly with sugar elicited extension (Minnich 1926). Here was a preparation where single chemoreceptors could be isolated and gustation studied behaviorally without the usual confounding postingestional effects.

Early in 1953, a student, Grabowski, and I succeeded by microtopical application of sucrose to single hairs in demonstrating unequivocally that the gustatory end-organs of the tarsi were hairs (setae) similar in appearance to those on the labellum, which Minnich had identified as gustatory (Grabowski & Dethier 1954). We were able to prove that the tips were not covered with cuticle, that the hairs were innervated by three bipolar sensory neurons (later examination with an electron microscope revealed two additional neurons), and that each of the cells responded to a different sensory modality. It was immediately apparent that this preparation could serve as an excellent model of a gustatory apparatus because, unlike the vertebrate taste papillae, the receptors were primary neurons, the axons of which led directly without synapsing into the head ganglia.

During the next ten years it was possible with this preparation to examine the response spectra of the chemosensory cells, detail, separate, and measure peripheral and central adaptation, measure temporal and spatial summation, evaluate differential thresholds ($\Delta I/I$), relate responsiveness to the structural configuration of stimulating molecules, propose a provisional functional map of gustatory projections in the central nervous system, demonstrate and evaluate in an intact organism central excitatory and inhibitory states as Sherrington had done with the flexion reflex in spinal cats, advance promising hypotheses regarding the nature of molecular receptor sites and transduction, and lay a groundwork for elucidating mechanisms underlying hunger and satiation. All of this work was eventually summarized in the book, *The Hungry Fly* (Dethier 1976).

The goal throughout this period was to understand the neural mechanisms mediating chemosensory responses and related behavior. At the same time, it was the behavior that was providing insight to the mechanisms. The behavioral approach was an exciting game of wits. It proved to be a powerful tactic, and many of its findings were subsequently shown by electrophysiology to be gratifyingly accurate.

Through the period of the late 1940s to 1958 there had been no breakthrough in the impasse to successful electrophysiological recording from insect chemoreceptors. Investigation of vertebrate systems had been making considerable progress, as exemplified by Pfaffmann's beautiful recordings of single chorda tympani fiber responses in the cat. At the Johns Hopkins University, to which I had moved after the war, Bronk, Hartline and others of the group were fully engaged in electrophysiological work; however, the only experimentation in chemoreception was being conducted by Beidler (then a graduate student). He was studying the integrated response of multiple chorda tympani fibers of the rat. Most of the studies with vertebrates were concerned with the neural basis of the four classical taste modalities.

In Europe there was progress in the field of olfaction, beginning with Adrian & Ludwig's (1938) recording of electrical activity in the olfactory nerves of catfish and carp and Adrian's (1942) recording of activity in the olfactory projections in the brain of the hedgehog. In 1953, Boistel and Coraboeuf succeeded in recording mixed neural responses in insect antennae. It was Schneider in Germany, however, who finally made a breakthrough (1955). Realizing in 1952–1953 that in the silkworm moth's response to specific pheromones he had the perfect experimental animal and the perfect olfactory stimulus, he exploited that system fully and in 1955 recorded the first single unit olfactory responses to pheromones. This accomplishment paved the way for his own elegant work and that of Kaissling, Boeckh, and others (Boeckh, Kaissling & Schneider 1965). The insect gustatory system remained intransigent, but behavioral analysis continued.

In the course of measuring behavioral thresholds, a number of workers observed that sensitivity decreased as the duration of deprivation increased. The question of whether the sensitivity of the receptors themselves changed or some postingestional factors influenced response steered our research in a direction that was also being investigated in vertebrates. Physiologists and physiological psychologists were probing the central nervous system for answers. Hetherington & Ranson (1942) had discovered that hypothalamic lesions in the brain of the rat affected feeding. Anand & Brobeck (1951a,b) at Yale and Teitlebaum & Stellar (1954) at Hopkins were also lesioning the brain. At that time it was believed that the lateral hypothalamus was a hunger or feeding center and the ventromedial hypothalamus a satiety center. In time the mechanisms were discovered to be more complex. Eventually, in the years to follow, investigators of the vertebrate system gradually worked their way toward the periphery, while those of us studying the blowfly were working from the periphery inwards.

Still employing threshold as a monitor, my associates and I began to isolate by surgery, ligation, and intubation various parts of the digestive system as possible origins of signals modulating the response to chemosensory stimulation. Complementary tests involved injection of the haemocole and also parabiosis. One tremendous advantage of the fly over the rat was the possibility of removing the entire digestive tract together with the oral gustatory receptors and a bit of brain and studying ingestion in vitro. Thus, by a process of elimination we determined that one mechanism regulating behavioral threshold and therefore ingestion was resident in the stomatogastric system, the analogue of the vertebrate autonomic system.

At this point in the investigation I discussed the problem with Dietrich Bodenstein, who was then a civilian employee at the Army Chemical Center in Edgewood, Maryland. Bodenstein was a developmental biologist who worked both on amphibians and cockroaches. He was one of the pioneers in insect developmental biology and a superb microsurgeon. When I mentioned to him my plan to section the recurrent nerve, he assured me that the operation was simplicity itself and proceeded to demonstrate by operating successfully on some *Drosophila*. Following his technique I sectioned the nerve in blowflies; the result was spectacular, a rapid and extreme hyperphagia. The whole sequence of normal feeding then became explicable in terms of interaction between chemosensory excitation and central inhibition triggered by internal mechanoreceptors (Dethier & Bodenstein 1958). This work, together with elaboration in the 1960s by Gelperin (1966a,b, 1967) and more recent refinements in our laboratory, gave us one of the most complete pictures of neural mechanisms of feeding up to that time.

In 1957 I was able to return to the gustatory receptor itself because of a remarkable innovation in electrophysiological recording. Hodgson, who had left our laboratory to continue postdoctoral studies with Roeder, developed, together with Lettvin and Roeder (1955), a technique that solved the problem of recording from single chemosensory sensilla. By one of those odd coincidences of science, Morita and his associates at Kyushu University made the same discovery in 1957. A glass micropipette containing weak saline plus the stimulating compound to be tested was placed over the tip of the hair and served as a salt bridge to a Ag/AgCl wire. Shortly thereafter Morita (1959) developed an elegant refinement whereby he inserted an electrode through a hole drilled in the shaft of the hair.

At Hopkins, Wolbarsht, Evans, and I immediately applied these new techniques to studies of the sensitivity and action spectra of individual chemosensory cells. The early behavioral conclusion that there was a sugar-sensitive cell, a salt-sensitive cell, and a mechanoreceptor was confirmed. In addition, the existence of two other cells, a water receptor (Wolbarsht 1957) and a second but distinctive salt receptor, was revealed. The first electron microscopic studies by Larsen (1962) in our laboratory and Adams et al (1965) at Rutgers University confirmed the existence of these cells, thus correcting the early methylene blue evidence.

These technical developments now made possible pursuit of two problems that had constituted the crux of chemosensory physiology from the beginning, namely transduction and coding. A chance observation by Barnhard & Chadwick at the Army Chemical Center in 1953 had put us on the track. They had observed that bait frequented by flies was more attractive than bait that had been protected from visiting flies. The obvious reason seemed to be that regurgitation and defecation altered the material; however, flies with both proboscis and anus plugged also enhanced the attractiveness. On a hunch I eluted the legs of flies with water, added sucrose to the eluate, and tested for enzyme activity. The eluate contained an alpha-glucosidase. The idea that this enzyme might initiate the process of transduction was tempting; however, since it was already known that some sugars without alpha-glucosidase linkages and some pentoses and hexoses as well were adequate stimuli, we did not pursue the matter further. Hansen in Germany and Morita, Kijima, Koizumi, and their associates in Japan picked up the trial (Hansen & Kuhner 1972, Kijima et al 1973). They proved that the enzymes were intimately and exclusively associated with the receptors. Exactly what part the glucosidases play in the process of transduction is still a mystery.

The relation between the sensitivity of the sugar receptor and the configuration of carbohydrate molecules was also unknown. Neither von Frisch (1935) nor I (1955) had been able to make any sense of the comparative stimulatory effectiveness of the various sugars. The first clue appeared in 1955, when behavioral studies of mixtures revealed that some sugars synergized and others inhibited each other. Evans (1963) suggested that the sugar receptor cell contained multiple molecular sites, specifically one for pyranose and one for furanose sugars [the idea of different multiple sites had been proposed earlier by Biedler for the rat salt receptor (1957, 1962)]. Six years later electrophysiological support for this hypothesis was provided (Omand & Dethier 1969). The matter was finally settled by Morita & Shiraishi (1968) and Shimada et al (1974) by further electrophysiological and pharmacological studies. As the matter now stands, the sugar receptor of the blowfly is presumed to have four specific molecular sites.

Comparable studies of the nature of the salt receptor were stimulated by Beidler's work at Hopkins with the rat chorda tympani. His equation based on the Law of Mass Action was found by Evans & Mellon (1962) to apply to the salt receptor of the fly.

One remaining series of studies attempting to relate stimulating effectiveness and chemical structure is worth mentioning. Over the years, beginning in 1947, more than 200 aliphatic compounds were tested and found to cause flies to reject sugar solutions. It was believed that these compounds stimulated a receptor mediating rejection. The "rejection" threshold could be predicted with great accuracy from the structural formulae. Discussions with Brink at Hopkins prompted me to apply to these data the same thermodynamic analyses that he and Posternak had made of narcotics (Brink & Posternak 1948). It was not until later (1965) in our laboratory at the University of Pennsylvania that Hanson showed electrophysiologically that these aliphatic compounds, rather than stimulating a "rejection" receptor, were narcotizing the receptors mediating acceptance. Steinhardt et al (1966) came to a similar conclusion.

Despite these forays into the nature of transduction, the bearing of gustation on feeding continued to be a central theme in our work. Richter, at the peak of his studies of food preference by the rat and the relation between preference and nutritional need, often visited our laboratory and observed a "two-bottle" preference apparatus that a student (Rhoades) and I had designed in emulation of his apparatus for rats (Dethier & Rhoades 1954). He encouraged us to investigate long-term ingestion and preference as they related to nutrition. Seven years earlier at the Army Chemical Center, a comprehensive survey of the nutritional adequacy of various carbohydrates had been undertaken (Hassett, Dethier & Gans 1950), so the stage was set for a study of preference. The final results indicated that insofar as sugars were concerned, taste preference was not an infallible guide to nutritional value.

After eleven stimulating and productive years at Hopkins, where I learned electrophysiology, electron microscopy, and physiological psychology, I moved to the University of Pennsylvania, where there was a strong multidisciplinary group studying many and varied aspects of feeding behavior at the Institute of Neurological Science. There the lines of work already described were continued and expanded. Another set of gustatory receptors was discovered in the oral cavity of flies; the sensory basis of water, alcohol, and protein ingestion was investigated, and the existence and characteristics of central excitatory and inhibitory states initiated by chemosensory input were established.

Advances in our knowledge of the chemoreceptors of the fly and the evolution of modern techniques prompted a return to the investigations on caterpillars begun 30 years earlier. Caterpillar gustatory receptors were discovered to be more complicated and versatile than those of the blowfly. They also appeared to be less specific, more broadly tuned. Schneider had classified the olfactory receptors of the silkworm moth as "specialists" and "generalists." The receptors of caterpillars fell somewhere between these two extremes of a response spectrum. Extensive recording of responses to plant saps led to the hypothesis that differential preferences were mediated by patterns of impulses from multiple receptors. Explanations previous to this, molded by Verschaffelt's (1910) discovery that special compounds triggered feeding and by the prevailing ethological concept of "sign" stimuli, emphasized the specificity of receptors. A theory stressing the importance of multineuronal afferent patterns (across-fiber patterns) proposed by Pfaffmann (1941) and eloborated by him and by Erickson (1963) seemed to fit the case of caterpillars. This work continued for many years at Pennsylvania and then at Princeton. Eventually some progress was made in decoding chemosensory messages with the help of information theory and analysis of spike-interval distributions by autocorrelograms (Dethier & Crnjar 1982).

The investigative journey in this selected field of inquiry has come full circle in the period spanning the very early development of neurophysiology in the first third of the century to the sophisticated armamentarium of neuroscience of the present. The character of the pursuit of knowledge in the field of chemoreception and related behavior reflects the general nature and evolution of sensory physiological investigation in this period of transition. The enlistment of ingenuity and indirect approaches that characterized general physiology in the 1920s when direct approaches were technically impossible carried sensory physiology forward. Awareness of progress in apparently unrelated fields facilitated advance toward focused goals. A contrapuntal relation gradually developed between ideation and technology.

In the broad search for and understanding of behavior, the investigation had advanced contripetally from sense organs to central phenomena, from a proximate goal of understanding the mechanics of stimulation at the receptor level to decoding sensory spike trains. Furthermore, it moved from information transmission to the meaning of all this for behavior. I do not presume to imply that all the questions addressed had been answered. It is clear, however, that what began as a specialized, one might even say parochial, interest, expanded along the way to contribute some measure of insight to broader issues relating neural machinery to behavior.

At the beginning of this prefatory chapter I referred to Harrison's *Masks of the Universe*. In introducing universes that are impermanent cosmic belief-systems of societies, he alluded also to private world pictures. Among the data from which these private worlds are constructed are sensory data.

Though our perception of the world changes with our intellectual and introspective knowledge, it still is limited by our senses. Considering that advanced technologies enormously extend our senses, one might well wonder whether biologically evolved sensory systems any longer faithfully serve our needs. To what extent does the world as perceived directly by sense organs have reality and validity? The molecular and time/space world, which we are convinced exists, is obviously different from the biological world as perceived.

The perceived world of infrahuman animals is certainly a reality in that, insofar as they survive and evolve successfully in the physical world, they perceive it correctly and so attest its reality. We, on the other hand, can generate our own perceptions independently of what our sense organs tell us. When we know that there is a Chernobyl, we act as though we can indeed sense it (even without the benefit of aversion-learning).

Paramount though the mind/brain problem may be for understanding behavior, the brain is limited in its perfection by sensory input, even though it can stimulate itself by introspection. Studies on sensory deprivation prove the need for sensory input (Zubek 1969). Thus, sensory physiology contributes in a major way to human understanding. Sense organs provide our only *direct* contact with the universe. Granit (1955) touched the heart of the matter when he wrote that sensory physiology is "a branch of natural science which is actually capable of giving some meaning to 'meaning'."

*Literature Cited*

Adams, J. R., Holbert, P. E., Forgash, A. J. 1965. Electronmicroscopy of the contact chemoreceptors of the stable fly, *Stomoxys, calcitrans* (Diptera: Muscidae). *Ann. Ent. Soc. Amer.* 58: 909–17

Adrian, E. D. 1926. The impulses produced by sensory nerve endings. Part I. *J. Physiol.* 61: 49–72

Adrian, E. D. 1928. *The Basis of Sensation: The Action of the Sense Organs.* London: Christophers

Adrian, E. D. 1942. Olfactory reactions in the brain of the hedgehog. *J. Physiol.* 100: 459–73

Adrian, E. D., Forbes, A. 1922. The all-or-nothing response of sensory nerve fibers. *J. Physiol.* 56: 301–30

Adrian, E. D., Ludwig, C. 1938. Nervous discharges from the olfactory organs of fish. *J. Physiol.* 94: 441–60

Adrian, E. D., Zotterman, Y. 1926. The impulses produced by sensory nerve endings. Part 2. The response of a single end-organ. *J. Physiol.* 61: 151–71

Anand, B. K., Brobeck, J. R. 1951a. Hypothalamic control of food intake in rats and cats. *Yale J. Biol. Med.* 24: 123–40

Anand, B. K., Brobeck, J. R. 1951b. Localization of a "feeding center" in the hypothalamus of the rat. *Proc. Soc. Exp. Biol. NY* 1951: 323.

Barnhard, C. S., Chadwick, L. E. 1953. A "fly factor" in intractant studies. *Science* 117: 104–5

Beidler, L. M. 1954. A theory of taste stimulation. *J. Gen. Physiol.* 38: 133–39

Biedler, L. M. 1957. Physiological basis of taste psychophysics. *Fed. Proc.* 16: 9

Beidler, L. M. 1962. Taste receptor stimulation. In *Progress in Biophysics and Biophysical Chemistry*, ed. J. A. V. Butler, H. E. Huxley, R. E. Zirkle, 12: 107–51. Oxford: Pergamon

Beidler, L. M. 1987. Research directions in the chemical senses. In *Neurobiology of Taste and Smell*, ed. T. E. Finger, W. L. Silver, pp. 423–37. New York: Wiley

Boeckh, J., Kaissling, K. E., Schneider, D.

1965. Insect olfactory receptors. *Cold Spring Harbor Symp. Quant. Biol.* 30: 263–80

Boistel, J., Coraboeuf, E. 1953. L'activité électrique dans l'antenne isolée de Lepidoptère au cours de l'étude de l'olfaction. *CR Soc. Biol. Paris* 147: 1172–75

Brink, F., Posternak, J. M. 1948. Thermodynamic analyses of the relative effectiveness of narcotics. *J. Cell Comp. Physiol.* 32: 211–33

Cannon, W. B. 1939. *The Wisdom of the Body.* New York: Norton

Crozier, W. J. 1934. Chemoreception. In *A Handbook of Experimental Psychology*, ed. C. Murchison, pp. 987–1036. Worcester, Mass.: Clark Univ. Press

Dethier, V. G. 1937. Gustation and olfaction in lepidopterous larvae. *Biol. Bull.* 72: 7–23

Dethier, V. G. 1941. The function of the antennal receptors in lepidopterous larvae. *Biol. Bull.* 80: 403–14

Dethier, V. G. 1955. The physiology and histology of the contact chemoreceptors of the blowfly. *Q. Rev. biol.* 30: 348–71

Dethier, V. G. 1976. *The Hungry Fly.* Cambridge: Harvard Univ. Press

Dethier, V. G., Bodenstein, D. 1958. Hunger in the blowfly. *Z. Tierpsychol.* 15: 129–40

Dethier, V. G., Crnjar, R. M. 1982. Candidate codes in the gustatory system of caterpillars. *J. Gen. Physiol.* 79: 549–69

Dethier, V. G., Rhoades, M. V. 1954. Sugar preference—aversion functions for the blowfly. *J. Exp. Zool.* 126: 177–204

Erickson, R. P. 1963. Sensory neural patterns and gustation. In *Olfaction and Taste*, ed. Y. Zotterman, 1: 205–13. Oxford: Pergamon

Evans, D. R. 1963. Chemical structure and stimulation by carbohydrates. In *Olfaction and Taste*, ed. Y. Zotterman, 1: 165–92. Oxford: Pergamon

Evans, D. R., Mellon, DeF. 1962. Stimulation of a primary taste receptor by salts. *J. Gen. Physiol.* 4: 651–61

Finger, T. E., Silver, W. L. 1987. *Neurobiology of Taste and Small.* New York: Wiley

Gasser, H. A., Erlanger, J. 1922. A study of the action currents of nerve with the cathode ray oscillograph. *Am. J. Physiol.* 62: 496–524

Gasser, H. S., Newcomer, H. S. 1921. Physiological action currents in the phrenic nerve. An application of the thermionic vacuum tube to nerve physiology. *Am. J. Physiol.* 57: 1–26

Gelperin, A. 1966a. Control of crop emptying in the blowfly. *J. Insect Physiol.* 12: 331–45

Gelperin, A. 1966b. Investigation of a foregut receptor essential to taste threshold regulation in the blowfly. *J. Insect Physiol.* 12: 829–41

Gelperin, A. 1967. Stretch receptors in the foregut of the blowfly. *Science* 157: 208–10

Grabowski, C. T., Dethier, V. G. 1954. The structure of the tarsal chemoreceptors of the blowfly, *Phormia regina* Meigen. *J. Morph.* 94: 1–17

Granit, R. 1955. *Receptors and Sensory Perception.* New Haven: Yale Univ. Press

Hansen, K. 1968. *Untersuchungen über den Mechanismus der Zucker-Perzeption bei Fliegen.* Habilitationschrift der Universität Heidelberg

Hansen, K., Kuhner, J. 1972. Properties of a possible receptor protein of the fly's sugar receptor. In *Olfaction and Taste IV*, ed. D. Schneider, pp. 350–56. Stuttgart: Wissenshaftliche Verlagsgesellschaft MBM

Hanson, F. E. 1965. *Electrophysiological studies on chemoreceptors of the blowfly*, Phormia regina *Meigen.* Phd dissertation, Univ. Penna., Philadelphia

Harrison, E. 1985. *Masks of the Universe.* New York: Macmillan

Hassett, C. C., Dethier, V. G., Gans, J. 1950. A comparison of nutritive value and taste thresholds of carbohydrate for the blowfly. *Biol. Bull.* 99: 446–53

Hetherington, A. W., Ranson, S. W. 1942. The spontaneous activity and food intake of rats with hypothalamic lesions. *Am. J. Physiol.* 136: 609–17

Hoagland, H. 1933. Specific nerve impulses from gustatory and tactile receptors in catfish. *J. Gen. Physiol.* 16: 685–714

Hodgson, E. S., Lettvin, J. Y., Roeder, K. D. 1955. Physiology of a primary chemoreceptor unit. *Science* 122: 417–18

Kijima, H., Koizumi, O., Morita, H. 1973. α-Glucosidase at the tip of the contact chemosensory seta of the blowfly, *Phormia regina. J. Insect Physiol.* 19: 1351–62

Larsen, J. R. 1962. The fine structure of the labellar chemosensory hairs of the blowfly, *Phormia regina* Meigen. *J. Insect Physiol.* 8: 683–91

Minnich, D. E. 1922. The chemical sensitivity of the tarsi of the red admiral butterfly, *Pyrameis atalanta* Linn. *J. Exp. Zool.* 35: 57–81

Minnich, D. E. 1926. The organs of taste on the proboscis of the blowfly, *Phormia regina* Meigen. *Anat. Rec.* 34: 126

Morita, H. 1959. Initiation of spike potentials in contact chemosensory hairs of insects. III. D.C. stimulation and generator potential of labellar chemoreceptor of *Calliphora. J. Cell. Comp. Physiol.* 54: 182–204

Morita, H., Doira, S., Takeda, K., Kuwabara, M. 1957. Electrical responses of contact chemoreceptors on tarsus of the butterfly, *Vanessa indica. Mem. Fac. Sci. Kyushu Univ.* E2: 119–39

Morita, H., Shiraishi, A. 1968. Stimulation of the labellar sugar receptor of the fleshfly by mono and disaccharides. *J. Gen. Physiol.* 52: 559–83

Omand, E., Dethier, V. G. 1969. An electrophysiological analysis of the action of carbohydrates on the sugar receptor of the blowfly. *Proc. Natl. Acad. Sci. USA* 62: 136–43

Parker, G. H. 1922. *Smell, Taste, and Allied Senses in the Vertebrates.* Philadelphia: Lippincott

Pfaffmann, C. 1941. Gustatory afferent impulses. *J. Cell. Comp. Physiol.* 17: 243–58

Pumphrey, J. 1935. Nerve impulses from receptors in the mouth of the frog. *J. Cell. Comp. Physiol.* 6: 457–67

Richter, C. P. 1942. Total self regulatory functions in animals and human beings. *Harvey Lect. Ser.* 38: 63–103

Schneider, D. 1955. Mikro-Electroden registrieren die electrischen Impulse einzelner Sinnesnervenzellen der Schmetterlingsantenne. *Ind. Electron. Forsch. Ferfigung* 3(3/4): 3–7

Shimada, I., Shiraishi, A., Kijima, H.,

Morita, H. 1974. Separation of two receptor sites in a single labellar sugar receptor of the flesh-fly by treatment with *p*-chloromercuribenzoate. *J. Insect Physiol.* 20: 605–21

Steinhardt, R. A., Morita, H., Hodgson, E. S. 1966. Mode of action of straight chain hydrocarbons on primary chemoreceptors of the blowfly. *Phormia regina. J. Cell. Physiol.* 67: 53–62

Teitlebaum, P., Stellar, E. 1954. Recovery from failure to eat produced by hypothalamic lesions. *Science* 120: 894–95

Verschaffelt, E. 1910. The cause determining the selection of food in some herbivorous insects. *Proc. R. Acad. Amsterdam* 13: 536–42

von Frisch, K. 1921. Über den Sitz des Geruchsinnes bei Insekten. *Zool. Jahrb. Abt. Zool. Physiol.* 38: 449–516

von Frisch, K. 1935. Über den Geschmackssinn der Biene. *Z. Physiol.* 21: 1–156

Wolbarsht, M. L. 1957. Water taste in *Phormia. Science* 125: 1248

Zotterman, Y. 1935. Action potentials from the chorda tympani. *Skand. Arch. Physiol.* 72: 73–77

Zubek, J. P. 1969. *Sensory Deprivation: Fifteen Years of Research.* New York: Appelton-Century-Crofts

*Annu. Rev. Neurosci. 1990. 13:15–24*

# THE MECHANICAL MIND

*Horace Barlow*

Physiological Laboratory, Cambridge, CB2 3EG, England

Most neuroscientists accept the machine as a useful metaphor or model of the mind. It points our research in a direction that has been outstandingly successful for more than a century, namely the reductionist analysis of brain function in terms of simpler physical, chemical, and biological processes, and because we can understand all, or almost all, about machines, the metaphor encourages us to think we can discover all, or almost all, about the mind. I thought when I started writing this piece that the metaphor was sound and useful, though its uncritical acceptance made me uneasy, especially because I knew that many of my colleagues in subjects like mathematics and linguistics received it with something close to incredulity, while many others disliked it intensely. I therefore thought it would be worthwhile to examine the idea in more detail, to try to find if there was any justification for unease, incredulity, or dislike; if you read on you will find that I have been forced to the conclusion that the metaphor is misleading and potentially harmful, but that this results from prevalent ignorance and prejudice about machines rather than any gross defect of the metaphor.

## IN DEFENCE OF THE ANALOGY

I think the main purpose behind calling the mind a machine is to drive out demons. We are saying, in effect, "Look, there is no more to the working of the brain than the physics and chemistry of its componets, just as there is no more to a machine than the physics and chemistry of *its* components." This is admirable as an invocation not to waste time on mental spirits and to study the physics and chemistry instead, so what are the objections? First there are three minor ones, namely that we don't actually know all about machines, that brains are made of totally different materials, and that they have come into existence in a strikingly different manner. By considering these objections we see where to be cautious about the meta-

15

phor, but they are not fatal to it. However, there are more serious problems that do justify mistrust and dislike: first, minds do some things that no current machines do, and if one is mainly interested in these particular tasks the analogy is not much help and could be misleading; second, it is the minds of other people one interacts with, so to treat minds as machines valued mainly for their usefulness has unpleasant ethical implications.

## We Do Not Understand All About Machines

The person who understands most about a machine is its designer, and no designer of a complex machine would claim that everything about it was perfectly understood. Engineers are really very different from scientists, for they use a body of knowledge to create something new, and provided the goal is reached they are not too concerned if their creation has unforeseen or unknown properties. Scientists make use of the same body of knowledge, but they must pay particular attention to anything unforeseen or unknown, since their goal is to extend the body of knowledge rather than to exploit it. In engineering, intuitive leaps and creative solutions are rightly admired when they make a machine work, but as scientists, we should admire creativity and intuition that enable us to understand what was previously unknown; we do not want to import the attitude "If it works, that's good enough" along with the concept that mind is a product of engineering.

The current enthusiasm for neural networks may show that this niggle is partly justified. Here are techniques that enable machines to perform tasks that hitherto lay in the province of the mind, but whereas artificial intelligence and the previous styles of computer simulation led to a more detailed analytical knowledge of the requirements for performing the task, the network style of simulation does not; instead it uses some blind procedure, such as back-propagation, and the success of a simulation is taken as justification for the procedure, rather than for the programmer's analysis of the task requirements. To be fair, the major proponents of the network approach do not regard this as an advantage and claim the method gives insight by, for instance, showing what intermediate elements or "hidden units" enable a task to be done; but one suspects the avoidance of detailed analysis is nonetheless a factor in the popularity of the method.

## The Materials Are Very Different

The objection that the materials are totally different need not be taken very seriously since, whatever the materials are, they still obey the rules of physics and chemistry. It just means more surprises for the neuroscientist, since he will come across unexpected properties and methods; birds do not use propellers or jet engines, but they do obey the same laws of aerodynamics as machines that fly.

Perhaps one should be a little more cautious when it comes to computers and the brain, for the architecture as well as the materials are so different, but surely some at least of the problems encountered when performing a task on a serial computer will carry over to the performance of the same task by the brain. As long as that is so, we can learn from the analogy.

## Minds and Machines Differ in Origin

Does it matter that brains are the product of millions of years of genetic evolution combined with a few months' ontogenesis and several years of teaching and experience, whereas machines are designed by those brains and made by human hand? I can't see why it should, and it could be claimed that the explicit knowledge required for design and construction gives deep insight into the nature of the mind. But once again there are cautions.

Evolution is an unprincipled and conservative designer. As a result, one rarely finds neat theoretical solutions embodied in brains, and in particular they often use parallel mechanisms for achieving a single goal. This fact tends to make life difficult for the experimentalist, because it frustrates attempts to do simple and effective controls. For instance, one might naively expect that if binocular vision enables people to judge distance, then blocking one eye would seriously impair this capacity, but motion parallax, knowledge of the normal sizes of objects, and other cues leave good distance judgment in one-eyed people. The use of alternative sensory cues for balance can cause confusion (see below), and the presence of many parallel methods has caused untold difficulty in the analysis of homing and other navigational feats in animals. It seems to be hard for us to avoid the automatic assumption that just one method is used to perform some difficult task, and this may partly result from the metaphor of the mechanical mind, for multiple alternative methods are rare in machines.

## WHAT THE MIND DOES

Many of the things the mind and brain do are also done by machines, and when this is the case understanding the machine obviously helps. To understand how a man balances on one leg, one should understand the principles of servo-feedback, though to reinforce the point made above, one should also know that a man's ability to stand on one leg with his eyes shut does not prove that vision is irrelevant to the task, and the ability to remain upright after the destruction of his vestibular organs does not prove they are unimportant either. But the brain also does things that no machine does, and here we get into trouble.

## Things No Computer Does

To start with, consider the fact that we constantly model the inanimate world around us, and also monitor it for changes. We know thoroughly the route from home to office, and notice when a section of road is being repaired, or when the office door is newly painted. Much of this model-making is quite automatic and unconscious and we only know it has been done when something changes; we do not consciously monitor the spectral composition of sunlight or the pitch of the front door bell, but it's pretty certain we would notice if they changed.

Robots are of course beginning to record and model their environments in order to find their way around in it, but compared with us they are extraordinarily backward, and so far they have little to teach us. But it would be premature to say that, for this reason, the metaphor is wrong or misleading, because the need for modeling has only recently arisen. It is likely that the principles will soon be better understood, and when this happens they may give us new insight into the brain's methods of building useful models of the environment.

## Modeling People

More serious problems arise when one considers the human brain's propensity to model people. This starts at a very early age, and parents quickly realize that they are not the only ones trying to run the household; a baby quickly becomes pretty expert at parent-control. It seems to me that this really is a most un-machinelike process, and likely to remain so. One reason is that machines are designed to do something useful for their owners, whereas babies are not. No doubt one could incorporate a few tricks in a computer that would enable it to get the better of its user—in fact many of them appear to do this effortlessly, without deliberate design; but in such cases it is clear upon reflection that they are not really getting the better of us, they are simply failing to give the desired service. To improve the man–machine interface, an engineer might program a computer to learn about the user's behavior, just as a baby learns about its parents' behavior; but there the analogy ends, for the engineer's purpose would always be to diminish the conflict of wills, while that is not the baby's purpose at all. A machine is by definition intended to be useful, and therefore it must be designed so that its user's will is unopposed as far as possible.

Here, then, we have reached a point where the machine is not a good metaphor for the mind, simply because no machine has been designed to do what the mind does. I am nor saying there is a theoretical reason that they should not be so designed, only that engineers are not likely to explore the means of doing something that it would be useless or counter-productive to do, and still less likely actually to do it. We, as neuro-scientists, can explore the problems of one brain out-guessing another, but we shall not at present get much help in doing so by regarding the mind as a machine, simply because that is not at present the sort of thing machines do.

Notice that this argument is only valid for the present; comparative and competitive interactions between computers can be modeled, and this may be very instructive for understanding human interactions. Hence the implications of the metaphor are not static, and it may mean something very different to future generations that have more understanding of the complexity of purely mechanistic interactions. All the same, for the present the metaphor is, at best, unhelpful on such problems.

## TAKING THE METAPHOR TOO SERIOUSLY

Perhaps this discussion has brought us to the reason for the unease and dislike aroused by the mind-machine metaphor. If we took it seriously, would we not regard other people's minds as machines, and hence be interested in them primarily for their utility to us? Machines do not and should not oppose our wills, and nor should other people, according to the metaphor.

This would all be less worrying if the newspapers were not full of the deeds and misdeeds of individuals who behave as though the metaphor was their gospel, and such an attitude is not entirely unfamiliar even in academic circles. Of course the metaphor is not intended to be a moral exhortation saying: "Treat other people's minds as machines that could be useful to you." But when we use it aren't we in effect saying: "It's alright to treat minds as machines, because actually they are"?

Thus the metaphor may not be entirely innocuous, since it might influence how people think about minds and consequently how they treat other people. To follow up this thought we need to see how people actually use the idea of the mind.

### Distinguishing Minds from Brains

So far, encouraged by the mind-machine metaphor, I have used the words "mind" and "brain" more or less interchangeably, but now we see that a distinction might be useful. The brain is what the metaphor properly applies to, while the mind is something different: it is the concept we use to describe the source of other people's, and our own, behavior. Since the concept of a particular person's mind is fashioned out of the observed behavior of that individual, it is a model one's own brain makes of the other individual's brain. Thus minds are the brain's models of itself and other brains, and the important thing is that the vast majority of people attribute behavior to such mind-models. As neuroscientists we believe it is the brain that controls behavior, but this is the belief and terminology of a small minority of experts; others attribute a person's behavior to his mind and care nothing for the beautiful nerve cells that we dedicate our lives to. That may be a pity, but in justification of the majority's attitude, pause to think how many of the facts you learn from this volume will alter the way you treat your colleagues, bring up your childern, talk to the janitor, or vote at the next election. Your beliefs about other people's minds, on the other hand, clearly do influence your way of life, and it is because the metaphor may modify your attitude to minds that it is potentially obnoxious.

One can be thoroughly mechanistic in believing that the brain controls behavior strictly within the laws of physics and chemistry without this distinction between mind and brain becoming a mockery. In most cases one has knowledge of only a minute fraction of the physical and chemical factors that are actually (we mechanistically believe) controlling the brain's output, but this does not stop us from making quite good and reliable judgments about the actions our own and other minds will initiate. There is nothing unusual in a model having such predictive power in spite of its use of incomplete data; in fact, good models stringently select the data they represent, both in the case of one's mental models of the physical environment with the people in it, and in the case of accurate scientific theories such as thermodynamics. The predictive power of a model depends on its correct identification of the dominant controlling factors and their influence, not upon its completeness. An incomplete model is often more generally useful than a more accurate one, as for example with Newtonian laws of motion.

## Minds Take Over the Control of Behavior

Figure 1 is an attempt to persuade neuroscientists that minds are actually more important than brains, and to show how this comes about. In stage 1, the mechanistic brain is in sole command, but brain A observes other brains, B for example, also controlling their own behavior. Occasionally they interact, or their behaviors conflict, so brain A builds a model of the way brain B controls what B says and does, based on what A has seen and heard. This model of brain B inside the brain of A is B's mind, shown in step 2; brain A uses his idea of B's mind to predict what B will say and do, and this should be beneficial to A when living in the same environment as B. B of course has done the same and now has its internal model of A's brain, A's mind.

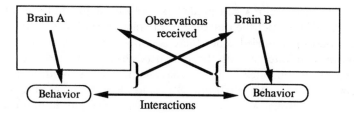

Stage 1. Brain A alone controls its own behavior, but observes B also controlling its own behavior. Brain B does likewise.

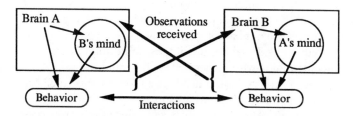

Stage 2. Brain A has made *B's mind*, an internal model of B's brain based on what A has observed B doing. This helps A predict B's behavior. Brain B has done the same.

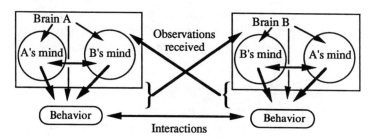

Stage 3. Brain A creates a model of itself, *A's mind*; the behavioral interactions between A and B can now be modeled within A's brain by the interactions of A's mind and B's mind. Minds now control social behavior.

*Figure 1*   Minds control behavior. We attribute other people's actions to their minds, which it therefore seems appropriate to regard as the models our own brains make of the source of other people's and our own behavior. This diagram shows three stages in the development within the brain of minds that take control of its social behavior, leaving the brain in charge of the unconscious, automatic actions we regard as mindless.

At this stage brain A can predict B's solo behavior, but this will not enable him to predict the outcome of interactions between A and B, for these cannot be modeled, since there is nothing inside A's brain suitable for B's mind to interact with. Brain A needs to make a model of the way its own brain generates its own behavior; this is A's mind, and the interactions between A's mind and B's mind allow brain A to model the interactions between A and B. By this time the two minds in brain A have become dominant in controlling much of the behavior of A, especially the verbal and other exchanges with other individuals. On the other hand, nonverbal and nonsocial behavior may still be generated predominantly by brain A, uninfluenced by the minds it has created within itself—these are, as we would say, mindless actions.

Of course all the time similar processes will have been going on in brain B, so A's models will not have it all their own way when it comes to making accurate predictions. Furthermore, if these mind-models are realistic, they should include minds within themselves; B's mind in brain A should, for example, have a little A's mind inside itself, for by this stage brain B will have its own version of A's mind, as shown in the right half of the figure. This infinite regress can presumably be cut short after a few cycles without seriously affecting the predictive power of the models, but it makes the brain into something more like a hall of mirrors than any currently understood machine. Surely, however, this hall of mirrors accurately portrays the fact that two people's minds interact with each other in an extraordinarily complex and intimate manner. At the very least, the interactions that this figure attempts to portray will have to be taken into account in any mechanistic accounts of higher human behavior, and I do not see how this can be done without giving the concept of the mind a very dominant role.

## Why Minds are More Interesting

Now that we have a distinction between mind and brain, we can see why most people are much more interested in one than the other. The mathematician, linguist, historian, or literary critic is not concerned with the nuts and bolts of brain mechanisms, but with mind—the abstract model that appears to be responsible for the works that he studies and perhaps produces himself. It is the same with cars, domestic appliances, word processors, and so on: except for the expert, people simply do not care how they work as long as they do the job expected. Of course when minds fail to work as expected we may call in the brain specialist, but until then it is minds we deal with.

I think this distinction between the mechanistic brain and the mind-models it makes has interesting implications for our understanding of pain, pleasure, ethics, and consciousness itself. As I have argued elsewhere, the part of the brain's functioning of which we are consciously aware seems to be confined to the inputs, outputs, and interactions of the minds of Figure 1. Just as ordinary people are more interested in minds than brains, so should be philosophers, for it is the source of peoples' actions rather than their mechanism that is their primary concern. Minds are the conceptual sources of this behavior, and although minds are the product of a mechanistic brain, their actions and interactions are not going to be understood in terms of physics and chemistry alone, even if they are ultimately determined by physics and chemistry alone.

Notice that within this entirely mechanistic framework one can ask interesting questions about minds: How is the stability of social interestion affected by how good a model of B's brain A has in the form of B's mind? Are inaccurate models unethical? Or might it be unethical to have too good a model? How about the same questions applied to A's mind, the model of his own brain? How far will taught precepts affect the form of these models? Do we also have mind-models of deities? Surely these are the sort of questions appropriate for theoretical analysis by philosophers and others, and the mechanistic nature of the brain hardly affects them. It is worth adding that such ideas might also form the basis for experiments.

Thus, to neuroscientists, the machine-like aspects of the brain are immediately important, because they firmly direct our attention to the physics and chemistry of the brain. But the metaphor should not be taken to imply that there is *no more* to the brain than can be described in terms of physics and chemistry, and the concept of mind is certain to be important in any satisfactory account of the way the brain controls social behavior. For this reason, philosphers and ordinary people are properly more concerned with minds than brains.

## CONCLUSIONS

To conclude, for most people in most circumstances the physical and chemical causation of the brain's output is not as important as the observable behavior that actually occurs; minds are what we attribute this behavior to, and our common-sense understanding of them will almost always give more useful predictions than knowledge of the physical and chemical processes that underlie them. For neuroscientists trying to refine such mechanistic accounts, the metaphor is appropriate and encouraging, but the dislike it arouses in others seems fully justified by its implication that other people's minds are mere machines and can be treated as such. What is needed to prevent the metaphor of the mechanical mind sanctioning unbridled egocentric behavior is to get rid of the prejudice that machines are essentially simple and deterministic, and to gain an appreciation of the complexity and difficulties in predicting behavior produced by two or more minds interacting in the manner shown in Figure 1. There is no reason to doubt that the brain is entirely mechanical, but it is a wonderful mechanism that can generate and use the concept of mind.

### ACKNOWLEDGEMENTS

I would like to acknowledge considerable help toward clarifying these ideas from Ian Glynn, Graeme Mitchison, Kathy Mullen, and Miranda Weston-Smith.

*Annu. Rev. Neurosci. 1990. 13:25–42*

# THE ATTENTION SYSTEM OF THE HUMAN BRAIN

*Michael I. Posner*

Department of Psychology, University of Oregon, Eugene, Oregon 97403

*Steven E. Petersen*

Department of Neurology and Neurological Surgery,
Washington University, School of Medicine, St. Louis, Missouri 63110

## INTRODUCTION

The concept of attention as central to human performance extends back to the start of experimental psychology (James 1890), yet even a few years ago, it would not have been possible to outline in even a preliminary form a functional anatomy of the human attentional system. New developments in neuroscience (Hillyard & Picton 1987, Raichle 1983, Wurtz et al 1980) have opened the study of higher cognition to physiological analysis, and have revealed a system of anatomical areas that appear to be basic to the selection of information for focal (conscious) processing.

The importance of attention is its unique role in connecting the mental level of description of processes used in cognitive science with the anatomical level common in neuroscience. Sperry (1988, p. 609) describes the central role that mental concepts play in understanding brain function as follows:

> Control from below upward is retained but is claimed to not furnish the whole story. The full explanation requires that one take into account new, previously nonexistent, emergent properties, including the mental, that interact causally at their own higher level and also exert causal control from above downward.

If there is hope of exploring causal control of brain systems by mental states, it must lie through an understanding of how voluntary control is exerted over more automatic brain systems. We argue that this can be

25

0147–006X/90/0301–0025$02.00

approached through understanding the human attentional system at the levels of both cognitive operations and neuronal activity.

As is the case for sensory and motor systems of the brain, our knowledge of the anatomy of attention is incomplete. Nevertheless, we can now begin to identify some principles of organization that allow attention to function as a unified system for the control of mental processing. Although many of our points are still speculative and controversial, we believe they constitute a basis for more detailed studies of attention from a cognitive-neuroscience viewpoint. Perhaps even more important for furthering future studies, multiple methods of mental chronometry, brain lesions, electrophysiology, and several types of neuroimaging have converged on common findings.

Three fundamental findings are basic to this chapter. First, the attention system of the brain is anatomically separate from the data processing systems that perform operations on specific inputs even when attention is oriented elsewhere. In this sense, the attention system is like other sensory and motor systems. It interacts with other parts of the brain, but maintains its own identity. Second, attention is carried out by a network of anatomical areas. It is neither the property of a single center, nor a general function of the brain operating as a whole (Mesulam 1981, Rizzolatti et al 1985). Third, the areas involved in attention carry out different functions, and these specific computations can be specified in cognitive terms (Posner et al 1988).

To illustrate these principles, it is important to divide the attention system into subsystems that perform different but interrelated functions. In this chapter, we consider three major functions that have been prominent in cognitive accounts of attention (Kahneman 1973, Posner & Boies 1971): (a) orienting to sensory events; (b) detecting signals for focal (conscious) processing, and (c) maintaining a vigilant or alert state.

For each of these subsystems, we adopt an approach that organizes the known information around a particular example. For orienting, we use visual locations as the model, because of the large amount of work done with this system. For detecting, we focus on reporting the presence of a target event. We think this system is a general one that is important for detection of information from sensory processing systems as well as information stored in memory. The extant data, however, concern primarily the detection of visual locations and processing of auditory and visual words. For alerting, we discuss situations in which one is required to prepare for processing of high priority target events (Posner 1978).

For the subsystems of orienting, detecting, and alerting, we review the known anatomy, the operations performed, and the relationship of attention to data processing systems (e.g. visual word forms, semantic memory) upon which that attentional subsystem is thought to operate. Thus, for orienting, we review the visual attention system in relationship to the data processing systems of the ventral occipital lobe. For detecting, we examine an anterior attention system in relationship to networks that subserve semantic associations. For alerting, we examine arousal systems in relationship to the selective aspects of attention. Insofar as possible, we draw together evidence from a wide variety of methods, rather than arguing for the primacy of a particular method.

## ORIENTING

### Visual Locations

Visual orienting is usually defined in terms of the foveation of a stimulus (overt). Foveating a stimulus improves efficiency of processing targets in terms of acuity, but it is also possible to change the priority given a stimulus by attending to its location covertly without any change in eye or head position (Posner 1988).

If a person or monkey attends to a location, events occurring at that location are responded to more rapidly (Eriksen & Hoffman 1972, Posner 1988), give rise to enhanced scalp electrical activity (Mangoun & Hillyard 1987), and can be reported at a lower threshold (Bashinski & Bachrach 1984, Downing 1988). This improvement in efficiency is found within the first 150 ms after an event occurs at the attended location. Similarly, if people are asked to move their eyes to a target, an improvement in efficiency at the target location begins well before the eyes move (Remington 1980). This covert shift of attention appears to function as a way of guiding the eye to an appropriate area of the visual field (Fischer & Breitmeyer 1987, Posner & Cohen 1984).

The sensory responses of neurons in several areas of the brain have been shown to have a greater discharge rate when a monkey attends to the location of the stimulus than when the monkey attends to some other spatial location. Three areas particularly identified with this enhancement effect are the posterior parietal lobe (Mountcastle 1978, Wurtz et al 1980), the lateral pulvinar nucleus of the postereolateral thalamus (Petersen et al 1987), and the superior colliculus. Similar effects in the parietal cortex have been shown in normal humans with positron emission tomography (Petersen et al 1988a).

Although brain injuries to any of these three areas in human subjects will cause a reduction in the ability to shift attention covertly (Posner 1988), each area seems to produce a somewhat different type of deficit. Damage to the posterior parietal lobe has its greatest effect on the ability to disengage from an attentional focus to a target located in a direction opposite to the side of the lesion (Posner et al 1984).

Patients with a progressive deterioration in the superior colliculus and/or surrounding areas also show a deficit in the ability to shift attention. In this case, the shift is slowed whether or not attention is first engaged elsewhere. This finding suggests that a computation involved in moving attention to the target is impaired. Patients with this damage also return to former target locations as readily as to fresh locations that have not recently been attended. Normal subjects and patients with parietal and other cortical lesions have a reduced probability of returning attention to already examined locations (Posner 1988, Posner & Cohen 1984). These two deficits appear to be those most closely tied to the mechanisms involved with saccadic eye movements.

Patients with lesions of the thalamus and monkeys with chemical injections into the lateral pulvinar also show difficulty in covert orienting (Petersen et al 1987, Posner 1988). This difficulty appears to be in engaging attention on a target on the side opposite the lesion so as to avoid being distracted by events at other locations. A study of patients with unilateral thalamic lesions showed slowing of responses to a cued target on the side opposite the lesion even when the subject had plenty of time to orient there. This contrasted with the results found with parietal and midbrain lesions, where responses are nearly normal on both sides once attention has been cued to that location. Alert monkeys with chemical lesions of this area made faster than normal responses when cued to the side opposite the lesion and given a target on the side of the lesion, as though the contralateral cue was not effective in engaging their attention (Petersen et al 1987). They were also worse than normal when given a target on the side opposite the lesion, irrespective of the side of the cue. It appears difficult for thalamic-lesioned animals to respond to a contralateral target when another competing event is also present in the ipsilateral field (R. Desimone, personal communication). Data from normal human subjects required to filter out irrelevancies, showed selective metabolic increases in the pulvinar contralateral to the field required to do the filtering (LaBerge & Buchsbaum 1988). Thalamic lesions appear to give problems in engaging the target location in a way that allows responding to be fully selective.

These findings make two important points. First, they confirm the idea that anatomical areas carry out quite specific cognitive operations. Second, they suggest a hypothesis about the circuitry involved in covert visual attention shifts to spatial locations. The parietal lobe first disengages attention from its present focus, then the midbrain area acts to move the index of attention to the area of the target, and the pulvinar is involved in reading out data from the indexed locations. Further studies of alert monkeys should provide ways of testing and modifying this hypothesis.

## Hemispheric Differences

The most accepted form of cognitive localization, resulting from studies of split brain patients (Gazzaniga 1970), is the view that the two hemispheres perform different functions. Unfortunately, in the absence of methods to study more detailed localization, the literature has tended to divide cognition into various dichotomies, assigning one to each hemisphere. As we develop a better understanding of how cognitive systems (e.g. attention) are localized, hemispheric dominance may be treated in a more differentiated manner.

Just as we can attend to locations in visual space, it is also possible to concentrate attention on a narrow area or to spread it over a wider area (Eriksen & Yeh 1985). To study this issue, Navon (1987) formed large letters out of smaller ones. It has been found in many studies that one can concentrate attention on either the small or large letters and that the attended stimulus controls the output even though the unattended letter still influences performance. The use of small and large letters as a method of directing local and global attention turns out to be related to allocation of visual channels to different spatial frequencies. Shulman & Wilson (1987) showed that when attending to the large letters, subjects are relatively more accurate in the perception of probe grating of low spatial frequency, and this reverses when attending to the small letters.

There is evidence from the study of patients that the right hemisphere is biased toward global processing (low spatial frequencies) and the left for local processing (high spatial frequencies) (Robertson & Delis 1986, Sergent 1982). Right-hemisphere patients may copy the small letters but miss the overall form, while those with left hemisphere lesions copy the overall form but miscopy the constituent small letters. Detailed chronometric studies of parietal patients reveal difficulties in attentional allocation so that right-hemisphere patients attend poorly to the global aspects and left-hemisphere patients to the local aspects (Robertson et al 1988).

These studies support a form of hemispheric specialization within the overall structure of the attention system. The left and right hemispheres both carry out the operations needed for shifts of attention in the contralateral direction, but they have more specialized functions in the level of detail to which attention is allocated. There is controversy over the existence (Grabowska et al 1989) and the nature (Kosslyn 1988) of these lateralization effects. It seems likely that these hemispheric specializations are neither absolute nor innate, but may instead develop over time, perhaps in conjunction with the development of literacy. Although the role of literacy in lateralization is not clear, there is some evidence that the degree of lateralization found in nonliterate normals and patients differs from that found in literate populations (Lecours et al 1988).

The general anatomy of the attention system that we have been describing lies in the dorsal visual pathway that has its primary cortical projection area in V1 and extends into the parietal lobe. The black areas on the lateral surface of Figure 1 indicate the parietal projection of this posterior attention system as shown in PET studies (Petersen et al 1988a). The parietal PET activation during visual orienting fits well with the lesion and single cell recording results discussed above. PET studies of blood flow also reveal prestriate areas related to visual word processing. For example, an area of the left ventral occipital lobe (gray area in Figure 1) is active during processing of visual words but not for letter-like forms (Snyder et al 1989). The posterior attention system is thought to operate upon the

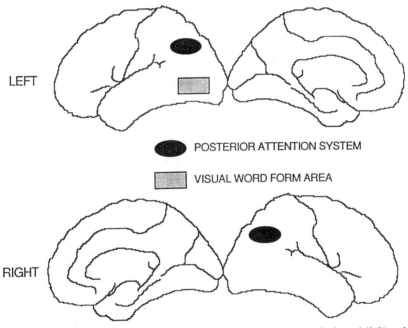

*Figure 1*   The posterior attention system. The *upper two drawings* are the lateral (*left*) and medial (*right*) surfaces of the left hemisphere. The *lower two drawings* are the medial (*left*) and lateral (*right*) surfaces of the right hemisphere. The location of the posterior visual spatial attention system is shown on the lateral surface of each hemisphere as determined by blood flow studies (Petersen et al 1988a). The location of the visual word form area on the lateral surface of the left hemisphere is from Snyder et al (1989).

ventral pathway during tasks requiring detailed processing of objects (e.g. during the visual search tasks discussed in the next section).

A major aspect of the study of attention is to see how attention could influence the operations of other cognitive systems such as those involved in the recognition of visual patterns. The visual pattern recognition system is thought to involve a ventral pathway, stretching from V1 to the infra-temporal cortex. Anatomically, these two areas of the brain can be coordinated through the thalamus (pulvinar) (Petersen et al 1987), or through other pathways (Zeki & Shipp 1988). Functionally, attention might be involved in various levels of pattern recognition, from the initial registration of the features to the storage of new visual patterns.

## Pattern Recognition

VISUAL SEARCH All neurons are selective in the range of activation to which they will respond. The role of the attention system is to modulate this selection for those types of stimuli that might be most important at a given moment. To understand how this form of modulation operates, it is important to know how a stimulus would be processed without the special effects of attention. In cognition, unattended processing is called "automatic" to distinguish it from the special processing that becomes available with attention.

We have learned quite a bit about the automatic processing that occurs in humans along the ventral pathway during recognition of visual objects (Posner 1988, Treisman & Gormican 1988). Treisman has shown that search of complex visual displays for single features can take place in parallel with relatively little effect of the number of distractors. When a target is defined as a conjunction of attributes (e.g. red triangle) and appears in a background of nontargets that are similar to the target (e.g. red squares and blue triangles), the search process becomes slow, attention demanding, and serial (Duncan & Humphreys 1989).

We know from cognitive studies (LaBerge & Brown 1989, Treisman & Gormican 1988) that cueing people to locations influences a number of aspects of visual perception. Treisman has shown that subjects use attention when attempting to conjoin features, and it has also been shown that spreading focal attention among several objects leads to a tendency for misconjoining features within those objects, regardless of the physical distance between them (Cohen & Ivry 1989). Thus, attention not only provides a high priority to attended features, but does so in a way that overrides even the physical distance between objects in a display.

While these reaction time results are by no means definitive markers of attention, there is also evidence from studies with brain lesioned patients that support a role of the visual spatial attention system. These clinical studies examine the ability of patients to bisect lines (Riddoch & Humphreys 1983), search complex visual patterns (Riddoch & Humphreys 1987), or report strings of letters (Friedrich et al 1985, Sieroff et al 1988). Damage to the posterior parietal lobe appears to have specific influences on these tasks. Patients with right parietal lesions frequently bisect lines too far to the right and fail to report the left-most letters of a random letter string (Sieroff et al 1988). However, these effects are attentional not in the recognition process itself. Evidence for this is that they can frequently be corrected by cueing the person to attend covertly to the neglected side (Riddoch & Humphreys 1983, Sieroff et al 1988). The cues appear to provide time for the damaged parietal lobe to disengage attention and thus compensates for the damage. It is also possible to compensate by substituting a word for a random letter string. Patients who fail to report the left-most letters of a random string will often report correctly when the letters make a word. If cues work by directing attention, they should also influence normal performance. Cues presented prior to a letter string do improve the performance of normals for nearby letters, but cues have little or no influence on the report of letters making words (Sieroff & Posner 1988). Blood flow studies of normal humans show that an area of the left ventral occipital lobe is unique to strings of letters that are either words or orthographically regular nonwords (Snyder et al 1989). This visual word form area (see gray area of Figure 1) appears to operate without attention, and this confirms other data that recognition of a word may be so automated as not to require spatial attention, whereas the related tasks of searching for a single letter, forming a conjunction, or reporting letters from a random string do appear to rely upon attention.

Studies of recording from individual cells in alert monkeys confirm that attention can play a role in the operation of the ventral pattern recognition system (Wise & Desimone 1988). It appears likely that the pathway by which the posterior attention system interacts with the pattern recognition system is through the thalamus (Petersen et al 1987). This interaction appears to require about 90 ms, since cells in V4 begin to respond to unattended items within their receptive field but shut these unattended areas off after 90 ms (Wise & Desimone 1988). Detailed models of the nature of the interaction between attention and pattern recognition are just beginning to appear (Crick 1984, LaBerge & Brown 1989).

IMAGERY  In most studies of pattern recognition, the sensory event begins the process. However, it is possible to instruct human subjects to take information from their long-term memories and construct a visual representation (image) that they might then inspect (Kosslyn 1988). This higher level visual function is called imagery. The importance of imagery as a means of studying mechanisms of high-level vision has not been well recognized in neuroscience. Imagery, when employed as a means of studying vision, allows more direct access to the higher levels of information processing without contamination from lower levels. There is by now considerable evidence that some of the same anatomical mechanisms are used in imagery as are involved in some aspects of pattern recognition (Farah 1988, Kosslyn 1988). Patients with right parietal lesions, who show deficits in visual orienting of the type that we have described above, also fail to report the contralesional side of visual images (Bisiach et al 1981). When asked to imagine a familiar scene, they make elaborate reports of the right side but not the left. The parts of the image that are reported when the patient is facing in one direction are neglected when facing in the other. This suggests that the deficit arises at the time of scanning the image.

When normal subjects imagine themselves walking on a familiar route, blood flow studies show activation of the superior parietal lobe on both sides (Roland 1985). Although many other areas of the brain are also active in this study, most of them are common to other verbal and arithmetical thoughts, but activation of the superior parietal lobe seems more unique to imagery. As discussed above, the parietal lobe seems to be central to spatial attention to external locations. Thus, it appears likely that the neural systems involved in attending to an external location are closely related to those used when subjects scan a visual image.

## TARGET DETECTION

In her paper on the topography of cognition, Goldman-Rakic (1988) describes the strong connections between the posterior parietal lobe and areas of the lateral and medial frontal cortex. This anatomical organization is appealing as a basis for relating what has been called involuntary orienting by Luria (1973), and what we have called the posterior attention system, to focal or conscious attention.

Cognitive studies of attention have often shown that detecting a target produces widespread interference with most other cognitive operations (Posner 1978). It has been shown that monitoring many spatial locations or modalities produces little or no interference over monitoring a single modality, unless a target occurs (Duncan 1980). This finding supports the distinction between a general alert state and one in which attention is clearly oriented and engaged in processing information. In the alert but disengaged state, any target of sufficient intensity has little trouble in summoning the mechanisms that produce detection. Thus monitoring multiple modalities or locations produces only small amounts of interference. The importance of engaging the focal attention system in the production of widespread interference between signals supports the idea that there is a unified system involved in detection of signals regardless of their source. As a consequence of detection of a signal by this system, we can produce a wide range of arbitrary responses to it. We take this ability to produce arbitrary responses as evidence that the person is aware of the signal.

Evidence that there are attentional systems common to spatial orienting as well as orienting to language comes from studies of cerebral blood flow during cognitive tasks. Roland (1985) has reported a lateral superior frontal area that is active both during tasks involving language and in spatial imagery tasks. However, these studies do not provide any clear evidence that such common areas are part of an attentional system. More compelling is evidence that midline frontal areas, including the anterior cingulate gyrus and the supplementary motor area, are active during semantic processing of words (Petersen et al 1988b), and that the degree of blood flow in the anterior cingulate increases as the number of targets to be detected increases (Posner et al 1988). Thus, the anterior cingulate seems to be particularly sensitive to the operations involved in target detection. (See Figure 2).

The anterior cingulate gyrus is an area reported by Goldman-Rakic (1988) to have alternating bands of cells that are labeled by injections into the posterior parietal lobe and the dorsolateral prefrontal cortex. These findings suggest that the anterior cingulate should be shown to be important in tasks requiring the posterior attention system as well as in language tasks. It has often been argued from lesion data that the anterior cingulate plays an important role in aspects of attention, including neglect (Mesulam 1981, Mirsky 1987).

Does attention involve a single unified system, or should we think of its functioning as being executed by separate independent systems? One way to test this idea is to determine whether attention in one domain (e.g. language) affects the ability of mechanisms in another domain (e.g. orienting toward a visual location). If the anterior cingulate system is important in both domains, there should be a specific interaction between even remote domains such as these two. Studies of patients with parietal lesions (Posner et al 1987) showed that when patients were required to monitor a stream of auditory information for a sound, they were slowed in their ability to orient toward a visual cue. The effect of the language task was rather different from engaging attention at a visual location because its effects were bilateral rather than being mainly on the side opposite the lesion. Thus, the language task appeared to involve some but not all of the same mechanisms that were used in visual orienting.

This result is compatible with the view that visual orienting involves systems separate but interconnected with those used for language processing. A similar result was found with normal subjects when they were given visual cues while shadowing an auditory message (Posner et al 1989). Here, the effects of the language task were most marked for cues in the right visual field, as though the common system might have involved lateralized mechanisms of the left hemisphere. These findings fit with the close anatomical links between the anterior cingulate and the posterior parietal lobe on the one hand and language areas of the lateral frontal lobe on the other. They suggest to us a possible hierarchy of attention systems in which the anterior system can pass control to the posterior system when it is not occupied with processing other material.

A spotlight analogy has often been used to describe the selection of information from the ventral pattern recognition system by the posterior attention system (Treisman & Gormican 1988). A spotlight is a very crude analogy but it does capture some of the dynamics involved in disengaging, moving, and engaging attention. This analogy can be stretched still further to consider aspects of the interaction between the anterior attention system and the associative network shown to be active during processing of semantic associates and categories by studies of cerebral blood flow (Petersen et al 1988a). The temporal dynamics of this type of interaction between attention and semantic activation have been studied in some detail (see Posner 1978, 1982, for review).

## ALERTING

An important attentional function is the ability to prepare and sustain alertness to process high priority signals. The relationship between the alert state and other aspects of information processing has been worked out in some detail for letter and word matching experiments (Posner 1978). The passive activation of internal units representing the physical form of a familiar letter, its name, and even its semantic classification (e.g. vowel) appears to take place at about the same rate, whether subjects are alert and expecting a target, or whether they are at a lower level of alertness because the target occurs without warning. The alert state produces more rapid responding, but this increase is accompanied by a higher error rate. It is as though the build-up of information about the classification of the

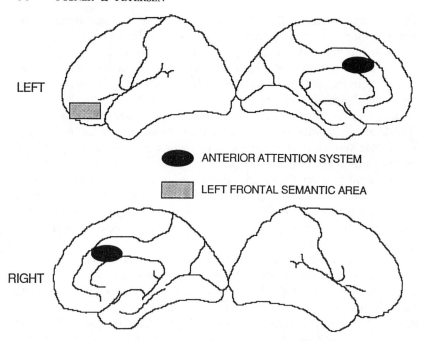

LEFT

RIGHT

● ANTERIOR ATTENTION SYSTEM

▨ LEFT FRONTAL SEMANTIC AREA

*Figure 2*  The anterior attention system. The *upper two drawings* are the lateral (*left*) and medial (*right*) surface of the left hemisphere. The *lower two drawings* are the medial (*left*) and lateral (*right*) surfaces of the right hemisphere. The semantic association area on the lateral aspect of the left hemisphere is determined by blood flow studies (Petersen et al 1988b). The anterior attention area is also from blood flow studies (Petersen et al 1988b, Posner et al 1988).

target occurs at the same rate regardless of alertness, but in states of high alertness, the selection of a response occurs more quickly, based upon a lower quality of information, thus resulting in an increase in errors. These results led to the conclusion that alertness does not affect the build-up of information in the sensory or memory systems but does affect the rate at which attention can respond to that stimulus (Posner 1978).

Anatomical evidence has accumulated on the nature of the systems producing a change in the alert state. One consistent finding is that the ability to develop and maintain the alert state depends heavily upon the integrity of the right cerebral hemisphere (Heilman et al 1985). This finding fits very well with the clinical observation that patients with right-hemisphere lesions more often show signs of neglect, and it has sometimes led to the notion that all of spatial attention is controlled by the right hemisphere. However, the bulk of the evidence discussed below seems to associate right-hemisphere dominance with tasks dependent upon the alert state.

Lesions of the right cerebral hemisphere cause difficulty with alerting. This has been shown with measurement of galvanic skin responses in humans and monkeys (Heilman et al 1985) and with heart rate responses to warning signals (Yokoyama et al 1987). Performance in vigilance tasks is also more impaired with right rather than left lesions (Coslett et al 1987, Wilkins et al 1987). It has also been observed in split-brain patients that vigilance is poor when information is presented to the isolated left hemisphere, but is relatively good when presented to the isolated right hemisphere (Dimond & Beaumont 1973). In summary, the isolated right hemisphere appears to contain the mechanism needed to maintain the alert state so that when lesioned, it reduces performance of the whole organism.

Studies of cerebral blood flow and metabolism involving vigilance tasks have also uniformly shown the importance of areas of the right cerebral hemisphere (Cohen et al 1988, Deutsch et al 1988; J. Pardo, P. T. Fox, M. E. Raichle, personal communication). Other attention demanding activity, e.g. semantic tasks and even imagery tasks, do not uniformly show greater activation of the right hemisphere (Petersen et al 1988b, Roland 1985). Thus, blood flow and metabolic studies also argue for a tie between the right cerebral hemisphere and alerting. Some of these studies provide somewhat better localization. Cohen et al found an area of the midfrontal cortex that appears to be the most active during their auditory discrimination task. This is an area also found to be active in both visual and somatosensory vigilance conditions (J. Pardo et al, personal communication). Of special interest is that Cohen et al report that the higher metabolic activation they found in the right prefrontal cortex was accompanied by reduced activation in the anterior cingulate. If one views the anterior cingulate as related to target detection, this makes sense. In tasks for which one needs to suspend activity while waiting for low probability signals, it is important not to interfere with detecting the external signal. Subjectively, one feels empty headed, due to the effort to avoid any thinking that will reduce the ability to detect the next signal.

There is evidence that the maintenance of the alert state is dependent upon right-hemisphere mechanisms, and also that it is closely tied with attention. These two facts both suggest the hypothesis that the norepinephrine (NE) system arising in the locus coeruleus may play a crucial role in the alert state. In a review of animal studies, Aston-Jones et al (1984) argue that NE cells play a role in changes in arousal or vigilance. Moreover, Robinson (1985) has shown in rats that lesions of the right cerebral hemisphere but not of the left hemisphere lead to depletion of NE on both sides, and that the effects are strongest with lesions near the frontal pole. These findings are consistent with the idea that NE pathways course through frontal areas, dividing as they go backward toward posterior areas. Thus, an anterior lesion would have a larger effect.

Morrison & Foote (1986) have studied the parts of the posterior visual system that are most strongly innervated by NE pathways. They find that in monkeys, NE innervation is most strongly present in the posterior parietal lobe, pulvinar, and superior colliculus. These are the areas related to the posterior attention system. Much weaker innervation was found in the geniculo-striate pathway and along the ventral pattern recognition pathway. These findings support the ideas that NE pathways provide the basis for maintaining alertness, and that they act most strongly on the posterior attention systems of the right cerebral hemisphere. In accord with these ideas, Posner et al (1987) found that patients with right parietal lesions were greatly affected when a warning signal was omitted before a target, while those with left parietal lesions were not. Clark et al (1989) have found that manipulation of NE levels by drugs had specific effects on attention shifting.

In summary, alertness involves a specific subsystem of attention that acts on the posterior attention system to support visual orienting and probably also influences other attentional subsystems. Physiologically, this system depends upon the NE pathways that arise in the LC and that are more strongly lateralized in the right hemisphere. Functionally, activation of NE works through the posterior attention system to increase the rate at which high priority visual information can be selected for further processing. This more rapid selection is often at the expense of lower quality information and produces a higher error rate.

## CONSEQUENCES

Study of attention from a neuroscience viewpoint has been impeded because attention has been thought of as a vague, almost vitalistic capacity, rather than as the operation of a separate set of neural areas whose interaction with domain-specific systems (e.g. visual word form, or semantic association) is the proper subject for empirical investigation. Even a crude knowledge of the anatomy of the selective attention system has a number of important consequences for research. It allows closer coordination between brain imaging studies using human subjects and animal studies involving recording from individual cells. In the case of the posterior attention system, we have outlined hypotheses about the connections between neural systems that can best be tested and expanded by studies designed to work out the connections at the cellular level. At higher levels, coordinated studies of PET and ERP imaging may tell us more details about communication between posterior visual word form systems and anterior semantics, and how attention is involved in this form of information transfer. A systems level analysis provides a framework for the more detailed studies that must follow.

A number of recent observations depend upon a better understanding of how attention relates to semantic activation. The psychological literature reflects a continuing effort to understand the limits to automatic priming of semantic systems (Posner 1982). In the study of sleep, we find challenging new hypotheses that tell us that during sleep, ongoing neural activity may be interpreted semantically by networks primed by daily activity (Hobson 1988). Similarly, research on split brain subjects (Gazzaniga 1970) has led to the idea of an interpreter system present in the left hemisphere that attempts to impose explanations for our behavior. Patients with lesions of the hippocampus, who show no memory that can be retrieved consciously, are able to demonstrate detailed storage by their performance (Squire 1986). This implies that for memory, as for performance, the distinction between automatic and conscious processing marks different neural mechanisms.

Finally, many disorders of higher level cognition are said to be due to deficits of attention. These include neglect, schizophrenia, closed head injury, and attention-deficit disorder, among others. The concept of an attentional system of the brain with specific operations allocated to distinct anatomical areas allows new approaches to these pathologies. One such example is the proposal that a core deficit in schizophrenia is a failure of the anterior attention system of the left hemisphere to impose the normal inhibitory pattern on the left lateralized semantic network (Early et al 1989). This proposal provides specific ideas on integration at the level of neurotransmission, anatomy, and cognition. Similar ideas may link attention-deficit disorder to the right hemisphere mechanisms that control sustaining of attention. A combined cognitive and anatomical approach may be useful in integrating the long separate physiological and psychosocial influences on psychopathology.

ACKNOWLEDGMENTS

This research was supported by Office of Naval Research Contract N-0014-86-0289 and by the Center for Higher Brain Function of Washington University School of Medicine. We acknowledge special appreciation to Drs. J. Pardo, P. Fox, M. Raichle, and A. Snyder for allowing citation of ongoing experiments. Dr. Pardo contributed heavily to our analysis of the alerting literature. Drs. Mary K. Rothbart, Asher Cohen, and Gordon Shulman were helpful in the presentation of this analysis.

## Literature Cited

Aston-Jones, G., Foote, S. L., Bloom, F. E. 1984. Anatomy and physiology of locus coeruleus neurons: Functional implications. In *Frontiers of Clinical Neuroscience*, Vol. 2, ed. M. G. Ziegler. Baltimore: Williams & Wilkins

Bashinski, H. S., Bachrach, R. T. 1984. Enhancement of perceptual sensitivity as the result of selectively attending to spatial locations. *Percept. Psychophys.* 28: 241–48

Bisiach, E., Luzzatti, C., Perani, D. 1981. Unilateral neglect, representational schema and consciousness. *Brain* 102: 757–65

Clark, C. R., Geffen, G. M., Geffen, L. B. 1989. Catecholamines and the covert orienting of attention. *Neuropsychol.* 27: 131–40

Cohen, A., Ivry, R. 1989. Illusory conjuctions inside and outside the focus of attention. *J. Exp. Psychol. Hum. Percept. Perf.* In press

Cohen, R. M., Semple, W. E., Gross, M., Holcomb, H. J., Dowling, S. M., Nordahl, T. E. 1988. Functional localization of sustained attention. *Neuropsych. Neuropsychol. Behav. Neurol.* 1: 3–20

Coslett, H. B., Bowers, D., Heilman, K. M. 1987. Reduction in cerebral activation after right hemisphere stroke. *Neurology* 37: 957–62

Crick, F. 1984. Function of the thalamic reticular complex: The searchlight hypothesis. *Proc. Natl. Acad. Sci.* 81: 4586–90

Deutsch, G., Papanicolaou, A. C., Bourbon, T., Eisenberg, H. M. 1988. Cerebral blood flow evidence of right cerebral activation in attention demanding tasks. *Int. J. Neurosci.* 36: 23–28

Dimond, S. J., Beaumont, J. G. 1973. Difference in the vigilance performance of the right and left hemisphere. *Cortex* 9: 259–65

Downing, C. J. 1988. Expectancy and visual-spatial attention effects on vision. *J. Exp. Psychol. Hum. Percept. Perf.* 14: 188–97

Duncan, J. 1980. The locus of interference in the perception of simultaneous stimuli. *Psychol. Rev.* 87: 272–300

Duncan, J., Humphreys, G. W. 1989. Visual search and stimulus similarity. *Psychol. Rev.* 96: 433–58

Early, T., Posner, M. I., Reiman, E., Raichle, M. E. 1989. Left striato-pallidal hyperactivity in schizophrenia. *Psychiat. Develop.* In press

Eriksen, C. W., Hoffman, J. E. 1972. Temporal and spatial characteristics of selective encoding from visual displays. *Percept. Psychophys.* 12: 201–4

Eriksen, C. W., Yeh, Y. 1985. Allocation of attention in the visual field. *J. Exp. Psychol. Hum Percept. Perf.* 11: 583–97

Farah, M. J. 1988. Is visual imagery really visual? Overlooked evidence from neuropsychology. *Psych Rev.* 95: 307–17

Fischer, B., Breitmeyer, B. 1987. Mechanisms of visual attention revealed by saccadic eye movements. *Neuropsychology* 25(1A): 73–84

Friedrich, F. J., Walker, J., Posner, M. I. 1985. Effects of parietal lesions on visual matching. *Cog. Neuropsychol.* 2: 253–64

Gazzaniga, M. S. 1970. *The Bisected Brain.* New York: Appleton. 171 pp.

Goldman-Rakic, P. S. 1988. Topography of cognition: Parallel distributed networks in primate association cortex. *Annu. Rev. Neurosci.* 11: 137–56

Grabowska, A., Semenza, C., Denes, G., Testa, S. 1989. Impaired grating discrimination following right hemisphere damage. *Neuropsychologia* 27(2): 259–64

Heilman, K. M., Watson, R. T., Valenstein, E. 1985. Neglect and related disorders. In *Clinical Neuropsychology*, ed. K. M. Heilman, E. Valenstein. pp. 243–93. New York: Oxford

Hillyard, S. A., Picton, T. W. 1987. Electrophysiology of cognition. *Handb. Physiol.* (pt. 2): 519–84

Hobson, J. A. 1988. *The Dreaming Brain.* New York: Basic. 319 pp.

James, W. 1890. *Principles of Psychology*, Vol. 1. New York: Holt

Kahneman, D. 1973. *Attention and Effort.* Englewood Cliffs, NJ: Prentice Hall. 246 pp.

Kosslyn, S. M. 1988. Aspects of a cognitive neuroscience of mental imagery. *Science* 240: 1621–26

LaBerge, D., Brown, V. 1989. Theory of attentional operations in shape identification. *Psychol. Rev.* 96: 101–24

LaBerge, D., Buchsbaum, M. S. 1988. Attention filtering and the pulvinar: Evidence from PET scan measures. *Program 29th Annu. Meet. Psychonom. Soc.*, p. 4. (Abstr.)

Lecours, A. R., Mehler, J., Parente, M. A. 1988. Illiteracy and brain damage. *Neuropsychologia* 26(4): 575–89

Luria, A. R. 1973. *The Working Brain.* New York: Basic. 398 pp.

Mangoun, G. R., Hillyard, S. A. 1987. The spatial allocation of attention as indexed by event-related brain potentials. *Hum. Factors* 29: 195–211

Mesulam, M. M. 1981. A cortical network for directed attention and unilateral neglect. *Ann. Neurol.* 10: 309–25

Mirsky, A. F. 1987. Behavioral and psychophysiological markers of disordered attention. *Environ. Health Persp.* 74: 191–99

Morrison, J. H., Foote, S. L. 1986. Noradrenergic and serotoninergic innervation of cortical, thalamic and tectal visual structures in old and new world monkeys. *J. Comp. Neurol.* 243: 117–28

Mountcastle, V. B. 1978. Brain mechanisms of directed attention. *J. R. Soc. Med.* 71: 14–27

Navon, D. 1977. Forest before trees: The precedence of global features in visual perception. *Cog. Psychol.* 9: 353–83

Petersen, S. E., Fox, P. T., Miezin, F. M., Raichle, M. E. 1988a. Modulation of cortical visual responses by direction of spatial attention measured by PET. *Assoc. Res. Vision Ophthal.*, p. 22 (Abstr.)

Petersen, S. E., Fox, P. T., Posner, M. I., Mintun, M., Raichle, M. E. 1988b. Positron emission tomographic studies of the cortical anatomy of single word processing. *Nature* 331: 585–89

Petersen, S. E., Robinson, D. L., Morris, J. D. 1987. Contributions of the pulvinar to visual spatial attention. *Neuropsychology* 25: 97–105

Posner, M. I. 1978. *Chronometric Explorations of Mind.* Englewood Heights, NJ: Erlbaum. 271 pp.

Posner, M. I. 1982. Cumulative development of attentional theory. *Am. Psychol.* 32: 53–64

Posner, M. I. 1988. Structures and functions of selective attention. In *Master Lectures in Clinical Neuropsychology*, ed. T. Boll, B. Bryant. 173–202 pp. Washington, DC: Am. Psych. Assoc.

Posner, M. I., Boies, S. J. 1971. Components of attention. *Psychol Rev.* 78: 391–408

Posner, M. I., Cohen, Y. 1984. Components of performance. In *Attention and Performance X*, ed. H. Bouma, D. Bowhuis. 531–56 pp. Hillsdale, NJ: Erlbaum

Posner, M. I., Inhoff, A., Friedrich, F. J., Cohen, A. 1987. Isolating attentional systems: A cognitive-anatomical analysis. *Psychobiology* 15: 107–21

Posner, M. I., Petersen, S. E., Fox, P. T., Raichle, M. E. 1988. Localization of cognitive operations in the human brain. *Science* 240: 1627–31

Posner, M. I., Sandson, J., Dhawan, M., Shulman, G. L. 1989. Is word recognition automatic? A cognitive-anatomical approach. *J. Cog. Neurosci.* 1: 50–60

Posner, M. I., Walker, J. A., Friedrich, F. J., Rafal, R. D. 1984. Effects of parietal lobe injury on covert orienting of visual attention. *J. Neurosci.* 4: 1863–74

Raichle, M. E. 1983. Positron emission tomography. *Annu. Rev. Neurosci.* 6: 249–67

Remington, R. 1980. Attention and saccadic eye movements. *J. Exp. Psychol. Hum. Percept. Perf.* 6: 726–44

Riddoch, M. J., Humphreys, G. W. 1983. The effect of cueing on unilateral neglect. *Neuropsychology* 21: 589–99

Riddoch, M. J., Humphreys, G. W. 1987. Perceptual and action systems in unilateral neglect. In *Neurophysiological and Neuropsychological Aspects of Spatial Neglect*, ed. M. Jeannerod, pp. 151–181. Amsterdam: North Holland

Rizzolatti, G., Gentilucci, M., Matelli, M. 1985. Selective spatial attention: One center, one circuit or many circuits. In *Attention and Performance XI*, ed. M. Posner, O. S. M. Marin, pp. 251–65. Hillsdale, NJ: Erlbaum

Robertson, L., Delis, D. C. 1986. Part-whole processing in unilateral brain damaged patients: Dysfunction of hierarchical organization. *Neuropsychology* 24: 363–70

Robertson, L., Lamb, M. R., Knight, R. T. 1988. Effects of lesions of temporal-parietal junction on perceptual and attentional processing in humans. *J. Neurosci.* 8(10): 3757–69

Robinson, R. G. 1985. Lateralized behavioral and neurochemical consequences of unilateral brain injury in rats. In *Cerebral Lateralization in Nonhuman Species*, ed. S. G. Glick, pp. 135–56. Orlando: Academic

Roland, P. E. 1985. Cortical organization of voluntary behavior in man. *Hum. Neurobiol.* 4: 155–67

Sergent, J. 1982. The cerebral balance of power: Confrontation or cooperation? *J. Exp. Psychol. Hum. Percept. Perf.* 8: 253–72

Shulman, G. L., Wilson, J. 1987. Spatial frequency and selective attention to local and global structure. *Perception* 16: 89–101

Sieroff, E., Pollatsek, A., Posner, M. I. 1988. Recognition of visual letter strings following damage to the posterior visual spatial attention system. *Cog. Neuropsychol.* 5: 427–49

Sieroff, E., Posner, M. I. 1988. Cueing spatial attention during processing of words and letter strings in normals. *Cog. Neuropsychol.* 5: 451–72

Snyder, A. Z., Petersen, S., Fox, P., Raichle, M. E. 1989. PET studies of visual word recognition. *J. Cerebral Blood Flow Metab.* 9: Suppl. 1–S576. (Abstr.)

Sperry, R. L. 1988. Psychology's mentalist paradigm and the religion/science tension. *Am. Psychol.* 43: 607–13

Squire, L. R. 1986. Mechanisms of memory. *Science* 232: 1612–19

Treisman, A. M., Gormican, S. 1988. Feature analysis in early vision: Evidence from search asymmetries. *Psychol. Rev.* 95: 15–48

Wilkins, A. J., Shallice, T., McCarthy, R. 1987. Frontal lesions and sustained attention. *Neuropsychology* 25: 359–66

Wise, S. P., Desimone, R. 1988. Behavioral neurophysiology: Insights into seeing and grasping. *Science* 242: 736–41

Wurtz, R. H., Goldberg, M. E., Robinson, D. L. 1980. Behavioral modulation of visual responses in monkeys. *Prog. Psychobiol. Physiol. Psychol.* 9: 42–83

Yokoyama, K., Jennings, R., Ackles, P., Hood, P., Boller, F. 1987. Lack of heart rate changes during an attention-demanding task after right hemisphere lesions. *Neurology* 37: 624–30

Zeki, S., Shipp, S. 1988. The functional logic of cortical connections. *Nature* 335: 311–17

*Annu. Rev. Neurosci. 1990. 13:43–60*

# PERIPHERAL NERVE REGENERATION

*James W. Fawcett† and Roger J. Keynes*

†The Physiological Laboratory, and *Department of Anatomy,
The University of Cambridge, Downing Street, Cambridge CB2 3DY,
England

## INTRODUCTION

Under normal circumstances peripheral nerves maintain stable connections with their targets throughout the life of an animal. Despite this appearance of stability, many of the cellular behaviors associated with growth persist in mature neurons: They may continuously remodel their connections (Purves & Voyvodic 1987), produce and transport to the nerve terminal many of the molecules needed for nerve growth (Lasek et al 1984); and they elongate considerably during body growth. On injury, nerve fibers distal to the lesion degenerate by the process known as Wallerian degeneration, and the neurons reinitiate axonal growth with its attendant metabolic changes. The cellular processes underlying degeneration and regeneration are the subject of this review.

## WALLERIAN DEGENERATION

An extensive literature now exists detailing the various molecular and morphological changes that take place in the distal stump following axotomy, and it is not intended to reiterate all these here (for review see Hallpike 1976, Allt 1976). Wallerian degeneration leads to the removal and recycling of axonal and myelin-derived material, and prepares the environment through which regenerating axons grow. Both the axon and the myelin degenerate, leaving behind dividing Schwann cells inside the basal lamina tube that surrounded the original nerve fiber; these columns of Schwann cells surrounded by basal lamina are known as endoneurial tubes or bands of Büngner. A similar process occurs in unmyelinated nerve fibers.

43

0147–006X/90/0301–0043$02.00

For many years there has been uncertainty about the relative roles of Schwann cells and macrophages in removing degenerating debris: Recent experiments by Friede and associates (Beuche & Friede 1984, Scheidt & Friede 1987) have resolved this question somewhat. In one experiment, they placed segments of mouse nerve inside Millipore diffusion chambers with a pore size sufficiently small (0.22 microns) to prevent cell ingress, and implanted the segments into the peritoneal cavity. For periods as long as 8 wk, they found that the Schwann cells did not proliferate, and that the myelin was discarded but not phagocytosed. On the other hand, when the chamber pore size was large enough (5 microns) to allow cells into the chambers, macrophages actively phagocytosed myelin. These results, assuming they are applicable to the normal degenerating nerve, suggest that macrophages are needed to trigger the clearance of myelin and its breakdown products. However, Schwann cells in degenerating nerves ingest myelin debris (Allt 1976), and Schwann cell phagocytosis of myelin has also been demonstrated in vitro (Bigbee et al 1987), although this appears to be a minor pathway for myelin removal (Crang & Blakemore 1987).

Many studies have documented Schwann cell proliferation during Wallerian degeneration in vivo (Abercrombie & Johnson 1946, Thomas 1966, Bradley & Asbury 1970, Allt 1976). In vitro it has been shown that axonal membrane and myelin debris stimulate Schwann cell mitosis (Salzer & Bunge 1980, Salzer et al 1980), and that macrophages that have phagocytosed myelin produce a conditioned medium that is mitogenic for Schwann cells (Baichwal et al 1988). Consistent with their possible role in triggering change in Schwann cell behavior, macrophages begin to accumulate in the distal stump just before the period of Schwann cell proliferation (Thomas 1966, Williams & Hall 1971, Perry et al 1987); and this correlates with the appearance in the distal stump of the macrophage product apolipoprotein E, which may be involved in the recycling of lipids (Snipes et al 1986, Ignatius et al 1986, 1987).

A recent study by Lunn et al (Lunn et al 1989) has placed the process of Wallerian degeneration in an interesting new perspective. These authors describe a strain of mouse in which, following sciatic nerve section, invasion of the distal stump by myelomonocytic cells is much reduced, axonal and myelin removal are very slow, and distal stump axons continue to conduct normal action potentials for up to 14 days. Despite the absence of Wallerian degeneration, sciatic nerve regeneration following crush proceeds at a normal rate (see below). The precise defect in this mutant has yet to be characterized, and it will be interesting to see whether it resides in the macrophages or in their recruitment by the cells of the degenerating nerve.

# CELL BODY RESPONSES TO AXOTOMY

Following axotomy, most surviving cell bodies undergo a variety of anatomical changes and modifications in gene expression and cellular metabolism. The main morphological event is "chromatolysis": the dispersal of Nissl substance due to the disintegration of large granular condensations of rough endoplasmic reticulum. Precisely what changes occur, how quickly and for how long, depend on factors such as age and distance of the lesion from the cell body. We do not detail all these changes (in many cases their functional significance is obscure). The interested reader is referred to the following comprehensive reviews (Cragg 1970, Grafstein & McQuarrie 1978, Lieberman 1971, Watson 1974).

The onset of regeneration is associated with expression of new genes and proteins. In general, the proteins produced during regeneration are the same as those associated with axonal growth in embryos. Substances that are abundant in developing axons, such as growth associated proteins (GAPs), tubulin, and actin, have their synthesis enhanced, whereas neurofilament protein, which mainly appears in development when axons have connected with their targets and are expanding radially, is decreased. GAP-43 is an axonally transported phosphoprotein found on the inside of the membrane of regenerating and growing axons, particularly near the growth cone (Meiri et al 1988). It is induced very rapidly following axotomy, the level of the protein increasing up to 100 times, and it then decreases again on reinnervation (Skene et al 1986, Verhaagen et al 1988, Karns et al 1987). It is a substrate for protein kinase C, and may play an important role in nerve fiber growth.

The pattern of synthesis of cytoskeletal components also recapitulates that found in development. Synthesis of neurofilament protein and expression of neurofilament mRNA decrease during axonal regeneration (Hoffman et al 1987, Oblinger & Lasek 1988). This is associated with a decrease in the quantity of neurofilament and a decrease in diameter of the regenerating nerve fibers. However, some neurons exhibit a paradoxical build up of phosphorylated neurofilament in the soma (whereas normally it is found largely in the axon) (Goldstein et al 1987, Sinicropi & McIlwain 1983). In dorsal root ganglion neurons the decrease in axonal neurofilament transport is restricted to the axonal branch connected to the periphery (Greenberg & Lasek 1988). Both these observations suggest that there is selective transport of neurofilament during regeneration. Two of the tubulin genes, T alpha 1 (Miller et al 1987) and class II beta, are expressed at a high level during axon development, but little in the adult, and both are re-induced during regeneration (Miller et al 1989, Hoffman & Cleveland 1988). Overall, tubulin synthesis and transport are usually increased in regeneration, but there are variations depending on the species, the nerve, and the position of the injury (Hall 1982, Giulian et al 1980, Sinicropi & McIlwain 1983, Oblinger & Lasek 1988).

## MORPHOLOGICAL RESPONSE OF THE AXON

Axons begin to regenerate within a few hours of axotomy. The first sprouts in myelinated axons are generally seen coming from the terminal nodes of Ranvier, through the gap left by partial retraction of the Schwann cells (Friede & Bischhausen 1980, Meller 1987, McQuarrie 1985, Morris et al 1972); unmyelinated axons sprout equally rapidly (Bray et al 1972). While these sprouts are forming, the cut tip of the axon swells, inflated with smooth endoplasmic reticulum, mitochondria, and eventually micro-tubules. The regenerating sprouts, of which there may be several from each axon, grow down the endoneurial tubes; the growth cones are usually in contact with the Schwann cell basal lamina on one side, and with the Schwann cell membrane on the other (Nathaniel & Pease 1963, Haftek & Thomas 1968, Scherer & Easter 1984): Axons continue to grow in this fashion back toward their targets. Both myelinated and unmyelinated axons may branch distal to the site of transection (Bray & Aguayo 1974, Jenq et al 1987). Thus each axon proximal to the cut may give rise to several axons distal to it. However, not all these branches survive: Over the months following a nerve repair, some axons will enlarge and return to an approximately normal diameter (Cragg & Thomas 1964), whereas others will disappear. Subsequent axonal enlargement and maturation is dependent on connection with a target, and branches that disappear have presumably failed to do this (Aitken 1949).

What changes in the cell body are essential for successful axonal regeneration? In view of the rapid onset of growth and the relative slowness of axonal transport, it is clear that the axon can begin to sprout without intervention from the cell body (see below). However, the regenerative response is considerably more energetic in axons that have received a conditioning crush two days or more previously; more axons sprout, and regenerating axons grow faster (McQuarrie 1985) (although these effects are generally much more marked in lower vertebrates than mammals). Axons also regenerate in vitro much more vigorously from explants taken during the period of embryonic axonal growth, or from older explants whose axons have previously been induced to regenerate (Collins & Lee 1982). The enhanced regenerative response following a conditioning lesion may be due to the early availability in the axon of molecules associated with regeneration. However, neither chromatolysis nor some of the changes in neuronal metabolism described above may be indispensable for successful axonal regeneration. After section of sensory axons between cell body and spinal cord (rhizotomy), DRG cells show little or no histological reaction and little or no change in synthesis of various proteins and cytoskeletal components (Hare & Hinsey 1940, Hall 1982, Perry et al 1983, Greenberg & Lasek 1988). Despite this minimal response, regeneration occurs after rhizotomy, at least as far as the entry to the spinal cord (Bignami et al 1984, Liuzzi & Lasek 1987), which for lumbar and sacral ganglia can be a substantial distance.

On the available evidence one cannot make any definitive pronouncement on which cell body changes are necessary for axon regeneration. The changes that occur in general lead to an increased supply of substances that will be used to rebuild the growing nerve fiber, and also to increases in Gap-43, which may play a central role in the process of axonogenesis. This must at the very least have a facilitatory effect on regeneration.

It is not clear what signals between cell body and axon tip regulate the transition to regenerative growth, nor indeed whether events at the axon tip are controlled in the same way as those at the cell body. A generally accepted model for the induction of the regenerative response in the cell body invokes changes in a trophic factor normally derived from the terminal/target and conveyed to the soma by axonal transport (Cragg 1970). A number of experiments are consistent with this hypothesis. In the adult guinea pig, local application of NGF to a sympathetic ganglion prevents the cell body reaction due to post-ganglionic axotomy, whereas application of NGF antiserum induces some of these changes (Nja & Purves 1978). Local application of colchicine to block axonal transport in post-ganglionic nerves also induces chromatolysis (Purves 1976), although this latter finding is not general to all neurons, where the effects of colchicine are rather complex, and may even delay the onset of chromatolysis (Singer et al 1982). A number of observations are inconsistent with the view that loss of a retrograde signal to the cell body is the only way in which neuronal responses to axotomy are initiated. The earliest regenerative sprouting at the axon tip can occur within a few hours of axotomy, too rapidly for the cell body to have been informed and to have sent its response, even by fast axonal transport. Indeed, sprouting at mouse motor nerve terminals can occur even when axons are disconnected from their cell bodies (Brown & Lunn 1988), and an axon completely disconnected from its cell body can elaborate a new growth cone at the site of the cut (Mason & Muller 1982, Shaw & Bray 1977). These observations suggest that there must be control mechanisms acting rapidly and locally at the axon tip. Similar considerations apply during development, during which it would be impossible for the frequent changes in the rate of axonal growth to be regulated sufficiently rapidly from the cell body. Growth cone motility in some neurons has been shown to be affected by neurotransmitters and by electrical events, both of which may operate by modulating the calcium concentration inside the growth cone (Mattson et al 1988, Patel & Poo 1984, Kater et al 1988), and it is possible that the cut axon tip is subject to similar influences.

## INTERACTIONS BETWEEN REGENERATING AXONS AND THEIR ENVIRONMENT

The environment through which axons regenerate in the PNS consists of Schwann cells and their basal lamina, fibroblasts, and collagen, and, earlier in regeneration, axonal debris, degenerating myelin, and phagocytic cells. Of these, the Schwann cell is the critical factor promoting axon regeneration. Whenever live Schwann cells are absent from the terrain confronting regenerating axons, the axons either fail to grow, or their growth is much reduced. The other components of the damaged nerve, fibroblasts (perineurial cells) and basal lamina, are not sufficient by themselves. A variety of experiments demonstrate this. If a length of peripheral nerve is frozen, the Schwann cells will be killed, but the basal lamina surrounding them will remain intact. When a nerve treated in this way is grafted to the proximal stump in a host animal, axons will regenerate through the graft, but only when accompanied by Schwann cells migrating in from the host proximal stump (Gulati 1988, Hall 1986a, Ide et al 1983); the same is true for nerve grafts immunologically rejected by their host (Zalewski et al 1982). Similarly, axons will regenerate through grafts of basal lamina and extracellular matrix derived from muscle (Fawcett & Keynes 1986, Glasby et al 1986), but each regenerating axon is accompanied by Schwann cells (Fawcett & Keynes 1988). If the invasion of live Schwann cells into killed nerve grafts is prevented by cytotoxic agents, axons fail to regenerate (Hall 1986b).

It has recently been possible to characterize the molecules responsible for many of the interactions between axons and Schwann cells. The Schwann cell basal lamina is similar to basal laminae elsewhere, in that it contains molecules such as laminin and fibronectin, which are potent promoters of neurite growth in culture (Rogers et al 1983, Bozyczko & Horwitz 1986). Since regenerating axons are in contact with the basal lamina, it would be surprising if it played no role in axon regeneration. In support of such a role, an antibody to a laminin-heparan sulphate proteoglycan complex is able to inhibit the growth of neurites on sections of peripheral nerve (Sandrock & Matthew 1987), but the regeneration-promoting properties of Schwann cells cannot be attributed entirely to basal lamina. Two experiments demonstrate this: First, as described above, nerve grafts in which the Schwann cells have been killed are relatively ineffective unless invaded by live Schwann cells; and second, Schwann cells grown in serum free medium have little or no basal lamina, yet they promote axonal growth effectively (Ard et al 1987). By using antibodies to known cell adhesion molecules, it has been possible to inhibit much of the neurite-promoting activity of Schwann cells. Antibodies to three identified cell surface adhesion molecules, L1/Ng-CAM, N-cadherin, and integrins, when applied together inhibit virtually all axonal growth on Schwann cells (Bixby et al 1988, Seilheimer & Schachner 1988), but none completely prevents growth by itself. L1/Ng-CAM is present on both axons and Schwann cells in greatly increased quantities following nerve section (Daniloff et al 1986, Martini & Schachner 1988), and binds homophilically, although it may have other receptors (Grumet & Edelman 1988). N-Cadherin also binds homophilically, and has been demonstrated on neurons (Hatta et al 1988, 1987). NCAM is present on both regenerating growth cones and Schwann cells, and may also play a role in axon growth (Martini & Schachner 1988).

There are changes in the quantities of nerve growth factor (NGF) and its receptor in regenerating nerves. Very little NGF or NGF receptor is found in a normal peripheral nerve. However, if the nerve is cut or crushed, the level of both molecules and their respective mRNAs in the region distal to the injury increases enormously (Taniuchi et al 1986, Raivich & Kreutzberg 1987, Heumann et al 1987). The increase in NGF production is probably caused by interleukin-1 secreted by macrophages (Lindholm et al 1987). The expression of NGF receptor by Schwann cells is probably controlled by axonal contact; NGF receptor levels increase in the absence of axonal contact, and decrease when contact is restored (Taniuchi et al 1988). The receptor on Schwann cells is not internalized, and is probably of the type II low-affinity variety (Taniuchi et al 1988). The precise role of NGF and its receptor in nerve regeneration is unclear. Sensory neurons require NGF for survival at embryonic stages but retain little sensitivity to its removal in adulthood (Yip et al 1984, Lindsay 1988); motoneurons are apparently insensitive to the presence or absence of NGF, although in embryos they have receptors for it, internalize it, and retrogradely transport it (Yan et al 1988). Moreover, NGF has no discernable effect on Schwann cells. Johnson and his colleagues hypothesize that the NGF receptor may act as a cell surface adhesion molecule, the NGF forming a bridge between NGF receptors on axon and Schwann cell and also giving trophic support to sensory neurons (Johnson et al 1988).

In addition to surface interactions, Schwann cells are able to influence regenerating axons from a distance. If a nerve is cut and the proximal and distal stump left less than 1 cm apart, axons will grow towards and regenerate into the distal stump (Cajal 1928, Politis et al 1982). This phenomenon has been studied in some detail by Kuffler (1986), who has shown that the attraction depends on the presence of live cells in the distal nerve segment. Neurite-promoting activity is secreted by cut nerves into chambers inserted between their ends, as it is by Schwann cells in culture (Longo et al 1984, Cornbrooks et al 1983), and this activity may be responsible for attracting axons to a distal stump. Much of this neurite-promoting activity can be inhibited or precipitated from Schwann cell conditioned media by antibodies to laminin (Chiu et al 1986, Lander et al 1985).

The Schwann cell plays a crucial role in regeneration, and its multiplication, stimulated by macrophages (Baichwal et al 1988), myelin debris, and axonal membrane (Salzer & Bunge 1980, Salzer et al 1980, Bigbee et al 1987), is critical for allowing axons to regenerate across gaps (Scaravilli et al 1986, Jenq & Coggeshall 1987a, Le Beau et al 1988). However, some of the other changes described above may not be essential for successful regeneration: This is suggested by recent observations on a strain of mouse in which normal Wallerian degeneration does not occur (see above). In the distal stumps of cut nerves in these mice, axons can be seen regenerating in association with Schwann cells, even though the Schwann cells are still forming anatomically normal myelin sheaths around axons that have been cut but that have failed to degenerate (Lunn et al 1989). How many of the changes in Schwann cells and neurons that are detailed above occur in these animals will be interesting to see.

Remyelination of regenerated axons starts after about eight days. In many ways remyelination recapitulates development. Regenerating myelinated axons are initially embedded in the Schwann cells of their endoneurial tube, which then wrap around them and form myelin. This process may be inhibited by antibodies to N-CAM (Rieger et al 1988). As a result of the division of Schwann cells following injury, the distance between internodes is smaller than normal, usually around 400 $\mu$m, and the relation between axon diameter and internodal distance is less marked (Minwegen & Friede 1985, Cragg & Thomas 1964, Vizoso & Young 1948, Ghabriel & Allt 1977). There follows a period in which myelin is remodeled, and some Schwann cells lose contact with axons (Hildebrand et al 1986); internodal distances remain short, possibly because there is no period of passive stretching of the nerve as occurs with normal body growth. The persistence of short internodal length and abnormally thin myelin sheaths is probably the reason for the slow conduction velocities of regenerated nerves (Cragg & Thomas 1964, Minwegen & Friede 1985). Unmyelinated axons regenerate in contact with Schwann cells, often with large groups of axons associated with each cell. In time, each unmyelinated axon becomes enveloped individually in Schwann cell processes, and the proportion of axons to Schwann cells decreases (Aguayo et al 1973). The information that determines whether axons will be myelinated or not is carried by the axons; myelinated axons will be myelinated on regeneration, regardless of whether the Schwann cells were originally associated with myelinated or unmyelinated axons (Weinberg & Spencer 1975, Aguayo et al 1976).

## ACCURACY OF REGENERATION—GUIDANCE AND SPECIFICITY

For optimal recovery of function after peripheral nerve regeneration, axons need to reconnect with their original targets. Nerve fibers regenerate much more accurately following crush rather than cut injuries (Sunderland 1978); if the endoneurial tubes and Schwann cell basal lamina are left intact, as is often the case with crush injuries (Haftek & Thomas 1968), regenerating axons usually remain in their parent tubes and are guided directly back to their targets. If the tubes are disrupted, however, regenerating sprouts may enter inappropriate tubes in the distal stump, and so be guided to inappropriate targets. The accuracy of regeneration following cut injuries varies according to body region, age, and species, and these variations have potentially interesting implications for guidance mechanisms.

The occurrence of aberrant reinnervation after axotomy is well established in the neuromuscular systems of adult mammals and anurans (Sperry 1945, Westerfield & Powell 1983, Bernstein & Guth 1961, Brushart & Mesulam 1980), and the same has now been demonstrated for the goldfish oculomotor system by Scherer (Scherer 1986). This is in contrast to the results following transection of motor axons in neonatal rats (Aldskogius & Thomander 1986, Hardman & Brown 1987), larval anurans (Farel & Bemelmans 1986) and adult urodeles (Wigston 1986, Grimm 1971, Holder et al 1982), where regeneration can restore appropriate neuromuscular connections in the periphery. Indeed, in the case of urodeles, accurate reinnervation of an individual limb muscle can take place despite surgical displacement of the proximal nerve stump (Holder et al 1984) or the muscle (Wigston & Kennedy 1987). A degree of positional selectivity has also been shown in the reinnervation of the adult rat diaphragm and serratus anterior muscles (Laskowski & Sanes 1988).

The mechanisms responsible for the restoration of original innervation patterns in these examples are by no means clear. There are perhaps two broad possibilities. In one, specificity would originate in the target muscle cells, which could provide diffusible and/or surface-bound cues for motor axons. The segmentally selective reinnervation of transplanted rat inter-costal muscles by preganglionic autonomic axons could be interpreted in this way (Wigston & Sanes 1982). A further possibility is that Schwann cells or fibroblasts might provide guidance cues for axons (Wigston & Donohue 1988). There is increasing evidence that regenerating rat per-ipheral axons can to some extent identify and direct their growth into appropriate branches of nerve trunks (Politis 1985, Brushart 1988). This raises the possibility that specific axon guidance cues operative during embryonic development might function to a limited extent during later regeneration (Keynes 1987).

There is a well-known precedent for the restoration of functionally correct connections after regeneration—the peripheral autonomic nervous system. The classical experiments of Langley (1895) showed segmental selectivity, since confirmed (Guth & Bernstein 1961, Nja & Purves 1977, Purves et al 1981), in the reconnection of regenerating preganglionic axons with postganglionic nerve cells after axotomy. Regeneration to peripheral target tissues after postganglionic axotomy is, by contrast, nonselective (Langley 1897, Purves & Thompson 1979, Hendry et al 1986). Positional nonselectivity is also found in the re-innervation of skin by sensory axons after cutaneous nerve lesions in the cat (Horch 1979) and rat (Jackson & Diamond (1984).

A further issue, of some functional significance, is whether axons that have regenerated to the periphery can identify appropriate types of target organ. At the level of gross mismatch, sensory axons will not form synapses on muscle fibers. In muscle, motor axons exhibit little specificity when selecting among fast and slow twitch muscle fibers and spindles (Miledi & Stefani 1969, Brown & Butler 1974, Foehring et al 1986, Gillespie et al 1986, Hoh 1975, Scott 1987). In skin, regenerating cutaneous axons frequently reinnervate appropriate receptors; alternatively, if receptors have atrophied, axons may influence them to regenerate in their previous location (Zelena 1981, Ide 1986). The accuracy of reconnection may depend on the degree to which anatomical factors allow competition to take place between correct and incorrect regenerating terminals. Close proximity between competing terminals may allow a successful outcome in the case of reinnervation of autonomic ganglia. The molecular nature of the factors determining such reinnervation specificity is an important area for future research. Some molecules that allow regenerating motor axons to recognize endplates are located in the basal lamina (Sanes et al 1978, Sanes et al 1986).

# NERVE REPAIR

Injuries to peripheral nerves are common, and the prolonged paralysis and anaesthesia that result are particularly disabling. The aim of the surgeon is to repair a damaged nerve so that the maximum number of axons regenerate through the site of injury, and so that the maximum proportion of these grow back to appropriate targets. The causes of unsuccessful repair are twofold: failure of axons to regenerate, and the innervation of incorrect targets. The first leads to weakness and poor sensory recovery; the second to poor muscular control, since muscles will receive inputs from axons originally connected to different muscles, and to poor sensory recovery, since axons connect to inappropriate regions of skin, leading to qualitatively abnormal and poorly localized sensation (Sunderland 1978, Terzis 1981).

The surgical technique that will enable the maximum number of axons to regenerate depends on the nature of the injury: the basic requirement is to appose the cut ends of the nerve trunk in such a way as to ensure the minimum of scar formation and the best possible blood supply. If a nerve is cut cleanly, this can be done by sewing the epineurium together, using optimized surgical technique (De Medinaceli et al 1983). However, often a considerable length of the nerve is damaged, and unless the damaged segment is removed it will be replaced by scar tissue. If this length is excised, the nerve ends usually cannot be rejoined unless they are pulled together with considerable tension, or unless the nerve is mobilized free from the surrounding tissue to the detriment of its blood supply. Tension at the suture line results in scarring and poor regeneration (Samii & Wallenborn 1972). In order to avoid these problems, it is usual to insert an autograft of the patient's own sural nerve to bridge the gap (Millesi 1977, Jenq & Coggeshall 1987b). This is not always possible, however, and there is considerable interest in alternative graft materials. A promising contender for this purpose is muscle basal lamina: Since muscle fibers are long cylinders that run parallel to one another, removal of the myoplasm leaves behind extracellular matrix in the form of many fine parallel tubes (Keynes et al 1984). This material promotes good axon regeneration (Fawcett & Keynes 1986, Glasby et al 1986); axons grow into the tubes accompanied by Schwann cells (Fawcett & Keynes 1988). Entubulation repairs using collagenous tubes are also reported to be successful, and filling tubes with laminin gel is reported to improve regeneration (Madison et al 1987, Madison 1987).

Since regenerating axons show little ability to navigate, their target is largely determined by the endoneurial tubes into which they grow. It is obviously impossible to match each proximal stump axon to its correct tube, but some gross matching is possible. Peripheral nerves are divided into fascicles; and some distance before a branch point, axons heading for a particular target group together in a single fascicle (Ueyama 1978). Repairs in which proximal stump fascicles are joined to the matching distal stump fascicle result in improved accuracy of regeneration and better function, as long as the repair site is near the target of the axons (Young et al 1981, Horch & Burgess 1980, Grabb et al 1970). Axons going to different targets are intermingled further proximally to such a degree that fascicular repair is of less benefit. Further improvement in results might come from control of the Schwann cell response to axotomy, control of scarring at the suture line, development of better nerve grafts, but above all from improving the accuracy with which axons regenerate to their targets.

Acknowledgments

We thank Dennis Bray and Michael Sofroniew for their help in writing this article.

*Literature Cited*

Abercrombie, M., Johnson, M. L. 1946. Quantitative histology of Wallerian degeneration. 1. Nuclear population in rabbit sciatic nerve. *J. Anat.* 80: 37–50

Aguayo, A. J., Peyronnard, J. M., Bray, G. M. 1973. A quantitative ultrastructural study of regeneration from isolated proximal stumps of transected unmyelinated nerves. *J. Neuropath. Exp. Neurol.* 32: 256–70

Aguayo, A. J., Epps, J., Charron, L., Bray, G. M. 1976. Multipotentiality of Schwann cells in cross anastomosed and grafted myelinated and unmyelinated nerves: Quantitative microscopy and radioautography. *Brain Res.* 104: 1–20

Aitken, J. 1949. The effect of peripheral connexions on the maturation of regenerating nerve fibres. *J. Anat.* 83: 32–43

Aldskogius, H., Thomander, L. 1986. Selective re-innervation of somatotopically appropriate muscles after facial nerve transection and regeneration in the neonatal rat. *Brain Res.* 375: 126–34

Allt, G. 1976. Pathology of the peripheral nerve. In *The Peripheral Nerve*, ed. D. N. Landon, pp. 666–739. London: Chapman & Hall

Ard, M. D., Bunge, R. P., Bunge, M. B. 1987. Comparison of the Schwann cell surface and Schwann cell extracellular matrix as promoters of neurite growth. *J. Neurocytol.* 16: 539–55

Baichwal, R. R., Bigbee, J. W., DeVries, G. H. 1988. Macrophage-mediated myelin-related mitogenic factor for cultured Schwann cells. *Proc. Natl. Acad. Sci. USA* 85: 1701–5

Bernstein, J. J., Guth, L. 1961. Non-selectivity in establishment of neuromuscular connections following nerve regeneration in the rat. *Exp. Neurol.* 4: 262–75

Beuche, W., Friede, R. L. 1984. The role of non-resident cells in Wallerian degeneration. *J. Neurocytol.* 13: 767–96

Bigbee, J. W., Yoshino, J. E., DeVries, G. H. 1987. Morphological and proliferative responses of cultured Schwann cells following rapid phagocytosis of a myelin-enriched fraction. *J. Neurocytol.* 16: 487–96

Bignami, A., Chi, N. M., Dahl, D. 1984. Regenerating dorsal roots and the nerve entry zone: An immunofluoroescence study with neurofilament and laminin antisera. *Exp. Neurol.* 85: 426–36

Bixby, J. L., Lilien, J., Reichardt, L. F. 1988. Identification of the major proteins that promote neuronal process outgrowth on Schwann cells in vitro. *J. Cell Biol.* 107: 353–61

Bozyczko, D., Horwitz, A. F. 1986. The participation of a putative cell surface receptor for laminin and fibronectin in peripheral neurite extension. *J. Neurosci.* 6: 1241–51

Bradley, W. G., Asbury, A. K. 1970. Duration of synthesis phase in neurilemma cells in mouse sciatic nerve during degeneration. *Exp. Neurol.* 26: 275–82

Bray, G. M., Peyronnard, J. M., Aguayo, A. J. 1972. Reactions of unmyelinated nerve fibres to injury—an ultrastructural study. *Brain Res.* 42: 297–309

Bray, G. M., Aguayo, A. J. 1974. Regeneration of peripheral unmyelinated nerves. Fate of the axonal sprouts which develop after injury. *J. Anat.* 117: 517–29

Brown, M. C., Butler, R. G. 1974. Evidence for innervation of muscle spindle intrafusal fibres by branches of alpha motoneurons following nerve injury. *J. Physiol.* 238: 41–2

Brown, M. C., Lunn, E. R. 1988. Mechanism of interaction between motoneurons and muscles. In *Plasticity of the Neuromuscular System, Ciba Symp.* 138: 78–96. Chichester: Wiley

Brushart, T. M. 1988. Preferential reinnervation of motor nerves by regenerating motor axons. *J. Neurosci.* 8: 1026–31

Brushart, T. M., Mesulam, M. M. 1980. Alteration in connections between muscle and anterior horn neurons. *Science* 208: 603–5

Cajal, S. Ramon y. 1928. In *Degeneration and Regeneration of the Nervous System*, ed. R. M. May. Oxford: Oxford Univ. Press

Chiu, A. Y., Matthew, W. D., Patterson, P. H. 1986. A monoclonal antibody that blocks the activity of a neurite regeneration-promoting factor: Studies on the binding site and its localization in vivo. *J. Cell Biol.* 103: 1383–98

Collins, F., Lee, M. R. 1982. A reversible developmental change in the ability of ciliary ganglion neurons to extend neurites in culture. *J. Neurosci.* 2: 424–30

Cornbrooks, C. J., Carey, D. J., McDonald, J. A., Timpl, R., Bunge, R. P. 1983. In vivo and in vitro observations on laminin production by Schwann cells. *Proc. Natl. Acad. Sci. USA* 80: 3850–54

Cragg, B. G. 1970. What is the signal for chromatolysis. *Brain Res.* 23: 1–21

Cragg, B. G., Thomas, P. K. 1964. The conduction velocity of regenerated peripheral nerve fibres. *J. Physiol.* 171: 164–75

Crang, A. J., Blakemore, W. F. 1987. Observations on the migratory behaviour of Schwann cells from adult peripheral nerve explant cultures. *J. Neurocytol.* 16: 423–31

Daniloff, J. K., Levi, G., Grumet, M., Rieger, F., Edelman, G. M. 1986. Altered expression of neuronal cell adhesion molecules induced by nerve injury and repair. *J. Cell Biol.* 103: 929–45

De Medinaceli, L., Wyatt, R. J., Freed, W. J. 1983. Peripheral nerve reconnection: Mechanical, thermal and ionic conditions that promote the return of function. *Exp. Neurol.* 81: 469–87

Farel, P. B., Bemelmans, S. E. 1986. Restoration of neuromuscular specificity following ventral rhizotomy in the bullfrog tadpole, *Rana catesbeiana*. *J. Comp. Neurol.* 254: 125–32

Fawcett, J. W., Keynes, R. J. 1986. Muscle basal lamina: A new graft material for peripheral nerve repair. *J. Neurosurg.* 65: 354–63

Fawcett, J. W., Keynes, R. J. 1988. Axonal growth and Schwann cell migration into basal lamina grafts to the rat sciatic nerve. *Soc. Neurosci. Abstr.* 14: 164

Foehring, R. C., Sypert, G. W., Munson, J. B. 1986. Properties of self-reinnervated motor units of medial gastrocnemius of cat. 1. Long-term reinnervation. *J. Neurophysiol.* 55: 931–65

Friede, R. L., Bischhausen, R. 1980. The fine structure of stumps of transected nerve fibers in subserial sections. *J. Neurol. Sci.* 44: 181–203

Ghabriel, M. N., Allt, G. 1977. Regeneration of the node of Ranvier: A light and electron microscope study. *Acta Neuropath.* 37: 153–63

Gillespie, M. J., Gordon, T., Murphy, P. R. 1986. Reinnervation of the lateral gastrocnemius and soleus muscles in the rat by their common nerve. *J. Physiol.* 372: 485–500

Giulian, D., Des Ruisseaux, H., Cowburn, D. 1980. Biosynthesis and intra-axonal transport of proteins during neuronal regeneration. *J. Biol. Chem.* 255: 6494–6501

Glasby, M. A., Gschmeissner, S. G., Hitchcock, R. J., Huang, C. L. 1986. The dependence of nerve regeneration through muscle grafts in the rat on the availability and orientation of basement membrane. *J. Neurocytol.* 15: 497–510

Goldstein, M. E., Cooper, H. S., Bruce, J., Carden, M. J., Lee, V. M. Y., Schlaepfer, W. W. 1987. Phosphorylation of neurofilament proteins and chromatolysis following transection of rat sciatic nerve. *J. Neurosci.* 7: 1586–94

Grabb, W. C., Bement, S. L., Koepke, G. H., Green, R. A. 1970. Comparison of methods of peripheral nerve suturing in monkeys. *Plast. Reconst. Surg.* 46: 31–38

Grafstein, B., McQuarrie, I. G. 1978. In *Neuronal Plasticity*, ed. C. W. Cotman, pp. 155–95. New York: Raven

Greenberg, S. G., Lasek, R. J. 1988. Neurofilament protein synthesis in DRG neurons decreases more after peripheral axotomy than after central axotomy. *J. Neurosci.* 8: 1739–46

Grimm, L. M. 1971. An evaluation of myotypic respecification in axolotls. *J. Exp. Zool.* 178: 479–96

Grumet, M., Edelman, G. M. 1988. Neuron-glia cell adhesion molecule interacts with neurons and astroglia via different binding mechanisms. *J. Cell Biol.* 106: 487–504

Gulati, A. K. 1988. Evaluation of acellular and cellular nerve grafts in repair of rat peripheral nerve. *J. Neurosurg.* 68: 117–23

Guth, L., Bernstein, J. J. 1961. Selectivity in the re-establishment of synapses in the superior cervical ganglion of the cat. *Exp. Neurol.* 4: 59–69

Haftek, J., Thomas, P. K. 1968. Electron microscope obsevations on the effects of localised crush injuries on the connective tissues of peripheral nerves. *J. Anat.* 103: 233–43

Hall, M. E. 1982. Changes in synthesis of specific proteins in axotomized dorsal root ganglia. *Exp. Neurol.* 76: 83–93

Hall, S. M. 1986a. Regeneration in cellular and acellular autografts in the peripheral nervous system. *Neuropathol. Appl. Neurobiol.* 12: 27–46

Hall, S. M. 1986b. The effect of inhibiting Schwann cell mitosis on the re-innervation of acellular autografts in the peripheral nervous system of the mouse. *Neuropathol. Appl. Neurobiol.* 12: 401–14

Hallpike, J. F. 1976. Histochemistry of peripheral nerves and nerve terminals. In *The Peripheral Nerve*, ed. D. N. Landon, pp. 605–65. London: Chapman & Hall

Hardman, V. J., Brown, M. C. 1987. Accuracy of reinnervation of rat internal intercostal muscles by their own segmental nerves. *J. Neurosci.* 7: 1031–36

Hare, W. K., Hinsey, J. C. 1940. Reactions of dorsal root ganglion cells to section of peripheral and central processes. *J. Comp. Neurol.* 73: 489–502

Hatta, K., Takagi, S., Fujisawa, H., Takeichi, M. 1987. Spatial and temporal expression pattern of N-cadherin cell adhesion molecules correlated with morphogenetic processes of chicken embryos. *Dev. Biol.* 120: 215–27

Hatta, K., Nose, A., Nagafuchi, A., Takeichi, M. 1988. Cloning and expression of cDNA encoding a neural calcium-dependent cell adhesion molecule: Its identity to the cadherin gene family. *J. Cell Biol.* 106: 873–81

Hendry, I. A., Hill, C. E., Watters, D. J. 1986. Long-term retention of Fast Blue in sympathetic neurones after axotomy and regeneration—demonstration of incorrect reconnections. *Brain. Res.* 376: 292–298

Heumann, R., Lindholm, D., Bandtlow, C., Meyer, M., Radeke, M. J., Misko, T. P., Shooter, E., Thoenen, H. 1987. Differential regulation of mRNA encoding nerve growth factor and its receptor in rat sciatic nerve during development, degeneration, and regeneration: Role of macrophages. *Proc. Natl. Acad. Sci. USA* 84: 8735–39

Hildebrand, C., Mustafa, G. Y., Waxman, S. G. 1986. Remodelling of internodes in regenerated rat sciatic nerve: Electron microscopic observations. *J. Neurocytol.* 15: 681–92

Hoffman, P. N., Cleveland, D. W., Griffin, J. W., Landes, P. W., Cowan, N. J., Price, D. L. 1987. Neurofilament gene expression: A major determinant of axonal caliber. *Proc. Natl. Acad. Sci. USA* 84: 3472–76

Hoffman, P. N., Cleveland, D. W. 1988. Neurofilament and tubulin expression recapitulates the development program during axonal regeneration: Induction of a specific beta-tubulin isotype. *Proc. Natl. Acad. Sci. USA* 85: 4530–33

Hoh, J. F. 1975. Selective and non-selective reinnervation of fast-twitch and slow-twitch rat skeletal muscle. *J. Physiol.* 251: 791–801

Holder, N., Mills, J., Tonge, D. A. 1982. Selective reinnervation of skeletal muscle in the newt. *Triturus cristatus*. *J. Physiol.* 326: 371–84

Holder, N., Tonge, D. A., Jesani, P. 1984. Directed regrowth of axons from a misrouted nerve to their correct muscles in the limb of the adult newt. *Proc. R. Soc. London Ser. B* 222: 477–89

Horch, K. 1979. Guidance of regrowing sensory axons after cutaneous nerve lesions in the cat. *J. Neurophysiol.* 42: 1437–49

Horch, K. W., Burgess, P. R. 1980. Functional specificity and somatotopic organization during peripheral nerve regeneration. In *Nerve Repair and Regeneration*, ed. J. L. Jewett, H. R. McCarrol. St. Louis, Mo.: Mosby

Ide, C. 1986. Basal laminae and Meissner corpuscle regeneration. *Brain. Res.* 384: 311–22

Ide, C., Tohyama, K., Yokota, R., Nitatori, T., Onodepa, H. 1983. Schwann cell basal lamina and nerve regeneration. *Brain Res.* 288: 61–65

Ignatius, M. J., Gebicke-Harter, P. J., Skene, J. H., Schilling, J. W., Weisgraber, K. H., Mahley, R. W., Shooter, E. M. 1986. Expression of apolipoprotein E during nerve degeneration and regeneration. *Proc. Natl. Acad. Sci. USA* 83: 1125–29

Ignatius, M. J., Shooter, E. M., Pitas, R. E., Mahley, R. W. 1987. Lipoprotein uptake by neuronal growth cones in vitro. *Science* 236: 959–62

Jackson, P. C., Diamond, J. 1984. Temporal and spatial constraints on the collateral sprouting of low-threshold mechanosensory nerves in the skin of rats. *J. Comp. Neurol.* 226: 336–45

Jenq, C. B., Coggeshall, R. E. 1987a. Permeable tubes increase the length of the gap that regenerating axons can span. *Brain Res.* 408: 239–42

Jenq, C. B., Coggeshall, R. E. 1987b. Sciatic nerve regeneration after autologous sural nerve transplantation in the rat. *Brain. Res.* 406: 52–61

Jenq, C. B., Jenq, L. L., Coggeshall, R. E. 1987. Numerical patterns of axon regeneration that follow sciatic nerve crush in the neonatal rat. *Exp. Neurol.* 95: 492–99

Johnson, E. M., Taniuchi, M., DiStefano, P. S. 1988. Expression and possible function of nerve growth factor receptors on Schwann cells. *Trends Neurosci.* 11: 299–304

Karns, L. R., Ng, S. C., Freeman, J. A., Fishman, M. C. 1987. Cloning of complementary DNA for GAP-43, a neuronal growth-related protein. *Science* 236: 597–600

Kater, S. B., Mattson, M. P., Cohan, C., Connor, J. 1988. Calcium regulation of the neuronal growth cone. *Trends Neurosci.* 11: 315–21

Keynes, R. J. 1987. Schwann cells during neural development and regeneration: Leaders or followers? *Trends Neurosci.* 10: 137–39

Keynes, R. J., Hopkins, W. G., Huang, L. H. 1984. Regeneration of mouse peripheral nerves in degenerating skeletal muscle: Guidance by residual muscle fibre basement membrane. *Brain Res.* 295: 275–81

Kuffler, D. P. 1986. Isolated satellite cells of a peripheral nerve direct the growth of regenerating frog axons. *J. Comp. Neurol.* 249: 57–64

Lander, A. D., Fujii, D. K., Reichardt, L. F. 1985. Laminin is associated with the neurite outgrowth promoting factors found in conditioned media. *Proc. Natl. Acad. Sci. USA* 82: 2183–87

Langley, J. N. 1895. Note on regeneration of preganglionic fibres of the sympathetic. *J. Physiol.* 18: 280–84

Langley, J. N. 1897. On the regeneration of preganglionic and of post-ganglionic visceral nerve fibres. *J. Physiol.* 22: 215–30

Lasek, R. J., Garner, J. A., Brady, S. T. 1984. Axonal transport of the cytoplasmic matrix. *J. Cell Biol.* 99: 212–21

Laskowski, M. B., Sanes, J. R. 1988. Topographically selective reinnervation of adult mammalian skeletal muscles. *J. Neurosci.* 8: 3094–99

Le Beau, J. M., Ellisman, M. H., Powell, H. C. 1988. Ultrastructural and morphometric analysis of long term peripheral nerve regeneration through silicone tubes. *J. Neurocytol.* 17: 161–72

Lieberman, A. R. 1971. The axon reaction: A review of the principal features of perikaryal responses to axon injury. *Int. Rev. Neurobiol.* 14: 49–124

Lindholm, D., Heumann, R., Meyer, M., Thoenen, H. 1987. Interleukin-1 regulates synthesis of nerve growth factor in non-neuronal cells of rat sciatic nerve. *Nature* 330: 658–59

Lindsay, R. M. 1988. Nerve growth factors (NGF, BDNF) enhance axonal regeneration but are not required for survival of adult sensory neurons. *J. Neurosci.* 8: 2394–2405

Liuzzi, F. J., Lasek, R. J. 1987. Astrocytes block axonal regeneration in mammals by activating the physiological stop pathway. *Science* 237: 642–45

Longo, F. M., Hayman, E. G., Davis, G. E., Ruoslahti, E., Engvall, E., Manthorpe, M., Varon, S. 1984. Neurite-promoting factors and extracellular matrix components accumulating in vivo within nerve regeneration chambers. *Brain Res.* 309: 105–17

Lunn, E. R., Perry, V. H., Brown, M. C., Rosen, H., Gordon, S. 1989. Absence of Wallerian degeneration does not hinder regeneration in peripheral nerve. *Eur. J. Neurosci.* 1: 27–33

Madison, R. 1987. A tubular prosthesis for nerve repair: Effects of basement membrane materials, laminin and porosity. *Soc. Neurosci. Abstr.* 13: 1042

Madison, R. D., da Silva, C., Dikkes, P., Sidman, R. L., Chiu, T. H. 1987. Peripheral nerve regeneration with entubulation repair: Comparison of biodegradeable nerve guides versus polyethylene tubes and the effects of a laminin-containing gel. *Exp. Neurol.* 95: 378–90

Martini, R., Schachner, M. 1988. Immunoelectron microscopic localization of neural cell adhesion molecules. (L1, N-CAM, and myelin-associated glycoprotein) in regenerating adult mouse sciatic nerve. *J. Cell Biol.* 106: 1735–46

Mason, A., Muller, K. J. 1982. Axon segments sprout at both ends: Tracking growth with fluorescent D-peptides. *Nature* 296: 655–57

## 58 FAWCETT & KEYNES

Mattson, M. P., Lee, R. E., Adams, M. E., Guthrie, P. B., Kater, S. B. 1988. Interactions between entorhinal axons and target hippocampal neurons: A role for glutamate in the development of hippocampal circuitry. *Neuron* 1: 865–76

McQuarrie, I. G. 1985. Effect of a conditioning lesion on axonal sprout formation at nodes of Ranvier. *J. Comp. Neurol.* 231: 239–49

Meiri, K. F., Willard, M., Johnson, M. I. 1988. Distribution and phosphorylation of the growth associated protein GAP 43 in regenerating sympathetic neurons in culture. *J. Neurosci.* 8: 2571–81

Meller, K. 1987. Early structural changes in the axoplasmic cytoskeleton after axotomy studied by cryofixation. *Cell Tissue Res.* 250: 663–72

Miledi, R., Stefani, E. 1969. Non-selective reinnervation of slow and fast muscle fibres in the rat. *Nature* 222: 569–71

Miller, F. D., Naus, C. C. G., Durand, M., Bloom, F. E., Milner, R. J. 1987. Isotypes of alpha tubulin are differentially regulated during neuronal maturation. *J. Cell Biol.* 105: 3065–73

Miller, F. D., Tetzlaff, W., Bisby, M. A., Fawcett, J. W., Milner, R. J. 1989. Rapid induction of the major embryonic alpha-tubulin mRNA, T alpha 1, during nerve regeneration in adult rats. *J. Neurosci.* 9: 1452–63

Millesi, H. 1977. Nerve grafting. In *Reconstructive Plastic Surgery*, ed. J. M. Converse, pp. 3227–42. Philadelphia: Saunders

Minwegen, P., Friede, R. L. 1985. A correlative study of internode proportions and sensitivity to procaine in regenerated frog sciatic nerves. *Exp. Neurol.* 87: 147–64

Morris, J. H., Hudson, A. R., Weddell, G. 1972. A study of degeneration and regeneration in divided rat sciatic nerve based on electron microscopy. *Z. Zellforsch.* 124: 103–30

Nathaniel, E. J., Pease, D. C. 1963. Regenerative changes in rat dorsal roots following Wallerian degeneration. *J. Ultrastruct. Res.* 9: 533–49

Nja, A., Purves, D. 1977. Specific reinnervation of guinea pig superior cervical ganglion cells by preganglionic fibers arising from different levels of the spinal cord. *J. Physiol.* 264: 565–83

Nja, A., Purves, D. 1978. The effects of nerve growth factor and its antiserum on synapses in the superior cervical ganglion of the guinea pig. *J. Physiol.* 277: 55–75

Oblinger, M. M., Lasek, R. J. 1988. Axotomy-induced alterations in the synthesis and transport of neurofilaments and microtubules in dorsal root ganglion cells. *J. Neurosci.* 8: 1747–58

Patel, N. B., Poo, M.-m. 1984. Perturbation of the direction of neurite growth by pulsed and focal electric fields. *J. Neurosci.* 4: 2939–47

Perry, G. W., Kryanek, S. R., Wilson, D. L. 1983. Protein synthesis and fast axonal transport during regeneration of dorsal roots. *J. Neurochem.* 40: 1590–98

Perry, V. H., Brown, M. C., Gordon, S. 1987. The macrophage response to central and peripheral nerve injury. A possible role for macrophages in regeneration. *J. Exp. Med.* 165: 1218–23

Politis, M. J., Ederle, K., Spencer, P. S. 1982. Tropism in nerve regeneration in vivo. Attraction of regenerating axons by diffusible factors derived from cells in distal nerve stumps of transected peripheral nerves. *Brain Res.* 253: 1–12

Politis, M. J. 1985. Specificity in mammalian peripheral nerve regeneration at the level of the nerve trunk. *Brain Res.* 328: 271–76

Purves, D. 1976. Functional and structural changes in mammalian sympathetic neurones following colchicine application to postganglionic nerves. *J. Physiol.* 259: 159–75

Purves, D., Thompson, W. 1979. The effects of post-ganglionic axotomy on selective synaptic connections in the superior cervical ganglion of the guinea pig. *J. Physiol.* 297: 95–110

Purves, D., Thompson, W., Yip, J. W. 1981. Reinnervation of ganglia transplanted to the neck from different levels of the guinea pig sympathetic chain. *J. Physiol.* 313: 49–63

Purves, D., Voyvodic, J. T. 1987. Imaging mammalian nerve cells and their connections over time in living animals. *Trends Neurosci.* 10: 398–404

Raivich, G., Kreutzberg, G. W. 1987. Expression of growth factor receptors in injured nervous tissue. I. Axotomy leads to a shift in the cellular distribution of specific beta-nerve growth factor binding in the injured and regenerating PNS. *J. Neurocytol.* 16: 689–700

Rieger, F., Nicolet, M., Pincon-Raymond, M., Murawsky, M., Levi, G., Edelman, G. M. 1988. Distribution and role in regeneration of N-CAM in the basal laminae of muscle and Schwann cells. *J. Cell Biol.* 107: 707–19

Rogers, S. L., Letourneau, P. C., Palm, S. L., McCarthy, J., Furcht, L. T. 1983. Neurite extension by peripheral and central nervous system neurons in response to substratum-bound fibronectin and laminin. *Dev. Biol.* 98: 212–20

Salzer, J. L., Bunge, R. P. 1980. Studies of

Schwann cell proliferation. 1. An analysis in tissue culture of proliferation during development, Wallerian degeneration and direct injury. *J. Cell Biol.* 84: 739–51

Salzer, J. L., Williams, A. K., Glaser, L., Bunge, R. P. 1980. Studies of Schwann cell proliferation. II. Characterization of the stimulation and specificity of the response to a neurite membrane fraction. *J. Cell Biol.* 84: 753–66

Samii, M., Wallenborn, R. 1972. Tierexperimentelle Untersuchunger über den Einfluss der Spannung auf den Regenerationserfolg nach Nervenaht. *Acta Neurochir.* 27: 87–110

Sandrock, A. W., Jr., Matthew, W. D. 1987. Identification of a peripheral nerve neurite growth-promoting activity by development and use of an in vitro bioassay. *Proc. Natl. Acad. Sci. USA* 84: 6934–38

Sanes, J. R., Marshall, L. M., McMahan, U. J. 1978. Reinnervation of muscle fiber basal lamina after removal of myofibers. *J. Cell Biol.* 78: 176–98

Sanes, J. R., Schachner, M., Covault, J. 1986. Expression of several adhesive macromolecules (N-CAM, L1, J1, NILE, uvomorulin, laminin, fibronectin, and a heparan sulfate proteoglycan) in embryonic, adult, and denervated adult skeletal muscle. *J. Cell Biol.* 102: 420–31

Scaravilli, F., Love, S., Myers, R. 1986. X-irradiation impairs regeneration of peripheral nerve across a gap. *J. Neurocytol.* 15: 439–49

Scheidt, P., Friede, R. L. 1987. Myelin phagocytosis in Wallerian degeneration. Properties of millipore diffusion chambers and immunohistochemical identification of cell populations. *Acta Neuropathol.* 75: 77–84

Scherer, S. S. 1986. Reinnervation of the extraocular muscles in goldfish is nonselective. *J. Neurosci.* 6: 764–73

Scherer, S. S., Easter, S. S. 1984. Degenerative and regenerative changes in the trochlear nerve of the goldfish. *J. Neurocytol.* 13: 519–65

Scott, J. J. 1987. The reinnervation of cat muscle spindles by skeletofusimotor axons. *Brain Res.* 401: 152–54

Seilheimer, B., Schachner, M. 1988. Studies of adhesion molecules mediating interactions between cells of peripheral nervous system indicate a major role for L1 in mediating sensory neuron growth on Schwann cells in culture. *J. Cell Biol.* 107: 341–51

Shaw, G., Bray, D. 1977. Movement and extension of isolated growth cones. *Exp. Cell Res.* 104: 55–62

Singer, P. A., Mehler, S., Fernandez, H. L. 1982. Blockade of retrograde axonal transport delays the onset of metabolic and morphological changes induced by axotomy. *J. Neurosci.* 2: 1299–1306

Sinicropi, D. V., McIlwain, D. L. 1983. Changes in the amounts of cytoskeletal proteins within the perikarya and axons of regenerating frog motoneurons. *J. Cell Biol.* 96: 240–47

Skene, J. H., Jacobson, R. D., Snipes, G. J., McGuire, C. B., Norden, J. J., Freeman, J. A. 1986. A protein induced during nerve growth (GAP-43) is a major component of growth-cone membranes. *Science* 233: 783–86

Snipes, G. J., McGuire, C. B., Norden, J. J., Freeman, J. A. 1986. Nerve injury stimulates the secretion of apolipoprotein E by nonneuronal cells. *Proc. Natl. Acad. Sci. USA* 1130–34

Sperry, R. W. 1945. The problem of central nervous system reorganization after nerve regeneration and muscle transposition. *Q. Rev. Biol.* 20: 311–69

Sunderland, S. 1978. In *Nerves and Nerve Injuries.* Edinburgh: Churchill Livingstone. 2nd ed.

Taniuchi, M., Clark, H. B., Johnson, E. M. Jr. 1986. Induction of nerve growth factor receptor in Schwann cells after axotomy. *Proc. Natl. Acad. Sci. USA* 83: 4094–98

Taniuchi, M., Clark, H. B., Schweitzer, J. B., Johnson, E.M. Jr. 1988. Expression of nerve growth factor receptors by Schwann cells of axotomized peripheral nerves: Ultrastructural location, suppression by axonal contact, and binding properties. *J. Neurosci.* 8: 664–81

Terzis, J. 1981. Patterns of cutaneous innervation and reinnervation following nerve transection. In *Posttraumatic Peripheral Nerve Regeneration*, ed. A. Gorio, H. Millesi, S. Mingrino, pp. 591–610, New York: Raven

Thomas, P. K. 1966. The cellular response to injury. 1. The cellular outgrowth from the distal stump of transected nerve. *J. Anat.* 100: 287–303

Ueyama, T. 1978. Course of fibres from different roots in dog sciatic nerve. *J. Anat.* 127: 277–89

Verhaagen, J., Oestreicher, A. B., Edwards, P. M., Veldman, H., Jennekens, F. G. I., Gispen, W. H. 1988. Light and electron-microscopical study of phosphoprotein B-50 following denervation and reinnervation of rat soleus muscle. *J. Neurosci.* 8: 1759–66

Vizoso, A. D., Young, J. Z. 1948. Internode length and fibre diameter in developing and regenerating nerves. *J. Anat.* 82: 110–34

Watson, W. E. 1974. Cellular responses to

axotomy and to related procedures. *Br. Med. Bull.* 30: 112–15

Weinberg, H. J., Spencer, P. S. 1975. Studies on the control of myelinogenesis. 1. Myelination of regenerating axons after entry into a foreign unmyelinated nerve. *J. Neurocytol.* 4: 395–418

Westerfield, M., Powell, S. L. 1983. Selective re-innervation of limb muscles of regenerating frog motor axons. *Dev. Brain Res.* 10: 301–4

Wigston, D. J. 1986. Selective innervation of transplanted limb muscles by regenerating motor axons in the axolotl. *J. Neurosci.* 6: 2757–63

Wigston, D. J., Donohue, S. P. 1988. The location of cues promoting selective reinnervation of axolotl muscles. *J. Neurosci.* 8: 3451–58

Wigston, D. J., Kennedy, P. R. 1987. Selective reinnervation of transplanted muscles by their original motoneurons in the axolotl. *J. Neurosci.* 7: 1857–65

Wigston, D. J., Sanes, J. R. 1982. Selective reinnervation of adult mammalian muscles by axons from different segmental levels. *Nature* 299: 464–67

Williams, P. L., Hall, S. M. 1971. Chronic Wallerian degeneration—an in vivo and ultrastructural study. *J. Anat.* 109: 487–503

Yan, Q., Snider, W. D., Pinzone, J. J., Johnson, E. M. 1988. Retrograde transport of nerve growth factor in motoneurons of developing rats: Assessment of potential neurotrophic effects. *Neuron* 1: 335–43

Yip, H. K., Rich, K. M., Lampe, P. A., Johnson, E. M. Jr. 1984. The effects of nerve growth factor and its antiserum on the postnatal development and survival after injury of sensory neurons in rat dorsal root ganglia. *J. Neurosci.* 4: 2986–92

Young, L., Wray, R., Weeks, P. 1981. A randomized prospective comparison of fascicular and epineural digital nerve repairs. *Plast. Reconstr. Surg.* 68: 89–93

Zalewski, A. A., Silvers, W. K., Gulati, A. K. 1982. Failure of host axons to regenerate through a once successful but later rejected long nerve allograft. *J. Comp. Neurol.* 209: 347–51

Zelena, J. 1981. The fate of Pacinian corpuscles after denervation and renervation. See Terzis 1981, pp. 563–71

*Annu. Rev. Neurosci. 1990. 13:61–73*

# ONTOGENY OF THE SOMATOSENSORY SYSTEM: Origins and Early Development of Primary Sensory Neurons

*Alun M. Davies*

Department of Anatomy, St. George's Hospital Medical School, London SW17 0RE, United Kingdom

*Andrew Lumsden*

Department of Anatomy, United Medical and Dental Schools, Guy's Hospital, London SE1 9RT, United Kingdom

## Introduction

Primary sensory neurons convey information to the CNS from a variety of sensory receptors in the periphery. In development, these neurons originate from progenitors that migrate from the neural crest and certain ectodermal placodes to the sites of developing sensory ganglia where they differentiate. Two axonal processes grow in opposite directions from the cell bodies of these early neurons to reach their peripheral and central target fields. The innervation of these target fields is, as elsewhere, associated with a period of neuronal death during which superfluous neurons are eliminated, followed by a period of modification and refinement of connections.

The accessibility of sensory ganglia from the earliest stages of their development is the main reason that much of our understanding of the cellular and molecular basis of neuronal development is founded on sensory neurons. The developing target fields of certain sensory ganglia are also well-defined and accessible for experimental studies, and this has permitted direct investigation of the regulatory influence of the target field on neuronal development. In this review we provide a chronological

61

0147–006X/90/0301–0061$02.00

account of sensory neuron development, with an emphasis on some fundamental principles that have emerged from studying these neurons.

## Origins of Sensory Neurons

Primary sensory neurons originate from thickenings of the embryonic ectoderm. The principal site of origin is the crest at the lateral margins of the neural plate, where cells detach from the neural epithelium and form all of the sensory neurons of spinal nerves and some of those of cranial sensory nerves. At a number of discrete locations in the head, sensory neurons also develop from the neurogenic placodes (D'Amico-Martel & Noden 1983); such ectodermal thickenings contribute neurons to all of the branchial nerves (V, VII, IX, X) and to the statoacoustic nerve (VIII). In the ganglion of the Vth nerve (trigeminal), neural crest-derived neurons occupy a dorso-medial position, whereas those of placodal origin lie ventro-lateral. In the VIIth, IXth, and Xth nerves, neural crest-derived neurons and placode-derived neurons occupy distinct proximal and distal ganglia.

In higher vertebrates, the only primary sensory neurons not associated with either dorsal root ganglia (DRG) or cranial sensory ganglia (CSG) are those of the mesencephalic nucleus of the trigeminal nerve. This neuronal population lies within the CNS, and is derived from neural crest cells (Narayanan & Narayanan 1978) whose migration is centripetal. The satellite cells and Schwann cells associated with both DRG and CSG arise exclusively from the neural crest.

## Multipotency of Neural Crest Precursors

Although the crest origins of both CSG and DRG neurons have been mapped in detail, little is known about the potentials of individual crest cells or the mechanisms whereby neuronal derivatives become distinct from others. At the time of initial migration from the neural epithelium, all crest cells look alike and there are few markers that distinguish any subsets (Barald 1982, Weston et al 1984, Girdlestone & Weston 1985).

Heterotopic grafting of various crest regions suggests that cell *populations* contain the full range of developmental potentials; for example, forebrain crest (which normally does not produce neural derivatives) gives rise to normal trigeminal ganglia when transposed to the metencephalic level and to DRG when transposed to the trunk level (Le Douarin et al 1986). Such studies demonstrate plasticity in crest populations and indicate that the specific environment through which (or into which) they migrate is important in the determination of cell fate. These studies do not show, however, whether the range of differentiation potentials is held in a single pluripotent precursor or is distributed among several partly or completely committed precursors that could all be present at each axial level.

Lineage analysis, both in situ and in vitro, reveals that pluripotent cells do exist in the neural crest but that the number of potencies available to any one cell becomes progressively restricted as the cells emerge from the margins of the neural plate (or tube) and disperse into the periphery. Single dorsal neural tube cells in chick embryos, labeled with an intracellular fluorescent dye, give rise to progeny in the majority of crest-derived tissues of the trunk: DRG, the pathways of peripheral nerves (probably Schwann cells), sympathetic ganglia, adrenal medulla, and sub-ectodermal spaces (Bronner-Fraser & Fraser 1988). Crest founder cells (or their daughters) may thus have the full range of potentials, including both sensory and sympathetic neuron phenotypes. Later, as the crest migrates away from the neural tube, heterogeneities of developmental potential emerge among its cells. Analysis of the clonal progeny of single migrating crest cells in vitro (Baroffio et al 1988) reveals that the large majority still have more than one potential but that a variety of partly committed precursors exists. Many clones contain both non-neurons and neurons that express either an adrenergic neurotransmitter-related molecule (tyrosine hydroxylase) or a neuropeptide (Substance P, VIP) or both. Taken together, the above findings suggest the existence of an early dual sensory-autonomic presursor that gives rise to dual neuronal-glial precursors of either the sensory or autonomic lineages.

## Neural Crest Migration and Dorsal Root Ganglion Formation

During their initial migration from the dorsal neural tube, crest cells emerge in a rostral to caudal sequence along the entire length of the axis (Weston 1963). Adhesive interactions between extracellular matrix (ECM) molecules (e.g. fibronectin, laminin) and cell surface receptors (integrins) are involved in crest cell migration. Blocking either the receptor (Bronner-Fraser 1985) or its ligand (Bronner-Fraser & Lallier 1988) with specific antibodies results in gross perturbations of migration.

The segmented disposition of spinal nerves is consequent on the formation of somites (Lehman 1927, Detwiler 1934) and the subdivision of these paraxially repeated mesodermal units into anterior (A) and posterior (P) halves. As each epithelial somite breaks down and sclerotome cells dissociate from its ventro-medial aspect, crest cells follow defined paths into the A-sclerotome (Rickmann et al 1985). Crest cells opposite P-half somites migrate longitudinally into the A-halves of the same somite or the posteriorly-adjacent somite, whichever is closer (Teillet et al 1987). Avoidance of P-halves by crest cells is mirrored closely by the growth cones of motor axons forming the ventral roots. Although a number of molecular differences have been found between A- and P- somite halves (Tan et al 1987, Mackie et al 1988, Layer et al 1988), only the peanut lectin receptor on P-sclerotome cells is detectable sufficiently early in development for it to qualify for a role in the segmental patterning of the PNS (Stern et al 1986, Keynes et al 1989). Detergent-extracted somite material, incorporated into liposomes, causes the rapid collapse of growth cones in vitro, whereas prior incubation of the solubilized material with immobilized peanut lectin abolishes the collapsing activity (J. Davies, G. M. W. Cooke, R. J. Keynes, and C. D. Stern, in preparation).

So far none of the ECM components implicated in crest migration has been shown to have a preferential distribution in A-halves of somites (Rickmann et al 1985, Krotoski et al 1986, Mackie et al 1988). It is therefore possible that the segmental array of DRG results from a combination of an overall facilitation of crest migration given by interactions with the comparatively uniform ECM and a segmentally reiterated inhibitory interaction.

Within the A-half of the sclerotome, some of the migrating crest cells stop alongside the mid dorso-ventral level of the spinal cord, whereas others continue their ventrad movement toward the dorsal aorta. These two groups of cells will form the DRG and sympathetic chain ganglia, respectively. NCAM expression is a feature of sympathetic chain formation but appears not to be associated with the arrest of migration and aggregation of DRG-producing crest (Lallier & Bronner-Fraser 1988). Cytotactin, an ECM molecule that has an inhibitory effect on crest translocation in vitro, is expressed in A-half sclerotome shortly after the crest enters it (Tan et al 1987, Mackie et al 1988) and may therefore be involved in the accumulation of cells to form DRG.

## Sensory Neuron Differentiation and Its Control

Although the potential to form both sensory and autonomic neurons is held in separate pools of precursor cells in the migrating crest (Le Douarin 1984, Ziller et al 1987), these precursors do not become completely segregated during migration. The technique of back-transplantation, in which quail ganglia are inserted into the crest migration path of younger chick embryos and allowed to migrate a second time (Le Douarin 1986), reveals not only the progress of commitment along the sensory line but also the existence of autonomic precursors in DRG. Early (E5) DRG contain precursors of both the sensory and the autonomic line as well as committed sensory neurons and glia. Later (E7–10) DRG can give rise to sympathetic neurons but not sensory neurons. Sensory precursors, therefore, are short-lived compared to autonomic precursors.

Sensory neurons develop provided their precursors are in proximity to the neural tube (Le Douarin 1984). If carried by migration into regions ventral to the neural tube, they die, and only cells of the autonomic line develop. A factor that influences DRG cell survival and differentiation has been demonstrated in the neural tube: Separation of the migrating crest from the neural tube by an impermeable membrane causes the rapid selective death of sensory line precursors. At least some of these cells survive when the inserted membrane is first impregnated with a neural tube extract (Kalcheim & Le Douarin 1986) or with both brain-derived neurotrophic factor and laminin (Kalcheim et al 1987).

Placode-derived neurons are initially larger than neural crest-derived neurons and are born earlier; in some cases neurons differentiate and extend neurofilament-positive processes while still emerging from the placode (D'Amico-Martel & Noden 1983).

## Process Outgrowth and Guidance

The distance sensory axons have to grow to reach their targets varies markedly among different CSG. For example, nodose ganglion axons grow over ten times further than vestibular ganglion axons. A comparative study of the rates at which the axons of these and other CSG grow to their targets in vivo has revealed a direct relationship between growth rate and target distance: The further the targets, the faster the axons grow. These differences in growth rate are intrinsically determined properties of early neurons because single isolated neurons from each ganglion grow at comparable rates in vitro (Davies 1989). It remains to be established whether this relationship between growth rate and target distance is a ubiquitous feature of neural development.

In peripheral sensory nerves, as in other systems, the earliest axons grow directly to their targets without sprouting or growth in aberrant directions. During the development of lumbosacral spinal nerves, for example, cutaneous sensory axons from DRG grow to their target skin along a stereotyped set of pathways and establish their dermatome at a precise location (Scott 1982). Sensory growth cones emerge after those of motoneurons, and, for the proximal part of their course at least, their pathways are pioneered by motor axons. It is possible that the spatially defined substratum provided by motor axons acts as a stereotropic guidance system for sensory growth cones (Tosney & Landmesser 1985) but this mechanical influence is not an absolute requirement for their successful navigation. When motoneuron precursors are ablated and motor axons consequently depleted, sensory nerves retain the ability to map out a typical pattern of main nerve trunks and branches (Swanson & Lewis 1986, Landmesser & Honig 1986) and can establish normal cutaneous innervation patterns (Scott 1988). These findings suggest that sensory neurons, like motoneurons, are specified with respect to their targets and that specific guidance cues exist that influence the distal (cutaneous) part of their course. They can also act as pioneers and respond correctly to the local, nonspecific guidance cues that determine the more proximal (mixed nerve) part of their course (Landmesser 1984).

Sensory neurons can be respecified; when the A-P order of a short stretch of DRG is reversed by the early rotation of the dorsal half of the neural tube (incorporating the neural crest), the cutaneous innervation patterns established by DRG are consonant with their new position. When the entire section of neural tube is rotated, the DRG tend to establish innervation patterns according to their original position (Scott 1986). This finding suggests that the ventral cord can influence the specification of DRG neurons to project along a particular pathway.

Homing behavior by displaced or misdirected axons (Lance-Jones & Landmesser 1980, Ferguson 1983, Harris 1986) indicates that specific cues are widely distributed in the embryo and can be detected away from normal pathways. The molecular basis of a distributed guidance mechanism is presently unknown but the tropic response of growth cones to a specific attractant diffusing from the target field is one possibility. That sensory growth cones can detect and respond by directional movement to gradients and that gradients can be effective over an adequate range in vivo have both been demonstrated by experiments using nerve growth factor (NGF) (Menesini-Chen et al 1978, Gundersen & Barrett 1979). To produce the highly stereotyped patterns of peripheral nerves, however, attractants would be expected to be regionally specific, to be available during initial outgrowth, and to be produced at a level sufficient for free diffusion and the establishment of gradients. NGF meets none of these requirements (see below).

Evidence for specific chemotropism has come from co-culture studies of the effects of mouse embryonic maxillary process target tissue on the growth and directionality of trigeminal ganglion axons. Here, neurites grow exclusively toward their target under the influence of an attractant produced by the target epithelial tissue that ultimately produces the sensory receptor cells of the whisker field (Lumsden & Davies 1983, 1986, Davies 1987, Lumsden 1988). The factor, which is immunochemically distinct from NGF (and other known growth factors) and laminin, is produced during the period of normal outgrowth, declines after normal growth cone-target encounter, and is not produced by the adjoining cutaneous field. Regions of the epidermis that receive a dense sensory innervation may thus be specified to attract appropriate growth cones from a distance. Chemotropism may be an especially significant guidance mechanism for peripheral nerves that lack a motor component throughout their course, such as the maxillary division of the trigeminal.

## Neuronal Death and Trophic Factor Dependence

From 20 to 80% of the neurons generated in sensory ganglia die shortly after they innervate their targets. An important function of neuronal death is to adjust the number of neurons to the requirements of their target fields. Experimental manipulation of target field size (Oppenheim 1981) and disruption of target field innervation (Yip & Johnson 1984) have demonstrated that both the peripheral and central target fields of sensory neurons play a role in regulating neuronal number.

Work on NGF has substantiated the hypothesis that developing target fields regulate the number of innervating neurons by the production of a limited quantity of a neurotrophic factor that the neurons require for their survival (Davies 1988; E. M. Shooter and H. Thoenen, in preparation). The most important evidence is that developing neurons whose survival is promoted by NGF in vitro, namely sympathetic neurons and a subset of sensory neurons, are also dependent on NGF in vivo. Anti-NGF antibodies eliminate these neurons during the phase of target field innervation, whereas exogenous NGF rescues neurons that would otherwise die.

Sensory neurons obtain NGF predominantly, if not exclusively, from their peripheral target fields. NGF mRNA is present in appreciable amounts in skin (Davies et al 1987a) but only low levels are present in the spinal cord (Shelton & Reichardt 1984). Accordingly, NGF is present in developing skin (Davies et al 1987a) and in the peripheral branches of sensory neurons but is undetectable in spinal cord and in the central branches of these neurons (Korsching & Thoenen 1985).

Elucidation of the spatial and temporal aspects of NGF synthesis and NGF receptor expression during cutaneous innervation (Davies et al 1987a) has clarified the role of NGF in development and has provided some understanding of how NGF regulates the number and distribution of nerve terminals. Studies of the most densely innervated cutaneous target field of the mouse embryo (the maxillary process) have shown that NGF synthesis commences with the arrival of the earliest sensory axons. The expression of NGF receptors on the innervating neurons also begins when their axons reach the target field. The survival and growth of sensory neurons is independent of NGF and other target-derived neurotrophic factors prior to target field innervation (Davies & Lumsden 1984, Ernsberger & Rohrer 1988). The demonstration that NGF is not present in the target field prior to its innervation and that sensory axons lack NGF receptors when they are growing to their targets discounts the view that target-derived NGF plays a role in long-range chemotropic guidance of early sensory axons (Levi-Montalcini 1982).

Assay of NGF mRNA in isolated cutaneous epithelium and mesenchyme together with localization by in situ hybridization has shown that its concentration correlates with innervation density. It is highest in epithelium (presumptive epidermis), lower in subjacent mesenchyme (presumptive dermis), and lowest in deep mesenchyme (presumptive subcutaneous tissue). The concentration of NGF mRNA in the epithelium of different cutaneous target fields during development is also related to their innervation density (S. Harper and A. M. Davies, unpublished observations). These findings suggest that the local availability of NGF regulates innervation density. They also discount the view, based on immunohistochemical studies of NGF in the denervated iris, that NGF is synthesized exclusively by Schwann cells (Rush 1984).

Although studies of NGF mRNA localization in developing cutaneous target fields indicate the kinds of cells that synthesize NGF, they provide no direct information on whether newly synthesized NGF is diffusible in the target field or whether its distribution and availability is restricted to specific sites. This information is important for understanding the nature of the competition between neurons for NGF. The finding that the NGF receptor gene is expressed in developing cutaneous target fields (S. Wyatt and A. M. Davies, unpublished observations) raises the possibility that low-affinity receptors may immobilize NGF to the surfaces of specific cells in the target field. This would provide a mechanism for selectively supporting the survival of neurons in accordance with the distribution of their axons in the target field, since only neurons whose axons contact these specific target cells would be able to obtain a supply of NGF.

NGF is not the only neurotrophic factor that promotes the survival of sensory neurons. Extracts of a variety of peripheral tissues contain factors distinct from NGF that promote the survival of embryonic sensory neurons in culture. This finding raises the issue of the kinds of sensory neurons that are supported by NGF and other peripheral target-derived factors. Several in vivo and in vitro studies of the effects of NGF have given rise to the view that NGF-dependence is directly related to sensory neuron ontogeny, in that all neural crest–derived sensory neurons require NGF for survival whereas placode-derived neurons do not (Pearson et al 1983, Davies & Lindsay 1985, Lindsay & Rohrer 1985). Recent work, however, has clearly shown that NGF-dependence is not a feature of all neural crest–derived sensory neurons. Trigeminal mesencephalic nucleus (TMN) neurons, a population of neural crest–derived proprioceptive neurons, are not supported by NGF in culture (Davies et al 1987b). Likewise, several experimental findings are consistent with the idea that the neural crest–derived proprioceptive neurons of DRG are also independent of NGF for survival (Davies 1987). Although TMN neurons do not survive in the presence of NGF, they are supported by a factor present in their peripheral target tissue, skeletal muscle (Davies 1986). In contrast to NGF, this muscle-derived factor has little or no effect on the survival of cutaneous sensory neurons. These findings suggest that the neurotrophic factor requirements of developing sensory neurons are related to the structures they innervate. For a detailed discussion of neurotrophic factor specificity in the sensory nervous system, see Davies (1987).

A candidate for a neurotrophic factor that mediates the trophic support of the CNS on developing sensory neurons is brain-derived neutrotrophic factor (BDNF; Barde et al 1982). This protein promotes the survival of embryonic DRG neurons in culture (Lindsay et al 1985) and rescues DRG neurons if administered to embryos during the period of natural neuronal death (Hofer & Barde 1988). Although BDNF-dependent neurons are present in all populations of sensory neurons, their proportion varies from 10 to 80% or more (Davies et al 1986a). In DRG, the majority of neurons respond to both BDNF and NGF in vitro in the early stages of their development (Ernsberger & Rohrer 1988). As development proceeds, BDNF-dependent and NGF-dependent neurons become largely distinct (Lindsay et al 1985). This suggests that the neurotrophic factor requirements of at least some sensory neurons become more restricted as they mature.

TMN neurons have been useful in resolving the issue of whether the survival of sensory neurons is regulated by the same or by different factors from the periphery and CNS. During the stage of target field innervation, the survival of the great majority of these neurons in culture is promoted by either of two neurotrophic factors: BDNF or a distinct factor present in skeletal muscle (Davies et al 1986b). There is no additional survival in the presence of saturating levels of both factors, thus indicating that each neuron responds to both factors. The combined effect of both factors is additive at concentrations that promote half-maximal survival alone and is greater than additive at very low concentrations. This suggests that peripheral and central neutrotrophic factors may potentiate each other at low concentrations. The finding that the responsiveness of TMN neurons to each factor is maximal during the period of natural neuronal death is further evidence that both factors cooperate in regulating sensory neuron survival during development.

## Target Recognition and the Specification of Connectivity

Neurons of a particular sensory modality connect with specific second order neurons in the various laminae of the spinal cord or equivalent nuclei in the brainstem. Several findings raise the possibility that certain oligosaccharide moieties associated with cell-surface glycoproteins and glycolipids play a role in target cell recognition. In the developing rat, subsets of DRG neurons that project to particular laminae in the spinal cord are labeled by antibodies that recognize either lactosyl or globosyl oligosaccharides (Dodd & Jessell 1985, 1986). Furthermore, DRG neurons that express lactosyl oligosaccharides make and release the complementary lactosyl-binding lectins (Regan et al 1986). Although there is as yet no direct evidence that these oligosaccharides are involved in target cell recognition in the developing sensory nervous system, it has been clearly demonstrated that lactosyl oligosaccharides play an important role in cell-cell recognition and adhesion in early embryonic development (Rutishauser & Jessell 1988).

An overriding question is whether the specific peripheral and central connections that primary sensory neurons make is determined prior to target field innervation or whether connectivity is governed by the particular targets these neurons encounter. The finding that sensory axons of the same modality tend to be clustered together in cutaneous nerves (Roberts & Elardo 1986) suggests that connectivity is specified prior to innervation rather than induced by various target cells encountered by chance. On the other hand, it is clear that the environment of developing sensory neurons can modify their peripheral and central connections. When thoracic DRG (predominantly cutaneous sensory) are transplated to the brachial region in tadpoles, a proportion of the neurons in these ganglia innervate stretch receptors in muscle and make appropriate terminations in the spinal cord (Smith & Frank 1987). The sequential timing of peripheral target field innervation and synaptogenesis in the spinal cord raises the possibility that the peripheral target may play a role in specifying central terminations (Smith & Frank 1988, Davis et al 1989). The periphery can also affect the phenotype of sensory neurons. Cross-anastomosis of cutaneous and muscle afferent nerves in adult rats results in changes in the neuropeptide content of sensory neurons that accord with their new targets (McMahon & Gibson 1987).

ACKNOWLEDGMENTS

The authors research is supported by the Cancer Research Campaign, Medical Research Council, and Wellcome Trust.

*Literature Cited*

Barald, K. F. 1982. Monoclonal antibodies to embryonic neurons: Cell specific markers for chick ciliary ganglion. In *Neuronal Development*, ed. N. Spitzer, pp. 110–19. New York: Plenum

Barde, Y. A., Edgar, D., Thoenen, H. 1982. Purification of a new neurotrophic factor from mammalian brain. *EMBO J*. 1: 549–53

Baroffio, A., Dupin, E., Le Douarin, N. M.

1988. Clone-forming ability and differentiation potential of migratory crest cells. *Proc. Natl. Acad. Sci. USA* 85: 5325–29

Bronner-Fraser, M. 1985. Alterations in neural crest migration by a monoclonal antibody that affects cell adhesion. *J. Cell Biol.* 101: 610–17

Bronner-Fraser, M., Fraser, S. E. 1988. Cell lineage analysis reveals multipotency of some avian neural crest cells. *Nature* 335: 161–64

Bronner-Fraser, M., Lallier, T. 1988. A monoclonal antibody against a laminin-heparan sulfate proteoglycan complex perturbs cranial neural crest migration *in vivo. J. Cell Biol.* 106: 1321–29

D'Amico-Martel, A., Noden, D. M. 1983. Contributions of placodal and neural crest cells to avian cranial peripheral ganglia. *Am. J. Anat.* 166: 445–68

Davies, A. M. 1986. The survival and growth of embryonic proprioceptive neurons is promoted by a factor present in skeletal muscle. *Dev. Biol.* 115: 56–67

Davies, A. M. 1987. Molecular and cellular aspects of patterning sensory neurone connections in the vertebrate nervous system. *Development* 101: 185–208

Davies, A. M. 1988. Role of neurotrophic factors in development. *Trends Genet.* 4: 139–43

Davies, A. M. 1989. Intrinsic differences in the growth rate of early nerve fibres related to target distance. *Nature* 337: 553–55

Davies, A. M., Bandtlow, C., Heumann, R., Korsching, S., Rohrer, H., Thoenen, H. 1987a. Timing and site of nerve growth factor synthesis in developing skin in relation to innervation and expression of the receptor. *Nature* 326: 353–58

Davies, A. M., Lindsay, R. M. 1985. The avian cranial sensory ganglia in culture: Differences in the response of placode-derived and neural crest-derived neurons to nerve growth factor. *Dev. Biol.* 111: 62–72

Davies, A. M., Lumsden, A. G. S. 1984. Relation of target encounter and neuronal death to nerve growth factor responsiveness in the developing mouse trigeminal ganglion. *J. Comp. Neurol.* 253: 13–24

Davies, A. M., Lumsden, A. G. S., Rohrer, H. 1987b. Neural crest-derived proprioceptive neurons express NGF receptors but are not supported by NGF in culture. *Neuroscience* 20: 37–46

Davies, A. M., Thoenen, H., Barde, Y. A. 1986a. The response of chick sensory neurons to brain-derived neurotrophic factor. *J. Neurosci.* 6: 1897–1904

Davies, A. M., Thoenen, H., Barde, Y. A. 1986b. Different factors from the central nervous system and periphery regulate the survival of sensory neurones. *Nature* 319: 497–99

Davis, B. M., Frank, E., Johnson, F. A., Scott, S. A. 1989. Development of central projections of lumbosacral sensory neurons in the chick. *J. Comp. Neurol.* 279: 556–66

Detwiler, S. R. 1934. An experimental study of spinal nerve segmentation in *Amblystoma* with reference to the plurisegmental contribution to the brachial plexus. *J. Exp. Zool.* 67: 395–441

Dodd, J., Jessell, T. M. 1985. Lactoseries carbohydrates specify subsets of sensory dorsal root ganglion neurons projecting to the superficial dorsal horn of the rat spinal cord. *J. Neurosci.* 5: 3278–94

Dodd, J., Jessell, T. M. 1986. Cell surface glycoconjugates and carbohydrate binding proteins: Possible recognition signals in sensory neurone development. *J. Exp. Biol.* 124: 225–38

Ernsberger, U., Rohrer, H. 1988. Neuronal precursor cells in chick dorsal root ganglia: Differentiation and survival *in vitro. Dev. Biol.* 126: 420–32

Ferguson, B. A. 1983. Development of motor innervation of the chick following dorso-ventral limb bud rotations. *J. Neurosci.* 3: 1760–72

Girdlestone, J., Weston, J. A. 1985. Identification of early neuronal subpopulations in avian neural crest cell cultures. *Dev. Biol.* 109: 274–87

Gundersen, R., Barrett, J. N. 1979. Neuronal chemotaxis: Chick dorsal root axons turn towards high concentrations of nerve growth factor. *Science* 206: 1079–80

Harris, W. A. 1986. Homing behaviour of axons in the embryonic vertebrate brain. *Nature* 320: 266–69

Hofer, M. M., Barde, Y. A. 1988. Brain-derived neutrophic factor prevents neuronal death *in vivo. Nature* 331: 261–62

Kalcheim, C., Barde, Y. A., Thoenen, H., Le Douarin, N. M. 1987. *In vivo* effect of brain-derived neutrophic factor on the survival of developing dorsal root ganglion cells. *EMBO J.* 6: 2871–73

Kalcheim, C., Le Douarin, N. M. 1986. Requirement of a neural tube signal for the differentiation of neutral crest cells into dorsal root ganglia. *Dev. Biol.* 116: 451–66

Keynes, R., Cook, G., Davies, J., Lumsden, A., Norris, W., Stern, C. D. 1989. Segmentation and the development of the vertebrate nervous system. *J. Physiol. Paris.* In press

Korsching, S., Thoenen, H. 1985. Nerve growth factor supply for sensory neurons: Site of origin and competition with the sympathetic nervous system. *Neurosci. Lett.* 39: 1–4

Krotoski, D. M., Domingo, C., Bronner-Fraser, M. 1986. Distribution of a putative cell surface receptor for fibronectin and laminin in the avian embryo. *J. Cell Biol.* 103: 1061–71

Lallier, T. E., Bronner-Fraser, M. 1988. A spatial and temporal analysis of dorsal root and sympathetic ganglion formation in the avian embryo. *Dev. Biol.* 127: 99–112

Lance-Jones, C., Landmesser, L. 1980. Motoneuron projection patterns in the chick hind limb following early partial spinal cord reversals. *J. Physiol. London* 302: 581–602

Landmesser, L. 1984. The development of specific motor pathways in the chick embryo. *Trends Neurosci.* 7: 336–39

Landmesser, L., Honig, M. 1986. Altered sensory projections in the chick hindlimb following early removal of motoneurons. *Dev. Biol.* 118: 511–31

Layer, P. G., Alber, A., Rathjen, F. G. 1988. Sequential activation of butyrylcholinesterase in rostral half somites and acetylcholinesterase in motoneurones and myotomes preceding growth of motor axons. *Development* 102: 387–96

Le Douarin, N. M. 1984. A model for cell line divergence in the ontogeny of the peripheral nervous system. In *Cellular and Molecular Biology of Neuronal Development*, ed. I. B. Black, pp. 3–28. New York: Plenum

Le Douarin, N. M. 1986. Cell line segregation during peripheral nervous system ontogeny. *Science* 231: 1515–22

Le Douarin, N. M., Fontaine-Perus, J., Couly, G. 1986. Cephalic ectodermal placodes and neurogenesis. *Trends Neurosci.* 9: 175–80

Lehman, F. 1927. Further studies on the morphogenetic role of the somites in the development of the nervous system of amphibians. The differentiation and arrangement of the spinal ganglia in *Pleurodeles waltlii. J. Exp. Zool.* 49: 93–131

Levi-Montalcini, R. 1982. Developmental neurobiology and the natural history of nerve growth factor. *Annu. Rev. Neurosci.* 5: 341–62

Lindsay, R. M., Rohrer, H. 1985. Placodal sensory neurons in culture: Nodose ganglion neurons are unresponsive to NGF, lack NGF receptors but are supported by a liver-derived neurotrophic factor. *Dev. Biol.* 112: 30–48

Lindsay, R. M., Thoenen, H., Barde, Y. A. 1985. Placode and neural crest-derived sensory neurons are responsive at early developmental stages to brain-derived neurotrophic factor. *Dev. Biol.* 112: 319–28

Lumsden, A. G. S. 1988. Diffusible factors and chemotropism in the development of the peripheral nervous system. In *The Making of the Nervous System*, ed. J. G. Parnavelas, C. D. Stern, R. V. Stirling, pp. 166–87. London: Oxford Univ. Press

Lumsden, A. G. S., Davies, A. M. 1983. Earliest sensory nerve fibres are guided to peripheral targets by attractants other than nerve growth factor. *Nature* 306: 786–88

Lumsden, A. G. S., Davies, A. M. 1986. Chemotropic effect of specific target epithelium in development of the mammalian nervous system. *Nature* 323: 538–39

Mackie, E. J., Tucker, R. P., Halfter, W., Chiquet-Ehrismann, R. 1988. The distribution of tenascin coincides with pathways of neural crest cell migration. *Development* 102: 237–50

McMahon, S. B., Gibson, S. 1987. Peptide expression is altered when afferent nerves reinnervate inappropriate tissue. *Neurosci. Lett.* 73: 9–15

Menesini-Chen, M. G., Chen, J. S., Levi-Montalcini, R. 1978. Sympathetic nerve fiber ingrowth in the central nervous system of neonatal rodents upon intracerebral NGF injections. *Arch Ital. Biol.* 116: 53–84

Narayanan, C. H., Narayanan, Y. 1978. Determination of the embryonic origin of the mesencephalic nucleus of the trigeminal nerve in birds. *J. Embryol. Exp. Morphol.* 43: 85–105

Oppenheim, R. W. 1981. Neuronal death and some related phenomena during neurogenesis: A selected historical review and progress report. In *Studies in Developmental Biology*, ed. W. M. Cowan, pp. 74–133. London: Oxford Univ. Press

Pearson, J., Johnson, E. M., Brandeis, L. 1983. Effects of antibodies to nerve growth factor on intra uterine development of derivatives of cranial neural crest and placode in the guinea pig. *Dev. Biol.* 96: 32–36

Regan, L., Dodd, J., Barondes, S., Jessell, T. M. 1986. Selective expression of endogenous lactose-binding and lactoseries glycoconjugates in subsets of rat sensory neurons. *Proc. Natl. Acad. Sci. USA* 83: 2248–52

Rickmann, M., Fawcett, J. W., Keynes, R. J. 1985. The migration of neural crest cells and the growth of motor axons through the rostral half of the chick somite. *J. Embryol. Exp.Morphol.* 90: 437–55

Roberts, W. J., Elardo, S. M. 1986. Clustering of primary afferent fibers in peripheral nerve fasicles by sensory modality. *Brain Res.* 370: 149–52

Rush, R. A. 1984. Immunohistochemical localization of endogenous nerve growth factor. *Nature* 312: 364–67

Rutishauser, U., Jessell, T. M. 1988. Cell adhesion molecules in vertebrate neural development. *Physiol. Rev.* 68: 819–57

Scott, S. A. 1982. The development of the segmental pattern of skin sensory innervation in embryonic chick hind limb. *J. Physiol.* 330: 203–30

Scott, S. A. 1986. Skin sensory innervation patterns in embryonic chick hindlimb following dorsal root ganglion reversals. *J. Neurobiol.* 17: 649–68

Scott, S. A. 1988. Skin sensory innervation patterns in embryonic chick hindlimbs deprived of motoneurons. *Dev. Biol.* 126: 362–74

Shelton, D. L., Reichardt, L. F. 1984. Expression of the nerve growth factor gene correlates with the density of sympathetic innervation in effector organs. *Proc. Natl. Acad. Sci. USA* 81: 7951–55

Smith, C. L., Frank, E. 1987. Peripheral specification of sensory neurons transplated to novel locations along the neuraxis. *J. Neurosci.* 7: 1537–49

Smith, C. L., Frank, E. 1988. Specificity of sensory projections to the spinal cord during development in bullfrogs. *J. Comp. Neurol.* 269: 96–108

Stern, C. D., Sisodiya, S. M., Keynes, R. J. 1986. Interactions between neurites and somite cells: Inhibition and stimulation of nerve growth in the chick embryo. *J. Embryol. Exp. Morphol.* 91: 209–26

Swanson, G. J., Lewis, J. H. 1986. Sensory nerve routes in chick wing buds deprived of motor innervation. *J. Embryol. Exp. Morphol.* 95: 37–52

Tan, S.-S., Crossin, K. L., Hoffman, H., Edelman, G. M. 1987. Asymmetric expression in somites of cytotactin and its proteoglycan ligand is correlated with neural crest cell distribution. *Proc. Natl. Acad. Sci. USA* 84: 7977–81

Teillet, M.-A., Kalcheim, C., Le Douarin, N. M. 1987. Formation of the dorsal root ganglia in the avian embryo: Segmental origin and migratory behaviour of neural crest progenitor cells. *Dev. Biol.* 120: 329–47

Tosney, K., Landmesser, L. 1985. Growth cone morphology and trajectory in the lumbosacral region of the chick embryo. *J. Neurosci.* 5: 2345–58

Weston, J. A. 1963. An autoradiographic analysis of the migration and localization of trunk neural crest cells in the chick. *Dev. Biol.* 6: 279–310

Weston, J. A., Girdlestone, J., Ciment, G. 1984. Heterogeneity in neural crest cell populations. In *Cellular and Molecular Biology of Neuronal Development*, ed. I. B. Black, pp. 51–62. New York: Plenum

Yip, H. K., Johnson, E. M. 1984. Developing dorsal root ganglion neurons require trophic support from their central processes: Evidence for a role of retrogradely transported nerve growth factor from the central nervous system to the periphery. *Proc. Natl. Acad. Sci. USA* 81: 6245–49

Ziller, C., Fauquet, M., Kalcheim, C., Smith, J., Le Douarin, N. M. 1987. Cell lineages in peripheral nervous system ontogeny: Medium-induced modulation of neuronal phenotypic expression in neural crest cell cultures. *Dev. Biol.* 120: 101–11

*Annu. Rev. Neurosci. 1990. 13:75–87*

# RNA AND PROTEIN METABOLISM IN THE AGING BRAIN

*Caleb E. Finch and David G. Morgan*

Andrus Gerontology Center, and Department of Biological Sciences, University of Southern California, Los Angeles, California 90089-0191

## INTRODUCTION

Substantial age-related changes in RNA and protein synthesis are reported in liver, immune system cells, and elsewhere, as well as in the brain. Our purpose is to discuss examples from this large and often controversial literature that bear on mechanisms of age-related neurodegeneration, including Alzheimer disease (AD) and Down's syndrome (DS); the discussion is not comprehensive, and we cite reviews where possible. Increased exchanges between neurobiologists studying aging or AD and biogerontologists studying other cells could increase the understanding of molecular mechanisms of aging that may be common to many cell types. We consider the following: slowed macromolecular biosynthesis; accumulation of abnormal proteins, including a discussion of whether amyloid and neurofibrillary tangles (NFT) of AD, DS, and normal brain aging are another aspect of abnormal protein accumulation; and age-related cell loss vs. shifts in the ratios of neurons and glial cell types. The focus is on the mammalian brain, since little is known about cellular aging in the nervous systems of lower vertebrates and invertebrates.

A major concern in studies of aging is the threat of confounds from diverse diseases that generally increase exponentially with advancing age and that have a high potential for idiosyncratically biasing specimens and increasing data heterogeneity. Dysfunctions of peripheral organs differ widely among older individuals, even in inbred rodent populations. At the least, in rodents these issues require data on the health status, which may be indicated by body weight changes before sampling, and postmortem

75

0147–006X/90/0301–0075$02.00

survey for gross pathology (Finch & Foster, 1973). Postmortem analysis of two-year-old male mice, for example, showed low plasma testosterone in subgroups with lymphoid tumors or lung disease, while healthy subgroups showed no decline in testosterone (Nelson et al 1975). Aging rodents also commonly develop kidney diseases that cause proteinuria and compensatorily increased synthesis of hepatic albumin mRNA and albumin (Wellinger & Guigoz 1986), as well as increased parathyroid hormone and calcitonin that increase bone resorption and reduces intestinal absorption of calcium (Kalu et al 1988). Changes in hepatic, kidney, and endocrine functions and the common lactotrophic pituitary tumors might well interact with age changes in brain mineral metabolism and other functions. Nutritionally balanced diet restriction reduces and delays many age-related diseases, as well as extending lifespan (Weindruch & Walford 1988, Kalu et al 1988).

## GENE ACTIVITY AND RNA METABOLISM

A major focus in gerontology concerns the extent of changes in gene activity. Although theories holding that aging results from randomly accumulated errors in DNA or other macromolecules have been hotly debated for decades, most evidence indicates that changes in gene activity during aging, when they occur, are selective for the gene locus and cell type (Reff 1985). Most cells retain their differentiated characteristics throughout life, e.g. the histological hallmarks of neurons, glia, hepatocytes. Consistent with this, few molecules change by >30% in levels, nor do the activities of isozymes and other enzymes (Finch 1972, Rogers & Bloom 1985), nor the abundant proteins in several neural tissues as resolved by 2-d gel electrophoresis (Cosgrove et al 1987, Wilson et al 1978). These simple observations indicate that the machinery for selective gene expression, upon which differentiated cell characteristics depend, changes little during aging. However, as described below, important changes may occur in the rates of RNA and protein synthesis.

Even in the absence of overt neuropathology, changes accompanying brain aging vary importantly between different cell types. Some neurons apparently remain intact, e.g. LHRH-neurons (Hoffman & Finch 1986), whereas others show hypertrophy of dendritic processes and increased perikaryal size, e.g. dentate granule neurons (Coleman & Flood 1987), or show increased perikaryal RNA content, e.g. subicular neurons (Uemura & Hartmann 1979). Nonetheless, many other neurons atrophy, e.g. melanin-containing neurons of substantia nigra (Mann & Yates 1983) and pyramidal neurons of the hippocampus (Ringborg 1966) and cerebral cortex (Peters et al 1987), which show reduced perikaryal RNA content and decreased size of their perikarya, nuclei, or nucleoli. Their volume loss is in the range of 5–30%, and again shows selectivity, e.g. nucleolar shrinkage is much less in the human locus ceruleus than substantia nigra (Mann & Yates 1979). The atrophy of large neurons may be intensified in AD (Mann et al 1981, Doebler et al 1987). On the other hand, cortical neuron atrophy was not ameliorated in four-year-old rats maintained on a life-long restricted diet that allowed them to live a year beyond the usual lifespan (Peters et al 1987). The link between neuron atrophy and neuron death is unclear.

The nucleolar shrinkage noted above probably represents decreased transcription of ribosomal RNA (rRNA) cistrons and assembly of ribosomes, i.e. a change in gene regulation. A different mechanism, suggested by a series of studies from Strehler (reviewed 1986), could be a 30–50% age-related loss of rRNA genes in DNA from brain and other tissues. Tandemly repeated sequences like rRNA cistrons are hypothesized to be at risk for excision, as classically observed for the Bar locus of *Drosophila*. This might be the first example of sequence-specific changes in mammalian DNA with aging, but rigorous verification is needed.

A different phenomenon, described so far in nonneural cells, is that certain repressed genes spontaneously reactive during aging, e.g. several genes on the genetically repressed ("Lyonized") X chromosome (Wareham et al 1987, Cattenach 1974). Other X-linked genes that might lead to gender differences of neural aging by increased dose of expression in females include steroid sulfatase, DNA α-polymerase, monoamine oxidase A, and glucose-6-phosphate dehydrogenase. The scope of this phenomenon in autosomes is unknown. We also note the growing evidence for progressive, but slow age-related demethylation of 5-methyldeoxycytidine in different DNA sequences in brain and other tissues (Mays-Hoopes 1989, Wilson et al 1987); demethylation of DNA is often correlated with increased transcription in eukaryotic cells, although the causality remains controversial. These findings suggest the occurrence of nonprogrammed changes of gene activity during brain aging.

The effect of aging on differential processing of HnRNA is an intriguing question. An unexpected phenomenon is shown by the spontaneous reversion of the phenotype for *analbuminemia*. This mutation in rats is a 7 nucleotide deletion that blocks processing of the primary hnRNA transcript; although albumin-HnRNA is present in hepatocyte nuclei of homozygotes, little albumin mRNA reaches the cytoplasm (Esumi et al 1982). Neonates (they are viable) have only traces of plasma albumin, which is produced by occasional hepatocytes that show intense immunocytochemistry (ICC) staining (Esumi et al 1985). Gradually, immunoassayable plasma albumin increases with age (Makino et al 1982), and by 12 months, the number of ICC+ cells increases 50-fold (Makino et al

1986). The mechanisms are unknown, but evidence points to epigenetic changes not involving the primary sequence, possibly through alternate splicing pathways. There is yet no example of age-related changes in splicing in neural tissues. (See Note [a] added in proof, p. 84.) Differential splicing of the $\beta$-amyloid precursor protein mRNA (APP), which has three alternately spliced forms (Selkoe et al 1988), might be considered in the delayed accumulation of amyloid $\beta$-protein in aging, AD, or DS. Twofold increases in the ratio of mRNA APP-751: APP-695 (Johnson et al 1988) in AD hippocampus could, however, be explained by other mechanisms.

The inventory of RNA does not change grossly in most tissues throughout the lifespan, as judged by content of total or polysomal poly(A)RNA and the polysomal poly(A)RNA sequence complexity (an estimate of the number of different RNA transcripts that includes low abundance mRNA) in whole monkey brain (Farquhar et al 1979), whole rodent brain (Colman et al 1980, Anzai et al 1983), most brain regions (Chaconas & Finch 1973), and other tissues (Richardson & Birchenall-Sparks, 1983). We note that the proportion of brain RNA in neurons vs. glia is unresolved by such studies, even though a particular sequence may be cell-type specific. Specific mRNA also show the selectivity of age changes in proteins, e.g. the whole-brain content of identifier mRNA (brain specific), did not change with age (Anzai & Goto 1987), whereas the hypothalamic content of pro-opiomelanocortin mRNA decreased 30% (Nelson et al 1988). Whole tissue analysis will soon yield to individual cell analysis of specific mRNA through *in situ* hybridization.

We are exploring a rat model showing atrophic changes in the substantia nigra that are similar to those in aging humans. Nine months after unilateral 6-hydroxydopamine lesions, the remaining dopaminergic nigral neurons had smaller perikarya and nucleoli (Pasinetti et al 1989). Moreover, mRNA prevalence in tyrosine hydroxylase (TH) immunopositive neurons was selectively altered, as determined by *in situ* hybridization measurements: The grain density corresponding to TH-mRNA was decreased by > 50%, while that for $\beta$-tubulin mRNA was unaltered. Because the remaining neurons were smaller, there was a net loss of both mRNA per cell, though the loss of TH-mRNA was much larger. These results suggest mechanisms in the atrophy of other cells with aging, e.g. modulation of inhibitory afferents, a chronic impact of neurotoxins, or alterations in trophic or growth factors.

Tracer studies of RNA synthesis rates in a few tissues indicate decreases by middle age in rodents that continue into later life; most changes reported are small, $\leq 30\%$ (Richardson & Burchenall-Sparks 1983). Nuclear run-on studies of $\alpha_{2u}$ globulin indicate a decreased transcription that parallels decreased mRNA prevalence in livers and reduced plasma levels in old rats. Diet restriction also restored the transcription of $\alpha_{2u}$ globulin mRNA (Richardson et al 1987). How aging alters RNA turnover is unresolved. Two-year-old rats showed no changes in the $t_{1/2}$ of bulk RNA from brain, of which most is rRNA (Menzies & Gold 1972) or in the $t_{1/2}$ of polysomal poly(A)RNA in liver (Gold et al 1980); because precursor pool turnover was slower in old rats of the latter study interpretation of the results is difficult.

## PROTEIN METABOLISM

We preface this discussion by noting that coordinate changes in protein synthesis and degradation are required to maintain the steady state levels of brain proteins throughout the lifespan (see above). Age-related declines in the rates of protein synthesis are widely described in brain and other rodent organs (Dwyer et al 1980, Rothstein 1982, Richardson et al 1987), whereas other studies show no changes (Cosgrove & Rapaport 1987). These divergent results between careful studies are difficult to interpret. In studies of the perfused rat liver with corrections of the specific activity of valyl-tRNA, Ward (1988) showed 40% decreases between 3 and 24 mo in the synthesis rate of hepatic total protein; diet restriction partly reversed these decrements. In rat brain, an autoradiographic study of leucine incorporation showed < 15% decreases in most regions up to 24 mo, while no region increased; larger decreases of 20–25% were found, e.g. in inferior olive, red nucleus, and locus ceruleus (Ingvar et al 1985). Although plasma pools were corrected for, there could be cell-type specific age changes in tracer metabolism. Slowed protein synthesis is consistent with the modest slowing of peripheral nerve axoplasmic flow in very old rats (Stromska & Ochs 1982). (See Note [b], p. 84.) Correlations between the changes of aging in RNA and protein synthesis have not been done in the same study.

The accumulation of amyloid and neurofibrillary tangles (NFT) are hallmarks of age-related neurodegeneration in humans (Selkoe 1988). Although brain-type amyloid and NFT are not found outside of the brain, many labs have described what we suspect could be a related phenomenon of aging, the "altered enzymes." Diverse tissues accumulate catalytically inactive enzymes, as detected by immunoenzymatic titration (Gershon & Gershon 1970, Rothstein 1982, Stadtman 1988). Select enzymes in many tissues of aging organisms show a decreased ratio of enzyme activity to immunoassayable protein. Enzymes vary between tissues during aging in the amount of these inactive forms, e.g. altered phosphoglycerate kinase is accumulated in brain, among several other tissues of old rats (Sharma & Rothstein 1984), whereas altered supraoxide dismutase accumulated only in liver (Rothstein 1982). TH, however, did not accumulate inactive forms in the adrenal of old rats, the only tissue examined, by immuno-enzymatic titration (Reis et al 1977). The few altered enzymes that have been analyzed in depth indicate altered conformations, not changes of amino acid sequence. Remarkably, full enzyme activity may be recoverable by cycles of denaturation-renaturation (Sharma & Rothstein 1980, Rothstein 1982, Yuh & Gafni 1987). A major hypothesis is that altered enzymes are intermediates on a degradative pathway, and accumulate because of slowed protein degradation and turnover during aging (Lavie et al 1982, Rothstein 1982). Thus the accumulation of NFT or amyloid could represent a more general phenomenon of slowed protein turnover during aging, which, like the altered enzymes, could vary between cell types. The regulation of protein turnover is not well understood, even in young cells, and involves multiple mechanisms and pathways (Dice 1987). Ubiquitinylation of NFT and certain other filamentous inclusions of degenerating neurons links these abnormal proteins to the nonlysozomal proteolytic pathway for degradation of damaged proteins in which ubiquitinylation is a key regulator of degradation; however, other abnormal proteins of AD are not ubiquitinylated, e.g. filaments in Hirano bodies (Selkoe 1988, Manetto et al 1988).

The extent to which changes in macromolecular biosynthesis with age limit neuronal responses is a major open question. Several studies indicate eventual decreases in plasticity, e.g. reactive synaptogenesis is slower in hippocampal neurons of old rats but eventually reaches the same extent as in young rats (Anderson et al 1986, Hoff et al 1982). Although several neuronal types in normal humans and rodents show increased dendritic arbors by the average lifespan, at later ages some previously expanded dendritic fields then regress (Coleman & Flood 1987, Rafols et al 1989). The increased dendritic arbor in dentate gyrus granule cells, e.g. during AD, is reasonably interpreted as a deafferentiation response to perforant path deterioration (Geddes et al 1985). Another example of decreased plasticity is the failure of sympathetic gangliar explants from two-year-old rats to increase substance P, in contrast to those from young adults (Adler & Black 1984). These decreases in plasticity could arise from general impairments in biosynthesis or from altered regulation of specific genes.

## CELL LOSS AND/OR SHIFTS IN CELL TYPES WITH AGING

This section considers how changes in brain cell populations may contribute to age-related changes of numerous purported neurochemical markers (Rogers & Bloom 1985, Morgan & Finch 1988). According to the traditional belief that extensive neuron loss occurs with aging, even without neurological disease (Brody 1955), the decline in many neurochemical markers may be secondary to neuron loss. Alternatively, neurochemical decreases may result from selective changes in protein synthesis, or in the enzymes that regulate levels of these markers. These two hypotheses are not mutually exclusive.

The notion that neuron loss with age is inexorable even in the absence of neurological disease is being challenged. Haug's (1984) prodigious morphometric analyses of over 100 human specimens with ages extending beyond 100 years indicated that neuron number in cortex is stable with age. The initial impression of neuron loss was attributed to an artifact from the greater shrinkage of young brains during fixation. Stable numbers of cortical neurons across the human lifespan were also found by Terry et al (1987) and Leuba & Garey (1987). In rodents, cortical neuron number seems stable through the average two to three year lifespan (Curcio & Coleman 1982, Heumann & Leuba 1983, Peters et al 1983, Hornberger et al 1985). Nonetheless, loss may occur at greater ages (Peters et al 1986). Coleman & Flood (1987) also conclude that neuron loss normally occurs during aging in human cerebral cortex. Both Haug (1984) and Terry et al (1987) report age-related neuronal shrinkage (10–30%) and a decreased thickness of the human cerebral cortex, which is consistent with the literature on pyramidal neuron atrophy. Neuron atrophy could be caused, as described above, by slowed macromolecular biosynthesis.

A third phenomenon, the increased volume of astrocytes, could underlie changes in neurochemical markers with aging. Thus, when normalized for protein or tissue mass, markers found primarily on neurons decline, those found in similar concentrations on both neurons and glia remain unchanged, while those found primarily on glia increase with age. This hypothesis requires neither a loss of neurons, nor a reduction in the steady state levels of markers in neurons (relative to their volume), but proposes that the neuronal compartment shrinks while the glial compartment increases with age. For example, the well-characterized age-related decrease in striatal D2 dopamine receptor density (Morgan & Finch 1988) is consistent with the presumed neuronal localization of this receptor. Conversely, the stability of cortical $\beta$-adrenergic receptor (Rogers & Bloom 1985) is consistent with localization on astrocytes and blood vessels, as well as on neurons.

The literature, mainly of anatomical studies, favoring an age-related increase in the fractional volume of glial cells in brain, while not yet detailed, is quite consistent. Several labs report 10–30% increases in glial cell density with age in rodent brain (Vaughan & Peters 1974, Sturrock 1980, Heumann & Leuba 1983, Hughes & Lantos 1987). In human frontal cortex, Hansen et al (1987) observed a five-fold increase in the density of glial fibrillary acidic protein (GFAP) positive astrocytes in the cellular laminae, but no significant change in lamina I, which has the highest density of astrocytes. However, Diamond et al (1977) found no age changes in neuron or glial densities in the rat occipital cortex.

Others found an increase in the size and/or fibrous character of astrocytes with aging. A summary of work by Landfield, Lynch and others concluded that aging in the rat dentate gyrus was accompanied by a major increase in the density of hypertrophied astrocytes by gold sublimate staining, but no change in the total astrocyte density (Lindsey et al 1979). Aging was associated with a conversion of astrocytes from a protoplasmic appearance to a fibrous or reactive appearance, a transition that was accelerated by adrenal steroids. De La Roza et al (1985) observed an increased filamentous appearance of astrocytes with age in the rat lateral geniculate nucleus and noted its similarity to that of reactive astrocytes observed after brain injury. In studies of GFAP-ICC+ astrocytes in rat brain smears (to allow visualization of the whole cell), Bjorklund et al (1985) reported that the astrocyte area increased during aging by 50% in cerebral cortex and hippocampus, and by 25% in cerebellum. Moreover, measurements of astrocytic and neuronal profiles in electron micrographs of dentate gyrus showed that the astrocyte fractional area increased slightly in old rats, from 10% to 15%, with a corresponding decrease of the neuronal area from 40% to 35% (Geinisman et al 1978).

These anatomical data are also consistent with the age-related 20–35% increase in the activity of glutamine synthetase (Danh et al 1985), an astrocyte specific enzyme (Norenberg & Martinez-Hernandez 1979). In addition, we find a selective 80% age-related increase in mRNA for the astrocyte-specific intermediate filament protein GFAP, in the cerebral cortex of healthy two-year-old mice (Goss et al 1988). Old mice enduring premorbid wasting conditions show even greater elevations of GFAP-mRNA (Goss et al 1987), which again emphasizes the impact of peripheral conditions on the CNS.

Taken together, these data suggest a considerable increase in the glial compartment of brain with aging due primarily to an increase in the volume occupied by astrocytes, as well as the number of astrocytes. Besides the changes in astrocytes, two laboratories reported increased numbers of microglia presenting HLA-DR (a MHC type II antigen) with AD and aging (McGeer et al 1988, Rogers et al 1988). A major question for the future concerns whether glial hyperactivity is always just a secondary response to degeneration or regression of neurons, or whether glia have a more aggressive primary role, as suggested by microglial HLA-DR reactivity (see above).

We also note the surprising number of neuron-like properties of astro-cytes in culture: neurotransmitter receptor-like binding sites (Hertz et al 1984; reviewed by Murphy & Pearce 1987), and intracellular responses to neurotransmitters such as changes in membrane potential (Hosli et al 1987, Kettenmann et al 1987; and many others), cyclic AMP stimulation (Northam & Mobley 1985), and receptor-mediated inositol phospholipid turnover (Pearce et al 1986). Astrocytes even have voltage-dependent potassium channels (Bevan & Raff 1985). However, the contribution of astrocytes to these purported neuronal markers is hard to assess in vivo. This derives partly from difficulty in estimating the relative number of astrocytes to neurons, and the dependency of many purported neuronal markers in astrocyte cultures on the culture conditions (Yu & Hertz 1982, Trimmer & McCarthy 1986, Whitaker-Azmitia & Azmitia 1986).

This discussion suggests that age-related changes in brain gene expression may result not only from changes in synthesis, the accumulation of abnormal proteins, or cell loss, but also from an increase in the fractional volume of one cell type (astrocytes) at the expense of another (neuronal shrinkage). Hence, aged neural cells need not modify their synthetic rate relative to cell volume to account for the wide variety of neurochemical changes.

## SUMMARY AND CONCLUSIONS

Much work in the molecular neurobiology of aging is still necessarily descriptive, pending more data on the relative contributions of neuron loss vs. neuron atrophy and glial hyperactivity during normal and neuro-pathological aging. Major loss of neuronal cells is not the rule, even in AD. A working hypothesis holds that increased astrocytic volumes and decreased neuronal volumes are a major factor in age-related neuro-chemical decreases. Resolution of these questions is needed to understand how regional RNA and protein synthesis change during aging.

We have presented alternative mechanisms for changes in gene expression that may be cell specific. The literature, mainly from non-neural tissues, suggests derepression of normally silent genes, possibly in association with DNA demethylation; decreased synthesis of neuronal rRNA due to deletion of rRNA genes; epigenetic changes in hnRNA splicing; reduced turnover rates and the accumulation of nonfunctional proteins. Although global qualitative changes in the inventory of mRNA and proteins are not found in neural tissues at advanced ages, the synthesis and turnover of RNA and protein may be slowed. Impaired protein turn-over or catabolism could contribute to the accumulation of NFT and brain amyloid. In sum, several different types of mechanisms appear to be important in RNA and protein metabolism of the brain during aging, but selectivity for brain region and cell type seems to prevail throughout these myriad changes.

ACKNOWLEDGMENTS

D. G. M. is supported by the Anna Greenwall Award of the American Federation for Aging Research. C. E. F. is supported by the grants from the NIA (P50-AG05142) and by the John D. and Katherine T. MacArthur Foundation Program on Successful Aging. We would like to thank Mark Hess, Jeff Masters, and Patrick May for their comments on drafts of this manuscript.

NOTES ADDED IN PROOF

[a] In a remarkably parallel example, the Brattleboro rat, which is homozygous for a mutation causing a frame shift and the synthesis of aberrant vasopressin precursor, shows a progressive age-related increase in the number of solitary neurons that make normal vasopressin as well as the mutant type from 0.1% at 1 month to 3% of the total vasopressin neurons at 20 months. This is attributed to somatic intrachromosomal gene conversion with the nearby oxytoxin gene (van Leeuen, F., van der Beek, E., Seger, M., Burbach, P., Ivell, R. 1989. Age-related development of a heterozygous phenotype in solitary neurons of the homozygous Brattleboro rat. *Proc. Natl. Acad. Sci. USA* 86: 6417–20). We note that the forthcoming inventories of DNA sequences from sequencing of human and other genomes might predict other neighboring loci that are vulnerable to sporadic age-related gene conversion events causing altered proteins.

[b] A detailed study of slow axonal flow at three ages across the adult lifespan in Fischer 344 rats shows progressive and substantial (40%) slowing of transport for cytoskeletal elements, without change in the transport of membranous vesicles (McQuarrie, I. G., Brady, S. T., Lasek, R. J. 1989. Retardation in the slow axonal transport of cytoskeletal elements during maturation and aging. *Neurobiol. Aging* 10: 359–65).

*Literature Cited*

Adler, J. E., Black, I. B. 1984. Plasticity of substance P in mature and aged sympathetic neurons in culture. *Science* 225: 1499–1500

Anderson, K. J., Scheff, S. W., DeKosky, S. T. 1986. Reactive synaptogenesis in hippocampal area CA1 of aged and young adult rats. *J. Comp. Neurol.* 252: 374–84

Anzai, K., Imazato, C., Goto, S. 1983.

mRNA population in the liver, kidney and brain of young and senescent mice. *Mech. Ageing Dev.* 23: 137–50

Anzai, K., Goto, S. 1987. Brain-specific small RNA during development and ageing of mice. *Mech. Aging Devel.* 39: 129–35

Bevan, S., Raff, M. 1985. Voltage-dependent potassium currents in cultured astrocytes. *Nature* 315: 229–32

Bjorklund, H., Eriksdotter-Nilsson, M., Dahl, D., Roxe, G., Hoffer, B., Olson, L. 1985. Image analysis of GFA-positive astrocytes from adolescence to senescence. *Exp. Brain Res.* 58: 163–70

Brody, H. 1955. Organization of the cerebral cortex. III. A study of aging in the human cerebral cortex. *J. Comp. Neurol.* 102: 511–56

Cattenach, B. M. 1974. Position effect variegation in the mouse. *Genetics Res.* 23: 291–306

Chaconas, G., Finch, C. E. 1973. The effect of ageing on RNA/DNA ratios in brain regions of the C57BL/6J male mouse. *J. Neurochem.* 21: 1469–73

Coleman, P. D., Flood, D. G. 1987. Neuron numbers and dendritic extent in normal aging and Alzheimer's disease. *Neurobiol. Aging* 8: 521–45

Coleman, P. D., Kaplan, B. B., Osterburg, H. H., Finch, C. E. 1980. Brain poly(A)RNA during aging: Stability of yield and sequence complexity in two rat stains. *J. Neurochem.* 34: 335–45

Cosgrove, J. W., Rapoport, S. I. 1987. Absence of age differences in protein synthesis by rat brain, measured with an initiating cell-free system. *Neurobiol. Aging* 8: 27–34

Cosgrove, J. W., Atack, J. R., Rapoport, S. I. 1987. Regional analysis of rat brain proteins during senescence. *Exp. Gerontol.* 22: 187–98

Curcio, C. A., Coleman, P. D. 1982. Stability of neuron number in cortical bands of aging mice. *J. Comp. Neurol.* 212: 158–72

Danh, H. C., Benedetti, M. S., Dostert, P. 1985. Age-related changes in glutamine synthetase activity of rat brain, liver and heart. *Gerontology* 31: 95–100

De La Roza, C., Cano, J., Reinoso-Suarez, F. 1985. An electron microscopic study of astroglia and oligodendroglia in the lateral geniculate nucleus of aged rats. *Mech. Ageing Devel.* 29: 267–81

Diamond, M. C., Johnson, R. E., Gold, M. W. 1977. Changes in neuron number and size and glia number in the young, adult, and aging rat medial occipital cortex. *Behav. Biology* 20: 409–18

Dice, J. F. 1987. Molecular determinants of protein half-lives in eukaryotic cells. *FASEB J.* 1: 349–57

Doebler, J. A., Markesberry, W. R., Anthony, A., Rhoads, R. E. 1987. Neuronal RNA in relation to neuronal loss and neurofibrillary pathology in the hippocampus in Alzheimer's disease. *J. Neuropathol. Exp. Neurol.* 46: 28–39

Dwyer, B. E., Fando, J. L., Wasterlain, C. G. 1980. Rat brain protein synthesis declines during postdevelopmental aging. *J. Neurochem.* 35: 746–49

Esumi, H., Takahashi, Y., Sekiya, T., Sato, S., Nagase, S., Sugimura, T. 1982. Presence of albumin mRNA precursors in nuclei of analbuminemic rat liver lacking cytoplasmic albumin mRNA. *Proc. Natl. Acad. Sci.* 79: 734–38

Esumi, H., Takahashi, Y., Makino, R., Sato, S., Sugimura, T. 1985. Appearance of albumin-producing cells in the liver of analbuminemic rats on aging and administration of carcinogens. In *Werner's Syndrome and Human Aging*, ed. D. Salk, Y. Fujiwara, G. M. Martin. *Adv. Exp. Biol. Med.* 190: 637–50

Farquhar, M. N., Kosky, K. J., Omenn, G. S. 1979. Gene expression in brain. In *Aging in Nonhuman Primates*, ed. D. M. Bowden, pp. 71–79. New York: Van Nostrand

Finch, C. E. 1972. Enzyme activities, gene function and ageing in mammals (review). *Exp. Gerontol.* 7: 53–67

Finch, C. E., Foster, J. R. 1973. Hematologic and serum electrolyte values of the C57BL/6J male mouse immaturity and senescence. *Lab. Animal Sci.* 23: 339–49

Geddes, J. W., Monaghan, D. T., Cotman, C. W., Lott, I. T., Kim, R. C., Chui, H. C. 1985. Plasticity of hippocampal circuitry in Alzheimer's disease. *Science* 230: 1179–81

Geinisman, Y., Bondareff, W., Dodge, J. T. 1978. Hypertrophy of astroglial processes in the dentate gyrus of the senescent rat. *Am. J. Anat.* 153: 537–44

Gershon, D., Gershon, D. 1970. Detection of inactive molecules in aging of organisms. *Nature* 227: 1214–17

Goss, J. R., Morgan, D. G., Finch, C. E. 1987. Glial fibrillary acidic protein mRNA increases during a wasting agonal state in old mice. Implications for the increased GFAP mRNA in Alzheimer's Disease. *Soc. Neurosci. Abstr.* 13: 1325

Goss, J. R., Morgan, D. G., Finch, C. E. 1988. Glial fibrillary acidic protein mRNA increases with age in mouse. *Soc. Neurosci. Abstr.* 14: 1008

Hansen, L. A., Armstrong, D. M., Terry, R. D. 1987. An immunohistochemical quantification of fibrous astrocytes in the aging human cerebral cortex. *Neurobiol. Aging* 8: 1–6

Haug, H. 1984. Macroscopic and microscopic morphometry of the human brain and cortex. A survey in the light of new results. *Brain Pathol.* 1: 123–49

Hertz, L., Schousboe, I., Hertz, L., Schousboe, A. 1984. Receptor expression in primary cultures of neurons or astrocytes. *Prog. Neuropsychopharmocol. Biol. Psychiat.* 8: 521–27

Heumann, D., Leuba, G. 1983. Neuronal

death in the development and aging of the cerebral cortex of the mouse. *Neuropath. Appl. Neurobiol.* 9: 297–311

Hoff, S. F., Scheff, S. W., Benardo, L. S., Cotman, C. W. 1982. Lesion-induced synaptogenesis in the dentate gyrus of aged rats: I. loss and reacquisition of normal synaptic density. *J. Comp. Neurol.* 205: 246–52

Hoffman, G. E., Finch, C. E. 1986. LHRH neurons in the female C57BL/6J mouse brain during reproductive aging: No loss up to middle-age. *Neurobiol. Aging* 7: 45–48

Hornberger, J. C., Buell, S. J., Flood, D. G., McNeill, T. H., Coleman, P. D. 1985. Stability of numbers but not size of mouse forebrain cholinergic neurons to 53 months. *Neurobiol. Aging* 6: 269–75

Hosli, L., Hosli, E., Baggi, M., Bassetti, C., Uhr, M. 1987. Action of dopamine and serotonin on the membrane potential of cultured astrocytes. *Exp. Brain Res.* 65: 482–85

Hughes, C. C. W., Lantos, P. L. 1987. A morphometric study of blood vessel, neuron and glial cell distribution in young and old rat brain. *J. Neurol. Sci.* 79: 101–10

Ingvar, M. C., Maeder, P., Sokoloff, L., Smith, C. B. 1985. Effects of ageing on local rates of cerebral protein synthesis in Sprague-Dawley rats. *Brain* 108: 155–70

Johnson, S. A., Pasinetti, G. M., May, P. C., Ponte, P. A., Cordell, B., Finch, C. E. 1988. Selective reduction of mRNA for the B-amyloid precursor protein that lacks a Kunitz-type protease inhibitor motif in cortex from AD. *Exp. Neurol.* 102: 264–68

Kalu, D. N., Herbert, D. C., Hardin, R. R., Yu, B. P., Kaplan, G., Jacobs, J. W. 1988. Mechanisms of dietary modulation of calcitonin levels in Fischer rats. *J. Gerontol.* 43: B125–31

Kettenmann, H., Backus, K. H., Schachner, M. 1987. Gamma-aminobutyric acid opens Cl-channels in cultured astrocytes. *Brain Res.* 404: 1–9

Lavie, L., Reznick, A. Z., Gershon, D. 1982. Decreased protein and puromycinyl-peptide degradation in livers of senescent mice. *Biochem. J.* 202: 47–51

Leuba, G., Garey, L. J. 1987. Evolution of neuronal numerical density in the developing and aging human visual cortex. *Hum. Neurobiol.* 6: 11–18

Lindsey, J. D., Landfield, P. W., Lynch, G. 1979. Early onset and topographical distribution of hypertrophied astrocytes in hippocampus of aging rats: A quantitative study. *J. Gerontol.* 34: 661–71

Makino, R., Esumi, H., Sato, S., Takashi,

Y., Nagase, S., Sugimura, T. 1982. Elevation of serum albumin concentration in analbuminemic rats by administration of 3'-methyl-4-dimethylaminobenzene. *Biochem. Biophys. Res. Comm.* 106: 863–70

Makino, R., Sato, S., Esumi, H., Negishi, C., Takano, M., Sugimura, T., Nagase, S., Tanaka, H. 1986. Presence of albumin-positive cells in the liver of analbuminemic rats and their increase on treatment with hepatocarcinogens. *Jpn. J. Canc.* 77: 153–59

McGeer, P. L., Itagaki, S., Akiyama, H., McGeer, E. G. 1988. Immune system response in Alzheimer's disease. In *The Molecular Biology of Alzheimer's Disease*, ed. C. E. Finch, P. Davies, pp. 47–50. *Curr. Comm. Molec. Biol.*

Manetto, V., Perry, G., Tabaton, M., Mulvihill, P., Fried, V. A., Smith, H. T., Gambetti, P., Autilio-Gambetti, L. 1988. Ubiquitin is associated with abnormal cytoplasmic filaments characteristic of neurodegenerative diseases. *Proc. Natl. Acad. Sci. USA* 85: 4501–5

Mann, D. M. A., Yates, P. O. 1979. The effects of ageing on the pigmented nerve cells of the human locus cerulus and substantia nigra. *Acta Neuropathol.* 47: 93–97

Mann, D. M. A., Neary, M., Yates, P. O., Lincoln, J., Snowden, J. S., Stanworth, P. 1981. Neurofibrillary pathology and protein synthetic capability in nerve cells in Alzheimer's disease. *Neuropath. Appl. Neurobiol.* 7: 37–47

Mann, D. M. A., Yates, P. O. 1983. Possible role of neuromelanin in the pathogenesis of Parkinson's disease. *Mech. Ageing Devel.* 21: 193–203

Mays-Hoopes, L. L. 1989. Age-related changes in DNA methylation: Do they represent continued developmental changes. *Int. Rev. Cytol.* 114: 181–220

Menzies, R. A., Gold, P. H. 1972. The apparent turnover of mitochondria, ribosomes, and sRNA of the brain in young adult and aged rats. *J. Neurochem.* 19: 1671–83

Morgan, D. G., Finch, C. E. 1988. Dopaminergic changes in the basal ganglia. A generalized phenomenon of aging in mammals. In *Central Determinants of Age-Related Declines in Motor Function*, ed. J. A: Joseph. *Ann. NY Acad. Sci.* 515: 145–60

Murphy, S., Pearce, B. 1987. Functional receptors for neurotransmitters on astroglial cells. *Neuroscience* 22: 381–94

Nelson, J. F., Latham, K., Finch, C. E. 1975. Plasma testosterone levels in C57BL/6J male mice: Effects of age and disease. *Acta. Endocrinol.* 80: 744–50

Nelson, J., Bender, M., Schachter, B. S. 1988. Age related changes in proopiomelanocortin messenger ribonucleic acid levels in hypothalamus and pituitary of female C57BL/6J mice. *Endocrinology* 123: 340–44

Norenberg, M. D., Martinez-Hernandez, A. 1979. Fine structural localization of glutamine synthetase in astrocytes of rat brain. *Brain Res.* 161: 303–10

Northam, W. J., Mobley, P. 1985. Clonidine pretreatment enhances the sensitivity of the beta-noradrenergic receptor coupled adenylate cyclase system in astrocytes. *Eur. J. Pharmacol.* 113: 153–54

Pasinetti, G. M., Lerner, S. P., Johnson, S. A., Morgan, D. G., Telford, N. A., Finch, C. E. 1989. Chronic lesions differentially decrease tyrosine hydroxylase mRNA in dopaminergic neurons of the substantia nigra. *Mol. Brain Res.* 5: 203–9

Pearce, B., Albrecht, J., Morrow, C., Murphy, S. 1986. Astrocyte glutamate receptor activation promotes inositol phospholipid turnover and calcium flux. *Neurosci. Lett.* 72: 335–40

Peters, A., Feldman, M. L., Vaughan, D. W. 1983. The effect of aging on the neuronal population within area 17 of adult rat cerebral cortex. *Neurobiol. Aging* 4: 273–82

Peters, A., Harriman, K. M., West, C. D. 1987. The effect of increased longevity, produced by dietary restriction on the neuronal population and area 17 in rat cerebral cortex. *Neurobiol. Aging* 8: 7–20

Rafols, J. A., Cheng, H. W., McNeill, T. H. 1989. Golgi study of the mouse striatum: Age-related dendritic changes in different neuronal populations. *J. Comp. Neurol.* 279: 212–27

Reff, M. E. 1985. In *Handbook of the Biology of Aging*, ed. C. E. Finch, E. L. Schneider, pp. 225–49. New York: Van Nostrand Reinhold. 2nd ed.

Reis, D. J., Ross, R. A., Joh, T. H. 1977. Changes in the activity and amounts of enzymes synthesizing catecholamines and acetylcholine in brain, adrenal medulla, and sympathetic ganglia of aged rat and mouse. *Brain Res.* 136: 465–74

Richardson, A., Birchenall-Sparks, M. C. 1983. Age-related changes in protein synthesis. *Rev. Biol. Res. Aging* 1: 255–73

Richardson, A., Butler, J. A., Rutherford, M. S., et al. 1987. Effect of age and dietary restriction on the expression of alpha$_{2\mu}$-globulin. *J. Biol. Chem.* 262: 12821–25

Ringborg, U. 1966. Composition and content of RNA in neurons of rat hippocampus at different ages. *Brain Res.* 2: 296–98

Rogers, J., Bloom, F. E. 1985. Neurotransmitter metabolism and function in the aging central nervous system. See Reff 1985, pp. 645–91

Rogers, J., Luber-Narod, J., Mufson, E. J., Styren, S. D., Civin, W. H. 1988. Presence of immune system markers in human brain and their potential role as a primary pathogenetic mechanism in Alzheimer's disease. See McGeer et al 1988, pp. 51–56

Rothstein, M. 1982. In *Biochemical Approaches to Aging*. New York: Academic

Selkoe, D. J. 1989. Biochemistry of altered brain proteins in Alzheimer's disease. *Annu. Rev. Neurosci.* 12: 463–90

Sharma, H. K., Rothstein, M. 1980. Altered enolase in aged *Turbatrix aceta*. Results from conformational changes in the enzyme. *Proc. Natl. Acad. Sci. USA* 77: 5865–68

Sharma, H. K., Rothstein, M. 1984. Altered brain phosphoglycerate kinase from aging rats. *Mech. Ageing Devel.* 25: 285–96

Stadtman, E. R. 1988. Minireview: Protein modification in aging. *J. Gerontol.* 43: B112–20

Strehler, B. L. 1986. Genetic instability as the primary cause of human ageing. *Exp. Geront.* 21: 283–319

Stromska, D. P., Ochs, S. 1982. Axoplasmic transport in aged rats. *Exp. Neurol.* 77: 214–24

Sturrock, R. R. 1980. A comparative quantitative and morphological study of ageing in the mouse neostriatum, indusium griseum and anterior commissure. *Neuropath. Appl. Neurobiol.* 6: 51–68

Terry, R. D., DeTeresa, R., Hansen, L. A. 1987. Neocortical cell counts in normal human adult aging. *Ann. Neurol.* 21: 530–39

Trimmer, P. A., McCarthy, K. D. 1986. Fetal, newborn and young adult rats express beta-adrenergic receptors in vitro. *Devel. Brain Res.* 27: 151–65

Uemura, E., Hartmann, H. A. 1979. RNA content and volume of nerve cell bodies in human brain. *Exp. Neurol.* 65: 107–17

Vaughan, D. W., Peters, A. 1974. Neuroglial cells in the cerebral cortex of rats from young adulthood to old age: An electron microscope study. *J. Neurocytol.* 3: 405–29

Ward, W. 1988. Enhancement by food restriction of liver protein synthesis in the aging Fischer 344 rat. *J. Gerontol.* 43: B50–53

Wareham, K. A., Lyon, M. F., Glenister, P. H., Williams, E. D. 1987. Age-related reactivation of an X-linked gene. *Nature* 327: 725–27

Weindruch, R., Walford, R. L. 1988. *The Retardation of Aging and Disease by Diet Restriction.* Springfield, Ill.: Thomas

Wellinger, R., Guigoz, Y. 1986. The effect of age on the induction of tyrosine aminotransferase and tryptophan oxygenase genes by physiological stress. *Mech. Ageing Devel.* 34: 203–17

Whitaker-Azmitia, P. M., Azmitia, E. C. 1986. [³H]5-Hydroxytryptamine binding to brain astroglial cells: Differences between intact and homogenized preparations and mature and immature cultures. *J. Neurochem.* 46: 1186–89

Wilson, D. E., Hall, M. E., Stone, G. C. 1978. Test of some aging hypotheses using two-dimensional protein mapping. *Gerontology* 24: 426–33

Wilson, V. L., Smith, R. A., Ma, S., Cutler, R. G. 1987. Genomic 5-methyl-deoxycytidine decreases with age. *J. Biol. Chem.* 262: 9948–51

Yu, P. H., Hertz, L. 1982. Differential expression of type A and type B monoamine oxidase of mouse astrocytes in primary cultures. *J. Neurochem.* 39: 1492–95

Yuh, K. M., Gafni, A. 1987. Reversal of age-related effects in rat muscle phosphoglycerate kinase. *Proc. Natl. Acad. Sci. USA* 84: 7458–62

*Annu. Rev. Neurosci. 1990. 13:89–109*

# FACE AGNOSIA AND THE NEURAL SUBSTRATES OF MEMORY

*Antonio R. Damasio, Daniel Tranel, and Hanna Damasio*

Department of Neurology, Division of Behavioral Neurology and Cognitive Neuroscience, University of Iowa College of Medicine, Iowa City, Iowa 52242

Focal damage to selective regions of the human association cortices can impair the ability to recognize the identity of previously familiar faces, even when visual perception and intellect remain unaltered. In general, the impairment is accompanied by an inability to learn the identity of new faces. The phenomenon has been noted since the turn of the century (Wilbrand 1892), and is known as *prosopagnosia,* or *face agnosia.* Its bizarre and unseeming nature lent itself to doubts that it could be caused by specific neural dysfunction, and psychodynamic interpretations were even offered. Recently, however, face agnosia has become the focus of serious study (e.g. Lhermitte et al 1972, Meadows 1974, Newcombe 1979, Benton 1980, Damasio et al 1982). Face agnosia, along with the varied neuropsychological disturbances that may accompany it, can now be analyzed with experimental paradigms and correlated with neuro-anatomical loci of damage identified by neuroimaging methods. This affords a rare opportunity to elucidate cognitive and neural mechanisms of perception, learning, and memory in humans.

## NEUROPSYCHOLOGIC CHARACTERIZATION OF FACE AGNOSIA

### The Presentation of Face Agnosia

The flavor of face agnosia can be easily captured by a brief description of one of our typical subjects. A 65-year-old woman suddenly developed an inability to recognize the faces of her husband and daughter. She could

89

0147–006X/90/0301–0089$02.00

not even recognize her own face in the mirror—although she knew the face she was observing must belong to her, she did not experience a sense of familiarity when viewing it. The faces of other relatives and friends became equally meaningless. Yet she remained fully capable of identifying all those persons from voice. About two months prior to the onset of the prosopagnosia, she had noticed some difficulty in her perception of colors. Specifically, colors appeared "washed out" and "dirty," although she continued seeing details of shape, depth, and movement without any problem. She had no other history of neurological problems, and no history of psychiatric disease.

Detailed evaluation upon her admission to our department revealed that her difficulties were circumscribed to the visual domain and, even there, occurred in a most selective manner. She was entirely unable to recognize a single face of any relative or friend, either in person or from photographs. Furthermore, she did not even have a hint that the faces belonged to persons with whom she was well acquainted. She was equally impaired in the recognition of celebrities from photographs or movies. These findings are represented by the experimental data in Table 1, which come from our patient E.H. The patient was given a series of 50 faces to identify, 8 of which were well known to her (*targets*). The targets were mixed, in pseudorandom order, with 42 photographs of persons she had never encountered before (*nontargets*). Two series were administered, one in which the target faces were of relatives, friends, and herself, and a second in which the targets were famous actors and politicians. Each face was presented on a slide, in black and white, as a full frontal pose. Features below the neckline and in the background were masked. The nontargets were selected so as to produce a group similar to the target group in terms of age range and gender ratio. Photographic charcteristics, including brightness, contrast, resolution, etc, were comparable in the target and nontarget

**Table 1**  Identity recognition and familiarity ratings for target and nontarget faces (patient E.H.)

|  | N | Identity recognition (% correct) | Average familiarity rating (s.d. in parentheses) |
|---|---|---|---|
| Retrograde-family experiment |  |  |  |
| Target | 8 | 0 | 6.0 (0.0) |
| Nontarget | 42 | — | 6.0 (0.0) |
| Retrograde-famous experiment |  |  |  |
| Target | 8 | 0 | 6.0 (0.0) |
| Nontarget | 42 | — | 6.0 (0.0) |

faces. In addition to being asked to recognize identity, she was also asked to rate, using a six-point rating scale that ranged from "1" (definite familiarity) to "6" (definite unfamiliarity), the extent to which she was familiar with each face. Identity recognition was at the zero level for both family and famous faces. Also, not only did she rate all of the target faces as "unfamiliar," she assigned the highest level of confidence to her ratings, i.e. she rated as "definitely unfamiliar" all eight target faces in both sets of stimuli. A parallel defect was present in the anterograde compartment, i.e. she was never able to learn the faces of persons she encountered after the onset of her condition, even though she could learn to recognize those individuals from their voices and other non-face cues.

The profound face recognition defect of this patient occurred against a background of otherwise virtually normal neuropsychological capacities. She had normal intellect, intact speech and language, and normal memory and learning for both verbal and nonverbal material (outside the visual realm). Recognition and naming of objects were normal in the auditory and tactile modalities. (Her visual recognition and naming of non-face stimuli, although normal at first glance, showed remarkable impairments, which are discussed in the following section.) She could read normally and she could write, with normal spelling. Of particular importance, visual acuity was 20/20 in both eyes, and she performed normally on a wide range of visuoperceptual, visuospatial, and visuoconstruction tasks.

The exceptions to the intactness of basic visual perception were: (*a*) a defect in color perception (known as *achromatopsia*); and (*b*) small form vision scotomata in the right superior and left superior quadrants. The patient could not perceive colors in any part of the visual field and saw form always in shades of black and white. Achromatopsia frequently accompanies "pure" prosopagnosia, as a result of the contiguity of cortical regions related to color processing and to processes necessary for face recognition (Damasio et al 1989a). By itself, however, it does not contribute to prosopagnosia. The defects in this patient were caused by infarctions in ventral visual association cortices, bilaterally, in the inferior occipital/posterior temporal region (see section on *Neuroanatomical Correlates* below). The infarction on the right side occurred first and caused left hemiachromatopsia, as suggested by her complaints of color perception loss. The second infarct, on the left side, further impaired her ability to process color and, in combination with the first lesion, created the neuroanatomical setting for prosopagnosia.

This case exemplifies a complete face recognition defect, restricted to the visual modality, and largely conforming to the classic notion of agnosia, i.e. a normal percept stripped of its meaning. For practical classification purposes we refer to this type of presentation as "pure associative" face agnosia. Note, however, that we do not believe perceptual and recognition/ recall processes can ever be rigidly compartmentalized. Those processes are operated in a continuum. Furthermore, additional research in the type of subject described above may reveal subclinical, high-level integrative defects that might well be conceptualized as perceptual.

There are other intriguing presentations of face agnosia. One is a partial identity recognition defect that compromises recognition of most or all real faces, in real time, but spares the recognition of many or all faces presented in photographs. The problem remains restricted to the visual modality, and is "pure" in that sense. Unlike the subject in the preceding description, this type of patient has a patent and very substantial defect in visual perception. There is difficulty in perceiving all parts of a visual array simultaneously, and in generating the image of a whole entity given a part. For instance, when shown the drawing of a part of an airplane, the patient is unable to imagine the whole and thus fails to recognize the stimulus. The ability to assemble parts of a model in a meaningful ensemble is also impaired. For practical classification purposes, we refer to this type of face agnosia as "partial apperceptive," although we are just as cautious about the implications of this label as we are with the label "associative" applied above. Many patients with manifest perceptual disorders, e.g. visual disorientation, neglect, defective pattern discrimination, impaired copying ability, impaired constructional praxis, and so on, remain fully capable of recognizing familiar faces. Such patients, in fact, not only remain capable of recognizing identity from faces, but they also preserve the ability to recognize the meaning of facial expressions, and to make accurate judgments about gender and age based on facial information (see control subjects in Tranel et al 1988). In other words, visuoperceptual disturbance as detected by current neuropsychological probes does not necessarily cause face agnosia (Meier & French 1965, Orgass et al 1972). It appears that other components of the recognition process must be altered for agnosia to occur.

In yet another presentation of face agnosia, patients have entirely normal visual perception, but their inability to recognize identities is not confined to the visual modality. The patients are unable to recognize identity regardless of the sensory channel used, e.g. neither viewing a face nor listening to the appropriate voice conjures up knowledge about the person behind both. This type we refer to as "amnesic associative," considering that it appears in the setting of amnesia. The physiopathological significance of this type of face agnosia cannot be overemphasized. It indicates that compromise of recall for episodic level knowledge compromises the recognition of face identity, which, in turn, reveals that episodic level knowledge—the knowledge of contextual detail that defines the uniqueness of entities or events—is critical for face learning and recognition (Damasio 1988, Damasio et al 1989b).

## Impaired Recognition of Non-face Entities in Face Agnosia

Early descriptions of face agnosia suggested, erroneously, that recognition of non-face stimuli, e.g. objects, animals, houses, etc, was normal. The error is not surprising, considering that patients with face agnosia are likely to complain primarily or perhaps exclusively about the problem that most disrupts their social interactions. Closer investigation, however, reveals that other recognition defects are present. Patients with face agnosia are also impaired in the visual recognition of other unique stimuli, and even of non-unique stimuli. Typical examples of the former include houses, cars, pets, and even personal effects, such as articles of clothing (Bornstein 1963, Lhermitte 1972, Damasio et al 1982, Pallis 1955). The patient described above, for instance, had pronounced difficulties in the recognition of her clothes, her car, and her house. She never had trouble categorizing individual stimuli at the appropriate supraordinate level— i.e. she always identified faces as "faces," cars as "cars," and houses as "houses." The problem was in assigning individual identity to various exemplars. Although not as disruptive of social interaction as the face recognition disability, these other defects were not insignificant—to locate her car in a parking lot, for example, she was forced to rely upon license plate number. The meaning of the accompanying defects is clear. Although faces are a problem stimulus, the dysfunction that causes the face agnosia involves detecting, from visual stimuli, the identity of many categories of entities, not just faces.

Although patients with face agnosia recognize the broad category to which virtually any stimulus belongs, it is clear that not all stimuli can be properly categorized (Damasio et al 1982). Any such patient can decide whether a stimulus is animate or inanimate, living or man-made, e.g. decide that a dog or an elephant are "animals," that a chair is "furniture," and that flowers or a tree are "plants." At supraordinate level, recognition operates normally. Surprisingly, however, the recognition of certain particular exemplars fails at "basic object" *and* "subordinate levels" [to use Rosch et al's (1976) taxonomic level nomenclature]. For instance, one of our patients with "pure associative" face agnosia was unable to distinguish a cat from a tiger or lion, although she knew that they were animals. The patient was not being asked to recognize identity, but simply to recognize a category.

The existence of defects at different taxonomic levels is of special importance, and allows for the investigation of possible factors contributing to failure or success of recognition. Current findings suggest several relevant factors (Damasio 1989a): (*a*) similarity of physical structure among exemplars; (*b*) size of the class formed by exemplars with a similar structure; (*c*) operation of the exemplar; (*d*) type of sensory and motor interaction required to map the exemplar; (*e*) value of the exemplar to the perceiver; (*f*) acquaintance with the exemplar. Because many of these factors are shared by entities that belong to the same conceptual categories, the recognition performance of agnosic patients may superficially suggest that some categories are recognized normally while others are pervasively affected, e.g. normal recognition of inanimate, man-made objects such as tools, contrasted with impaired recognition with animate, natural kinds such as animals or plants. Careful analysis of the data, however, reveals that there are exceptions to the rule in both the "normal" and the "impaired" categories, and that success or failure ultimately depends on constraints governing the mapping of each individual entity (Damasio 1989a). Thus, the constraints related to some entities create a burden for learning and recognition at categorical level that is just as complex as the recognition of other entities at unique level. This distinction is not merely academic, but forms the basis for rejecting the notion that semantic systems are organized, cognitively and neurally, according to conceptual definitions and respecting conceptual boundaries.

In short, in virtually every instance, patients with face agnosia fail to recognize an exemplar as a unique individual. Faces and other visual stimuli alike cannot conjure up pertinent information on the basis of which identity can be recovered. In addition, some stimuli fail to be recognized even if mere categorical assignment is requested.

## Other Aspects of Face Processing

As mentioned above, although patients with face agnosia are often unable to recognize the identity behind any face, they have no difficulty recognizing faces as faces. Another aspect of face processing that is often intact is the recognition of facial expressions. Tranel et al (1988) demonstrated experimentally that many patients are entirely normal in their ability to recognize a whole array of facial expressions presented in static photographs of human faces. Six facial expressions were investigated: anger, fear, happiness, sadness, surprise, and disgust. For each picture on a 24-item test, subjects were asked to choose from a multiple choice list the name of the expression that was being modeled in the picture. The performances of four subjects with face agnosia were compared to normals and to a brain-damaged control group (see Table 2).

Three of the patients with face agnosia performed perfectly, achieving scores fully comparable to those of the controls. Only one patient, #2, performed defectively. This result accords with the clinical observation

**Table 2**    Recognition of the meaning of facial expressions

| Subject group | Score |
|---|---|
| Controls-normal ($n = 20$) | 20.9/24 |
| Controls-brain damaged ($n = 26$) | 18.0/24 |
| Face agnosic patients | |
| #1 | 17/24 |
| #2 | 11/24[a] |
| #3 | 18/24 |
| #4 | 21/24 |

[a] Defective performance.

that patients with face agnosia generally can tell the meaning of the facial expressions as they are produced by the persons with whom they interact. The findings also reveal that such recognition can be based on a non-moving stimulus, i.e. a stimulus that lacks movement unfolding in real time. Other face processing abilities preserved in such patients include the assignment of gender and the estimation of age (Tranel et al. 1988).

## Visual Recognition of Identity from Non-face Cues

Johansson (1973) has shown that normal individuals can recognize human patterns of movement, e.g. walking, jumping, dancing, from films depicting the motion of point-light sources placed on main joints of an actor dressed in black, moving against a black background (a situation in which the viewer sees nothing but movement of lighted dots against a dark background). When the display was shown as a static picture, observers were unable to recognize patterns of movement, or even to recognize that the point-light sources represented human figures. For the dynamic patterns, the seemingly difficult recognition task can be performed in as little as 200 msec (Johansson 1976). Other investigators have used this method to show that persons can identify gender (Kozlowski & Cutting 1977), and can even recognize unique identity (e.g. of the self or friends; Cutting & Kozlowski 1977), on the basis of gait information.

One intriguing finding in many patients with face agnosia is that they are able to recognize the identity of a person on the basis of non-face visual cues, such as movement and posture (Damasio et al 1982). Given a circumscribed context, in which only a limited number of identity hypotheses can be entertained, such patients will correctly guess the identity of a person with a distinctive posture of the head or trunk, or with a particular gait, or with a characteristic pattern of deportment. This finding reveals the selectivity of the visual recognition defect. Unique identity can still be retrieved through visual channels based on stimuli other than faces. This is important, from a physiological standpoint, because such an ability can only be explained by the existence of a multiplicity of visual representations, keyed to a single unique entity (in this case, a person). In our view (see discussion of model, below), any of those multiple representations can provide an interface to yet other representations, visual and nonvisual, related to the unique entity.

## Nonconscious Recognition of Faces

PSYCHOLOGICAL STUDIES   Most patients with severe face agnosia report a complete inability to recognize the identity of familiar faces, and they cannot decide, on an individual basis, whether a face is familiar or not. However, certain specialized probes and paradigms have been used to show that these patients do preserve, at a nonconscious level, some capacity to discriminate familiar from unfamiliar faces. A clear demonstration of this effect was reported by Tranel & Damasio (1985, 1988), who used the electrodermal skin conductance response (SCR) to investigate covert discrimination of familiar faces. Subjects viewed sets of facial stimuli that included a mix of *target* (individuals with whom the subject was well acquainted) and *nontarget* (individuals the subject had never seen before) faces. Each face was presented for 2 sec, at intervals of 20–25 sec. The subject was instructed to view each face carefully, but no verbal or motor response was called for. Skin conductance responses to the stimuli were measured throughout stimulus presentation. For each face, the amplitude of the largest SCR that began within 1 to 5 sec after stimulus onset was recorded, and then the average target SCR and nontarget SCR were calculated.

Relevant data are reproduced in Table 3. Despite their severe defects in familiar face recognition, four subjects generated larger and more frequent SCRs to familiar faces, as compared to unfamiliar ones. Subject A, for example, produced an average SCR of 0.934 $\mu$S to the eight target faces, which is significantly larger than her average nontarget response of 0.048 $\mu$S ($p < 0.001$). In another experiment, the same effect was obtained for "famous" faces. In the anterograde compartment (i.e. in the period since the onset of agnosia), where none of these subjects has shown any conscious learning of faces encountered since the onset of their defect, the SCRs also revealed significant nonconscious discrimination, although the effect was smaller than for previously known faces.

There is other evidence from the psychophysiological domain that individuals with face agnosia can show an appreciable amount of nonconscious familiar face discrimination. Bauer (Bauer 1984, Bauer & Verfaellie 1988), for example, used a paradigm in which SCRs were measured while subjects

**Table 3**  Skin conductance responses for target and nontarget faces: Retrograde-family experiment

| Subject | N | | Skin conductance responses | | | |
|---|---|---|---|---|---|---|
| | | | Frequency (in %) | | Amplitude (in $\mu$S) | |
| | Target | Nontarget | Target | Nontarget | Target | Nontarget |
| A | 8 | 42 | 71 | 12 | 0.934 (0.723) | 0.048 (0.134) |
| B | 8 | 30 | 75 | 33 | 0.170 (0.200) | 0.039 (0.069) |
| C | 8 | 42 | 100 | 36 | 1.660 (1.110) | 0.146 (0.317) |
| D | 14 | 56 | 100 | 30 | 0.528 (0.547) | 0.011 (0.034) |

viewed and listened to correct and incorrect pairs of face-name matches. In these experiments, subjects more often showed larger SCRs to the correct face-name pairs. This effect provides convergent support for the notion that patients with face agnosia can discriminate familiar faces at a nonconscious level.

OTHER PARADIGMS    In a study of eye movements in two subjects with face agnosia, Rizzo et al (1987) found that the subjects scanned facial stimuli in the same manner as normal controls. For example, the patients showed the greatest number and duration of fixations on regions of the face that contained essential facial features, e.g. the eyes, nose, mouth, chin and hairline, just as controls did. Quantitative analysis of the scanpaths also documented a high degree of similarity between the patients and controls. The authors concluded that impaired facial recognition and learning were not associated with impaired scanning of faces.

An intriguing finding reported by Rizzo et al was that scanpath properties in face agnosic subjects varied in subtle but reliable ways depending upon whether the face being scanned was previously known to the subject. When the transition of scanpaths among quadrants was cast in a first-order Markov model transition matrix, it appeared that scanning patterns for familiar, personally meaningful faces, were markedly different from the patterns generated for unfamiliar faces. The key difference was in the *predictability* of the scanpaths, with previously familiar faces generating *less* predictable scanpaths than novel ones. Considering that the physical contours of faces activate neural structures that drive the scanpath and alter the sequential dependence of fixations, the results were taken to mean that scanpaths for familiar faces in face agnosic individuals were driven by a previously acquired and still accessible internal "schema."

Other investigators have used different experimental paradigms to demonstrate preserved "covert" familiar face processing in face agnosic subjects. de Haan et al (1987), for example, used a reaction time paradigm in which subjects had to decide whether two photographs "matched" (were of the same individual) or did not match (were of different individuals). A patient with face agnosia was faster in matching the identities of familiar than unfamiliar faces. This effect, which is also manifest in normal subjects, occurred despite the patient's complete inability to recognize face identity, or even to classify familiar faces as "familiar" or "unfamiliar" (Young & de Haan 1988). In another example, Bruyer et al (1983) demonstrated that their agnosic subject could accurately classify previously familiar faces as "known faces," despite his inability to recognize identity. These behavioral paradigms provide corroborative support for the psychophysiological and psychophysical studies described above, in the conclusion that patients with severe inability to recognize facial identity can show reliable "covert recognition" of familiar faces.

## NEUROANATOMICAL CORRELATES OF FACE AGNOSIA

Face agnosia of the "associative" type is caused by bilateral damage in inferior occipital and temporal visual association cortices, i.e. in the inferior component of cytoarchitectonic fields 18 and 19, and part of the nearby cytoarchitectonic field 37. The superior component of fields 18 and 19 and the inferior parietal region (field 39) are generally intact (Figure 1a,b).

Face agnosia of the "amnesic associative" type is also caused by bilateral damage but in anterior temporal regions, i.e. hippocampal system structures and the surrounding higher order cortices with which the hippocampus is interconnected (Figure 1a,c). The posterior occipito-temporal cortices are intact.

By contrast, face agnosia of the "apperceptive" type is associated with damage in right visual association cortices within the occipital and parietal regions. Based on current evidence from our laboratory, we believe there must also be damage to inferior *and* superior component of fields 18 and 19 on the right, mesially and laterally, along with damage to part of fields 39 and 37 on the right (Figure 1a,d).

The specificity of these loci of damage can be judged by contrasting face agnosic patients with controls who have lesions in adjoining fields but who

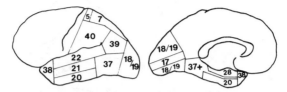

*Figure 1a*   Diagram of human cytoarchitectonic fields in regions whose damage is associated with face agnosia. Note that the disposition of human visual cortices is remarkably different from that of the monkey. The primary visual cortex (field 17 or V1) is placed entirely on the mesial brain surface. It occupies the depths and banks of the calcarine fissure. Fields 18 and 19 contain several functional regions that have been defined neurophysiologically, e.g. V2, V3. Note also the remarkable size of the human fields 39 and 40 in the inferior parietal lobule and of field 37 in the posterior temporal region. The label [37+] designates the combination of fields 37, 36, and 35 in the mesial temporal region.

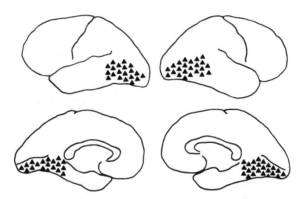

*Figure 1b*   Regions of damage correlated with face agnosia of the "associative" type. The lesions are bilateral and are located below the plane defined by the calcarine fissure. The lesions compromise the inferior component of cytoarchitectonic fields 18 and 19 and part of the nearby cytoarchitectonic field 37. Note that the superior component of fields 18 and 19 as well as the cytoarchitectonic fields in the parietal region are not involved.

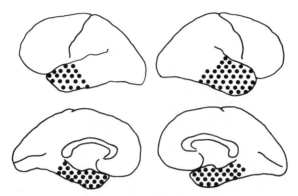

*Figure 1c*   Distribution of damage found in cases of face agnosia of the "amnesic associative" type. The lesions are bilateral and involve the anterior temporal region but not the posterior occipitotemporal cortices. The damage compromises the hippocampal system (entorhinal cortex, hippocampal formation, and amygdala), as well as paralimbic and neocortical fields in cytoarchitectonic fields 38, 20, 21, and 22.

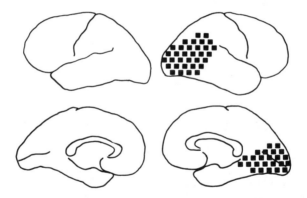

*Figure 1d*   Damage found in face agnosia of the "apperceptive" type. Note that the lesions are in the right hemisphere and that no lesion in left cortices appears necessary. Current evidence suggests that damage to both inferior and superior components of fields 18 and 19, both mesially and laterally, are necessary. Damage to part of fields 39 and 37, on the right, has been found in all the cases we studied.

do not have face agnosia. For instance, bilateral damage of visual association cortices in the superior occipital region does not cause face agnosia.
Unilateral damage to parietal or to occipito-parietal association cortices
does not cause face agnosia either. In fact, unilateral damage to right
occipito-parietal cortices can cause a host of visuoperceptive defects and
yet spare face recognition (Meier & French 1965, Orgass et al 1972).
Unilateral left parietal damage causes neither visuoperceptive defects nor
face agnosia. Thus, it appears that combined superior *and* inferior damage
in right posterior association cortices is a requisite for face agnosia. Our
patient J.W. # 1324 and a recent patient described by Landis et al (1988)
support this view.

Unilateral damage to inferior occipito-temporal cortices does not cause
full-fledged prosopagnosia. Some unilateral left occipital lesions, in
addition to causing alexia, also cause partial defects of facial recognition,
which we have termed "deep prosopagnosia" (Damasio et al 1988). Rather
than recognizing the precise target, the patients recognize someone whose
biographical characteristics are quite close in terms of gender, age, activity,
and so on (e.g. recognizing Betty Grable as Marilyn Monroe). Unilateral
right occipital lesions may cause slow and erratic defects in face recognition
but, again, not pervasive face agnosia.

In brief, face agnosia is generally correlated with bilateral lesions located
either posteriorly in the inferior occipital region or anteriorly in the temporal region. However, unilateral posterior lesions, especially those located
in the right hemisphere and involving cortices in both occipital and parietal
regions, can cause partial defects of face recognition. The evidence indicates unequivocally that face recognition can be performed by *both* right
and left neural networks (as Levy et al 1972 had proposed based on their
study of split-brain patients). But it is also apparent that left and right
structures do not contribute equally to face recognition. The right visual
cortices have a definite advantage in face processing.

# COGNITIVE AND NEURAL ASPECTS OF FACE LEARNING AND RECOGNITION

We have used the profiles of neuropsychological characteristics described
above, along with their anatomical correlates, to hypothesize cognitive
and neural mechanisms behind face recognition and, by extension, visual
recognition. In the sections below, we discuss the status of faces as a
class of visual stimuli, and the constraints imposed on their learning and
recognition. We then outline cognitive mechanisms and neural substrates
for the process of face recognition.

## *The Nature of Faces as Visual Stimuli*

In the course of a lifetime, humans encounter many thousands of faces, in person and through the media. Behind each face there is a unique individual with distinctive physical and biographical characteristics. Yet, given the frequency, ease, and efficiency with which we carry out facial recognition, there is little to remind us of the complexity of the task and of the burden it must pose for the brain. A number of reasons help explain the demands of the problem.

Most obvious is the sheer volume of the task. We are required to recognize a huge number of faces. And the problem is compounded because those *many different* faces must be recognized at the level of *unique*, individual identity. Most stimuli that humans are called upon to identify need only be recognized as members of a conceptual class, and do not require recognition as unique individuals. We can interact satisfactorily with the objects we manipulate daily, the vehicles we see or use for transport, or the animals we encounter in the environment, by merely recognizing the general categories to which they belong. In some instances, subcategorizations are required, e.g. "jet plane," "robin," but few exemplars require unique identification (exceptions include one's own car, house, or pet).

Although recognition of identity from faces is just one among several means of determining unique meaning, and other means (e.g. voice, gait, context) can partially or wholly support the recognition process, it is important to note that with few exceptions, the non-face routes are less powerful, efficacious, and specific than the face route. The presentation of a face, without any additional clues, can lead to effective and instant identity recognition, and even fairly radical alterations in viewing angle, presence or absence of appendages (e.g. hat, earrings, mustache), context, and other such factors will normally produce only small delays and increased chance for error. By contrast, recognition based on other representations of an entity requires a considerable amount of adjuvant context in order for maximal accuracy to be achieved. These routes are also more vulnerable to perceptual distortion. Consider, for example, the difficulty imposed on recognition from voice, when attempted over the telephone. Face processing is thus a burden with rewards. Its special requirements— for instance, the need to map fine physical details—pay off in terms of speed and accuracy, and permit a system capable of meeting the challenges described above.

Faces are not only *many*, and *different*, and *unique*, but they are also *similar*. In spite of their physical differences, faces respect such strict constraints on the variation of physical structure of components and on the spatial arrangement of those components that they are configurationally more alike than different. Finally, there are social demands on the process of face recognition that call for both high accuracy and high speed. In short, it is difficult to imagine another domain of knowledge controlled by an equivalent collection of constraints and demands. In that sense, faces are indeed special. It must be noted, however, that certain professional activities can create somewhat similar constraints and demands, which are not shared by all humans. An example is the task of recognizing aircraft visually, or analyzing histological material under the microscope, or judging the type and quality of dog or horse specimens. The latter task was used by Diamond & Carey (1986) to illustrate the correct point that faces are both special and not special, since there are some stimuli that under particular circumstances may pose comparable problems.

## Face Recognition Processes from a Cognitive Standpoint

From a cognitive standpoint, face recognition is a process of relating *some* face records specific to a face, to a set of *some* non-face records specific to the entity behind the face. At the identity level, the process of face recognition is one of conjuring up meaning for a familiar face in the form of a unique set of coactivated memoranda, i.e. recognition is subsidiary to "recall" of a pertinent set of memories. But not all levels and types of face processing are aimed at the same cognitive demands. Recognizing a facial expression among a limited set of possible facial expressions, while of great relevance in social interactions, requires evocation of a specific but non-unique set of memoranda.

In our model, recognition of identity from faces, depends on (*a*) the establishment of *face records* (records of the physical characteristics of a unique face, apprehended by vision); (*b*) the establishment of linkages between *face records* and *non-face* records (non-face records inscribe other characteristics pertinent to the entity); and (*c*) access to (*a*) and (*b*), leading to simultaneous reactivation ("recall"), at conscious level, of a set of non-face records sufficient for unequivocal identification of the possessor of the face.

The process described in (*a*) is necessary because the separation of extremely similar face patterns requires detailed mapping of face features and configurational arrangements in order to render records distinctive (Damasio 1988). As we mentioned, parallel sources of information (e.g. tell-tale appendages, movement, special environmental context) do assist with disambiguation. However, the information available in the face alone is often enough, and thus there is a clear advantage to record as much of it as possible for each individual face. In other words, even if it can be argued that recognizing the identity of a given person can be successfully achieved without recording or accessing *all* physical details of a face, a perceiver is likely to recognize more quickly and accurately if the details are indeed recorded and accessible. Incidentally, such detailed records are indeed laid down for faces of significant individuals, or it would not be possible for normal perceivers to recall the extraordinary amount of physical structure features that they do, and to inspect these features and ensembles with their mind's eyes.

The process described in (*b*) is necessary because visual records of face structure do not contain information relative to other characteristics of the entity behind the face (Damasio et al 1982). It is necessary to link face records to non-face records, first during learning and later during recognition. Non-face records are both nonverbal and verbal. They include visual records such as body parts, typical attire, typical context, characteristic motion, etc and verbal information such as names, certain demographic information, and other characteristic lexical descriptors of the entity.

For face recognition, then, the critical ingredient is the partial reconstruction of previous experiences related to a target face, which in turn depends on the simultaneous reconstruction of key sensory and motor components of those experiences. Viewing a familiar face leads to the activation of face records learned during previous exposures to the face, e.g. shape of face components, face contour, face motion during facial expressions and face turning, scanpaths (transition of fixations) used to perceive all of the above, texture, and color. In turn, based on those many sources, combined or in isolation, non-face records acquired during previous exposures and pertinently associated with the familiar face also become activated. The success of recognition depends on the activation of a sufficiently comprehensive set of *non-face* records *and* face records, in synchronous fashion. In other words, within approximately the same time window, the perceiver must not only see aspects of the face, but also have an internally recalled experience of information that pertains uniquely to that face.

In cognitive terms, face agnosia can be caused by a disturbance at any point in this multicomponent process. For instance, some perceptual defects may preclude the activation of some types of face records. Or the process that mediates activation of non-face records may be defective and fail to evoke a sufficient amount of pertinent information relative to an otherwise normal percept. Or the process may generate activations that fail to reach a level commensurate with conscious experience and thus go unattended. Such activations would not produce *evocations* but rather *covert* activity sufficient to influence behavior in some experimental paradigms but not enough to generate conscious mental contents. In traditional psychological teminology, one might say that they do not produce true recall, or that they produce only partial recall.

## Neural Substrates of Face Recognition

Our account of the neural substrates of facial recognition is based on the application of a systems level model of learning and memory discussed elsewhere (Damasio 1989b,c). In brief, the model posits: (*a*) that the representation of the physical properties of any entity occurs at feature level, in fragmented fashion and in anatomically separate regions of early sensory cortices and motor cortices. The patterns of neural activity corresponding to different features and dimensions of entities are thus recorded in a distributed manner in the same neural ensembles engaged during perception; (*b*) that the integration of the varied aspects of external and internal reality, in both perception and in recall, depends on the timelocked activation of those geographically separate sites of neural activity (the model rejects the notion of a single site, spatial integration of different representations in higher-order cortices); (*c*) that the combinatorial arrangements that bind features into entities, and entities into events, i.e. their spatial and temporal coincidences, are recorded in separate neural ensembles, called convergence zones; and (*d*) that the separate cortical regions that record featural fragments and the convergence zones that record the combinatorial binding codes that correspond to their linkage in previous perceptual or recorded experiences are interconnected by feedforward and feedback projections. This permits convergence zones to trigger and synchronize neural activity in a way that attempts to reproduce patterns that were pertinently associated in previous experience. Convergence zones are located throughout the telencephalon in association cortices, limbic cortices, and nonlimbic subcortical nuclei such as the basal ganglia, and form hierarchical and heterarchical networks. In general, convergence zones located near early cortices are called "local" and bind featural components of entities. Convergence zones located further downstream in the system are called "non-local" and bind progressively more complex entities and events.

Drawing on the above framework, and taking into account the lesion method data presented above, we propose the following:

1. Face records are made up of a variety of fragmentary representations of the physical structure of faces and of processes utilized by the brain during repeated perception of such physical structures. They include unique shapes of face contour and face components, linkage codes for their spatial assembly, their spatial transformations as viewing perspectives

change, and scanpaths (transition of fixations) used and modified by repeated exposure.

2. Face records are bound by "local" convergence zones, located in association cortices near and directly downstream from the cortices where separate face components or scanpaths are represented. Such convergence zones thus subsume sets of descriptive characteristics for different faces (i.e. they represent amodally a combinatorial arrangement of characteristics), although no single convergence zone is presumed to subsume *all* traits of one face. "Local" convergence zones interact with other "local" convergence zones and also with "non-local" convergence zones located downstream in the system. The latter bind events in which a specific face has participated. When signaled by feedforward projections, nonlocal convergence zones project back to a multiplicity of sensory cortices, visual and not, verbal and nonverbal, where the components of events pertinent to a face (nonface records) can be simultaneously coactivated.

3. Face records are contained in posterior visual association cortices bilaterally (in functional regions within cytoarchitectonic fields 18 and 19, and 39 and 7), but are not evenly or symmetrically distributed in those cortices. This is a consequence of the particular anatomical and physiological arrangement of the visual cortices. The processing of visual properties such as shape, texture, color, and motion depends on varied cellular channels and cortical regions within the visually related cortices and such a functional segregation imposes a separation of the corresponding face records (see Van Essen & Maunsell 1983, Livingstone & Hubel 1988, Damasio 1985).

The evidence suggests persuasively that shape-related face records are probably based on inferior visual association cortices, i.e. the records and the binding local convergence zones are preferentially located in the inferior visual cortices. This is in agreement with behavioral lesion data in both humans (see Damasio 1985) and non-human primates (Ungerleider & Mishkin 1982). There is also preliminary evidence that static records might be skewed toward the left sector of those cortices, whereas records in which shape would be recorded dynamically with transformations around vertical and anterior-posterior axes would be recorded in the right sector of the system. Our findings in patients with unilateral occipital lesions speak to this point and so do studies of face processing in normals (Sergent & Bindra 1981, Ellis 1983, Gazzaniga & Smylie 1983). The mappings of linkages among face components, as related to scanpaths over a unique face, are likely to depend upon superiorly located cortices in the right occipito-parietal region (fields 18 and 19, as well as fields 39 and perhaps 7), in keeping with the clear role of such cortices in motion detection, motion learning, and eye movement control.

4. "Non-local" convergence zones, on the basis of which event level information can be reconstituted, are located bilaterally in temporal neocortices of fields 20, 21, 22, 35, 36, 37, in paralimbic and limbic fields 28, 38, hippocampus and amygdala, in insular cortices, and in prefrontal cortices.

Recognition of a familiar face thus starts as face perception, in multiple visual association cortices, and terminates as synchronized multimodal recall, in multiple discrete cortical regions. The process starts in early cortices and returns to early cortices, recurrently and iteratively. By holding the record to pertinent combinatorial arrangements, convergence zones at all levels, local and nonlocal, in early, intermediate, and high-order cortices, guide the process of recurrence and iteration.

The testing and refinement of the account proposed here depends on further work in humans, based on cognitive experiments in subjects with small lesion probes, along with neurophysiological studies of face processing in nonhuman primates, building on the findings of Bruce et al (1981), Perrett et al (1982), and Baylis et al (1985).

ACKNOWLEDGMENT

Supported by NINDS Grant PO1 NS1 9632.

*Literature Cited*

Bauer, R. M. 1984. Autonomic recognition of names and faces in prosopagnosia: A neurophysiological application of the Guilty Knowledge Test. *Neuropsychologia* 22: 457–69

Bauer, R. M., Verfaellie, M. 1988. Electrodermal discrimination of familiar but not unfamiliar faces in prosopagnosia. *Brain Cogn.* 8: 240–52

Baylis, G. C., Rolls, E. T., Leonard, C. M. 1985. Selectivity between faces in the responses of a population of neurons in the cortex in the superior temporal sulcus of the monkey. *Brain Res.* 342: 91–102

Benton, A. L. 1980. The neuropsychology of facial recognition. *Am. Psychol.* 35: 176–86

Bornstein, B. 1963. Prosopagnosia. In *Problems of Dynamic Neurology*, ed. L. Halpern, pp. 283–318. Jerusalem: Hadassah Medical Organ

Bruce, C., Desimone, R., Gross, C. G. 1981. Visual properties of neurons in a polysensory area in superior temporal sulcus

of the macaque. *J. Neurophysiol.* 46: 369–84

Bruyer, R., Laterre, C., Seron, X., Feyereisen, P., Strypstein, E., Pierrard, E., Rectem, D. 1983. A case of prosopagnosia with some preserved covert remembrance of familiar faces. *Brain Cogn.* 2: 257–84

Cutting, J. E., Kozlowski, L. T. 1977. Recognizing friends by their walk: Gait perception without familiarity cues. *Bull. Psychom. Soc.* 9: 353–56

Damasio, A. R. 1985. Disorders of complex visual processing. In *Principles of Behavioral Neurology, Contemporary Neurology Series*, ed. M. M. Mesulam, pp. 259–88. Philadelphia: Davis

Damasio, A. R. 1988. Neural mechanisms. In *Handbook of Research on Face Processing*, ed. A. Young, H. Ellis. London: North-Holland

Damasio, A. R. 1989a. Category-related recognition defects and the organization of meaning systems. *Trends Neurosci.* In press

Damasio, A. R. 1989b. Time-locked multi-regional retroactivation: A systems level model for some neural substrates of recall and cognition. *Cognition.* In press

Damasio, A. R. 1989c. The brain binds entities and events by multiregional activation from convergence zones. *Neural Comp.* 1: 123–32

Damasio, A. R., Damasio, H., Van Hoesen, G. W. 1982. Prosopagnosia: Anatomic basis and behavioral mechanisms. *Neurology* 32: 331–41

Damasio, A. R., Tranel, D., Damasio, H. 1988. "Deep" prosopagnosia: A new form of acquired face recognition defect caused by left hemisphere damage. *Neurology* 38: 172

Damasio, A. R., Tranel, D., Damasio, H. 1989a. Disorders of visual recognition. In *Handbook of Neuropsychology*, Vol. II, ed. A. Damasio, pp. 317–32. Amsterdam: Elsevier

Damasio, A. R., Damasio, H., Tranel, D. 1989b. Impairments of visual recognition as clues to the processes of memory. In *Signal and Sense: Local and Global Order in Perceptual Maps*, ed. G. Edelman, E. Gall, M. Cowan. New York: Neurosci. Inst. Monogr., New York: Wiley. In press

de Haan, E. H. F., Young, A., Newcombe, F. 1987. Face recognition without awareness. *Cogn. Neuropsychol.* 4: 385–415

Diamond, R., Carey, S. 1986. Why faces are and are not special: An effect of expertise. *J. Exp. Psychol.* 115: 107–17

Ellis, H. D. 1983. The role of the right hemisphere in face perception. In *Functions of the Right Cerebral Hemisphere*, ed. A. Young. London: Academic

Gazzaniga, M. S., Smylie, C. S. 1983. Facial recognition and brain asymmetries: Clues to underlying mechanisms. *Ann. Neurol.* 13: 536–40

Johansson, G. 1976. Spatio-temporal differentiation and integration in visual motion perception. *Psychol. Res.* 38: 379–93

Johansson, G. 1973. Visual perception of biological motion and a model for its analysis. *Percept. Psychophys.* 14: 201–11

Kozlowski, L. T., Cutting, J. E. 1977. Recognizing the sex of walker from dynamic point-light displays. *Percept. Psychophys.* 21: 575–80

Landis, T., Regard, M., Bliestle, A., Kleihues, P. 1988. Prosopagnosia and agnosia for noncanonical views. *Brain* 111: 1287–97

Levy, J., Trevarthen, C., Sperry, R. W. 1972. Perception of bilateral chimeric figures following hemispheric disconnection. *Brain* 95: 61–78

Lhermitte, J., Chain, F., Escourolle, R., Ducarne, B., Pillon, B. 1972. Etude anatomo-clinique d'un cas de prosopagnosie. *Rev. Neurol.* 126: 329–46

Livingstone, M. S., Hubel, D. H. 1988. Segregation of form, color, movement, and depth: Anatomy, physiology, and perception. *Science* 240: 740–49

Meadows, J. C. 1974. The anatomical basis of prosopagnosia. *J. Neurol. Neurosurg. Psychiatry* 37: 489–501

Meier, M. J., French, L. A. 1965. Lateralized deficits in complex visual discrimination and bilateral transfer of reminiscence following unilateral temporal lobectomy. *Neuropsychologia* 3: 261–72

Newcombe, F. 1979. The processing of visual information in prosopagnosia and acquired dyslexia: Functional versus physiological interpretation. In *Research in Psychology and Medicine*, ed. D. J. Osborne, M. M. Gruneberg, J. R. Eiser, 1: 315–22. London: Academic

Orgass, B., Poeck, K., Kerchensteiner, M., Hartje, W. 1972. Visuocognitive performances in patients with unilateral hemispheric lesions. *Z. Neurol.* 202: 177–95

Pallis, C. A. 1955. Impaired identification of faces and places with agnosia for colours. *J. Neurol. Neurosurg. Psychiatry* 18: 218–24

Perrett, D. I., Rolls, E. T., Caan, W. 1982. Visual neurons responsive to faces in the monkey temporal cortex. *Exp. Brain Res.* 47: 329–42

Rizzo, M., Hurtig, R., Damasio, A. R. 1987. The role of scanpaths in facial learning and recognition. *Ann. Neurol.* 22: 41–45

Rosch, E., Mervis, C., Gray, W., Johnson, D., Boyes-Braem, P. 1976. Basic objects in natural categories. *Cogn. Psychol.* 8: 382–439

Sergent, J., Bindra, D. 1981. Differential hemispheric processing of faces: methodological considerations and reinterpretation. *Psychol. Bull.* 89: 541–54

Tranel, D., Damasio, A. R. 1988. Nonconscious face recognition in patients with face agnosia. *Behav. Brain Res.* 30: 235–49

Tranel, D., Damasio, A. R. 1985. Knowledge without awareness: An autonomic index of facial recognition by prosopagnosics. *Science* 228: 1453–54

Tranel, D., Damasio, A. R., Damasio, H. 1988. Intact recognition of facial expression, gender, and age in patients with impaired recognition of face identity. *Neurology* 38: 690–96

Ungerleider, L. G., Mishkin, M. 1982. Two cortical visual systems. In *The Analysis of*

*Visual Behavior*, ed. D. J. Ingle, R. J. W. Mansfield, M. A. Goodale. Cambridge: MIT Press

Van Essen, D. C., Maunsell, J. H. R. 1983. Hierarchical organization and functional streams in the visual cortex. *Trends Neurosci.* 6: 370–75

Wilbrand, H. 1892. Ein Fall von See-lenblindheit und hemianopsie mit Sectionsbefund. *Dtsch. Z. Nervenheilkd.* 2: 361–87

Young, A. W., de Haan, E. H. F. 1988. Boundaries of covert recognition in prosopagnosia. *Cogn. Neuropsychol.* 5: 317–36

*Annu. Rev. Neurosci. 1990. 13:111–27*

# REGULATION OF NEUROPEPTIDE GENE EXPRESSION

## Richard H. Goodman

Division of Molecular Medicine, New England Medical Center Hospital, Boston, Massachusetts 02111

## Introduction

Neuronal plasticity depends on the capacity of cells to make long-term adjustments to environmental stimuli. Neuropeptide genes provide a good model for examining how stimuli detected on the cell surface influence neuronal gene expression. This review focuses on the mechanisms that control the regulated expression of neuropeptide genes. These mechanisms use signal transduction pathways common to many types of eukaryotic cells, not only neurons. Signals received at the cell surface are transduced through G-protein coupled receptors to regulate the production of second messengers such as cyclic AMP (cAMP) and diacylglycerol. These second messengers activate protein kinases, which, in turn, induce the expression of selected genes. Certain proto-oncogene products can also function in normal cellular signal transduction processes. The pathways activated by these proto-oncogenes have provided clues to the mechanisms underlying neuropeptide gene regulation in normal cells.

## Elements Involved in Gene Regulation

The expression of eukaryotic genes is thought to be controlled through *cis*-acting DNA elements surrounding a gene's coding region. For genes transcribed by RNA polymerase II (Pol II), these elements include the promoter, located 25–30 base pairs upstream from the transcriptional start-site, upstream regulatory elements (UREs), and enhancers, which may be located in any position or orientation within the transcriptional unit (Dynan & Tjian 1985, Maniatis et al 1987). Promoters usually consist of AT-rich sequences (the TATA box) and are important for establishing the start-site of transcription (Breathnach & Chambon 1981). Promoters

111

provide a binding site for Pol II and a variety of TATA-binding factors (TFIIB, TFIID, TFIIE; Sawadogo & Roeder 1985). UREs increase the constitutive level of expression and can be considered part of the promoter. Examples of UREs include the CAAT box and the GC box (GGGCGGG; McKnight & Tjian 1986). Although promoters and UREs can provide some degree of tissue-specific expression (Edlund et al 1985), this role is primarily reserved for enhancers. Enhancer sequences can be categorized as regulated or constitutive. Like promoters and UREs, enhancers provide a binding-site for soluble proteins (*trans*-acting factors). *Trans*-acting factors limited to particular cell-types are important for tissue-specific expression. Presumably other enhancer-binding proteins are involved in signal transduction.

Enhancers are frequently located in the 5′ flanking region of eukaryotic genes. Consequently, fusion genes containing putative regulatory regions linked to easily assayed reporters, such as the bacterial enzyme, chloramphenicol acetyltransferase (CAT), can be used to monitor enhancer activity (Gorman et al 1982). Because enhancers function in a position- and orientation-independent manner, they are generally active even when separated from their natural promoter. Nonetheless, some enhancers appear to be fully functional only when paired with a specific promoter.

## Cyclic AMP-Mediated Gene Expression

Early studies on the regulation of neuropeptide gene expression focused on the mediator cAMP. Montminy et al (1986a) showed that somatostatin mRNA levels increased in primary diencephalic cultures after exposure to forskolin, a post-receptor activator of adenylate cyclase. Forskolin also induced expression of a 4.2 kilobase segment of the somatostatin gene that had been introduced into 3T3 fibroblast cells, suggesting that cAMP-regulated expression was a property of the somatostatin transcriptional unit.

The DNA element responsible for cAMP-regulated expression was characterized by deletional analysis of somatostatin-CAT fusion genes (Montminy et al 1986b). These studies identified a sequence, between 29 and 60 base pairs upstream from the transcriptional initiation site, that could confer cAMP-responsiveness to a heterologous promoter. This sequence was designated the "cAMP-responsive element" or CRE. The somatostatin CRE contains an 8 base pair palindrome, 5′-TGACGTCA-3′, which was noted to be conserved in other genes regulated by cAMP (Wynshaw-Boris et al 1984). Several other cellular and viral genes were found subsequently to contain a CRE-like sequence (Table 1). Most of these elements are found within 150 base pairs of the transcriptional initiation site.

**Table 1**  Cyclic AMP-regulated genetic elements

| Gene | Sequence | | Reference |
|---|---|---|---|
| Rat somatostatin | TGACGTCA | −41 | (Montminy et al 1986b) |
| Human proenkephalin | CTGCGTCA | −90 | (Comb et al 1986) |
| Rat phosphoenol-pyruvate carboxykinase | TTACGTCA | −83 | (Short et al 1986) |
| Human chorionic | TGACGTCA | −116 | (Silver et al 1987) |
| gonadotropin | TGACGTCA | −136 | |
| Rat corticotropin releasing hormone | TGACGTCA | −221 | (Seasholtz et al 1988) |
| Rat tyrosine hydroxylase | TGACGTCA | −37 | (Lewis et al 1987) |
| Human vasoactive intestinal peptide | CGTCAxxxxxTGACGTC | −70 | (Tsukada et al 1987) |
| c-*fos* | TGACGT | −61 | (Verma & Sassone-Corsi 1987) |
| Cytomegalovirus | TGACGTCA | −133 | (Ghazal et al 1987) |
| Adenovirus (E4) | TGACGT | −46 | (Lin & Green 1988) |
|  | TGACG | −140 | |
|  | TGACGT | −163 | |
| (E3) | TGACG | −58 | |
| (E2) | TGACGT | −72 | |

Montminy & Bilezikjian (1987) used DNA-affinity chromatography to purify a somatostatin CRE binding-protein from PC12 cells. This protein, designated CREB (CRE-binding protein), has a molecular weight of approximately 43 kD and appears to bind constitutively to the CRE. CREB is phosphorylated by the catalytic subunit of protein kinase A (PKA) in vitro and by forskolin in vivo, but paradoxically this phosphorylation does not alter its binding properties. It is possible, however, that the apparent constitutive binding of CREB to the CRE in vitro occurs through artifactual modifications that take place during the process of protein isolation.

Nonetheless, phosphorylation of CREB very likely is important for somatostatin gene regulation. Yamamoto et al (1988) showed that PKA stimulates somatostatin gene transcription in vitro 20-fold. They also showed that, like several other transcription factors, CREB exists in monomeric and dimeric forms. Phosphorylation of CREB by protein kinase C (PKC) induces formation of the dimer, which has a 10-fold higher affinity for the CRE than the monomeric form. Disruption of the palindromic sequence within the somatostatin CRE blocks dimer formation and subsequent transcriptional activation by cAMP. Furthermore, the ability of CREB to dimerize depends on its phosphorylation state. Treatment with alkaline phosphatase decreases the level of the dimer, supporting the concept that phosphorylation of CREB is involved in transcriptional activation. Dimerization of CREB was shown to be induced by PKC rather than PKA, however. Because somatostatin gene expression was not known to respond to activation of the PKC pathway, the significance of CREB dimerization for transcriptional regulation of the somatostatin gene was uncertain. Nonetheless, dimerization of CREB suggested an alternative kinase pathway for regulation of the somatostatin gene. Recent studies by Verhave et al (submitted) suggest that the somatostatin CRE can, in fact, be activated by PKC. The utilization of serum in the growth medium probably explains the failure of other investigators to observe phorbol ester–mediated regulation of somatostatin gene expression (Deutsch et al 1988a).

Roesler et al (1988) have proposed that rapid changes in the intracellular level of cAMP lead to modifications in the transcriptional activation domains of pre-bound CREB. Phosphorylation of CREB could stabilize the interactions of Pol II (or associated TATA-factors) with the promoter or could allow the participation of additional non-DNA-binding proteins. Although CREB is a phosphoprotein, CREB phosphorylation may not be directly involved in gene activation. It is conceivable that phosphorylation of another protein, such as c-*fos*, is primarily responsible for stimulating cAMP-responsive genes.

Hoeffler et al (1988) recently cloned a cDNA encoding a protein that binds to the human gonadotropin alpha subunit CRE. A fusion protein expressed from the cloned cDNA binds to several different CRE sequences, including that found in the rat somatostatin gene. Nonfunctional CREs and phorbol ester–responsive enhancers were not recognized by the placental CREB. The deduced primary sequence of the 326 amino acid protein revealed a region of 61% identity with a portion of the transcription factor AP-1, including a putative "leucine zipper" domain. This domain, consisting of four leucine residues spaced seven residues apart at the carboxy-terminus of the protein, is characteristic of transcription factors (Landschulz et al 1988). A highly homologous cDNA was subsequently isolated from a rat PC12 cDNA library (Gonzalez et al 1989). Human hypothalamus contains an mRNA encoding a protein identical to placental CREB (R. Rehfuss and R. H. Goodman, unpublished observation).

Through studies of the human proenkephalin gene, Comb et al (1986) first demonstrated that the CRE fulfills the criteria of an enhancer, in that it functions in an orientation- and position-independent manner. Further, the proekephalin CRE was found to respond to both forskolin and to activators of PKC such as phorbol 12-myristate 13-acetate (TPA). Interpretation of the transcriptional responsiveness to TPA was complicated by the requirement for co-treatment with phosphodiestrase inhibitors, however. The proenkephalin CRE is important for both basal and regulated expression, like the CRE sequences in other neuropeptide genes.

The human proenkephalin promoter contains two functionally distinct elements, designated EnkCRE-1 (5′-TGGCGTA-3′) and EnkCRE-2 (5′-CGTCA-3′) (Comb et al 1988, Hyman et al 1988). The upstream EnkCRE-1 element has little activity by itself but greatly augments cAMP-inducibility if combined with the downstream EnkCRE-2 element. EnkCRE-2 is partially active in the absence of EnkCRE-1. At least four distinct factors bind to the proenkephalin enhancer (Figure 1). These include EnkTF-1 (a novel factor that binds to the EnkCRE-1 element), AP-1, AP-4 and AP-2. AP-1, AP-2 and AP-4 are transcription factors that were characterized as activators of the SV40 enhancer and are described in greater detail below. The AP-1 and AP-4 binding sites overlie the EnkCRE-2 locus, and the AP-2 site lies 10 base pairs downstream. It has been proposed that EnkTF-1, binding to the upstream site, interacts with AP-1 and AP-4, which bind to EnkCRE-2. The spacing between the EnkCRE-1 and EnkCRE-2 sites represents one turn of the DNA helix. Consequently, proteins binding to the two elements should lie on the same side of the helix and may be favorably positioned to interact with each other. Mutation of the proenkephalin enhancer so that the two elements lie on opposite sides of the helix dramatically decreases basal and stimulated expression. Binding of AP-1 to EnkCRE-2 is directly correlated with cAMP-inducibility of the enhancer. These findings suggest a primary role for the transcription factor AP-1 in cAMP-mediated regulation. The inability of Imagawa et al (1987) to observe cAMP-regulation of genes containing AP-1 sites may have been due to differences in experimental conditions. Deutsch et al (1988a) and Fink et al (submitted) have also found that AP-1 recognition sites can mediate cAMP-responses. Additionally, Piette et al (1988) showed that binding of nuclear extracts to an AP-1 recognition site was stimulated two- to three-fold after treatment of cells with dibutyryl cAMP.

AP-2 sites have been proposed to mediate activation through both the PKA and PKC pathways (Imagawa et al 1987, Mitchell et al 1987). Like the other kinase-regulated enhancers, AP-2 sites are active in a position-independent manner but are generally found adjacent to the promoter. The importance of the AP-2 site for proenkephalin gene regulation has not been elucidated, and the mechanism of AP-2 activation is poorly under-

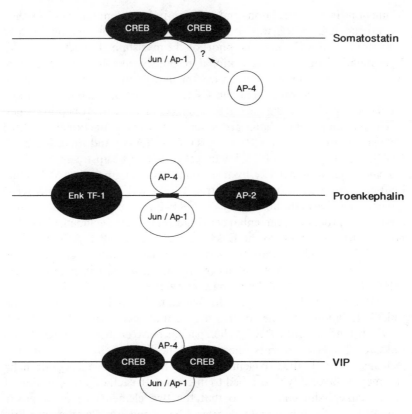

*Figure 1* Schematic representation of transcription factors interacting with the somatostatin, proenkephalin, and VIP enhancers. AP-4 binding to somatostatin, and CREB binding to VIP has not yet been shown directly.

stood. Unlike AP-1, AP-2 binding activity is not activated by either TPA or cAMP (Imagawa et al 1987).

The role of AP-4 in proenkephalin gene regulation is also unclear. Although regulation of the SV40 enhancer has been proposed to require synergism between AP-1 and AP-4 (Mermod et al 1988), it is also possible that these factors could compete with each other for binding.

Cyclic AMP-responsiveness of the vasoactive intestinal peptide (VIP) gene requires a 17 base pair sequence located between 86 and 70 base pairs upstream from the transcriptional initiation site (Tsukada et al 1987). This element contains two copies of a 5 base pair motif, CGTCA, in inverted orientations (Figure 1). Although the CRE sequences in different neuropeptide genes have varying structures, the CGTCA motif appears to be the one characteristic that is shared by all CREs. Mutational analysis indicated that the precise CGTCA motifs within the VIP enhancer, rather than their palindromic nature, are required for activity (Fink et al 1988, Hyman et al 1988). Mutations between the motifs have no effect on enhancer function. In contrast to the proenkephalin enhancer, separating the two motifs in the VIP enhancer by a half-turn of the DNA helix does not dramatically decrease activity. These findings suggest that the upstream motifs in the VIP and proenkephalin enhancers interact with functionally different proteins. DNase footprint competition studies support this hypothesis (Hardy & Shenk 1988, Hyman et al 1988).

Although the DNA sequences of different CREs are not identical, the proteins that bind to these elements are highly related (Hardy & Shenk 1988, Fink et al 1988, Lin & Green 1988, Hyman et al 1988, Deutsch et al 1988a,b). In general, there is a good correlation between the functional activity of CREs and their ability to bind nuclear proteins as assessed by gel mobility-shift assays (Bokar et al 1988). However, sequences flanking the core motifs are also important for binding and activity. For example, the bovine parathyroid hormone and rat glucagon genes contain CRE sequences near their promoters but are not regulated by cAMP; nor do they compete with functional elements for DNA binding proteins (Deutsch et al 1988b). This requirement for specific flanking sequences has complicated a detailed analysis of CRE-promoter interactions.

## Regulated and Basal Expression of the CRE

Several studies have suggested that maximal activity of the CRE requires PKA activity. Montminy et al (1986b) showed that the somatostatin CRE does not respond to cAMP in a PKA-deficient line of PC12 cells. Grove et al (1987) demonstrated by co-transfecting proenkephalin-CAT fusion genes with a vector encoding the Walsh inhibitor of PKA that the catalytic rather than regulatory subunit of PKA was required for activation of the CRE. These studies were extended by Riabowol et al (1988), who utilized cell lines containing a fusion gene consisting of the VIP CRE linked to a $\beta$-galactosidase reporter. Cells containing the reporter gene stain blue after treatment with forskolin or after injection with purified catalytic subunit of PKA. Co-injection of regulatory subunit blocks the response to the catalytic subunit. These studies provided the first direct evidence that CRE activity requires stimulation by the catalytic subunit.

Although maximal stimulation of the CRE clearly requires PKA activity, CRE sequences also appear to have a role in basal and tissue-specific expression. For example, fusion genes containing the somatostatin 5′ flanking region are expressed at a much higher level in CA77 cells (a somatostatin-producing medullary thyroid carcinoma cell line) than in control cells (Andrisani et al 1987). Deletional analysis indicates that the segment of the somatostatin 5′ flanking region responsible for high-level expression in CA77 cells coincides with the CRE. Similarly, the phosphoenolpyruvate carboxykinase CRE and CRE-elements within several viral enhancers may be required for basal (but not necessarily tissue-specific) expression (Quinn et al 1988, Ghazal et al 1987, Lee & Green 1987). Studies of the adenovirus E4 gene suggest that proteins binding to the CRE increase the binding of TATA factors to the promoter (Horikoshi et al 1988).

In other genes, the CRE is necessary, but not sufficient, for tissue-specific expression. Delegeane et al (1987) found that the human glycoprotein alpha subunit gene requires two DNA elements for high-level expression in choriocarcinoma cells. The first consists of an 18 base pair repeated sequence containing the CGTCA motif. This element confers cAMP-responsiveness in both placental and nonplacental cells (Silver et al 1987, Jameson et al 1987). An additional element upstream from the CRE has no independent activity but, in combination with the CRE, allows tissue-specific expression. This upstream element is functional when combined with CREs from the somatostatin or VIP genes as well (Deutsch et al 1988b).

## Regulation of Neuropeptide Gene Expression by Protein Kinase C

Regulation through the PKC pathway is commonly studied by using phorbol esters, such as TPA, that directly stimulate PKC activity. TPA-responsive sequences (TREs) were originally identified in the SV40 enhancer (Lee et al 1987a) and in the cellular metallothionein IIA and collagenase genes (Angel et al 1987a). Many neuropeptide genes are also regulated through this pathway (Ohsawa et al 1985, Anderson et al 1988, deBustros et al 1986). It has been proposed that the AP-1 family of proteins mediates transcriptional regulation of TREs by phorbol esters (Lee et al 1987b, Angel et al 1987b).

Studies of the proto-oncogene c-*jun* have dramatically increased the understanding of PKC-regulated gene expression. The linking of c-*jun* and TPA-regulation emerged from the unexpected finding that the carboxy-terminal sequence of the oncogene v-*jun* resembled the DNA-binding domain of the yeast transcriptional activator GCN4. The similarity of the GCN4 and AP-1 binding sites (5′-TGACTCA-3′) suggested that v-*jun*, GCN4, and AP-1 might be closely related (Vogt et al 1987). Indeed, v-*jun* expression vectors were shown to increase transcription of genes containing AP-1 sites (Angel et al 1988a), and the carboxy-terminus of v-*jun* was found to be functionally equivalent to the GCN4 DNA-binding domain (Struhl 1988). Structural, binding and functional studies confirmed that AP-1 and c-*jun* are identical (Bohmann et al 1987, Angel et al 1988a). Recently, other *jun*-related genes have been characterized and shown to have similar binding and functional properties (Ryder et al 1988). Characterization of *jun*/AP-1 provided conclusive evidence that oncogene products can interact with intracellular signal transduction mechanisms. These findings further suggested that the tumor-promoting properties of TPA depend on the activation of genes containing AP-1 sites.

The mechanism of *jun*/AP-1 activation by PKC is not completely understood. *Jun*/AP-1 is a phosphoprotein, but TPA does not markedly change the degree of its phosphorylation. Nonetheless, treatment of cells with TPA rapidly increases AP-1 binding activity (Chiu et al 1987). It is possible that phosphorylation of c-*fos* is important in mediating this effect (Barber & Verma 1987). The role of c-*fos* in TPA-regulation is discussed below. The *jun*/AP-1 gene has also been shown to contain a functional AP-1 site near its promoter, suggesting that it may induce its own synthesis (Angel et al 1988b). Thus, TPA may increase *jun*/AP-1 activity at both transcriptional and post-transcriptional levels.

The transcription factor *jun*/AP-1 is likely to be important in the regulation of neuropeptide genes by phorbol esters. Pre-prothyrotropin releasing hormone (Lee et al 1988) and peptide YY (S. Krazinski et al, unpublished observations) are two examples of TPA-regulated neuropeptide genes that contain AP-1 recognition sites near their promoters. The AP-1 site in the pre-prothyrotropin releasing hormone gene has been shown in DNase I footprinting assays to bind recombinant *jun*/AP-1.

CRE sequences are structurally related to TREs and, as discussed above, can bind *jun*/AP-1. Consequently, as originally proposed by Comb et al (1986), CREs may mediate TPA- as well as cAMP-regulation. Fink et al have recently shown that the VIP CRE is induced by TPA (without the requirement for phosphodiesterase inhibitors), that recombinant c-*jun* binds specifically to the CRE, and that v-*jun* expression vectors in association with c-*fos* can activate transcription of the VIP CRE in vivo (submitted). These findings confirm that CREs can mediate transcriptional activation through two separate signal transduction pathways. Other examples of DNA elements binding to multiple transcription factors include the octanucleotide motif (Fletcher et al 1987, Scheidereit et al 1987), the CCAAT box (Dorn et al 1987), and the NF-kB/H2TF1 site (Baldwin & Sharp 1988).

Other transcription factors have also been implicated in TPA-regulation, including AP-2 (Imagawa et al 1987), AP-3 (Chiu et al 1987), and NF-kB (Lenardo et al 1987). Of these additional factors, only AP-2 has been proposed to function in neuropeptide gene regulation (Comb et al 1988). C-

*fos* expression vectors activate genes containing TRE sequences (Sassone-Corsi et al 1988), but because *trans*-activation by c-*fos* occurs only in cells that contain *jun*/AP-1, this effect is probably indirect (Chiu et al 1988).

## Role of c-fos in Neuropeptide Gene Regulation

Several studies have suggested that the proto-oncogene product c-*fos* might participate as a "third messenger" in the regulation of neuropeptide genes. C-*fos* is particularly well-situated for transducing intracellular signals because cell-surface stimulation rapidly induces its synthesis, its intracellular half-life is relatively short, and its location is predominantly nuclear (Curran et al 1984, Curran & Morgan 1986). The suggestion that v-*fos* could activate expression of cellular genes was first proposed by Stetoyama et al (1986), who showed that v-*fos* expression vectors increased transcription of the type III collagen promoter. Distel et al (1987) subsequently showed that c-*fos* interacts with a nucleoprotein complex important for adipocyte gene regulation. The sequence of the DNA element involved in this interaction was identical to the AP-1 binding site, a finding that led these investigators to propose that c-*fos* might associate with AP-1 or that it might bind directly to AP-1 sites. Subsequent studies showed that c-*fos* associates with the transcription factor AP-1 rather than with AP-1 recognition sites (Franza et al 1988, Rauscher et al 1988, Chiu et al 1988). Association of *jun*/AP-1 and c-*fos* does not require DNA (Chiu et al 1988) and is thought to occur through the "leucine zipper" domains (Kouzarides & Ziff 1988). The model for this interaction, which is predicated upon the hydrophobic interaction of leucine residues aligned on each face of two alpha-helices, was originally proposed by Landschulz et al (1988). Like AP-1, several discrete forms of c-*fos* exist, presumably arising from separate genes (Rauscher et al 1988). Current thinking is that cell-surface stimulation activates the expression of both *jun*/AP-1 and c-*fos*, which subsequently act in concert to stimulate the expression of genes containing AP-1 recognition sites (Chiu et al 1988). Inhibition of c-*fos* expression by antisense RNA prevents the induction of genes by TPA (Schonthal et al 1988).

The ability of *jun*/AP-1 to stimulate its own transcription and the participation of c-*fos* in AP-1 function provides several potential mechanisms for amplifying and terminating signals from cell-surface receptors. The initial step in TPA-regulation probably involves post-transcriptional modifications of pre-existing *jun*/AP-1 or changes in the association of *jun*/AP-1 with *fos*. Activated *jun*/AP-1 may then induce expression of the *jun*/AP-1 gene through the AP-1 recognition site near the promoter. Increased *jun*/AP-1 transcription should be self-sustaining, but *jun*/AP-1 in association with the c-*fos* protein can also down-regulate c-*fos* gene expression

(Schonthal et al 1988, Sassone-Corsi et al 1988). Because of the short half-life of *fos*, the decrease in *fos* expression may diminish the binding activity of *jun*/AP-1 and thereby decrease *jun*/AP-1 gene expression.

It is likely that c-*fos* also contributes to the regulation of genes containing CRE sequences. As discussed above, CREs can mediate responses to stimulation of both the PKA and PKC pathways. Verhave et al (submitted) have shown that v-*jun* expression vectors stimulate the expression of some CRE-containing genes, but not others. For example, transcription from the somatostatin CRE is induced over 20-fold by v-*jun* but reporter genes containing the VIP CRE are barely affected. Co-transfection of c-*fos* expression vectors allows the VIP CRE to respond to v-*jun* and increases the response of somatostatin still further. Because *jun*/AP-1 lacks a consensus phosphorylation site for PKA, c-*fos* may provide a mechanism for stimulating AP-1 activity through the PKA pathway. This stimulation is likely to occur through transcriptional pathways, since phosphorylation of c-*fos* is not altered by cAMP (Barber & Verma 1987). Whether c-*fos* associates with CREB has not been determined, but the presence of a "leucine zipper" domain within the sequence of placental CREB makes this interaction plausible.

## Relationship of CREs to Viral Enhancers

The utilization of common transcriptional elements by cellular and viral genes is a frequent theme in molecular biology. The coexistence of CRE sequences in viral and cellular genes may provide insights into the evolution of these elements and may suggest approaches to elucidating the pathways of enhancer activation.

Several laboratories have noted the similarities between cellular CREs and the E1A-inducible adenoviral early gene enhancers (Hardy & Shenk 1988, Sassone-Corsi 1988, Garcia et al 1987, Lee & Green 1987). A cellular transcription factor, designated ATF (Lee et al 1987), is thought to mediate the activation of adenoviral early genes by E1A. Sassone-Corsi (1988) has recently shown that the adenoviral E2A and E3 genes are induced by cAMP in PC12 cells and that this activation requires type 2 PKA. Although E1A activates the adenoviral enhancers through a CRE-like sequence, stimulation by E1A nonetheless occurs in PC12 cells that lack PKA. Consequently, it has been proposed that E1A can bypass the requirement of the CRE for PKA activity. In support of this hypothesis, E1A was shown to activate expression of a fusion gene in which the adenovirus E4 enhancer had been replaced by the VIP CRE (Lee et al, submitted). These findings indicate that E1A might interact with the cAMP signaling system (Hardy & Shenk 1988). This effect appears to be promoter-specific, however, because the intact VIP promoter is not regulated by E1A.

The relationship between the adenoviral enhancers and the cellular CREs has also been demonstrated with DNase I footprint and gel-shift competition assays (Sassone-Corsi 1988, Fink et al 1988, Hardy & Shenk 1988). Additionally, CREB and ATF have been shown to be similar in size (Hurst & Jones 1987) and to react to some of the same antisera (Hai et al 1988), thus leading to the hypothesis that CREB and ATF are related, if not identical, proteins. Other investigators, using functional complementation assays, have suggested that the early adenoviral enhancers are activated by a 65–72 kD protein, however (Cortes et al 1988). Whether E1A increases binding activity of adenoviral transcription factors depends on the particular enhancer being tested: Binding to the E2 promoter is increased by E1A (Kovesdi et al 1986) but binding to E4 is not (Lee & Green 1987). It is possible, however, that the E1A-mediated activation of transcription factor binding cannot be determined accurately in vitro. It is also unknown whether E1A and PKA cause similar modifications in their target proteins.

Another viral system that utilizes CRE sequences for *trans*-activation is the human T-cell leukemia virus (HTLV-1). HTLV-1 activates its own expression through a protein known as *tax* or $p40^x$ (Sodroski et al 1984). *Tax* interacts with a region of the HTLV-1 5′ LTR that contains three copies of the CRE core motif (Park et al 1988). Because *tax* does not directly interact with DNA, it has been proposed that activation depends on a pre-existing cellular protein. CREB/ATF is a reasonable candidate for this cellular factor for two reasons. First, the HTLV-1 LTR can be induced by E1A and cAMP (Tan & Roeder 1988). Second, the predominant *tax*-associated protein has a molecular weight of 43 kD, identical to the predicted size of CREB (Slamon et al 1988). It is possible, therefore, that *tax*, E1A and PKA activate the same transcription factor. Recent studies have suggested that c-*myc* activates the adenovirus E4 promoter through the same target sequence as E1A (Onclercq et al 1988). Thus c-*myc* may also utilize the CREB/ATF pathway.

The CRE is clearly important for basal, regulated, and tissue-specific expression of neuropeptide genes. It is possible that this element may also be involved in the pathogenesis of ectopic hormone syndromes in certain types of neuroendocrine tumors. CRE activity depends on interactions with soluble cellular factors that can be activated by membrane receptors or by the products of specific transforming genes (E1A, *tax*, c-*myc*). Certain oncogenes may have the potential, therefore, to induce cellular transformation and to enhance the expression of CRE-containing neuropeptide genes. This hypothesis predicts that many different neuropeptides would be produced simultaneously by certain tumors, an occurrence that is characteristic of ectopic hormonal syndromes. It is possible that E1A, *tax*, or some as yet undiscovered transforming gene activates proteins with CRE-binding activity, or that CREB itself represents the cellular form of an unknown oncogene.

## Conclusions

Neuropeptide genes have provided a very useful model for studies of regulated gene expression, but many questions are still unanswered. First, it is necessary to determine whether cAMP-mediated gene regulation involves single or multiple forms of CREB. Second, the relationship among CREB, ATF, and the *tax*-associated proteins must be determined. Presumably both of these issues will be elucidated after the appropriate cDNAs have been cloned and expressed. Third, the individual contributions of CREB and AP-1 in regulated CRE activity need to be examined, as does the contribution of these factors to tissue-specific gene expression. The possibility that CRE or AP-1 recognition sites might provide a link between oncogenesis and ectopic neuropeptide production obviously requires further study. Finally, the mechanism of CREB activation by protein kinase systems, the significance of CREB dimerization, and the role of c-*fos* or other associated proteins in cAMP-regulated transcription remain unclear. Progress in elucidating these issues will greatly increase the understanding of neuropeptide gene regulation in normal cells and in neuroendocrine tumors.

*Literature Cited*

Anderson, B., Milsted, A., Kennedy, G., Nilson, J. H. 1988. Cyclic AMP and phorbol esters interact synergistically to regulate expression of the chorionic gonadotrophin genes. *J. Biol. Chem.* 263: 15578–83

Andrisani, O. M., Hayes, T. E., Roos, B., Dixon, J. E. 1987. Identification of the promoter sequences involved in the cell specific expression of the rat somatostatin gene. *Nucleic Acids Res.* 15: 5715–27

Angel, P., Allegretto, E. A., Okino, S. T., Hattori, K., Boyle, W. J., Hunter, T., Karin, M. 1988a. Oncogene *jun* encodes a sequence-specific *trans*-activator similar to AP-1. *Nature* 332: 166–71

Angel, P., Baumann, I., Stein, B., Delius, H., Rahmsdorf, H. J., Herrlich, P. 1987a. 12-O-tetradecanoyl-phorbol-13-acetate induction of the human collagenase gene is mediated by an inducible enhancer element located in the 5′-flanking region. *Mol. Cell. Biol.* 7: 2256–66

Angel, P., Hattori, K., Smeal, T., Karin, M.

1988b. The *jun* proto-oncogene is positively autoregulated by its product, *jun*/AP-1. *Cell* 55: 875–85

Angel, P., Imagawa, M., Chiu, R., Stein, B., Imbra, R. J., Rahmsdorf, H. J., Jonat, C., Herrlich, P., Karin, M. 1987b. Phorbol ester-inducible genes contain a common *cis* element recognized by a TPA-modulated *trans*-acting factor. *Cell* 49: 729–39

Baldwin, A. S., Sharp, P. A. 1988. Two transcription factors NF-kB and H2TF1, interact with a single regulatory sequence in the class 1 major histocompatibility complex promoter. *Proc. Natl. Acad. Sci. USA* 85: 723–27

Barber, J. R., Verma, I. M. 1987. Modification of *fos* proteins: Phosphorylation of c-*fos*, but not v-*fos*, is stimulated by 12-tetradecanoyl-phorbol-13-acetate and serum. *Mol. Cell. Biol.* 7: 2201–11

Bohmann, D., Bos, T. J., Admon, A., Nishimura, T., Vogt, P. K., Tjian, R. 1987. Human proto-oncogene c-*jun* encodes a DNA binding protein with structural and functional properties of transcription factor AP-1. *Science* 238: 1386–92

Bokar, J. A., Roesler, W. J., Vandenbark, G. R., Kaetzel, D. M., Hanson, R. W., Nilson, J. H. 1988. Characterization of the cAMP responsive elements from the genes for the α-subunit of glycoprotein hormones and phosphoenolpyruvate carboxykinase (GTP). *J. Biol. Chem.* 263: 19740–47

Breathnach, R., Chambon, P. 1981. Organization and expression of eukaryotic split genes coding for proteins. *Annu. Rev. Biochem.* 50: 349–83

Chiu, R., Boyle, W. J., Meek, J., Smeal, T., Hunter, T., Karin, M. 1988. The c-*fos* protein interacts with c-*jun*/AP-1 to stimulate transcription of AP-1 responsive genes. *Cell* 54: 541–52

Chiu, R., Imagawa, M., Imbra, R. J., Bockoven, J. R., Karin, M. 1987. Multiple *cis*- and *trans*-acting elements mediate the transcriptional response to phorbol esters. *Nature* 329: 648–51

Comb, M., Birnberg, N. C., Seasholtz, A., Herbert, E., Goodman, H. M. 1986. A cyclic AMP- and phorbol ester-inducible DNA element. *Nature* 323: 353–55

Comb, M., Mermod, N., Hyman, S. E., Pearlberg, J., Ross, M. E., Goodman, H. M. 1988. Proteins bound at adjacent DNA elements act synergistically to regulate human proenkephalin cAMP inducible transcription. *EMBO J.* 7: 3793–805

Cortes, P., Buckbinder, L., Leza, M. A., Rak, N., Hearing, P., Merino, A., Reinberg, D. 1988. EivF, a factor required for transcription of the adenovirus EIV promoter, binds to an element involved in EIa-dependent activation and cAMP induction. *Genes Dev.* 2: 975–90

Curran, T., Miller, A. D., Zokas, L., Verma, I. M. 1984. Viral and cellular *fos* proteins: A comparative analysis. *Cell* 36: 259–68

Curran, T., Morgan, J. I. 1986. Superinduction of c-*fos* by nerve growth factor in the presence of peripherally active benzodiazephines. *Science* 229: 1265–68

deBustros, A., Baylin, S. B., Levine, M. A., Nelkin, B. D. 1986. Cyclic AMP and phorbol esters separately induce growth inhibition, calcitonin secretion and calcitonin gene transcription in cultured human medullary thyroid carcinoma. *J. Biol. Chem.* 261: 8036–41

Delegeane, A. M., Ferland, L. H., Mellon, P. L. 1987. Tissue-specific enhancer of the human glycoprotein hormone α-subunit gene: dependence on cyclic AMP-inducible elements. *Mol. Cell. Biol.* 7: 3994–4002

Deutsch, P. J., Hoeffler, J. P., Jameson, J. L., Habener, J. F. 1988a. Cyclic AMP and phorbol ester-stimulated transcription mediated by similar DNA elements that bind distinct proteins. *Proc. Natl. Acad. Sci. USA* 85: 7922–26

Deutsch, P. J., Hoeffler, J. P., Jameson, J. L., Lin, J. C., Habener, J. F. 1988b. Structural determinants for transcriptional activation by cAMP-responsive DNA elements. *J. Biol. Chem.* 263: 18466–72

Distel, R. J., Ro, H.-S., Rosen, B. S., Groves, D. L., Spiegelman, B. M. 1987. Nucleoprotein complexes that regulate gene expression in adipocyte differentiation: Direct participation of c-*fos*. *Cell* 49: 835–44

Dorn, A., Bollekens, J., Staub, A., Benoist, C., Mathis, D. 1987. A multiplicity of CCAAT box-binding proteins. *Cell* 50: 863–72

Dynan, W. S., Tjian, R. 1965. Control of eukaryotic messenger RNA synthesis by sequence-specific DNA-binding proteins. *Nature* 316: 774–78

Edlund, T., Walker, M. D., Barr, P. J., Rutter, W. J. 1985. Cell-specific expression of the rat insulin gene: Evidence for role of two distinct 5′ flanking elements. *Science* 230: 912–16

Fink, J. S., Verhave, M., Kasper, S., Tsukada, T., Mandel, G., Goodman, R. H. 1988. The CGTCA sequence motif is essential for biological activity of the vasoactive intestinal peptide gene cAMP-regulated enhancer. *Proc. Natl. Acad. Sci. USA* 85: 6662–66

Fletcher, C., Heintz, N., Roeder, R. G. 1987. Purification and characterization of OTF-1, a transcription factor regulating cell cycle expression of a human histone H2B gene. *Cell* 51: 773–81

Franza, B. R. Jr., Rauscher, F. J. III, Josephs, S. F., Curran, T. 1988. The *fos* complex and *fos*-related antigens recognize sequence elements that contain AP-1 binding sites. *Science* 239: 1150–53

Garcia, J., Wu, F., Gaynor, R. 1987. Upstream regulatory regions required to stabilize binding to the TATA sequence in an adenovirus early promoter. *Nucleic Acids Res.* 15: 8367–85

Ghazal, P., Lubon, H., Fleckenstein, B., Hennighausen, L. 1987. Binding of transcription factors and creation of a large nucleoprotein complex on the human cytomegalovirus enhancer. *Proc. Natl. Acad. Sci. USA* 84: 3658–62

Gonzalez, G. A., Yamamoto, K. K., Fischer, W. H., Karr, D., Menzel, P., Biggs, W., Vale, W. V., Montminy, M. R. 1989. A cluster of phosphorylation sites on the cyclic AMP-regulated nuclear factor CREB predicted by its sequence. *Nature* 337: 749–52

Gorman, C. M., Moffat, L. F., Howard, B. H. 1982. Recombinant genomes which express chloramphenicol acetyltransferase in mammalian cells. *Mol. Cell. Biol.* 2: 1044–51

Grove, J. R., Price, D. J., Goodman, H. R., Avruch, J. 1987. Recombinant fragment of protein kinase inhibitor blocks cyclic AMP-dependent gene transcription. *Science* 238: 530–33

Hai, T., Liu, F., Allegretto, E. A., Karin, M., Green, M. R. 1988. A family of immunologically related transcription factors that includes multiple forms of ATF and AP-1. *Genes Dev.* 2: 1216–26

Hardy, S., Shenk, T. 1988. Adenoviral control regions activated by E1A and the cAMP response element bind to the same factor. *Proc. Natl. Acad. Sci. USA* 85: 4171–75

Hoeffler, J. P., Meyer, T. E., Yun, Y., Jameson, J. L., Habener, J. F. 1988. Cyclic AMP-responsive DNA-binding protein: Structure based on a cloned placental cDNA. *Science* 242: 1430–33

Horikoshi, M., Hai, T., Lin, Y.-S., Green, M. R., Roeder, R. G. 1988. Transcription factor ATF interacts with the TATA factor to facilitate establishment of a preinitiation complex. *Cell* 54: 1003–42

Hurst, H. C., Jones, N. C. 1987. Identification of factors that interact with the E1A-inducible adenovirus E3 promoter. *Genes Dev.* 1: 1132–46

Hyman, S. E., Comb, M., Lin, Y.-S., Pearlberg, J., Green, M. R., Goodman, H. M. 1988. A common *trans*-acting factor is involved in transcriptional regulation of neurotransmitter genes by cyclic AMP. *Mol. Cell. Biol.* 8: 4225–33

Imagawa, M., Chiu, R., Karin, M. 1987. Transcription factor AP-2 mediates induction by two different signal-transduction pathways: Protein kinase C and cAMP. *Cell* 51: 251–60

Jameson, J. L., Deutsch, P. J., Gallagher, G. D., Jaffe, R. C., Habener, J. F. 1987. *Trans*-acting factors interact with a cyclic AMP response element to modulate expression of the human gonadotropin a gene. *Mol. Cell. Biol.* 7: 3032–40

Kouzarides, T., Ziff, E. 1988. The role of the leucine zipper in the *fos-jun* interaction. *Nature* 226: 646–51

Kovesdi, I., Reichel, R., Neving, J. R. 1986. Identification of a cellular transcription factor involved in E1A trans-activation. *Cell* 45: 219–28

Landschulz, W. H., Johnson, P. F., McKnight, S. L. 1988. The leucine zipper: A hypothetical structure common to a new class of DNA-binding proteins. *Science* 240: 1759–64

Lee, K. A. W., Green, M. 1987. A cellular transcription factor E4F1 interacts with an E1A-inducible enhancer and mediates constitutive enhancer function in vitro. *EMBO J.* 6: 1345–53

Lee, K. A. W., Hai, T. Y., Siva-Ramen, L., Thimmappaya, B., Hurst, H. C., Jones, N. C., Green, M. R. 1987. A cellular protein, activating transcriptional factor, activates transcription of multiple E1A-inducible adenovirus early promoters. *Proc. Natl. Acad. Sci. USA* 84: 8355–59

Lee, S. A., Stewart, K., Goodman, R. H. 1988. Structure and expression of the rat pro-TRH gene. *J. Biol. Chem.* 263: 16604–9

Lee, W., Haslinger, A., Karin, M., Tjian, R. 1987a. Activation of transcription by two factors that bind promoter and enhancer sequences of the human metallothionein gene and SV40. *Nature* 325: 368–72

Lee, W., Mitchell, P., Tjian, R. 1987b. Purified transcription factor AP-1 interacts with TPA-inducible enhancer elements. *Cell* 49: 741–52

Lenardo, M., Pierce, J. W., Baltimore, D. 1987. Protein-binding sites in Ig gene enhancers determine transcriptional activity and inducibility. *Science* 236: 1573–77

Lewis, E. J., Harrington, C. A., Chikaraishi, D. M. 1987. Transcriptional regulation of the tyrosine hydroxylase gene by glucocorticoid and cyclic AMP. *Proc. Natl. Acad. Sci. USA* 84: 3550–54

Lin, Y.-S., Green, M. R. 1988. Interaction of a common cellular transcription factor, ATF, with regulatory elements in both E1a- and cyclic AMP-inducible promoters. *Proc. Natl. Acad. Sci. USA* 85: 3396–400

Maniatis, T., Goodbourn, S., Fischer, J. A. 1987. Regulation of inducible and tissue-specific gene expression. *Science* 236: 1237–45

McKnight, S., Tjian, R. 1986. Transcriptional selectivity of viral genes in mammalian cells. *Cell* 46: 795–805

Mermod, N., Williams, T. J., Tjian, R. 1988. Enhancer binding factors AP-4 and AP-1 act in concert to activate SV40 late transcription in vitro. *Nature* 332: 557–61

Mitchell, P. J., Wang, C., Tjian, R. 1987. Positive and negative regulation of transcription in vitro: Enhancer-binding protein AP-2 is inhibited by SV40 T antigen. *Cell* 50: 847–61

Montminy, M. R., Bilezikjian, L. M. 1987. Binding of nuclear protein to the cyclic-AMP response element of the somatostatin gene. *Nature* 328: 175–78

Montminy, M. R., Low, M. J., Tapia-Arancibia, L., Reichlin, S., Mandel, G., Goodman, R. H. 1986a. Cyclic AMP regulates somatostatin mRNA accumulation in primary diencephalic cultures and in transfected fibroblast cells. *J. Neurosci.* 6: 1171–76

Montminy, M. R., Sevarino, K. A., Wagner, J. A., Mandel, G., Goodman, R. H. 1986b. Identification of a cyclic-AMP-responsive element within the rat somatostatin gene. *Proc. Natl. Acad. Sci. USA* 83: 6682–86

Ohsawa, K., Hayakawa, Y., Nishizawa, M., Yamagami, T., Yamamoto, H., Yanaihara, N., Okamoto, H. 1985. Synergistic stimulation of VIP/PHM-27 gene expression by cyclic AMP and phorbol esters in human neuroblastoma cells. *Biochem. Biophys. Res. Commun.* 132: 885–91

Onclercq, R., Gilardi, P., Lavenu, A., Cremisi, C. 1988. c-*myc* products *trans*-activate the adenovirus E4 promoter in EC stem cells by using the same target sequence as E1A products. *J. Virol.* 62: 4533–37

Park, R. E., Haseltine, W. A., Rosen, C. A. 1988. A nuclear factor is required for transactivation of HTLV-I gene expression. *Oncogene* 3: 275–79

Piette, J., Hiral, S.-I., Yaniv, M. 1988. Constitutive synthesis of activator protein 1 transcription factor after viral transformation of mouse fibroblasts. *Proc. Natl. Acad. Sci. USA* 85: 3401–5

Quinn, P. G., Wong, T. W., Magnuson, M. A., Shabb, J. B., Granner, D. K. 1988. Identification of basal and cyclic AMP regulatory elements in the promoter of the phosphoenolpyruvate carboxykinase gene. *Mol. Cell. Biol.* 8: 3467–75

Rauscher, F. J. III, Cohen, D. R., Curran, T., Bos, T. J., Vogt, P. K., Bohmann, D., Tjian, R., Franza, B. R. Jr. 1988. *Fos*-associated protein p39 is the product of the *jun* proto-oncogene. *Science* 240: 1010–16

Riabowol, K. T., Fink, J. S., Gilman, M. Z., Walsh, D. A., Goodman, R. H., Feramisco, J. R. 1988. The catalytic subunit of cAMP-dependent protein kinase induces expression of genes containing cAMP-responsive enhancer elements. *Nature* 336: 83–86

Roesler, W. J., Vandenbark, G. R., Hanson, R. W. 1988. Cyclic AMP and the induction of eukaryotic gene transcription. *J. Biol. Chem.* 263: 9063–66

Ryder, K., Lau, L. F., Nathans, D. 1988. A gene activated by growth factor is related to the oncogene v-*jun*. *Proc. Natl. Acad. Sci. USA* 85: 1487–91

Sassone-Corsi, P. 1988. Cyclic AMP induction of early adenovirus promoters involves sequences required for E1A

trans-activation. *Proc. Natl. Acad. Sci. USA* 85: 7192–96

Sassone-Corsi, P., Sisson, J. C., Verma, I. M. 1988. Transcriptional autoregulation of the proto-oncogene *fos*. *Nature* 334: 314–19

Sawadogo, M., Roeder, R. G. 1985. Factors involved in specific transcription by human RNA polymerase II: Analysis by a rapid and quantitative in vitro assay. *Proc. Natl. Acad. Sci. USA* 82: 4394–98

Scheidereit, C., Heguy, A., Roeder, R. G. 1987. Identification and purification of a human lymphoid-specific octamer binding protein (OTF-2) that activates transcription of an immunoglobulin promoter in vitro. *Cell* 51: 783–93

Schonthal, A., Herrlich, P., Rahmsdorf, H. J., Ponta, H. 1988. Requirement for *fos* gene expression in the transcriptional activation of collagenase by other oncogenes and phorbol esters. *Cell* 54: 325–34

Seasholtz, A. F., Thompson, R. C., Douglass, J. O. 1988. Identification of a cyclic adenosine monophosphate-responsive element in the rat corticotropin-releasing hormone gene. *Mol. Endocrinol.* 12: 1311–18

Setoyama, C., Frunzio, R., Liau, G., Mudryj, M., deCrombugghe, B. 1986. Transcription activation encoded by the v-*fos* gene. *Proc. Natl. Acad. Sci. USA* 83: 3213–17

Short, J. M., Wynshaw-Boris, A., Short, H. P., Hanson, R. W. 1986. Characterization of the phosphoenolpyruvate carboxykinase promoter regulatory region: II. Identification of cAMP and glucocorticoid regulatory domains. *J. Biol. Chem.* 261: 9721–26

Silver, B. J., Boker, J. A., Virgin, J. B., Vallen, E. A., Milsted, A., Nilson, J. H. 1987. Cyclic AMP regulation of the human glycoprotein hormone alpha subunit gene is mediated by an 18-base-pair element. *Proc. Natl. Acad. Sci. USA* 84: 2198–202

Slamon, D. J., Boyle, W. J., Keith, D. E., Press, M. F., Golde, D. W., Souza, L. M. 1988. Subnuclear localization of the *trans*-activating protein of human T-cell leukemia virus type I. *J. Virol.* 62: 680–86

Sodroski, J. G., Rosen, C. A., Hazeltine, W. A. 1984. *Trans*-acting transcriptional activation of the long terminal repeat of human T lymphotropic viruses in infected cells. *Science* 225: 381–85

Struhl, K. 1988. The JUN oncoprotein, a vertebrate transcription factor, activates transcription in yeast. *Nature* 332: 649–50

Tan, H. T., Roeder, R. G. 1988. Identification and purification of nuclear factors interacting with the HTLV-1 $tat_1$-responsive element within the HTLV-I LTR. In *RNA Tumor Viruses*, ed. S. Goff, N. Rosenberg. Cold Spring Harbor: Cold Spring Harbor Lab. 56 pp.

Tsukada, T., Fink, J. S., Mandel, G., Goodman, R. H. 1987. Identification of a region in the human vasoactive intestinal polypeptide gene responsible for regulation by cyclic AMP. *J. Biol. Chem.* 262: 8743–47

Verma, I. M., Sassone-Corsi, P. 1987. Protooncogene *fos*: Complex but versatile regulation. *Cell* 51: 513–14

Vogt, P. K., Bos, T. J., Doolittle, R. F. 1987. Homology between the DNA-binding domain of the GCN4 regulatory protein of yeast and the carboxy-terminal region

of a protein coded for by the oncogene *jun. Proc. Natl. Acad. Sci. USA* 84: 3316–19

Wynshaw-Boris, A. J., Lugo, T. G., Short, J. M., Fournier, R. E., Hanson, R. W. 1984. Identification of a cAMP regulatory region in the gene for rat cytosolic phosphoenolpyruvate carboxykinase (GTP). Use of chimeric genes transfected into hepatoma cells. *J. Biol. Chem.* 259: 12161–69

Yamamoto, K. K., Gonzalez, G. A., Briggs, W. H. III, Montminy, M. R. 1988. Phosphorylation-induced binding and transcriptional efficacy of nuclear factor CREB. *Nature* 334: 494–98

*Annu. Rev. Neurosci. 1990. 13:129–54*

# PATTERNED ACTIVITY, SYNAPTIC CONVERGENCE, AND THE NMDA RECEPTOR IN DEVELOPING VISUAL PATHWAYS

*Martha Constantine-Paton, Hollis T. Cline, and Elizabeth Debski*

Department of Biology, Yale University, New Haven, Connecticut 06511

## INTRODUCTION

Plasticity, activity-dependent competition, and Hebbian synapes are terms frequently associated with a set of cellular interactions that causally link early neural activity to the final stages of neural circuit differentiation. Similar terms and, in fact, similar interactions are used in discussions of learning and memory. This use of a common vocabulary reflects the hope of neurobiologists that the developmental mechanisms of plasticity will prove similar to those underlying learning and memory in the mature brain. The visual pathway has been the focus of continuous and intense experimental work in developmental plasticity for over three decades, and quite recent work in this area has suggested a molecular mechanism also found in hippocampal long-term potentiation (LTP). Extensive reviews are available in both fields (Movshon & Van Sluyters 1981, Sherman & Spear 1982, Fregnac & Imbert 1984, Shatz & Sretavan 1986, Stryker 1986, Collingridge & Bliss 1987, Teyler & DiSenna 1987, Nicoll 1988, Udin & Fawcett 1988, Brown et al 1990). Consequently, this treatment focuses only on the developing visual pathway and only on those visual system experiments that conceptually link plasticity in the differentiating and mature brain. It presents the arguments and evidence suggesting that temporal correlations in the action potential patterns of young visual synapses determine their relative positions within local regions of neuropil and their convergence onto the same sets of post-synaptic cells. It also

129

0147–006X/90/0301–0129$02.00

considers the issue of whether functionally detected developmental changes in synaptic effectiveness are isomorphic with activity-dependent changes detected anatomically. Finally, it reviews recent data on the involvement of the *N*-methyl D-aspartate (NMDA) subclass of excitatory amino acid receptor in use-dependent modifications of the developing visual pathway.

## ACTIVITY AND SYNAPTIC CONVERGENCE IN MAMMALS

In the mammalian geniculocortical visual pathway, inputs from the two retinas are segregated in the dorsal lateral geniculate nucleus (dLGN) and in the afferent layers of visual cortex, primarily layer IV. In the nonafferent layers of cortex, synapses from layer IV cells finally converge upon cortical neurons so that left and right retinal pathways are no longer segregated. The segregation of retinally driven inputs as well as their ability to converge on single cortical cells has been shown to be affected by the patterns of activity they convey. As outlined below, this activity dependence was first detected in experiments that perturbed binocular convergence as well as experiments that altered the balance of activity between the two eyes.

### Artificially Induced Strabismus: Experimental Deprivation of Binocular Convergence

Animals raised with one eye muscle cut so that their retinas cannot converge on the same point in visual space lose significant numbers of binocular cortical neurons. The changes occur in the apparent absence of sluggish or dying neurons, thus suggesting that the normal convergence of retinally driven inputs rather than cortical cell survival is the parameter that has been affected by the treatment (Hubel & Wiesel 1965). There have been many permutations of experiments in which animals are prevented from seeing the same stimuli through both eyes. All demonstrate the same eye-specific segregation phenomenon and share one principle in common: When visual stimulation does not allow near simultaneous delivery of similar patterns of action potential activity to the same cortical cells through the left and right eye visual pathways, the two pathways fail to maintain the ability to drive those cells. Instead the inputs from the two eyes functionally segregate onto two mutually exclusive sets of cortical neurons (Hirsch & Spinelli 1970, Leventhal & Hirsch 1977, Bruce et al 1981; see Fregnac & Imbert 1984 for review).

### Monocular Deprivation Experiments

In monocular deprivation studies one eye of a neonate kitten or monkey is sutured closed for periods ranging from weeks to months. After this period the animal is prepared for physiological recordings, the sutured eye is opened, and the ability of isolated cortical units to respond to either eye is tested. The consistent physiological result is that the previously sutured, deprived eye loses its ability to drive cortical units and all cortical neurons develop pronounced responses to the open eye (Wiesel & Hubel 1963, Hubel et al 1977). The degree of domination is related to the duration of the treatment, but even short exposures during a brief, most sensitive period can produce pronounced changes in eye dominance (Olson & Freeman 1975). If monocular deprivation is begun very early in the postnatal period, this "take-over" occurs in all cortical layers, including layers IV, which normally, in binocular regions of cortex, contains approximately equal numbers of monocular cells driven by either the right or left retina.

## The Issue of Structural Versus Functional Changes

It is frequently assumed that activity-dependent developmental changes involve structural "rewiring," whereas changes in the mature brain reflect alterations in the efficacy of existing synapses. In fact there is relatively little evidence to justify generalization for either case. The same issue of assigning an observed change to a structural versus functional cause also arises in the visual development literature. Specifically, structural rewiring in the dLGN and afferent cortical layers is detectable with physiological techniques at the affected level and higher. However, functionally detected changes in the relative efficacy of retinal inputs to cortical neurons do not necessarily imply actual losses or gains in the numbers of synapses to those cells.

In afferent layer IV of visual cortex the functionally detected takeover of a large proportion of neurons by the nondeprived eye has a firm structural correlate in the relative amount of termination space occupied by the geniculate inputs corresponding to the two eyes. Layer IV ocular dominance columns of the nondeprived eye fail to retract at the expense of those from the deprived eye (Hubel et al 1977, Shatz & Stryker 1978, LeVay et al 1980). Furthermore, the dLGN neurons receiving input from the deprived eye show reduced somal size relative to the corresponding cells receiving input from the nondeprived eye. These reductions are found only in binocular regions of the dLGN and they do not occur when animals are binocularly deprived of pattern vision (a procedure that does not unbalance the distribution of terminals in layer IV). Consequently, the somal shrinkage is not due to a loss of activity but probably reflects a loss of ocular dominance termination space in cortex (Guillery & Stelzner 1970, Guillery 1972; see for review Movshon & Van Sluyters 1981, Sherman & Spear 1982).

More recently there have been a number of investigations of deprivation treatment effects on the morphology of single neurons and their geniculate and cortical terminal arbors. The work has generally documented some changes in the structure of single cell arbors consistent with deprivation-induced alterations in function or frequencies of encountering units with particular responses (see Sherman & Spear 1982, Shatz & Sretavan 1986, Sur 1988 for reviews).

Despite this evidence of anatomical change at lower levels in the geniculocortical pathway, however, there is no evidence for actual structural changes in afferent convergence as a result of perturbations in normal activity patterns in nonafferent cortical layers. Thus, some of the physiological observations of developmental plasticity in cortex may reflect actual structural changes in the numbers or positions of synapses while others reflect functional changes in synaptic efficacy. For example rearing a kitten in the dark prolongs the period in which the nondeprived eye can functionally dominate cortical neurons when physiologically assayed. It does not prolong the period during which the ocular dominance columns of the nondeprived eye can show anatomical expansion in layer IV (Mower et al 1985). In addition, when, during a brief period of early life, the initially deprived eye of kittens or monkeys is opened and the open eye sutured shut, there is a rapid recovery of a population of cortical neurons responding to the initially deprived eye (Blakemore & Van Sluyters 1974, Kratz et al 1976, Blakemore et al 1978). The speed of this recovery suggests that some of the effects of monocular deprivation result from increased inhibitory suppression of the initially deprived eye's inputs by the open eye rather than competitive displacement of its terminals from post-synaptic membranes. However, the actual mechanisms involved remain controversial (Duffy et al 1976, Movshon & Van Sluyters 1981).

## The Hebbian Synapse Hypothesis

The correlated activity requirement for the developmental maintenance of binocular neurons and the activity-dependent interaction that allows the open eye to take over cortical cells under conditions of monocular deprivation are linked by the theoretical framework provided initially by D. O. Hebb (1949). Hebb's postulate for associative learning (1949) and its modern articulation by Stent (1973) and Changeaux & Danchin (1976) suggests a two-part rule for the use-dependent modification of young labile synapses:

1. Synaptic contacts between synchronously active pre- and post-synaptic neurons are selectively reinforced.

2. Synaptic contacts between asynchronously active pre- and post-synaptic neurons are selectively depressed or eliminated.

The ability of the post-synaptic membrane to temporally summate the synaptic potentials from converging synapses that are synchronously, or near synchronously, active means that inputs with correlated action potential patterns are more likely than asynchronous inputs to have their activity covary with activation of the post-synaptic cell. Hebb's rule, therefore, extended to an array of synapses converging on a single cell, predicts that synchronous inputs are likely to be reinforced whereas asynchronous inputs will be functionally eliminated even though they may carry the same average amount of activity (Stent 1973). In the monocular deprivation paradigm, greater activity in one eye compared to the other provides the opportunity for more frequent correlations of activity among the converging synapses from the active eye, and, consequently, more effective driving of cortical neurons by that eye. The deprived eye not only has a lower probability of driving synaptic events that will sum to activate cortical neurons, but, in addition, its synapses will frequently be silent while the cortical neurons are driven by the open eye. In short, Hebbian ideas explain the results of monocular deprivation as a functional disconnection of the deprived eye from all cortical cells because cortical cell activity is not correlated with synaptic activity from that eye.

In contrast to monocular deprivation, strabismus does not depress either retina's ability to drive cortical neurons, but it does eliminate interocularly correlated synaptic activity when inputs from the two pathways converge on the same cortical cells. Functional segregation of each retina's inputs to separate sets of these post-synaptic cells ensues, according to the Hebb postulate, because that situation allows synapses driven by each eye to maximize their association with synchronously active cells and minimize their association with cells "asynchronously" activated by the other eye.

A key element in the application of Hebbian ideas to normal visual system development as well as monocular domination or binocular segregation within cortex was the demonstration that highly correlated patterns of activity do, in fact, exist *within a retina*. Physiological recordings in goldfish, cats, and rabbits have revealed that action potential patterns of neighboring ganglion cells of the same response type are nearly identical, and this similarity decreases between increasingly distant pairs of neurons. Action potential patterns of neighboring ganglion cells of opposite response type are negatively correlated. Moreover, the correlations remain in the absence of pattern stimulation and even in the spontaneous activity of the cells under conditions of complete dark adaptation (Arnett 1978, Arnett & Spraker 1981, Mastronarde 1983a,b).

The presence of these spatially organized high correlations of activity within a single eye's projection allow a Hebbian explanation for the fact that ocular dominance columns still develop relatively normally under conditions of dark-rearing in cats (LeVay et al 1978, Mower et al 1985, Swindale 1988) or during intrauterine development in monkeys (Rakic 1976, 1977, Hubel et al 1977). High intraretinal correlations in activity and synapses that follow Hebbian rules also provide an explanation for the fact that left and right retinal pathways converge in nonafferent cortical layers and yet segregate in afferent layers. In layer IV, synapses from the left and right eye geniculate laminae constitute virtually all of the inputs to the cortical cells and these synapses begin to sort out before visual experiences and binocular convergence of the eyes are possible. Consequently, intraeye correlations of activity have an opportunity to dominate over the later onset and less frequent correlations of synaptic input between the two eyes. Compared to layer IV, synapses on neurons in nonafferent cortical layers are capable of sorting relatively later in development during periods when the two eyes become capable of good binocular alignment. These neurons have converging inputs from many layer IV cells, resulting in larger visual receptive fields and less tightly correlated intraretinal activity, and they also receive many intracortical and nongeniculate extracortical inputs only indirectly associated with retinal activity (Gilbert 1983). Thus intraretinal correlations in activity are a much smaller proportion of the total number of possible synchronous events on nonlayer IV cortical neurons, whereas events triggered through the retinas of both eyes simultaneously have the better chance of producing simultaneously active synaptic inputs, temporal summation, and effective post-synaptic activation. Several diverse experimental paradigms have been used to manipulate nonretinal activity converging on cortical cells during monocular deprivation. They have all supported the idea that nondeprived eye activity must be correlated with activation of some nonretinal inputs in order to functionally "take-over" cells in nonafferent cortical layers (Freeman & Bonds 1979, Raushecker & Singer 1981, Bienenstock et al 1982, Singer & Raushecker 1982, Singer et al 1982, Fregnac & Imbert 1984).

## TTX Blockade in Mammals

A large number of experiments have employed the voltage-dependent $Na^+$ channel blocker, tetrodotoxin (TTX) to eliminate "spontaneous" activity and demonstrate its importance in utero or before the eye is capable of pattern vision. The anatomical results of these studies are all consistent with Hebb's postulate. Left and right eye inputs do not segregate from each other in the dLGN (Shatz & Stryker 1988) or into cortical layer IV ocular dominance columns (Stryker & Harris 1986). Blocking of all activity, however, has many effects that could be interpreted as a simple retardation of normal development (Edwards & Grafstein 1984, Kalil et al 1986, Cohan & Kater 1986, Casagrande & Condo 1988), and only two laboratories have actually generated evidence specifically implicating the temporal pattern of synaptic activation as the important missing parameter under TTX blockade.

Recordings in the kitten geniculate after blockade of retinal activity have documented a pronounced disruption of functional segregation in the retinogeniculate pathway that would never have been predicted on the basis of structural data alone (Archer et al 1982, Dubin et al 1986). Neurons within the geniculate laminae of normal kittens generally have well-defined "on-center, off-surround" or "off-center, on-surround" receptive field structure. TTX-treated kittens, recorded from shortly after removal of the retinal blockade, have many geniculate cells that are unusual in that they can be driven by both eyes (Dubin et al 1986). The most significant observation in these experiments, however, is that a large proportion of neurons respond to both "light on" and "light off" throughout their receptive field. The concentric field organization of ganglion cells is not perturbed by TTX treatment (Archer et al 1982). In addition, Dubin et al (1986) were occasionally able to record simultaneously from three cells, a geniculate neuron, an "on-center" ganglion cell that drove it, and an "off-center" ganglion cell that drove it. Thus it appears quite clear from this study that disruption of geniculate neuron functional organization occurs at the level of retinal axon convergence onto geniculate dendrites.

Disruption of a mechanism that stabilizes ganglion cell synapses by virtue of their temporal synchrony and ability to drive geniculate neurons would produce exactly this result. Thus, during normal development, a particular geniculate neuron would by chance receive a majority of its inputs from "on-center" ganglion cells driven by stimuli in the same small region of the visual field. Once this bias is established, synapses from functionally different ganglion cells, responding to cessation of illumination in the same small region, cannot be stabilized on the same neuron: "Off-center" retinal input would never fire in synchrony with the majority of "on-center" inputs to that cell. The "off-center" ganglion cells would, instead, functionally segregate onto closely adjacent geniculate neurons that initially received a slightly larger complement of synapses from the "off-center" ganglion cell type. Elimination of these patterned activity cues through TTX retinal blockade allows both "on-center" and "off-center" ganglion cells to maintain roughly equal ability to drive the same post-synaptic neurons.

In kitten primary visual cortex, TTX retinal treatments in the neonatal period have demonstrated that monocular takeover of cortical cells can be accomplished by a vision-deprived eye in competition with an eye lacking spontaneous activity (Chapman et al 1986). Binocular treatment of retinas (Stryker & Harris 1986) or direct treatment of cortex with TTX (Reiter et al 1986) will eliminate monocular takeover.

Most significantly, however, fixed temporal patterns of activation in the pathways of the two eyes have been applied to kittens with binocular retinal blockades that would, by themselves, prevent ocular dominance segregation in layer IV. This was accomplished with chronically implanted stimulating electrodes that delivered either simultaneous or out-of-phase volleys of activity to the central visual pathways of both eyes for 1–2 hr per day. The majority of cortical neurons were driven through both eyes in the synchronously stimulated group. Very few neurons were binocular in the asynchronously treated group (Stryker & Strickland 1984, Stryker 1986). Anatomically, in kittens experiencing asynchronous pathway volleys, the TTX effects on blocking segregation were considerably mitigated, whereas synchronously stimulated animals showed no signs of ocular dominance segregation (M. P. Stryker, personal communication).

## ACTIVITY AND SYNAPTIC CONVERGENCE IN NONMAMMALIAN VERTEBRATES

Most evidence that temporal parameters in afferent activity patterns have structural effects on visual projections comes from work on the retinotectal systems of goldfish and frogs. These species show retinal ganglion cell response properties that are similar to those of mammals. They exhibit equally refined maps of visual space within their brains, and some tectal neurons receive converging projections from the binocular region of the visual field through both eyes. Nevertheless, the retinal ganglion cells of cold-blooded vertebrates are more resistant than mammalian cells to perturbations of their normal target contacts, and their synapses in tectum are continually broken and remade throughout the larval period and well into adult life (Gaze et al 1979, Reh & Constantine-Paton 1983, Easter & Stuermer 1984). The latter properties permit explorations of dynamic synaptic interactions that are difficult in the internally developing mammals. The three aspects of these cold-blooded vertebrate visual projections that have been used effectively to explore the role played by activity are summarized briefly below.

### Synapse Segregation Based On Presynaptic Cell Body Proximity

The first hints of neural activity's role in sculpting the visual pathways of lower vertebrates arose in optic nerve regeneration experiments on goldfish

(Levine & Jacobson 1975, Cronlly-Dillon & Glaizner 1974). In both fish and frogs the retinotectal projection is normally completely crossed. However, when axons from two retinas are normally caused to converge on a single tectal lobe, the inputs from the two eyes do not mix but rather remain segregated into eye-specific "clumps" of retinal terminals.

The similarity between eye-specific clumping in tectum and eye-specific segregation in mammals was first recognized when the tectal terminations of supernumerary amphibian eyes were examined anatomically. In embryonically created three-eyed frogs, two complete visual projections converge on the same tectum from the earliest developmental stages, and anterograde labeling experiments invariably reveal that the continuous retinal projections of both the host and third eye are disrupted into an alternating, periodic pattern highly reminiscent of mammalian ocular dominance stripes (Constantine-Paton & Law 1978). Using the amphibian preparations and embryonic microsurgery, a series of experiments rapidly ruled out right and left eye labels (Law & Constantine-Paton 1981), genotypic differences between the eyes (Ide et al 1983), differences in time of arrival of the two projections (Law & Constantine-Paton 1980), and selective fasciculation within the optic tract (Constantine-Paton et al 1983) as parameters that could provide for recognition and "aggregation" of synapses arising from the same retina. In fact, the only parameter that is invariably correlated with segregation is convergence of terminals from nonneighboring regions of retina within a small region of the target zone (Fawcett & Willshaw 1982). Consequently, given the evidence that retinal neighbors share action potential patterns, it seemed likely that a Hebb-like interaction was causing convergence of each eye's projection into retina-specific zones where nearest presynaptic neighbors have the maximum probability of driving or depolarizing the same tectal neurons. A recent study combining intracellular Lucifer Yellow filling of tectal neurons and simultaneous visualization of eye-specific stripes provided anatomical support for this idea. The majority of tectal neurons restrict either their entire dendritic tree or an entire primary dendritic branch to the terminal zones of only one eye (Katz & Constantine-Paton 1988).

That retinal activity is necessary for eye-specific segregation in fish and frogs has now been confirmed with bilateral TTX-blocking experiments in several different laboratories (Meyer 1982, Boss & Schmidt 1984, Reh & Constantine-Paton 1985). Monocular blockade is not sufficient to eliminate this eye-specific segregation in goldfish (Meyer 1982), although the borders of the terminal "clumps" seem to become more diffuse (Schmidt & Tieman 1985). In addition, in frog larvae with supernumerary eyes it has been possible to rule out the possibility that TTX treatment (i.e. lack of activity) merely delays normal development of the segregated pattern.

In three-eyed tadpoles, because of the normal turnover of retinotectal synapses (Reh & Constantine-Paton 1983), if an activity-dependent mechanism is crucial to the maintenance of segregation, it must remain continually active. In these animals it is possible to observe ocular-dominance stripes in various stages of desegregation as animals are examined after increasingly long durations of TTX retinal blockade (Reh & Constantine-Paton 1985).

## Normal Topographic Map Refinement

There is now general agreement that activity-independent differences, in cell surface properties and ability to "read" axon guidance cues, bring visual axons to the vicinity of their topographically appropriate target cells. It is the fine-tuning of these projections through the local ordering of synapses that is established through activity-dependent nearest-neighbor sorting (see Udin & Fawcett 1988 for review).

The prediction from this two-stage mapping hypothesis is that projections lacking retinal activity or lacking correlations in activity that are related to the retinal proximity of ganglion cells should show less refined representations of the retinal surface within the tectal lobe. Conversely, ganglion cells from a larger region of retina should maintain synapses in any given region of the tectal neuropil. Both of these predictions have been borne out by experiments using TTX blockade of retinal activity (Meyer 1983, Schmidt & Edwards 1983) or stroboscopic stimulation of regenerating retinal projections in goldfish (Schmidt & Eisele 1985, Cook & Rankin 1986, Cook 1988).

## Binocular Projections of Xenopus Frogs

Binocular maps within amphibian tecta arise through the nucleus isthmi (the homologue of the mammalian parabigeminal nucleus), which relays binocular visual field information from the contralateral tectum to the locations representing the same visual field positions in the ipsilateral tectum (Gruberg & Udin 1978). Numerous experiments over the past decade have shown that the development of binocular neurons in these tectal positions depends critically on similar patterns of action potential activity arriving from the contralateral eye and from the ipsilateral eye through the nucleus isthmi. For example, surgical rotation of one retina during tadpole stages so that the location of visual field positions in the contralateral tectum is correspondingly rotated, causes a physiologically detectable shift in the projection of the nucleus isthmi projection to that tectum. The ipsilateral eye projection through the isthmotectal relay therefore reattains registration with the perturbed contralateral eye projection, and inputs carrying information about the same point in the visual field are still able to converge on a small locus in the tectum (Keating 1974, Udin & Keating 1981). Patterned visual stimulation is crucial to the development of this convergence. Animals maintained in complete darkness during the critical metamorphic period attain diffuse, relatively weak, isthmotectal projections driven by the ipsilateral eye, and the ipsilateral map fails to come into register with the contralateral map (Keating 1975, Keating & Feldman 1975).

Significantly, these physiological observations have striking anatomical correlates. When the nucleus isthmi projection is filled with HRP, the axons carrying ipsilateral eye information can be followed to their termination sites in the tectal lobe. In animals with one eye rotated by 90 degrees, these axons approach the tectum normally and arrive at their normal arborization site. However, they then grow in apparently random fashion until encountering the displaced sites with functionally correlated contralateral eye activity (Udin 1983). They form morphologically normal arbors in these novel positions, leaving little more than a vestigial branch at the tectal position where dense arborization would normally occur (Udin 1985).

## CELLULAR MECHANISMS UNDERLYING THE CORRELATED ACTIVITY RULE

### Modulation of Synaptic Effectiveness Through Post-Synaptic Cell Excitability

A number of studies have investigated the relationship between action potential activity in single cortical neurons and the efficacy of particular inputs. The results suggest that post-synaptic action potentials produce an increase in the effectiveness of the correlated input (Barany & Feher 1981; see Fregnac & Imbert 1984 for review) and support the idea that near synchrony of pre- and post-synaptic activity is an important factor in modulating cortical synaptic strength. In the visual cortex, pairing of visual stimulation through one eye with post-synaptic cell firing by applied current causes an increase in the effectiveness of the paired eye's input more frequently in kittens than in adult cats. The changes can last from 15 minutes to several hours (Fregnac et al 1988). Similar results have been obtained in experiments in which iontophoresis of excitatory transmitters was paired with activation of one eye's inputs (Greuel et al 1988).

Decreased post-synaptic excitability or reduced sensory signal compared to background "noise" has also been implicated in cat visual cortical plasticity in an extensive literature of studies in which chronic drug treatments of cortex (Kasamatsu et al 1979, Daw et al 1983, 1985, Bear et al 1983, Paradiso et al 1983; see Gordon et al 1987 for review) or lesions of noncortical regions projecting to cortex (Adrien et al 1985, Daw et al 1984, Singer 1982, Bear & Singer 1986) have eliminated the domination of visual cortical cells by a nondeprived eye. The observations support the notion that a level of post-synaptic activation reflecting both retinal and non-retinal inputs must be reached during sensory stimulation, in order to trigger selective increases in synaptic efficacy (Bienenstock et al 1982, Singer et al 1982, Singer & Raushecker 1982, Bear et al 1987).

One very recent experiment, however, suggests that many of these observations may have to be reexamined for the potentially complicating effects of a competitive disadvantage inflicted upon active inputs that fail to fire post-synaptic cells. Reiter & Stryker (1988) depressed cortical excitability with chronic infusion of the $GABA_a$ receptor agonist muscimol. This treatment suppressed spontaneous activity and inhibited visually elicited responses in kitten cortical neurons. Under these conditions, in which cortical neurons were prevented from firing, the *deprived* (less active) eye came to dominate over the active eye in the monocular deprivation paradigm. This is the first experimental evidence in the visual development literature for a synaptic interaction in which less active inputs appear to have an advantage over more active neighbors (Cooper et al 1979, Levy & Desmond 1985). Synaptic activity seems to be detrimental to an eye's functional dominance of a cortical neuron if that cortical neuron cannot be driven (Reiter & Stryker 1988).

## NMDA-Mediated Plasticity in Visual Cortex

Many developmental studies of cortical plasticity have now focused on the properties of the *N*-methyl D-aspartate subclass of excitatory amino receptors as the long sought detectors for correlated synaptic events. This would be equivalent to the documented triggering function of this receptor system in LTP of hippocampal CA1 synapses. The essential idea is that temporal summation of EPSPs results in $Ca^{2+}$ influx through the NMDA receptor channel because the first, non-NMDA channel–mediated responses depolarize the post-synaptic membrane (Wigstrom & Gustafsson 1985). These initial EPSPs relieve the $Mg^{2+}$ block of the NMDA channel so that subsequent excitatory amino acid (EAA)–mediated synaptic events open the channel and let $Ca^{2+}$ into the post-synaptic cell (Mayer et al 1984, Nowak et al 1984).

There is general agreement that the LGN inputs to the visual cortex are predominantly glutaminergic (Hagihara et al 1988). Many of the inter-laminar and intracortical projections may also use EAAs (Hicks 1987). However, the particular role of the NMDA receptor in developmental use-dependent plasticity or in visual synaptic transmission is controversial.

EFFECTS OF MANIPULATION OF EAAs    The earliest study to demonstrate an effect on plasticity of manipulating EAAs used cortical glutamate infusions in monocularly deprived kittens to demonstrate failure of the nondeprived eye functionally to dominate cortical neurons (Shaw & Cynader 1984). This result was interpreted as demonstrating that any nonspecific imbalances of cortical activity could disrupt plasticity.

The study that actually focused developmental cortical work specifically on the NMDA receptor used chronic infusion of 2-amino-5-phosphono-valeric acid (APV), a specific antagonist of NMDA receptors (Harris et al 1984), into the cortex of neonate, monocularly deprived kittens. The treatment blocks the monocular takeover of cortical neurons by the non-deprived eye, prevents the acquisition or the maintenance of stimulus orientation selectivity, and produces sluggish cells with relatively large receptive fields (Kleinschmidt et al 1987). Singer and his colleagues suggested that the effects arise specifically from blocking the ability of cortical NMDA receptors to detect correlated events rather than a general effect of depression of cortical neuron excitability.

The problem with this interpretation is that there are no definitive experiments tying the level of post-synaptic excitability necessary to induce use-dependent modification of synapses to the threshold for NMDA receptor, or other high threshold $Ca^{2+}$ channel, activation. For example, in the studies such as those of Fregnac et al (1988) and Greuel et al (1988), in which stimulation through one eye is paired with electrical or transmitter stimulation of a binocular cortical cell, it would be extremely interesting to know whether low, NMDA-selective, doses of simultaneously applied APV selectively blocked the increase in the paired eye's effectiveness.

Data from several laboratories suggest that treatments that block cortical NMDA channels also depress neuronal excitability in kitten cortex at least temporarily. Tsumoto and his colleagues (Tsumoto et al 1987, Hagihara et al 1988) tested visually and electrically stimulated cortical responses to application of APV and kynurenate, an antagonist that blocks all EAA receptor types. They found that 70% of visual responses are APV-sensitive in kittens. This dropped to 30% in adult cats. Similar depressions in responsiveness have been observed by Fox et al (1989) and are described in more detail below. Thus, the question has become whether NMDA receptor activation has a unique function as a detector of correlated activity as opposed to having a more general role as one of several transmitter systems that modulate cortical cell excitability. The issue is difficult to address definitively in cortex. In in vivo studies, the absolute concentrations of applied APV at the post-synaptic membrane are never known. At high concentrations, APV loses its selectivity for NMDA receptors and suppresses activation of kainate and quisqualate receptors as well (Collingridge & Bliss 1987). Miller et al (1989) have tried to overcome this difficulty by infusing APV into cortex with an osmotic pump beginning one day before physiological recording in order to obtain stable concentrations of the drug at fixed distances from the cannulae. They have found a close correlation between sluggishness of cell responses to visual stimulation and selective elevation of responsiveness to applied NMDA but not to similarly applied quisqualate or kainate. They conclude that in normally reared adult cats, NMDA receptor activation is a normal component of the excitatory response of cortical neurons to visual stimulation. Equivalent studies are not yet available for kittens.

This controversy over the function of NMDA receptors in the visual cortex in developmental use-dependent changes will not be easily resolved. The NMDA receptor could still be a unique detector of correlated activity in cortex, even though APV infusions may be acting to decrease cortical excitability. Detection of correlated events via a specific ability to trigger post-synaptic $Ca^{2+}$ influx and participation in visual transmission as an excitatory post-synaptic receptor are not necessarily mutually exclusive functions. In addition, all of the perturbations that depress post-synaptic excitability in cortex and thus block cortical plasticity in the monocular deprivation paradigm would simultaneously decrease the probability of NMDA receptor activation.

SYNAPTIC WEIGHT CHANGES VERSUS ANATOMICAL REDISTRIBUTION OF SYNAPSES   Assuming that at least some of the experience dependent plasticity in developing visual cortex is attributable to blocking a specific correlated activity detector function of NMDA receptors, there remains one other major, unresolved, issue in the cortical studies dealing with this issue. Specifically, are the plastic properties attributable to NMDA receptors reflections of structural changes in synaptic convergence, or do they simply reflect long-term changes in synaptic efficacy? Changes in synaptic weight via increased EAA release (Errington et al 1987) or increased EAA receptors (Lynch & Baudry 1984) have been suggested as mechanisms underlying hippocampal LTP. Similarities between the developing cortex and hippocampus are supported by studies in cortical tissue slices that demonstrate an APV sensitive component of a potentiation of cortical synaptic transmission that can be induced with frequent stimulation of cortical white matter (Artola & Singer 1987). Several laboratories have also generated evidence suggesting that visual cortical potentiation may have a sensitive period that corresponds closely to the period of maximal cortical plasticity (Komatsu et al 1981, 1988). Structural changes in the sizes of synaptic contacts or in the number or shape of post-synaptic spines have been, to varying degrees (Coss & Perkel 1985), associated with both cortical (Perkins & Teyler 1988) and hippocampal LTP (Desmond & Levy

1986a,b). Nevertheless, the rapid onset of LTP is generally taken as evidence that most of the functional change results from alterations in the efficacy of existing synapses rather than from the actual loss or repositioning of contacts.

To date only two pieces of evidence even hint that structural rewiring may result from cortical NMDA receptor activation. Rauschecker & Hahn (1987) have detected retrograde inhibition of ocular dominance shifts produced in alert kittens by monocular occlusion. This inhibition was accomplished by intramuscular injections with ketamine-xylazine after each of 15 20-minute exposures to monocular visual experience during the height of the sensitive period for nondeprived eye domination of the cortex. Ketamine, a sigma opiod receptor agonist, is also an activity-dependent blocker of the NMDA channel. Neither saline injections nor xylazine injections alone produced this effect. The authors suggest that an ongoing process that may involve a structural change is implicated by the fact that ketamine treatment is effective after the experience has ceased (Rauschecker & Hahn 1987). If NMDA receptors are involved in a consolidation process necessary for an ocular dominance shift to occur, it is not clear that they are functioning as specific correlated activity detectors, or that the ongoing process involves structural changes.

The final evidence suggesting a structuring role for NMDA channel activation has been generated by physiological studies of cortical neuronal responsiveness to iontophoretically applied D-APV (Tsumoto et al 1987, Fox et al 1989). The reasoning here is that if NMDA receptor function is important as a normal excitatory receptor component of cortical neuron activation, D-APV at low doses should be able to suppress some component of the visual response and the effect should be relatively constant at all ages. However, if absence of NMDA receptor function is a critical factor in limiting plasticity, then *changes* in NMDA receptor sensitivity should correlate with independent measures of changes in plasticity. D-APV will suppress spontaneous activity and visually elicited activity in all visual cortical layers of young kittens. However, the ability of low doses of the drug to block visual activity in layers IV, V, and VI is gradually lost in successively older animals. In adults, visual driving of cells in these layers is insensitive to APV at relatively low doses that are, nevertheless, still successful at significantly suppressing spontaneous activity (Fox et al 1989). The potentially important point is that only in layer IV was the dropoff in APV sensitivity correlated with a sensitive period for plasticity, and this parallel was with the structural segregation of geniculocortical afferents (LeVay et al 1978). Significantly, in layers II and III, APV is able to suppress retinal activation of cortical neurons at all ages (Fox et al 1989).

## NMDA-Mediated Plasticity in the Retinotectal System

Evidence that excitatory amino acids are the predominant transmitter in the retinotectal pathway has been obtained, over the past several years, in quantitative autoradiographic and physiological studies of both the gold-fish (Langdon & Freeman 1986, 1987, Henley & Oswald 1988) and the frog (Debski et al 1987, Debski & Constantine-Paton 1988, Fox & Fraser 1987, McDonald et al 1989). This information has motivated investigations that capitalize on the ability, in these systems, directly to manipulate, record from, and visualize the cell processes (retinal ganglion cell terminals) believed to be the initiators of plastic synaptic interactions mediated by excitatory amino acids.

APV EXPERIMENTS    Chronic application of DL-APV to the doubly inner-vated tecta of tadpoles or frogs with a supernumerary eye causes complete, anatomically assayed desegregation of the two retinal inputs. This gradual breakdown of ocular dominance stripes is not observed when the biologi-cally inactive isomer L-APV is similarly applied. Furthermore, the effects of the active isomer are fully reversible (Cline et al 1987).

Pharmacological and physiological investigations of EAA transmission in the retinotectal system have been undertaken in normal tadpoles by using an intact, unanesthetized, brain preparation in which all nociceptive inputs have been severed. In this preparation the animals are maintained via vascular perfusion on oxygenated, glucose-supplemented saline solu-tion to which drugs of known concentration can be added and carefully controlled (Debski et al 1987). These studies indicate that the con-centrations of APV used in the chronic experiments do not depress action potential activity in retinal ganglion cell axons (Cline et al 1987) or the level of excitability of tectal neurons assessed as the post-synaptic component of the tectal field potential in response to electrical stimulation of the contralateral optic nerve (Debski & Constantine-Paton 1988). In addition, in a completely independent series of experiments, chronic application of APV to the tectum of juvenile *Xenopus*, in concentrations sufficient to block convergence of binocular maps (see below), does not depress retinotectal transmission of visual activity through a tectal relay to the nucleus isthmi (Scherer & Udin 1988). Thus, it appears that in the retinotectal pathway, application of APV at concentrations at which it is selective for the NMDA receptor subtype and at which it has minor, if any effects, on depressing retinal terminal or tectal cell excitability, can effectively block the selective stabilization of synapses from retinal neighbors on tectal dendrites.

In goldfish, chronic treatment of tecta with DL-APV during the final stages of optic nerve regeneration increases the size of multiunit receptive fields recorded in the tectal neuropil (Schmidt 1988). In addition, normal frog tadpoles chronically treated with APV for six to eight weeks have much larger areas of retina projecting to the same defined point in the tectal lobe as compared to sham-operated or normal tadpoles (Cline & Constantine-Paton 1989). Both of these observations are fully consistent with the idea that the Hebbian synaptic interaction underlying segregation is identical to that which normally increases the fidelity of continuous topographic maps and that the NMDA receptor is similarly critical to both processes.

Finally, Udin and her colleagues have shown that the matching of binocular tectal maps in *Xenopus laevis* with 90 degree rotations of the contralateral eye is completely blocked by chronic treatment of these tecta with DL-APV during the critical period for this plasticity (Scherer & Udin 1988). Thus, NMDA receptors appear to be mediators of binocular convergence in this lower vertebrate preparation as well. This finding is broadly significant. Convergence in the *Xenopus* binocular assay is dependent on the same type of binocular visual experience required for binocular convergence in mammalian cortex, and the system has a limited developmental critical period (Keating 1975). In *Xenopus*, binocular visual experience produces plastic changes by a pronounced structural relocation of synapses (Udin & Keating 1981, Udin 1985).

The most parsimonious explanation for all of the observations in lower vertebrates following APV application is that a functional or selectively activated NMDA channel is a critical trigger in whatever cascade of events ultimately increases the lifetime of visual synapses on the same post-synaptic membrane in response to covariance of pre- and post-synaptic activity. Moreover, in each of these preparations, there is a clear indication, if not direct evidence, that the plastic changes are structural relocations of synaptic contacts and not simply functional alterations in existing synapses.

The idea that a common mechanism underlies developmental and mature plasticity could be strengthened by the demonstration that functional changes in synaptic efficacy similar to hippocampal CA1 LTP accompany structural changes in visual pathways attributable to Hebbian synaptic interactions. Data on this point is only available for the goldfish visual projection. As mentioned above, topographic map refinement in the tectum, during the final stages of goldfish optic nerve regeneration, is blocked by chronic treatment with APV. During the time that this refinement is taking place, visual responses are capable of a potentiation resembling that found in the hippocampus. Schmidt (1987, 1988) has demonstrated that following a short train of low-frequency stimuli, the tectal response to optic nerve stimulation potentiates more quickly and to a greater extent than had been found previously in the mature goldfish optic nerve projection (Lewis & Teyler 1986). This potentiation can be blocked by APV and is, therefore, presumably, NMDA-mediated (Schmidt 1988).

SINGLE-CELL STRUCTURAL CHANGES UNDERLYING ACTIVITY-DEPENDENT PLASTICITY   In our own laboratory we have examined retinal and tectal cell responses to chronic application of NMDA itself. The results suggest a specific relationship between NMDA receptor activation and at least one of the processes that sculpt the morphology and connectivity of single retinal ganglion cell terminal arbors.

Chronic application of nontoxic doses of NMDA to doubly innervated frog tecta produce a pronounced increase in eye-specific segregation (Cline et al 1987, Cline & Constantine-Paton 1987). Stripe boundaries become sharper, and there are fewer fusions and forks in the striped pattern. Thus it appears that the continuous presence of exogenous excitatory ligand for the NMDA receptor improves the ability of the system to discriminate correlated from noncorrelated synaptic events. Although the responses of the stripe pattern to chronic exposure of NMDA or APV seem complementary, closer examination of the morphological and electrophysiological effects of the treatments suggest that the response to NMDA is not the simple opposite of the response of APV. Reconstructions of retinal ganglion cell terminal arbors from NMDA-treated three-eyed animals reveal a dramatic, 50% reduction in the number of terminal branches in the treated arbors compared to the untreated arbors (Cline & Constantine-Paton 1989). Electrophysiological studies in the cannulated tadpole brain preparation were carried out on animals chronically treated for 4 weeks with NMDA in the same concentrations used to produce the anatomical increase in segregation. We found that the response of the retinotectal pathway to applied NMDA is significantly decreased in chronically NMDA-treated tadpoles (E. A. Debski and H. T. Cline, unpublished). We do not yet know whether the decreased sensitivity is due to a down-regulation of receptor number, to a decrease in agonist binding affinity, or to a change in the tectal circuits in which NMDA receptors are involved. Receptor desensitization, however, seems unlikely because NMDA from the implant is washed out of the animal before NMDA sensitivity is assayed. (The receptor would be expected to recover from desensitization during this wash.)

It seems reasonable to suggest that a decreased sensitivity of the receptor system may increase the amount or degree of correlated activity necessary to activate NMDA-gated channels and ultimately stabilize individual retinal ganglion cell synapses. However, it is important to point out that NMDA treatment at any concentration tested does not cause stripe desegregation. Therefore, chronic NMDA treatment does not appear to inactivate the ability of the system to discriminate inputs from the two eyes (Cline & Constantine-Paton 1989).

Stripe boundaries are regions of relatively low correlations in synaptic activity because they are regions of neuropil in which noncorrelated inputs from the two eyes intermix. Quantitative EM analyses of single retinal terminals indicates that most retinotectal synapses are located on the highest order branches of the arbor (Yen & Constantine-Paton 1988). A mechanism in which the survival of these branches is dependent on the formation of some critical number of synapses stabilized via NMDA receptor activation would explain stripe sharpening with decreased NMDA sensitivity. The number of stabilized synapses in boundary regions would be the first to drop below the critical value necessary to sustain a branch. Furthermore, because in the tadpole and young frog the arbors of ganglion cells are constantly withdrawing branches in some regions and initiating new branches in others (Reh & Constantine-Paton 1983), this same reasoning explains the chronic NMDA-induced decreases in branches of individual arbors even within the stripes of one eye. Specifically, with chronic NMDA treatment and decreased NMDA receptor sensitivity, many of the new branches in ganglion cell arbors simply do not establish enough stabilized synapses to survive. The new branches are uniquely vulnerable to decreased NMDA receptor sensitivity because they have the fewest numbers of stabilized synapses to begin with: Initial contact with a post-synaptic process is likely to be by trial and error and only a few of a new branch's synapses can be expected to converge on post-synaptic processes already receiving inputs that are active simultaneously with them. In addition, for each synapse on a distal branch that succeeds in converging along with some nearly synchronized "other" input, the degree of correlation has a high probability of being low.

Over the past ten years it has become clear that the terminal arbors of projection neurons in the developing visual pathways of both cold- and warm-blooded vertebrates are dynamic structures capable of relatively extensive remodeling in response to perturbations of activity. Thus, chick retinal ganglion cells, like those of fish and frogs, appear to make their initial synapses in a region of tectum quite separate from their terminal sites in the mature brain (McLoon 1985). Geniculocortical terminals from the nondeprived eye of young macaque monkeys show some sprouting into the deprived eye's ocular dominance stripes even when monocular occlusion is begun relatively late in development, after segregation is established (LeVay et al 1980). In the kitten geniculocortical pathway, neonatal monocular deprivation or retinal TTX block has been shown to affect selectively and differently the morphology of X and Y ganglion cell arbors in a pattern that is consistent with the type or degree of activity deprivation to which the system was exposed (Sherman & Spear 1982, Friedlander et al 1982, Sur 1988). Finally, in fetal kittens, as had previously been observed for frog larvae (Reh & Constantine-Paton 1985), TTX blockade results in individual retinal ganglion cells that have greatly expanded central arborizations (Sretavan et al 1988). It is very likely that, in both these studies, the activity blocks suppress post-synaptic as well as presynaptic activity. Is this enlargement simply due to increased growth rates in silent ganglion cells as has been suggested for isolated mollusc neurons in culture (Cohan & Kater 1986)? Is it due to an increased sprouting stimulus from inactive post-synaptic neurons, as has been suggested for the neuromuscular system (Brown et al 1981)? Is it the result of inactivation of a Hebbian selective stabilization system in which no terminal branches are stabilized and none are trimmed because both pre- and post-synaptic processes are silent? Clearly, some of the phenomena described on a system level will soon be addressed at the level of cell biology. However, we know little about morphological changes in post-synaptic neurons and little about structural plasticity in nonprojection neurons. It is also clear that much more information has to be collected on the biochemistry of the changes produced by activity perturbations before the numerous newly raised questions can be answered.

## CONCLUSION

It has historically been the case in both the fields of developmental and mature brain plasticity that each new set of observations opens up a new host of questions without necessarily promising that a complete explanation lies immediately around the corner. An association between high sensitivity of NMDA-mediated synaptic transmission and viable "exploratory" contacts of young neurons could be a critical cellular difference between the pronounced structural plasticity of the developing brain and the profoundly less plastic properties of many regions in the mature brain. Only the correlation cited above between the timing of cortical layer IV ocular dominance column plasticity and high NMDA receptor sensitivity of layer IV neurons is currently available to support this hypothesis for the mammalian brain.

From a developmental biologist's point of view, the major unanswered question in all areas of neural plasticity is whether functionally detected changes in synaptic weight represent one end of the same spectrum of interactions that produce structural relocation of synapses at its other extreme. Even in those visual pathways for which data support a specific correlated activity detector function for the NMDA receptor, there is essentially no unambiguous information on the biochemical events subsequent to NMDA receptor activation. Apparently also little agreement has been attained about these events in the more intensively studied hippocampus (Collingridge 1987). Without documentation of biochemical similarity, the argument that brain evolution had adapted exactly the same activity-dependent mechanisms used during development to the mature functions of learning and memory is speculative at best. Nevertheless, for a few, but maybe not all developing visual projections, as for some, but not all hippocampal model systems for learning and memory (Brown 1990), it can now be said that activation of the NMDA subclass of glutamate receptors plays a critical role in modulating the long-term interactions between pre- and post-synaptic neurons.

## ACKNOWLEDGMENTS

We wish to thank Dr. Susan Udin, Dr. Michael Stryker, and Dr. Kenneth Miller for helpful comments on this manuscript.

*Literature Cited*

Adrien, J. Blanc, G., Buisseret, P., Fregnac, Y., Gary-Bobo, E., Imbert, M., Tassin, J. P., Trotter, Y. 1985. Noradrenaline and functional plasticity in kitten visual cortex: A re-examination. *J. Physiol.* 367: 73–98

Archer, S. M., Dubin, M. W., Stark, L. A. 1982. Abnormal development of kitten retino-geniculate connectivity in the absence of action potentials. *Science* 217: 743–45

Arnett, D. W. 1978. Statistical dependence between neighboring retinal ganglion cells in goldfish. *Exp. Brain Res.* 32: 49–53

Arnett, D., Spraker, T. E. 1981. Cross-correlation analysis of the maintained discharge of rabbit retinal ganglion cell. *J. Physiol.* 317: 29–47

Artola, A., Singer, W. 1987. Long-term potentiation and NMDA receptors in rat visual cortex. *Nature* 330: 649–52

Baranyi, A., Feher, O. 1981. Long-term facilitation of excitatory synaptic transmission in single cortical neurons of the cat produced by repetitive pairing of synaptic potentials following intracellular stimulation. *Neurosci. Lett.* 23: 303–8

Bear, M. F., Cooper, L. N., Ebner, F. F. 1987. A physiological basis for a theory of synapse modification. *Science* 237: 42–48

Bear, M. F., Paradiso, M. A., Schwartz, M., Nelson, S. B., Carnes, K. M., Daniels, J. D. 1983. Two methods of catecholamine depletion in kitten visual cortex yield different effects on plasticity. *Nature* 302: 245–47

Bear, M. F., Singer, W. 1986. Modulation of visual cortical plasticity by acetylcholine and noradrenaline. *Nature* 320: 172–76

Bienenstock, E., Cooper, L. N., Munro, P. 1982. Theory for the development of neuron selectivity: Orientation specificity and binocular interaction in visual cortex. *J. Neurosci.* 2: 32–48

Blakemore, C., Garey, L. J., Vital-Durand, F. 1978. The physiological effects of monocular deprivation and their reversal in the monkey's visual cortex. *J. Physiol.* 283: 223–62

Blakemore, C., Van Sluyters, R. C. 1974. Reversal of the physiological effects of monocular deprivation in kittens: Further evidence for a sensitive period. *J. Physiol.* 237: 195–216

Boss, V. C., Schmidt, J. T. 1984. Activity and the formation of ocular dominance patches in dually innervated tectum of goldfish. *J. Neurosci.* 4: 2891–2905

Brown, M. C., Holland, R. L., Hopkins, W. G. 1981. Motor nerve sprouting. *Annu. Rev. Neurosci.* 4: 17–42

Brown, T. H., Kairiss, E. W., Keenan, C. L. 1990. Hebbian synapses: Mechanisms and algorithms. *Annu. Rev. Neurosci.* 13: In press

Bruce, C. J., Isley, M. R., Shinkman, P. G. 1981. Visual experience and the development of interocular orientation disparity in visual cortex. *J. Neurophysiol.* 46: 215–28

Casagrande, V. A., Condo, G. J. 1988. The effect of altered neuronal activity on the development of layers in the lateral geniculate nucleus. *J. Neurosci.* 8: 395–416

Changeux, J. P., Danchin, A. 1976. Selective stabilization of developing synapses as a mechanism for the specification of neural networks. *Nature* 264: 705–12

Chapman, B., Jacobson, M. D., Reiter, H. O., Stryker, M. P. 1986. Ocular dominance shift in kitten visual cortex caused by imbalance in retinal electrical activity. *Nature* 324: 154–56

Cline, H. T., Constantine-Paton, M. 1989. NMDA receptor antagonists disrupt the retinotectal topographic map. *Neuron* 3: In press

Cline, H. T., Constantine-Paton, M. 1988. NMDA receptor antagonist, APV, disorganizes the retinotectal map. *Proc. Soc. Neurosci.* 14: 674

Cline, H. T., Debski, E., Constantine-Paton, M. 1987. NMDA receptor antagonist desegregates eye-specific stripes. *Proc. Natl. Acad. Sci. USA* 84: 4342–45

Cohan, C. S., Kater, S. B. 1986. Suppression of neurite elongation and growth cone motility by electrical activity. *Science* 232: 1638–40

Collingridge, G. 1987. The role of NMDA receptors in learning and memory. *Nature* 330: 604–5

Collingridge, G. L., Bliss, T. V. P. 1987. NMDA receptors—their role in long-term potentiation. *Trends Neurosci.* 10: 288–93

Constantine-Paton, M., Law, M. I. 1978. Eye-specific termination bands in tecta of three-eyed frogs. *Science* 202: 639–41

Constantine-Paton, M., Pitts, E. C., Reh, T. 1983. The organization of the optic pathway and its relationship to termination in the retinotectal system of *Rana pipiens. J. Comp. Neurol.* 218: 298–313

Cook, J. E. 1988. Topographic refinement of the goldfish retinotectal projection: Sensitivity to stroboscopic light at different periods during optic nerve regeneration. *Exp. Brain Res.* 70: 109–16

Cook, J. E., Rankin, E. C. C. 1986. Impaired refinement of the regenerated retinotectal projection of the goldfish in stroboscopic light: A quantitative WGA-HRP study. *Exp. Brain Res.* 63: 421–30

Cooper, L. N., Liberman, F., Oja, E. 1979. A theory for the acquisition and loss of neuron specificity in visual cortex. *Biol. Cybernet.* 33: 9–28

Coss, R. G., Perkel, D. H. 1985. The function of dendritic spines: A review of theoretical issues. *Behav. Neural Biol.* 44: 151–85

Cronlly-Dillon, J. R., Glaizner, B. 1974. Specificity of regenerating optic fibres for left and right optic tectum in goldfish. *Nature* 251: 505–7

Daw, N. W., Rader, R. K., Robertson, T. W., Ariel, M. 1983. The effects of 6-hydroxydopamine on visual deprivation in the kitten striate cortex. *J. Neurosci.* 3: 907–14

Daw, N. W., Robertson, T. W., Rader, R. W., Videen, T. O., Coscia, C. J. 1984. Effect of lesions of dorsal noradrenergic bundle on visual deprivation in kitten striate cortex. *J. Neurosci.* 4: 1354–60

Daw, N. W., Videen, T. O., Parkinson, D., Rader, R. K. 1985. DSP-4 (N-(2-chloroethyl)-*N*-ethyl-2-bromobenzylamine) depletes noradrenaline in kitten visual cortex without altering the effects of monocular deprivation. *J. Neurosci.* 5: 1925–33

Debski, E. A., Cline, H. T., Constantine-Paton, M. 1987. Kynurenic acid blocks retinal-tectal transmission in *Rana pipiens. Proc. Soc. Neurosci.* 13: 1691

Debski, E. A., Constantine-Paton, M. 1988. The effects of glutamate receptor agonists and antagonists on the evoked potential in *Rana pipiens. Proc. Soc. Neurosci.* 14: 674

Desmond, N. L., Levy, W. B. 1986a. Changes in the numerical density of synaptic contacts with long-term potentiation in the hippocampal dentate gyrus. *J. Comp. Neurol.* 253: 466–75

Desmond, N. L., Levy, W. B. 1986b. Changes in the post-synaptic density with long-term potentiation in the dentate gyrus. *J. Comp. Neurol.* 253: 476–82

Dubin, M. W., Stark, L. A., Archer, S. M. 1986. A role for action-potential activity in the development of neural connections in the kitten retinogeniculate pathway. *J. Neurosci.* 6: 1021–36

Duffy, F. H., Snodgrass, S. R., Burchfiel, J. L., Conway, J. L. 1976. Bicuculline reversal of deprivation amblyopia in the cat. *Nature* 260: 256–57

Easter, S. S. Jr., Stuermer, C. A. G. 1984. An evaluation of the hypothesis of shifting terminals in the goldfish optic tectum. *J. Neurosci.* 4: 1052–63

Edwards, D. L., Grafstein, B. 1984. Intraocular injection of tetrodotoxin in goldfish decreases fast axonal transport of $^3$H-glucosamine-labeled materials in optic axons. *Brain Res.* 299: 190–94

Errington, M. L., Lynch, M. A., Bliss, T. V. P. 1987. Long-term potentiation in the dentate gyrus: Induction and increased glutamate release are blocked by DL-aminophosphonovalerate. *Neuroscience* 20: 279–84

Fawcett, J. W., Willshaw, D. J. 1982. Compound eyes project stripes on the optic tectum in *Xenopus*. *Nature* 296: 350–52

Fox, B. E. S., Fraser, S. E. 1987. Excitatory amino acids in the retino-tectal system of *Xenopus laevis*. *Proc. Soc. Neurosci.* 13: 766

Fox, K., Sato, H., Daw, N. 1989. The location and function of NMDA receptors in cat and kitten visual cortex. *J. Neurosci.* 9: 2443–54

Freeman, R. D., Bonds, A. B. 1979. Cortical plasticity in monocularly deprived immobilized kittens depends on eye movement. *Science* 206: 1093–95

Fregnac, Y., Imbert, M. 1984. Development of neuronal selectivity in primary visual cortex of cat. *Physiol. Rev.* 64: 325–434

Fregnac, Y., Shulz, D., Thorpe, S., Bienenstock, E. 1988. A cellular analogue of visual cortical plasticity. *Nature* 333: 367–70

Friedlander, M. J., Stanford, L. R., Sherman, S. M. 1982. Effects of monocular deprivation on the structural function relationship of individual neurons in the cats lateral geniculate nucleus. *J. Neurosci.* 2: 321–30

Gaze, R. M., Keating, M. J., Ostberg, A., Chung, S. H. 1979. The relationship between retinal and tectal growth in larval *Xenopus*. Implications for the development of the retinotectal projection. *J. Embryol. Exp. Morphol.* 53: 103–43

Gilbert, C. D. 1983. Microcircuitry of the visual cortex. *Annu. Rev. Neurosci.* 6: 217–47

Gordon, B., Allen, E. E., Trombley, P. Q. 1987. The role of norepinephrine in plasticity of visual cortex. *Prog. Neurobiol.* 30: 171–91

Greuel, J. M., Luhmann, H. J., Singer, W. 1988. Pharmacological induction of use-dependent receptive field modifications in the visual cortex. *Science* 242: 74–77

Gruberg, E. R., Udin, S. B. 1978. Topographic projections between the nucleus isthmi and the tectum of the frog *Rana pipiens*. *J. Comp. Neurol.* 179: 487–500

Guillery, R. W. 1972. Binocular competition in the control of geniculate cell growth. *J. Comp. Neurol.* 144: 117–30

Guillery, R. W., Stelzner, D. J. 1970. The differential effects of unilateral lid closure upon the monocular and binocular segments of the dorsal lateral geniculate nucleus in the cat. *J. Comp. Neurol.* 139: 413–22

Hagihara, K., Tsumoto, T., Sato, H., Hata, Y. 1988. Actions of excitatory amino acid antagonists on geniculo-cortical transmission in the cat's visual cortex. *Exp. Brain Res.* 69: 407–16

Harris, E. W., Ganong, A. H., Cotman, C.

W. 1984. Long-term potentiation in the hippocampus involves activation of N-methyl-D-aspartate receptors. *Brain Res.* 323: 132–37

Hebb, D. O. 1949. *Organization of Behavior*. New York: Wiley

Henley, J. M., Oswald, R. E. 1988. Characterization and regional distribution of glutamatergic and cholinergic ligand binding sites in goldfish brain. *J. Neurosci.* 8: 2101–7

Hicks, T. P. 1987. Excitatory amino acid pathways in cerebral cortex. In *Excitatory Amino Acid Transmission*, ed. T. P. Hick, D. Lodge, H. McLennan, pp. 373–80. New York: Liss

Hirsch, H. V. B., Spinelli, D. N. 1970. Visual experience modifies distribution of horizontally and vertically oriented receptive fields in cats. *Science* 168: 869–71

Hubel, D. H., Wiesel, T. N. 1965. Binocular interaction in striate cortex of kittens reared with artificial squint. *J. Neurophysiol.* 28: 1041–59

Hubel, D. H., Wiesel, T. N., LeVay, S. 1977. Plasticity of ocular dominance columns in monkey striate cortex. *Philos. Trans. R. Soc. London Ser. B* 278: 377–409

Ide, C. R., Fraser, S. E., Meyer, R. L. 1983. Eye dominance columns formed by an isogenic double nasal frog eye. *Science* 221: 293–95

Kalil, R. E., Dubin, M. W., Scott, G., Stark, L. A. 1986. Elimination of action potentials blocks the structural development of retino-geniculate synapses. *Nature* 323: 156–58

Kasamatsu, T., Pettigrew, J. D., Ary, M. 1979. Restoration of visual cortical plasticity by local microperfusion of norepinephrine. *J. Comp. Neurol.* 185: 163–82

Katz, L. C., Constantine-Paton, M. 1988. Relationship between segregated afferents and post-synaptic neurons in the optic tectum of three-eyed frogs. *J. Neurosci.* 8: 3160–80

Keating, M. J. 1974. The role of visual function in the patterning of binocular visual connections. *Br. Med. Bull.* 30: 145–51

Keating, M. J. 1975. The time course of experience dependent synaptic switching of visual connections in *Xenopus laevis*. *Proc. R. Soc. London Ser. B* 189: 603–10

Keating, M. J., Feldman, J. 1975. Visual deprivation and intertectal neuronal connections in *Xenopus laevis*. *Proc. R. Soc. London Ser. B* 191: 467–74

Kleinschmidt, A., Bear, M. F., Singer, W. 1987. Blockade of NMDA receptors disrupts experience-dependent plasticity of kitten striate cortex. *Science* 238: 355–58

Komatsu, Y., Fujii, K., Maeda, J., Sakaguchi, H., Toyama, K. 1988. Long-term potentiation of synaptic transmission in kitten visual cortex. *J. Neurophysiol.* 59: 124–41

Komatsu, Y., Toyama, K., Maeda, J., Sakaguchi, H. 1981. Long-term potentiation investigated in a slice preparation of the visual cortex of young kittens. *Neurosci. Lett.* 26: 269–72

Kratz, K. E., Spear, P. D., Smith, D. C. 1976. Postcritical period reversal of effects of monocular deprivation on striate cortex cells in the cat. *J. Neurophysiol.* 39: 501–11

Langdon, R. B., Freeman, J. A. 1986. Antagonists of glutaminergic neurotransmission block retino-tectal transmission in goldfish. *Brain Res.* 398: 169–74

Langdon, R. B., Freeman, J. A. 1987. Pharmacology of retinotectal transmission in the goldfish: Effects of nicotinic ligands, strychnine, and kynurenic acid. *J. Neurosci.* 7: 760–73

Law, M. I., Constantine-Paton, M. 1980. Right and left eye bands in frogs with unilateral tectal ablations. *Proc. Natl. Acad. Sci. USA* 77: 2314–18

Law, M. I., Constantine-Paton, M. 1981. Anatomy and physiology of experimentally produced striped tecta. *J. Neurosci.* 1: 741–59

LeVay, S., Stryker, M. D., Shatz, C. J. 1978. Ocular dominance columns and their development in layer IV of the cat's visual cortex. *J. Comp. Neurol.* 179: 223–24

LeVay, S., Wiesel, T. N., Hubel, D. H. 1980. The development of ocular dominance columns in normal and visually deprived monkeys. *J. Comp. Neurol.* 191: 1–51

Leventhal, A. G., Hirsch, H. V. B. 1977. Effects of early experience upon orientation sensitivity and the binocularity of neurons in the cat's visual cortex. *Proc. Natl. Acad. Sci. USA* 74: 1272–76

Levine, R., Jacobson, M. 1975. Discontinuous mapping of retina on to tectum innervated by both eyes. *Brain Res.* 98: 172–76

Levy, W. B., Desmond, N. L. 1985. The rules of elemental synaptic plasticity. In *Synaptic Modification, Neuron Selectivity and Nervous System Organization*, ed. W. B. Levy, J. A. Anderson, S. Lehmkuhle, pp. 105–21. Hillsdale, NJ: Erlbaum

Lewis, D., Teyler, T. J. 1986. Long-term potentiation in the goldfish optic tectum. *Brain Res.* 375: 246–50

Lynch, G., Baudry, M. 1984. The biochemistry of memory: A new and specific hypothesis. *Science* 224: 1057–63

Mastronarde, D. N. 1983a. Correlated firing of cat retinal ganglion cells. I. Spontaneously active inputs to X- and Y-cells. *J. Neurophysiol.* 49: 303–24

Mastronarde, D. N. 1983b. Interactions between ganglion cells in cat retina. *J. Neurophysiol.* 49: 350–65

Mayer, M. L., Westbrook, G. L., Guthrie, P. B. 1984. Voltage-dependent block by $Mg^{++}$ of NMDA responses in spinal cord neurones. *Nature* 309: 261–63

McDonald, J. W., Cline, H. T., Constantine-Paton, M., Maragos, W. E., Johnston, M. V., Young, A. B. 1989. Quantitative autoradiographic localization of NMDA, quisqualate and PCP receptors in the frog tectum. *Brain Res.* 482: 155–59

McLoon, S. C. 1985. Evidence for shifting connections during the development of the chick retinotectal projection. *J. Neurosci.* 5: 2570–80

Meyer, R. L. 1982. Tetrodotoxin blocks the formation of ocular dominance columns in goldfish. *Science* 218: 589–91

Meyer, R. L. 1983. Tetrodotoxin inhibits the formation of refined retinotopography in goldfish. *Dev. Brain Res.* 6: 293–98

Miller, K. D., Chapman, B., Stryker, M. P. 1989. Visual responses in adult cat visual cortex depend on N-methyl-D-aspartate receptors. *Proc. Natl. Acad. Sci. USA* 86: 5183–87

Movshon, J. A., Van Sluyters, R. C. 1981. Visual neural development. *Annu. Rev. Psychol.* 32: 477–522

Mower, G. D., Caplan, C. J., Christin, W. G., Duffy, F. 1985. Dark rearing prolongs physiological but not anatomical plasticity of the cat visual cortex. *J. Comp. Neurol.* 235: 448–66

Nicoll, R. A., Kauer, J. A., Malenka, R. C. 1988. The current excitement in long-term potentiation. *Neuron* 1: 97–103

Nowak, L., Bregestovski, P., Ascher, P., Herbet, A., Prochiantz, A. 1984. Magnesium gates glutamate-activated channels in mouse central neurones. *Nature* 307: 462–65

Olson, C. R., Freeman, R. D. 1975. Progressive changes in kitten striate cortex during monocular vision. *J. Neurophysiol.* 38: 26–32

Paradiso, M. A., Bear, M. F., Daniels, J. D. 1983. Effects of intracortical infusion of 6-hydroxydopamine on the response of kitten visual cortex to monocular deprivation. *Exp. Brain Res.* 51: 413–22

Perkins, A. T., Teyler, T. J. 1988. A critical period for long-term potentiation in the developing rat visual cortex. *Brain Res.* 439: 222–29

Rakic, P. 1976. Prenatal genesis of connections subserving ocular dominance in the rhesus monkey. *Nature* 261: 589–91

Rakic, P. 1977. Prenatal development of the visual system in rhesus monkey. *Philos. Trans. R. Soc. London Ser. B* 278: 245–60

Rauschecker, J. P., Hahn, S. 1987. Keta-mine-xylazine anaesthesia blocks consolidation of ocular dominance changes in kitten visual cortex. *Nature* 326: 183–85

Rauschecker, J. P., Singer, W. 1981. The effects of early visual experience on the cat's visual cortex and their possible explanation by Hebb synapses. *J. Physiol.* 310: 215–39

Reh, T., Constantine-Paton, M. 1983. Retinal ganglion cells change their projection sites during larval development of *Rana pipiens. J. Neurosci.* 4: 442–57

Reh, T. A., Constantine-Paton, M. 1985. Eye-specific segregation requires neural activity in three-eyed *Rana pipiens. J. Neurosci.* 5: 1132–43

Reiter, H. O., Stryker, M. P. 1988. Neural plasticity without post-synaptic action potentials: Less-active inputs become dominant when kitten visual cortical cells are pharmacologically inhibited. *Proc. Natl. Acad. Sci. USA* 85: 3623–27

Reiter, H. O., Waitzman, D. M., Stryker, M. P. 1986. Cortical activity blockade prevents ocular dominance plasticity in kitten visual cortex. *Exp. Brain Res.* 65: 182–88

Scherer, W. S., Udin, S. B. 1988. The role of NMDA receptors in the development of binocular maps in *Xenopus* tectum. *Proc. Soc. Neurosci.* 14: 272.16

Schmidt, J. T. 1987. Increased potentiation of post-synaptic responses correlates with sensitive period during optic nerve regeneration in goldfish. *Proc. Soc. Neurosci.* 13: 241

Schmidt, J. T. 1988. NMDA blockers prevent both retinotopic sharpening and LTP in regenerating optic pathway of goldfish. *Proc. Soc. Neurosci.* 14: 675

Schmidt, J. T., Edwards, D. L. 1983. Activity sharpens the map during the regeneration of the retinotectal projection in goldfish. *Brain Res.* 209: 29–39

Schmidt, J. T., Eisele, L. E. 1985. Stroboscopic illumination and dark rearing block the sharpening of the regenerated retinotectal map in goldfish. *Neuroscience* 14: 535–46

Schmidt, J. T., Tieman, S. B. 1985. Eye-specific segregation of optic afferents in mammals, fish and frogs: The role of activity. *Cell Mol. Neurobiol.* 5: 5–34

Shatz, C. J., Sretavan, D. W. 1986. Interactions between retinal ganglion cells during the development of the mammalian visual system. *Annu. Rev. Neurosci.* 9: 171–207

Shatz, C. J., Styker, M. P. 1978. Ocular dominance in layer IV of the cats visual cortex and the effects of monocular deprivation. *J. Physiol.* 281: 267–83

Shatz, C. J., Stryker, M. P. 1988. Prenatal tetrodotoxin infusion blocks segregation of retinogeniculate afferents. *Science* 242: 87–89

Shaw, C., Cynader, M. 1984. Disruption of cortical activity prevents ocular dominance changes in monocularly deprived kittens. *Nature* 308: 731–34

Sherman, S. M., Spear, P. D. 1982. Organization of visual pathways in normal and visually deprived cats. *Physiol. Rev.* 62: 740–855

Singer, W. 1982. Central-core control of developmental plasticity in the kitten visual cortex. I. Diencephalic lesions. *Exp. Brain Res.* 47: 209–22

Singer, W., Rauschecker, J. P. 1982. Central core control of developmental plasticity in the kitten visual cortex. II. Electrical activation of mesencephalic and diencephalic projection. *Exp. Brain Res.* 47: 223–33

Singer, W., Tretter, F., Yinon, U. 1982. Central gating of developmental plasticity in kitten visual cortex. *J. Physiol.* 324: 221

Stent, G. S. 1973. A physiological mechanism for Hebb's postulate of learning. *Proc. Natl. Acad. Sci. USA* 70: 997–1001

Sretavan, D. W., Shatz, C. J., Stryker, M. P. 1988. Modification of retinal ganglion cell axon morphology by prenatal infusion of tetrodotoxin. *Nature* 336: 468–71

Stryker, M. P. 1986. The role of neural activity in rearranging connections in the central nervous system. In *The Biology of Change in Otolaryngology*, ed. R. W. Ruben, et al, pp. 211–24. Amsterdam: Elsevier

Stryker, M. P., Harris, W. A. 1986. Binocular impulse blockade prevents the formation of ocular dominance columns in cat visual cortex. *J. Neurosci.* 6: 2117–33

Stryker, M. P., Strickland, S. L. 1984. Physiological segregation of ocular dominance columns depends on patterns of afferent activity. *Invest. Opthalmol. Suppl.* 25: 278

Sur, M. 1988. Development and plasticity of retinal X and Y axon terminations in the cat's lateral geniculate nucleus. *Brain Behav. Evol.* 31: 243–51

Swindale, N. V. 1988. Role of visual experience in promoting segregation of eye dominance patches in the visual cortex of the cat. *J. Comp. Neurol.* 267: 472–88

Teyler, T. J., DiScenna, P. 1987. Long-term potentiation. *Annu. Rev. Neurosci.* 10: 131–61

Tsumoto, T., Hagihara, K., Sato, H., Hata, Y. 1987. NMDA receptors in the visual cortex of young kittens are more effective than those of adult cats. *Nature* 327: 513–14

Udin, S. B. 1983. Abnormal visual input

## 154 CONSTANTINE-PATON ET AL

leads to development of abnormal axon trajectories in frogs. *Nature* 301: 336–38

Udin, S. B. 1985. The role of visual experience in the formation of binocular projections in frogs. *Cell. Mol. Neurobiol.* 5: 85–102

Udin, S. B., Fawcett, J. W. 1988. Formation of topographic maps. *Annu. Rev. Neurosci.* 11: 289–327

Udin, S. B., Keating, M. J. 1981. Plasticity in a central nervous pathway in *Xenopus*: Anatomical changes in the isthmo-tectal projection after larval eye rotation. *J. Comp. Neurol.* 203: 575–94

Wiesel, T. N., Hubel, D. H. 1963. Single cell responses in striate cortex of kittens deprived of vision in one eye. *J. Neurophysiol.* 26: 1003–7

Wigstrom, H., Gustafsson, B. 1985. On long-lasting potentiation in the hippocampus: A proposed mechanism for its dependence on coincident pre- and post-synaptic activity. *Acta Physiol. Scand.* 123: 519–22

Yen, L.-H., Constantine-Paton, M. 1988. EM analysis of single retinal ganglion cell terminals in developing *Rana pipiens*. *Proc. Soc. Neurosci.* 14: 674

*Annu. Rev. Neurosci. 1990. 13:155–69*

# EARLY EVENTS IN THE EMBRYOGENESIS OF THE VERTEBRATE VISUAL SYSTEM:
## Cellular Determination and Pathfinding

*William A. Harris and Christine E. Holt*

Department of Biology and Center for Molecular Genetics, University of California, San Diego, La Jolla, California 92093

## Introduction

A sheet of undifferentiated epithelium folds into a tube with anterior swellings from which the vertebrate eyes and brain will arise. The neuro-epithelial cells divide and give rise to mature neurons with a stunning variety of shapes, chemistries, and functions. The neurons grow to inter-connect in a complex but highly ordered way, so that a light pattern falling on the retina can be processed into a useful mental image of the world with the speed and precision that far surpasses today's most advanced computers. How is this accomplished? Our understanding of the biological basis of this development is almost hopelessly sketchy. Below we outline some of the progress that has been made on two of the earliest processes that are part of this scenario: the acquisition of neuronal identity and axonal pathfinding.

## Neuronal Identity

In the retinas of most vertebrates, there are laminar and cell-type gradients of histogenesis. Ganglion cells are the first to be born, followed quickly by cone and horizontal cells, while rods are the last (Sidman 1961, Morest 1970, Kahn 1974, Hinds & Hinds 1983, Holt et al 1988). Retinal his-togenesis takes weeks in mammals and therefore ganglion cells are usually

155

0147–006X/90/0301–0155$02.00

all postmitotic before the first rod cells are born. In *Xenopus laevis*, the South African clawed-toed frog, there is tremendous temporal compression in histogenesis, so cells of all retinal layers are born nearly simultaneously. The central sector of the embryonic retina completes histogenesis within a 24 hour period (Jacobson 1968, Holt et al 1988).

Lineage tracing studies with retroviral markers in the rat (Turner & Cepko 1987) and intracellular injections in the frog (Holt et al 1988, Wetts & Fraser 1988) have shown that single progenitor cells give rise to radial clones of retinal neurons of many different types in all three retinal layers. Clones containing marked ganglion cells in the mouse must be generated early in retinal histogenesis, and such clones tend to be large (Turner et al 1988), whereas clones containing ganglion cells in the *Xenopus* can occur late in retinal histogenesis and may be very small with only two cells (Holt et al 1988). This difference is a direct consequence of the differing duration of histogenesis in these two species. In both systems, however, it is clear that the last division of a precursor can simultaneously give rise to cells of two different layers or types. The apparently random assortment of cell types in small clones in the frog is a persuasive argument that cell fate is determined at or after the final mitosis (Holt et al 1988).

Cell type determination in the retina could thus be the result of an internal stochastic mechanism with multiple states, in which case the internal decision, once made in a postmitotic cell, would drive the cell to its proper position in the retina. An alternative possibility is that a cell, soon after its final mitosis, is exposed to a specific array of microenvironmental cues, such as signaling molecules on neighboring cells or local gradients of extracellular materials. These signals instruct or limit the nascent cell to choose a particular fate. The latter model could explain the loose relationship between birthdate and determination in the retina, for a cell born at a particular time may be exposed to particular cues generated in part by the cells that have been born previously. This then would allow birthdate to be a "soft" predictor rather than a "hard" determiner of cell fate. One could imagine that a cell born after a number of ganglion cells had already been generated might be influenced by the fact that the only differentiating cells nearby are ganglion cells. Ganglion cells could reduce the probability of other cells' choosing a similar fate by an inhibitory cellular interaction. Alternatively, postmitotic ganglion cells might instruct the newly generated neurons to become, for example, horizontal cells.

Fish and amphibian retinas add new cells from the ciliary margin throughout most of the life of the animal (Hollyfield 1968, Straznicky & Gaze 1971, Johns 1977, Beach & Jacobson 1979). Clones of cells generated from the ciliary margin give rise to cells of all retinal types, thus indicating again that there are no determined mother cells in this growing zone (Wetts et al 1988). The cell types spun off by terminal divisions of ciliary margin cells can be altered experimentally. Reh & Tully (1986) have shown that a neurotoxin (6-OH dopamine), which kills the central retinal dopaminergic amacrine cells, leads temporarily to an increased production of this cell type from the ciliary margin. Similarly, killing the central glutaminergically stimulated cells with kainic acid leads to an increased production from the ciliary margin of cells to those layers most depleted by the toxin (Reh 1987). These results argue that the ciliary margin precursors, themselves uncommitted, throw off daughters that differentiate along particular pathways in response to signals from the more central retina (likely the locally differentiated retina that opposes the ciliary margin).

Reh and colleagues (Reh & Cljavin 1989 and personal communication) have taken this further with *in vitro* studies of the rat retina. When newly postmitotic cells are dissociated and forced to differentiate in culture, at embryonic day 15 (E15), a time when ganglion cells but not rods are usually born, ganglion cells but no rods differentiate. Similarly, newly postmitotic cells from rat retinas dissociated on postnatal day 3 (P3) produce rods but not ganglion cells. Interestingly, when dissociated P3 and E15 cells are mixed in culture, some newly born E15 cells differentiate into rods. This indicates that neuronal determination of postmitotic cells is indeed both flexible at a cellular level and responsive to extracellular cues that may be spatiotemporally regulated and at least partially due to cell–cell interactions between committed and uncommitted cells.

Some rod precursors in fish, and probably amphibians also, appear committed to give rise only to rods (Fernald 1989). The retina grows not only by addition of cells at the ciliary margin, as mentioned above, but also by stretching (Johns 1977). The stretching decreases the density of all cell types in the central retina, except the rods, which stay at a constant density. Therefore clearly rods, and only rods, must be added to the central retina throughout the life of the animal. In the outer nuclear layer, just inside the ciliary margin of fish there appears to be a second germinal zone of rod precursors (R. D. Fernald and J. Scholes, personal communication). Many of these cells divide as precursors and migrate peripherally. Some give off rods to the retina soon after they themselves are added to the outer nuclear layer. Some continue to divide at a low level throughout the life of the animal, each depositing rods at a particular site, so that new rods are intercalated evenly over the entire central retina as it stretches. In this second germinal zone, the rod precursors, the progeny of pluripotent marginal zone cells, may themselves be pluripotent. In ouabain-treated retinas, only neuroblasts are spared, and a zone of retina is regenerated from mitotic cells in the outer nuclear layer just central to the ciliary margin (Kurz-Isler & Wolburg 1982). These mitotic cells may have arisen from dedifferentiated rods (Kurz-Isler & Wolburg 1982), or they may be the rod precursors, since they are located at the site of the main density of this population (R. D. Fernald and J. Scholes, personal communication). If they are rod precursors, why do they produce only rods in the normal retina? It may be that only the extracellular signals for rod histogenesis exist in the central retinas of older fish. This idea fits with the basic notion presented here of uncommitted stem cells and of determination by local interactions.

Amphibians have the ability to regenerate their entire neural retina from the pigment epithelium (Stone 1950, Loposhov & Sologub 1972). This means that pigment epithelial cells must have the ability to divide and transdifferentiate under specific conditions into other cell types. Pigment epithelial cells grown in tissue culture can also transdifferentiate (Okada 1980, Reh et al 1987). When grown on laminin substrates, pigment epithelial cells lose their pigment and give rise to neuronal colonies, including retinal ganglion cells (Reh et al 1987). The inner limiting membrane of the retina is a laminin-rich membrane and perhaps holds the cues that lead to retinal ganglion cell differentiation in vivo. When grown on collagen, however, pigment epithelial cells usually transdifferentiate into glial cells (D. Sakaguchi, personal communication). In still other culture conditions, pigment epithelial cells may transdifferentiate into lens cells (Okada 1980). Thus, clearly as yet poorly defined extracellular factors can influence the fate of these retinal precursors. Whether these same factors are used during the development of the retina, or just during its regeneration, is unknown.

Cell type determination has also been studied by retroviral lineage tracers in the optic tectum of the chick (Gray et al 1988) and the cortex of the mouse (Luskin et al 1988) and rat (Walsh & Cepko 1988). Studies from these systems and from the retina verify the classical idea that neuroblasts, dividing at the ventricular zone, spin off a collection of postmitotic cells that migrate radially and form a pillar of clonally related but functionally distinct cells (Sauer 1935, Morest 1970, Rakic 1972, 1974). Although the radial alignment of cells in the chick optic tectum is spectacular (Gray et al 1988), there is no indication that such a pillar of related cells has any functional significance other than that the cells come to lie in radial register by a developmental restriction not strictly tied to function. The best evidence that this pattern has no functional significance is simply that the clonal pillars in the cortex are by no means perfect. Occasionally cells of a clone may be displaced horizontally more than the diameter of a functional column (Walsh & Cepko 1988). In the tectum, the cortex, and the retina, single precursors give rise to a heterogeneous and radial array of offspring, a finding that suggests very strongly that fundamentally similar mechanisms of cell type determination are at work in these different parts of the

CNS. Yet experiments on cellular determination in the cortex challenge this simple notion. We next discuss these experiments and then offer some models that may resolve the apparently conflicting findings.

Layers of cells in the cortex are much more clearly segregated by birth-date than cells in the retina (Rakic 1974). Layer VI cells are born first, and then, in order, layer V, IV, III, II and finally layer I cells are born. This is the classical inside-out pattern of histogenesis at its clearest (Angevine & Sidman 1961, Rakic 1974). Obviously then, cellular birthdate cannot be automatically ruled out as a mechanism of determination. In fact, three kinds of experiments support the notion that cell-type determination may be fixed at birth, and may be independent of local cell interactions in the cortical plate. First, the *reeler* mutant mouse has a defect in postmitotic neuronal migration such that the cortex and several other structures of the CNS are disorganized radially (Caviness 1976). The normally inside-out pattern of cortical histogenesis is reversed in *reeler* mice. Thus the earliest born cells are in layer I and the latest in layer VI, and this gradient is superimposed on a more helter-skelter arrangement of cell type than is common (Caviness 1982). Despite this severe disruption of the lamination of the cortex, all the cell types that have been looked for are present, though in abnormal laminae, and cortical cells have normal receptive fields, thus indicating normal local and long-range connections (Dräger 1981, Lemmon & Pearlman 1981). The interpretation generally offered to explain the normal cellular composition of the reeler cortex is that cell type must be determined before migration, possibly before birth, so that although a cell's route of migration and placement in the cortex may be perturbed, the cell, wherever it arrives, is capable of differentiating according to its programmed fate.

Studies in which methylazoxymethanol (MAM), a toxin for dividing cells, has been delivered to embryos during the histogenesis of particular cortical layers also point to the early determination of cell fate in the cortex (Jones et al 1982). For example, treating pregnant rats with MAM when their fetuses are generating layers II, III, and IV in the cortex leads to rat pups born without these layers. These studies suggest that cell-type determination is occurring by the birth of cells destined for the cortical plate, or that temporal factors govern these decisions. In addition, unlike the regulation for cell-type loss that occurs in the retina, the dividing ventricular zone of the cortex does not compensate for the absence of a particular cell type. This may be, however, simply because no cortical cells are born after the MAM effect wears off.

Finally, McConnell (1988) showed that when premigratory cortical cells in the ferret with layer V birthdates are transplanted to the proliferative zone of older animals such that these cells should end up in upper layers

(II and III), most cells simply refuse to migrate. Some do migrate, however, some to the layers predicted by their birthdates, others to layers II and III. This is a complex result, open to many interpretations. The one favored by McConnell is that cells at birth are largely committed to certain migratory routes and fates. If such a cell, so determined is in a strange environment, it may simply not migrate. Alternatively, many of the cells dissociated by McConnell may have been stem cells forced to differentiate before they were naturally ready, therefore, they could simply be incompetent to migrate. The heterochronically transplanted cells, when they do reach the cortex, may be layer II/III or layer V cells. Why only some cells are "reprogrammed" is unclear. It may be that, at the time of dissociation, some nascent cells are older (i.e. they have been post-mitotic longer) than others, and therefore are more committed to a particular fate, whereas the younger cells are more plastic.

In sum, cell fate in the retina seems to be determined late (probably postmitotically), appears to be dependent on cell birthdate but only insofar as birthdate correlates with a changing microenvironment for a postmitotic cell, and is clearly affected by local cellular interactions. Lineage is not a mechanism of determination. In the cortex, however, neuronal fate seems to be determined early (perhaps premitotically) and be highly dependent on cell birthdate if not lineage. There is little evidence of regulation or involvement of cellular interactions in cell determination. Yet in both systems there is evidence for an ependymal precursor giving rise to a radial pillar of assorted progeny, suggesting a fundamentally similar mechanism. Can we reconcile the apparent retinal-cortical discrepancies? There are at least some possibilities.

One possibility is that cell fate is determined by exactly the same kinds of mechanisms (i.e. spatially and temporally regulated extracellular determiner substances) but that it happens later in the retina (perhaps after the final division of a cell) and earlier in the cortex (e.g. before the final division of a cell). As a variation on this notion, we might remember that restriction of cell fate is gradual, some restrictions occurring well before neuroblasts give rise to their final products, some occurring later. Cells first become committed to a neural fate during gastrulation (Spemann 1918). In the neural plate, cells of different regions become committed to give rise to certain brain structures, like the retina, the cortex, and so on (Mangold 1933, Källen 1958, Corner 1964). In the retina, most choices of retinal neuron type are still open to a postmitotic cell, but in the cortex a postmitotic cell must choose among a more restricted set of fates. A third possibility is simply that cell type is determined at the same time in the life history of the cell in both systems (just around the birth of the cell), and that what influences cell fate is the local microenvironment around the

nascent cell (i.e. near the ependymal zone in the retina and in the proliferative zone in the cortex). McLoon & Barnes (1989) developed an
antibody to a ganglion cell–specific marker and used it to show that
ganglion cells in the chick express this marker immediately after their final
mitosis, well before the cells detach from the ependymal layer and come
to reside in the ganglion cell layer. Newly born migratory cells in the proliferative zone of the cortex have long processes that may sample the
more distal microenvironment (e.g. in the intermediate zone), where the
earlier born cells are still migrating. Thus in both the cortex and the retina,
newly postmitotic cells may interact only locally with more differentiated
cells to derive a cellular fate.

The steps in cellular commitment in the visual CNS are likely to be
many, and to be based on signals that begin with induction of the neural
plate and end with cell type induction of newly postmitotic neurons. This
scheme is reminiscent of what is now considered a reasonably hypothesis
for the generation of neural crest derivatives (Anderson 1989). Although
there is pluripotency among nascent crest cells, such cells may soon become
restricted by local environmental factors into subpopulations with more
restricted potencies (Anderson 1989). This scheme is also reminiscent of
cellular determination in the *Drosophila* retina (Tomlinson & Ready 1987).
Here cellular fate is assigned to a postmitotic cell based on what its
immediate neighbors are, i.e. where and when it becomes integrated into
the growing neurocrystalline array. The same types of signals and receptors
that are now being described in the *Drosophila* retina (Basler & Hafen
1988) may have their homologs in the vertebrate CNS and may be used
in various combinations throughout neurogenesis, from the formation of
the neural plate to the final determination of neuronal cell type.

## Pathfinding

Cells of the visual pathway of the CNS, once born and committed to
particular fates, must begin to send out processes, axons, and dendrites,
that will become responsible for the integrative functioning of the visual
system (Hinds & Hinds 1974, Holt 1989). Retinal ganglion cells, in which
this development has been most extensively studied, elaborate lengthy
axons, which exit the eye, enter the brain at the ventral diencephalon, and
grow up the optic tract to innervate their synaptic targets in the thalamus
or optic tectum. The first axons to grow along the primordial visual
pathway do so in a directed fashion following a stereotyped and error-free
course (Shatz & Kliot 1982, Thanos & Bonhoeffer 1983, Holt & Harris
1983, Godemont et al 1987, Stuermer 1988). Several recent studies have
yielded some new insights into this remarkable navigational feat.

Time-lapse video recordings of single growth cones traveling along the optic tract in *Xenopus* have shown that axons grow at a fairly constant rate without pausing or retracting until they reach the optic tectum, where they slow down and begin to branch (Harris et al 1987). Growth cones without cell bodies (following eye removal) continue to grow, to navigate correctly, and to respond to the target appropriately, thus indicating that local interactions of the growth cone are responsible for guidance, and that the soma does not need to be continuously consulted during pathfinding or axon extension. This implies that materials in the growth cone are sufficient for interpreting the pathway cues and responding appropriately to them.

Sometimes, and this has been noted in both the cat and the *Xenopus* embryo, a small fraction of ganglion cells send out more than one axon (Ramoa et al 1987, Holt 1989). In *Xenopus*, two axons from a single cell can travel out of the ventral fissure, along the optic stalk, and into the optic tract. Interestingly, the two axons from the same cell (or two axons from neighboring cells) show no indication of fasciculating, or even traveling very close to each other, in the retina, the optic nerve, or the optic tract, thus indicating that at these early stages in the frog there may not be much topographic order among the fibers (Holt 1989). This result also suggests that growth cones track to their targets independently of each other. Fasciculation of fibers with the same target address is not the general mechanism of guidance in the embryo, and their is no evidence for a subset of pioneering axons among these embryonic axons with which other embryonic axons fasciculate. In fact, when the first fibers that grow out of the retina (from the dorsal half) are delayed, and the ventral fibers are forced therefore to pioneer the pathway to the tectum, they show no signs of impaired navigation. Clearly then the dorsal pioneers are not critical for, or even likely to be involved in, the pathfinding of the later growing ventral axons (Holt 1984). The main conclusion from these studies is that each growth cone acts as an independent individual as it grows to the target.

Growth cones in a number of systems have been noted to display particularly complex morphologies at decision points, i.e. where sharp turns are made and where different tracts may be followed (Raper et al 1983, Tosney & Landmesser 1985, Bovolenta & Mason 1987). The morphology of *Xenopus* retinal ganglion cell growth cones are most complex at the optic nerve head. Here the axon makes an abrupt change in direction and begins to dive through the retina (Holt 1989) instead of continuing to grow along its surface. The complexity of the growth cone suggests that the ganglion cell axon is not just passively funneled into the optic nerve but makes an active choice at this point. The growth cone becomes simple again in the optic nerve in both frogs (Holt 1989) and mice (Bovolenta & Mason 1987), and then more complex when the growth cone reaches the junction between the optic nerve and the ventral thalamus. At this point in fish, these axons encounter a different type of glia, slow down, and rearrange themselves into a new topographic order (Maggs & Scholes 1986). In *Xenopus*, in which almost all the early axons travel contralaterally, the growth cones remain equally complex through the chiasm and up the contralateral optic tract. In the mouse, in which the chiasm represents a point of choice, many axons make U-turns and head to the ipsilateral tract; here the growth cones are particularly complex (Bovolenta & Mason 1987). What causes some axons to make abrupt turns at the chiasm and travel to the ipsilateral pathway is not clear. The fraction of axons making this choice, however, is clearly affected in mammalian mutants that reduce pigmentation in the pigment epithelium (Lund 1965, Guillery 1969, LaVail et al 1978). In the embryo, there is a clear pattern of pigmentation along the optic nerve, extending to the midline, and fibers seem to restrict themselves to areas of the nerve furthest from the pigment, a finding that suggests that optic axons are somehow pigmentophobic (Silver & Sapiro 1981, Webster et al 1988). The pigment also is a $Ca^{2+}$ ion chelator, so one hypothesis (Dräger 1985) is that pigment regulates local extracellular $Ca^{2+}$ concentration, which in turn may affect growth cone mobility and turning.

Growth cones are clearly dynamic sensory structures capable of responding to guidance cues in their environment (Letourneau 182, Bray 1982). Where are the directional signals located and how are they detected by the growth cone? One likely possibility involves surface receptor-mediated mechanisms. Another is that small molecules pass directly from the interiors of cells in the pathway to growth cones via transiently formed gap junctions, as has been suggested but not disproved in the leg of the grasshopper embryo. This second possibility has been directly examined by intracellular dye injection of early retinal ganglion cells. Lucifer Yellow is small enough (MW 457) to pass through gap junctions (Stewart 1978), thus cells that are dye-coupled after a single intracellular injection can transmit small molecules among each other (see Taghert et al 1982). When Lucifer Yellow was injected into retinal ganglion cells that had axons at varying stages *en route* to the tectum no instances were found of dye passing from the growth cone tip to adjacent cells (Holt 1989). This finding argues that growth cones use and respond to extracellular cues via receptor-mediated interactions.

In vitro experiments show that growth cones need to advance on a substrate and that not all substrates are equally effective at promoting or initiating outgrowth. Laminin (Rogers et al 1983, Manthorpe et al 1983, Smalheiser et al 1984) and laminin conjugated with a heparin sulfate proteoglycan (Chiu et al 1986) have been shown to be excellent substrates for retinal neurite outgrowth in culture. They are components of natural basement membranes. Studies in situ however, have shown that retinal growth cones in the brain rarely make contact with the basement membrane that bounds the surface of the brain. Rather, they are in contact with the endfeet of embryonic radial glial cells (Silver & Rutishauser 1984, Bork et al 1987). At early stages of chick retinal axon outgrowth, however, laminin may well be present on the endfeet of glial cells (Cohen et al 1987). Neural cell adhesion molecule (NCAM) is also on the endfeet of embryonic glial cells, and antibody perturbation studies of NCAM-mediated adhesion result in distortions of growth cone interactions with neuroepithelial cells leading to disorganized fiber tracts (Silver & Rutishauser 1984). Several laboratories are actively pursuing the molecular basis of glial-promoted neurite outgrowth. Astrocytic glial cells promote retinal neurite extension in vitro (Fallon 1985). Antibody perturbation studies have shown that astrocytes have a variety of substances on their surfaces that promote neurite outgrowth of retinal cells. N-cadherin, NCAM, and integrins are among the identified molecules that promote retinal growth cone extension on astrocytes in vitro (Neugenbauer et al 1988). Other, as yet unidentified, molecules are likely involved in glial endfeet (Sakaguchi et al 1989).

What the instructional cues are that guide growth cones to their central targets remains an open question. That any of the molecules mentioned above from the in vitro studies are responsible for long-range guidance seems unlikely, for none is known to be distributed on the neuroepithelium in a pattern that would lead optic fibers to the tectum. Rather, these may be growth-promoting factors present on the endfeet of all astrocytic glia in the young brain. Recent in vitro studies, however, suggest that similar types of glial cells from different parts of the brain have differing capacities to promote the differentiation and survival of neurons from different brain regions (Chamak et al 1987). Glial cells from target regions of the dissociated neurons are particularly effective in promoting their neurite growth (Leith et al 1988). This indicates that there may be regionally specific molecular differences among the glial population in the neuro-epithelium.

Experiments that shift the embryonic eye and stalk anteriorly or posteriorly show that retinal axons can orient to the tectum starting from a variety of brain entry points (Harris 1986). This suggests that the tectum may release a diffusible chemotropic factor. Such factors have been shown to exist in other parts of the nervous system (Lumsden & Davies 1983, Dodd & Jessell 1988). There is reason to believe, however, that the tectum is not the source of such a factor, for if the tectum is removed before retinal fibers begin to grow into the brain, axons still travel along their normal routes until they reach the point at which the operation has been done (Taylor 1987b). At this point, the fibers turn posteriorly and head off to the spinal cord. If, in a similar type of embryonic experiment, the tectum has been displaced anteriorly or posteriorly, the fibers will still travel along their normal pathways until they come very close to their misplaced targets. They will then turn in the appropriate directions and innervate it. Thus the tectum may only have a very short-range influence on retinal axon trajectory (Taylor 1987b). An alternative to chemotaxis for long-range guidance, then, is that retinal axons (and other projection fibers) use local positional cues in the neuroepithelium to orient to their targets. Thus, for example, if there were stable positional markers in the form of antero-posterior and dorsoventral gradients, or sectors of neuroepithelium with distinct combinations of positional markers, it would be possible for growth cones to use these markers to orient to a particular target, as if they were reading locations on a map of the embryonic brain. To test this idea, squares of presumptive optic tract tissue were rotated either clockwise or counterclockwise by about 90° (Harris 1989). When retinal fibers subsequently entered the rotated tissue, they were deflected in accordance with the direction of rotation. Upon exiting the rotated piece of tissue, the axons reoriented to the tectum. This result makes it clear that retinal axons are paying attention to relatively stable local cues and not long-range diffusible ones. This suggests that there are indeed positional markers on the neuroepithelium (perhaps on the endfeet of glial cells) that may be useful in guiding retinal axons and possibly other long-range projections. What these markers might be remains a mystery, but perhaps they are analogous to the highly spatially regulated gene products that give developmentally important positional information to the *Drosophila* embryo (Akam 1987).

## Conclusions

The vertebrate visual system has been and will continue to be a superb system in which to investigate issues of early neurogenesis and pathfinding because of its accessibility and because of the great wealth of neuroanatomical and physiological background in this system. We are beginning to understand the developmental strategies by which neurons become committed to particular fates. We are in a similar position with regard to axonal pathfinding in the CNS. In both cases we have only a few insights into the molecular basis of these early developmental events.

ACKNOWLEDGEMENTS

We would like to thank Tom Reh for his insightful discussions concerning cell fate. Many of his ideas have been incorporated in this manuscript.

W. A. H. and C. E. H. are supported by grants from the NIH and the McKnight Foundation.

## Literature Cited

Akam, M. 1987. The molecular basis for metameric pattern in pattern in the *Drosophila* embryo. *Development* 101: 1–22

Anderson, D. 1989. The neural crest cell lineage problem: Neuropoiesis? *Neuron.* In press

Angevine, J. B. Jr., Sidman, R. L. 1961. Autoradiographic study of cell migration during histogenesis of cerebral cortex in the mouse. *Nature* 192: 766–68

Basler, K., Hafen, E. 1988. *Sevenless* and *Drosophila* eye development: A tyrosine kinase controls cell fate. *Trends Genet.* 4: 74–79

Beach, D. H., Jacobson, M. 1979. Patterns of cell proliferation in the retina of the clawed frog during development. *J. Comp. Neurol.* 183: 603–14

Bork, T., Schabtach, E., Grant, P. 1987. Factors guiding optic fibres in developing *Xenopus* retina. *J. Comp. Neurol.* 264: 147–58

Bovolenta, P., Mason, C. 1987. Growth cone morphology varies with position in the developing mouse visual pathway from retina to first targets. *J. Neurosci.* 7(5): 1447–60

Bray, D. 1982. Filopodial contraction and growth cone guidance. In *Cell Behaviour*, ed. R. Bellairs, A. Curtis, G. Dunn, pp. 299–317. London: Cambridge Univ. Press

Caviness, V. S. Jr. 1976. Patterns of cell and fiber distribution in the neocortex of the reeler mutant mouse. *J. Comp. Neurol.* 170: 435–48

Caviness, V. S. Jr. 1982. Neocortical histogenesis in normal and reeler mice: A developmental study based on [$^3$H]thymidine autoradiography. *Dev. Brain Res.* 4: 293–302

Chamak, B., Fellous, A., Glowinski, J., Prochiantz, A. 1987. MAP2 expression and neuritic outgrowth and branching are coregulated through region-specific neuroastroglial interactions. *J. Neurosci.* 7: 3163–70

Chiu, A. Y., Matthew, W. D., Patterson, P. H. 1986. A monoclonal antibody that blocks the activity of a neurite regeneration-promoting factor: Studies on the binding site and its localization in vivo. *J. Cell Biol.* 103: 1383–98

Cohen, J., Burne, J. F., McKinlay, C., Winter, J. 1987. The role of laminin and the laminin/fibronectin receptor complex in the outgrowth of retinal ganglion cell axons. *Dev. Biol.* 122: 407–18

Corner, M. A. 1964. Localization of capacities for functional development in the neural plate of *Xenopus laevis*. *J. Comp. Neurol.* 123: 243–56

Dodd, J., Jessell, T. M. 1988. Axon guidance and the patterning of neuronal projections in vertebrates. *Science* 242: 692–99

Dräger, U. C. 1981. Observations on the organization of the visual cortex in the reeler mouse. *J. Comp. Neurol.* 201: 555–70

Dräger, U. C. 1985. Calcium binding in pigmented and albino eyes. *Proc. Natl. Acad. Sci. USA* 82: 6716–20

Fallon, J. 1985. Preferential outgrowth of central nervous system neurites on astrocytes and Schwann cells as compared with non-glial cells in vitro. *J. Cell Biol.* 100: 198–207

Fernald, R. D. 1989. Retinal rod neurogenesis. In *Development of the Vertebrate Retina*, ed. B. Finlay, D. Sengelaub. New York: Plenum. In press

Godement, P., Vanselow, J., Thanos, S., Bonhoeffer, F. 1987. A study in developing visual systems with a new method of staining neurones and their processes in fixed tissue. *Development* 101: 697–713

Gray, G. E., Glover, J. C., Majors, J., Sanes, J. R., 1988. Radial arrangement of clonally related cells in the chicken optic tectum: Lineage analysis with a recombinant retrovirus. *Proc. Natl. Acad. Sci. USA.* 85: 7356–60

Guillery, R. W. 1969. An abnormal retinogeniculate projection in Siamese cats. *Brain Res.* 14: 739–41

Harris, W. A. 1986. Homing behaviour of axons in the embryonic vertebrate brain. *Nature* 320: 266–69

Harris, W. A. 1989. Local positional cues in the neuroepithelium guide retinal axons in the embryonic *Xenopus* brain. *Nature* 339: 218–21

Harris, W. A., Holt, C. E., Bonhoeffer, F. 1987. Retinal axons with and without their somata, growing to and arborizing in the tectum of *Xenopus* embryos: A time-lapse video study of single fibres in vivo. *Development* 101: 123–33

Hinds, J. W., Hinds, P. L. 1974. Early ganglion cell differentiation in the mouse retina. An electron microscope analysis utilizing serial sections. *Dev. Biol.* 37: 381–416

Hinds, J. W., Hinds, P. L. 1983. Development of retinal amacrine cells in the mouse embryo: Evidence for two modes of formation. *J. Comp. Neurol.* 213: 1–23

Hollyfield, J. G. 1968. Differential addition of cells to the retina in *Rana pipiens* tadpoles. *Dev. Biol.* 18: 163–79

Holt, C. E. 1984. Does timing of axon outgrowth influence initial retinotectal topography in *Xenopus*? *J. Neurosci.* 4: 1130–52

Holt, C. E. 1989. A single cell analysis of early retinal ganglion cell differentiation in *Xenopus*: From soma to axon tip. *J. Neurosci.* In press

Holt, C. E., Harris, W. A. 1983. Order in the initial retinotectal map in *Xenopus*: A new technique for labelling growing nerve fibres. *Nature* 301: 150–52

Holt, C. E., Bertsch, T. W., Ellis, H. M., Harris, W. A. 1988. Cellular determination in the *Xenopus* retina is independent of lineage and birth date. *Neuron* 1: 15–26

Jacobson, M. 1968. Cessation of DNA synthesis in retinal ganglion cells is correlated with the time of specification of their central connections. *Dev. Biol.* 17: 219–32

Johns, P. R. 1977. Growth of the adult goldfish eye. III. Source of the new retinal cells. *J. Comp. Neurol.* 170: 343–57

Jones, E. G., Valentino, K. L., Fleshman, J. W. Jr. 1982. Adjustment of connectivity in rat neocortex after prenatal destruction of precursor cells of layer II–IV. *Dev. Brain Res.* 2: 425–31

Kahn, A. J. 1974. An autoradiographic analysis of the time of appearance of neurons in the developing chick neural retina. *Dev. Biol.* 38: 30–40

Källén, B. 1958. Studies on the differentiation capacity of neural epithelium cells in chick embryos. *Z. Zellforsch. Mikrosk. Anat. Abt. Histochem.* 47: 469–80

Kurz-Isler, G., Wolburg, H. 1982. Morphological study on the regeneration of the retina in the rainbow trout after ouabain-induced damage: Evidence for dedifferentiation of photoreceptors. *Cell Tissue Res.* 225: 165–78

LaVail, J. H., Nixon, R. A., Sidman, R. L. 1978. Genetic control of retinal ganglion cell projections. *J. Comp. Neurol.* 182: 399–422

Leith, E., McClay, D. R., Lander, J. M. 1988. Neuronal adhesion and neurite outgrowth on glia enriched substrates from target and non-target brain regions. *Neurosci. Abstr.* 14: 581

Lemmon, V., Pearlman, A. L. 1981. Does laminar position determine the receptive field properties of cortical neurons? A study of corticotectal cells in area 17 of the normal mouse and the reeler mutant. *J. Neurosci.* 1: 83–93

Letouneau, P. C. 1982. Nerve fibre growth and extrinsic factors. In *Neuronal Development*, ed. N. C. Spitzer, pp. 213–54. New York: Plenum

Loposhov, G. V., Sologub, A. A. 1972. Artificial metaplasia of pigmented epithelium into retina in tadpoles and adult frogs. *J. Embryol. Exp. Morphol.* 28: 521–47

Lumsden, A. G. S., Davies, A. M. 1983. Earliest sensory nerve fibres are guided to peripheral targets by attractants other than nerve growth factor. *Nature* 306: 786–88

Lund, R. D. 1965. Uncrossed visual pathways of hooded and albino rats. *Science* 149: 1506–7

Luskin, M. B., Pearlman, A. L., Sanes, J. R. 1988. Cell lineage in the cerebral cortex of the mouse studied in vivo and in vitro with a recombinant retrovirus. *Neuron* 1: 635–47

Maggs, A., Scholes, J. 1986. Glial domains and nerve fiber patterns in the fish retinotectal pathway. *J. Neurosci.* 6: 424–38

Mangold, O. 1933. Isolationsversuche zür Analyse der Entwicklung bestimmter Kopforgane. *Naturwissen* 21: 394–97

Manthorpe, M., Engvall, E., Rouslahti, E., Longo, F. M., Davis, G. E., Varon, S. 1983. Laminin promotes neuritic regeneration from cultured peripheral and central neurons. *J. Cell Biol.* 97: 1882–90

McConnell, S. K. 1988. Fates of visual cortical neurons in the ferret after isochronic and heterochronic transplantation. *J. Neurosci.* 8: 945–74

McLoon, S. C., Barnes, R. B. 1989. Early differentiation of retinal ganglion cells: An axonal protein expressed by premigratory and migrating retinal ganglion cells. *J. Neurosci.* 94: 1424–32

Morest, D. K. 1970. The pattern of neurogenesis in the retina of the rat. *Z. Anat. Entwicklungsgesch.* 131: 45–67

Neugebauer, K. M., Tomaselli, K. J., Lilien, J., Reichardt, L. F. 198. N-cadherin, N-CAM, and integrins promote retinal neurite outgrowth on astrocytes in vitro. *J. Cell Biol.* 107: 1177–87

Okada, T. S. 1980. Cellular metaplasia or transdifferentiation as a model for retinal cell differentiation. *Curr. Top. Dev. Biol.* 16: 349–80

Rakic, P. 1972. Mode of cell migration to the superficial layers of fetal monkey neocortex. *J. Comp. Neurol.* 145: 61–84

Rakic, P. 1974. Neurons in the rhesus monkey visual cortex: Systematic relation between time of origin and eventual deposition. *Science* 183: 425–27

Ramoa, A. S., Campbell, A., Shatz, C. J. 1987. Transient morphological features of identified ganglion cells in living fetal and neonatal retina. *Science* 237: 522–25

Ramoa, A. S., Campbell, A., Shatz, C. J. 1988. Dendritic growth and remodeling of cat retinal ganglion cells during fetal and postnatal development. *J. Neurosci.* 8: 4239–61

Raper, J. A., Bastianni, M. J., Goodman, C. S. 1983. Pathfinding by neuronal growth cones in grasshopper embryos. I. Divergent choices made by the growth cones of sibling neurons. *J. Neurosci.* 3: 20–30

Reh, T. A. 1987. Cell-specific regulation of neuronal production in the larval frog retina. *J. Neurosci.* 7: 3317–24

Reh, T. A., Cljavin, I. 1989. Age of differentiation determines neuronal phenotype in rat retinal germinal cells. *J. Neurosci.* In press

Reh, T. A., Nagy, T. 1987. A possible role for the vascular membrane in retinal regeneration in *Rana caterbienna* tadpoles. *Dev. Biol.* 122: 471–82

Reh, T. A., Nagy, T., Gretton, H. 1987. Retinal pigmented epithelial cells included to transdifferentiate to neurons by laminin. *Nature* 330: 68–71

Reh, T. A., Tully, T. 1986. Regulation of tyrosine hydroxylase-containing amacrine cell number in larval frog retina. *Dev. Biol.* 114: 463–89

Rogers, S. L., Letourneau, P. C., Palm, S. L., McCarthy, J., Furcht, L. T. 1983. Neurite extension by peripheral and central nervous system neurons in response to substratum-bound fibronectin and laminin. *Dev. Biol.* 98: 212–20

Sakaguchi, D. S., Moeller, J. F., Coffman, C. R., Gallenson, N., Harris, W. A. 1989. Growth cone interactions with a glial cell line from embryonic *Xenopus* retina. *Dev. Biol.* 134: 158–74

Sauer, F. C. 1935. Mitosis in the neural tube. *J. Comp. Neurol.* 62: 377–405

Shatz, C. J., Kliot, M. 1982. Prenatal misrouting of the retinogeniculate pathway in Siamese cats. *Nature* 300: 525–29

Sidman, R. L. 1961. Histogenesis of the mouse retina studied with thymidine-H³ in the structure of the eye. In *The Structure of the Eye*, ed. G. K. Smelser, pp. 487–506. New York: Academic

Silver, J., Rutishauser, U. 1984. Guidance of optic axons in vivo by a preformed adhesive pathway on neuroepithelial endfeet. *Dev. Biol.* 106: 485–99

Silver, J., Sapiro, J. 1981. Axonal guidance during development of the optic nerve: The role of pigmented epithelia and other

extrinsic factors. *J. Comp. Neurol.* 202: 521–38

Smalheiser, N. R., Crain, S. M., Reid, L. M. 1984. Laminin as a substrate for retinal axons in vitro. *Dev. Brain Res.* 12: 136–40

Spemann, H. 1918. Über die Determination der ersten Organlagen des Amphibienembryo. *Wilhelm Roux Arch. Entwicklungsmech. Org.* 43: 488–555

Stewart, W. W. 1978. Functional connections between cells as revealed by dye-coupling with a highly fluorescent naphthalimide tracer. *Cell* 14: 741–59

Stone, L. S. 1950. Neural retina degeneration followed by regeneration from surviving retinal pigment cells in grafted adult salamander eyes. *Anat. Rec.* 106: 89–109

Straznicky, K., Gaze, R. M. 1971. The growth of the retina in *Xenopus laevis*: An autoradiographic study. *J. Embryol. Exp. Morphol.* 26: 67–79

Stuermer, C: A. O. 1988. Retinotopic organization of the developing retinotectal projection in the Zebrafish embryo. *J. Neurosci.* 8: 4513–30

Taghert, P. H., Bastiani, M. J., Ho, R. K., Goodman, C. S. 1982. Guidance of pioneer growth cones: Filopodial contacts and coupling revealed with an antibody to Lucifer Yellow. *Dev. Biol.* 94: 391–99

Taylor, J. S. H. 1987a. Fibre organization and reorganization in the retinotectal projection of *Xenopus Development* 99: 393–410

Taylor, J. S. H. 1987b. Target recognition and pathway cues in the primary development of the retinotectal projection. *Soc. Neurosci. Abstr.* 14: 369

Thanos, S., Bonhoeffer, F. 1983. Investigations on the development of topographic order of retinotectal axons: Anterograde and retrograde staining of axons and perikarya with rhodamine in vivo. *J. Comp. Neurol.* 219: 420–30

Tomlinson, A., Ready, D. F. 1987. Cell fate in the *Drosophila* ommatidium. *Dev. Biol.* 123: 264–75

Tosney, K. W., Landmesser, L. T. 1985. Growth cone morphology and trajectory in the lumbosacral region of the chick embryo. *J. Neurosci.* 5: 2345–58

Turner, D. L., Cepko, C. L. 1987. A common progenitor for neurons and glia persists in rat retina late in development. *Nature* 238: 131-36

Turner, D. L., Snyder, E. Y., Cepko, C. L. 1988. Cell lineage analysis in the embryonic mouse retina by retroviral vector-mediated gene transfer. *Soc. Neurosci. Abstr.* 14: 892

Walsh, C., Cepko, C. L. 1988. Clonally related cortical cells show several migration patterns. *Science* 241: 1342–45

Webster, M. J., Shatz, C. J., Kliot, M., Silver, J. 1988. Abnormal pigmentation and unusual morphogenesis of the optic stalk may be correlated with retinal axon misguidance in embryonic Siamese cats. *J. Comp. Neurol.* 269: 592–611

Wetts, R., Fraser, S. E. 198. Multipotent precursors can give rise to all major cell types of the frog retina. *Science* 239: 1142–45

Wetts, R., Serbedzija, A., Fraser, S. E. 1988. Multipotent precursor cells in *Xenopus* ciliary margin: A cell lineage analysis. *Soc. Neurosci. Abstr.* 14: 1129

*Annu. Rev. Neurosci. 1990. 13:171–82*

# THE ROLE OF GLUTAMATE NEUROTOXICITY IN HYPOXIC– ISCHEMIC NEURONAL DEATH

*Dennis W. Choi*

Department of Neurology, Stanford University, Stanford, California 94305

*Steven M. Rothman*

Departments of Pediatrics, Neurology, and Anatomy and Neurobiology, Washington University, St. Louis, Missouri 63110

The human brain depends on its blood supply for a continuous supply of oxygen and glucose. Irreversible brain damage occurs if blood flow is reduced below about 10 ml/100 g tissue/min and if blood flow is completely interrupted, damage will occur in only a few minutes. Unfortunately, such reductions (ischemia) are common in disease states: either localized to individual vascular territories, as in stroke; or globally, as in cardiac arrest. Cerebral hypoxia can also occur in isolation, for example in respiratory arrest, carbon monoxide poisoning, or near-drowning; pure glucose deprivation can occur in insulin overdose or a variety of metabolic disorders. As a group, these disorders are a leading cause of neurological disability and death; stroke alone is the third most common cause of death in North America.

Despite its clinical importance, little is known about the cellular pathogenesis of hypoxic-ischemic brain damage, and at present there is no effective therapy. A critical question has been why brain, more than most other tissues, is so vulnerable to hypoxic-ischemic insults. In particular, certain neuronal subpopulations, such as hippocampal field CA1 and neocortical layers 3, 5, and 6, are characteristically destroyed after submaximal hypoxic-ischemic exposure. A possible answer has emerged in the last few years: At least some of this special vulnerability may be accounted for by the central neurotoxicity of the endogenous excitatory

171

0147–006X/90/0301–0171$02.00

amino acid neurotransmitter, glutamate, released into the extracellular space under hypoxic-ischemic conditions.

A link between glutamate and cerebral hypoxia was anticipated 30 years ago by Van Harreveld (1959), who was studying cortical spreading depression (SD) in the rabbit. He had shown that the spreading depression could be produced by applying glutamate to the cortical surface and suspected that it was related to neuronal hypoxia. He then suggested that the two were connected: "If glutamic (aspartic) acid is involved in the mechanism of SD, it is likely that it also plays a major part in the asphyxial cortical changes which have a striking resemblance to this phenomenon." Evidence reviewed here and previously (Meldrum 1985, Rothman & Olney 1986, Choi 1988b) supports Van Harreveld's speculation.

## Glutamate Neurotoxicity

Befitting its dominant role in central excitatory neurotransmission (Curtis & Johnston 1974), glutamate is present in excitatory presynaptic terminals throughout the brain, achieving millimolar whole tissue levels. It is somewhat counterintuitive—and unsettling—to consider that such a ubiquitous agent can lethally injure neurons, but the neurotoxicity of intense exposure to extracellular glutamate was established more than 30 years ago in the retina (Lucas & Newhouse 1957) and 20 years ago in the brain (Olney & Sharpe 1969). Olney went on to show that this neurotoxicity, which he later called "excitotoxicity," was a general property of excitatory amino acids on central neurons (Olney 1978).

Under normal conditions, powerful neuronal and glial uptake systems rapidly remove synaptically released glutamate from the extracellular space before toxicity occurs (Schousboe 1981). Cellular uptake also serves to mask the neurotoxicity of exogenously administered glutamate in experimental animals; as a result, many more studies have been done over the years with the plant excitotoxin, kainate, than with glutamate itself. More recently, it has been possible to examine directly glutamate neurotoxicity in neuronal cell cultures, where exposure can be controlled. These in vitro studies have suggested that glutamate neurotoxicity has several specific characteristics consistent with an important role in the pathogenesis of hypoxic-ischemic brain injury.

First, glutamate appears to be a remarkably potent and rapidly acting neurotoxin. Exposure to only 100 $\mu$M glutamate for 5 min sufficed to destroy large numbers of cultured cortical neurons (Choi et al 1987). Thus it is possible that the transient release of only a small fraction of the intracellular stores of glutamate into the extracellular space can damage neurons, a finding that places glutamate neurotoxicity early in the chain of lethal events that might ensue in the wake of hypoxia-ischemia.

Second, glutamate neurotoxicity may be largely mediated by a toxic influx of extracellular calcium. Intense glutamate exposure produces immediate neuronal swelling, which can be prevented by the removal of extracellular sodium or chloride (Rothman 1985, Olney et al 1986), and is probably due to the entry of sodium, chloride, and water into the cells. However, even without this acute swelling, most neurons exposed briefly to glutamate will still go on to degenerate in a delayed fashion, dependent on the presence of extracellular calcium (Choi 1987). Removal of extracellular calcium substantially attenuates excitatory amino acid-induced neuronal loss in cortical (Choi 1985) and hippocampal (Rothman et al 1987a) cultures, as well as in cerebellar slices (Garthwaite & Garthwaite 1986). Furthermore, glutamate-induced [45]calcium accumulation by cortical neurons is highly correlated with resultant neuronal degeneration (Marcoux et al 1988; M. Kurth and D. Choi, unpublished observations). These findings provide an attractive link between glutamate neurotoxicity and earlier data suggesting that a large calcium influx accompanies hypoxic-ischemic neuronal injury in vivo (Siesjo 1988).

Third, glutamate neurotoxicity may be blocked by antagonist compounds (Rothman 1984), in particular those effective against the N-methyl-D-aspartate (NMDA) subtype of glutamate receptor-ionophore complexes (Choi et al 1988). Glutamate activates several subtypes of receptor-ionophore complexes, named for their preferred pharmacological agonists: NMDA, kainate, and quisqualate (Watkins & Olverman 1987). Both NMDA and non-NMDA (kainate and quisqualate) receptors mediate the ability of glutamate to excite neurons, or to produce acute excitotoxic neuronal swelling. Selective antagonism of NMDA receptors alone, however, while inadequate to prevent either neuroexcitation or acute neuronal swelling, suffices to block the late neuronal degeneration induced by brief glutamate exposure (Choi et al 1988). This prominent role of NMDA receptors in glutamate neurotoxicity is consistent with the latter's dependence on extracellular calcium, since only the NMDA subtype of glutamate receptors opens a membrane channel that is highly permeable to calcium (MacDermott et al 1986). The NMDA receptor-activated channel may be the major route by which glutamate induces a toxic calcium influx, although one should keep in mind that additional calcium entry probably occurs through voltage-activated calcium channels, the sodium-calcium exchanger, and nonspecific membrane leakage (Choi 1988a), and also that calcium also may be released from intracellular stores by the action of glutamate on "metabotropic" quisqualate receptors (Sladeczek et al 1985, Nicoletti et al 1986). In any case, specific dependence on NMDA receptors provides a specific pharmacological link between glutamate neurotoxicity and hypoxic-ischemic injury (see below).

Fourth, glutamate neurotoxicity can be attenuated by antagonists added after glutamate exposure (Rothman et al 1987a, Choi et al 1988). This finding suggests that toxic exposure to glutamate is self-propagating: i.e. that an initial toxic exposure to glutamate subsequently triggers further neuronal injury, mediated by the excessive release or leakage of endogenous glutamate stores. The protective efficacy of late antagonist administration has been confirmed against excitatory amino acid neurotoxicity in vivo (Foster et al 1988), and may represent the cellular substrate for auspicious reports that glutamate antagonists can be neuroprotective when administered after the onset of hypoxic-ischemic injury both in vitro and in vivo (see below).

Finally, neuronal vulnerability to excitatory amino acid–induced injury is not uniform. Cortical neurons containing NADPH-diaphorase (which co-localizes with somatostatin) (Koh & Choi 1988a), or GABA (Tecoma & Choi 1989), and striatal neurons containing acetylcholinesterase (Koh & Choi 1988c), all possess some intrinsic resistance to injury by NMDA in vitro; the former are also characterized by heightened vulnerability to injury by kainate or quisqualate. Such differences in intrinsic neuronal vulnerability may not directly predict survival after in vivo hypoxia-ischemia, since several other factors [e.g. the density of glutamatergic inputs, or the co-release of zinc, which can alter the receptor distribution activated by glutamate (Koh & Choi 1988b)] also can influence resultant injury. However, these differences could contribute to the phenomenon of selective neuronal loss, a basic feature of hypoxic-ischemic neuronal damage. Of special note, is the fact that cortical NADPH-diaphorase-containing neurons appear to be selectively spared in a neonatal rat model of ischemic injury (Ferriero et al 1988).

## In Vitro Hypoxia

Observations on dispersed rodent hippocampal neurons in tissue culture provided the first direct indication that excitatory synaptic transmission might influence the sensitivity of neurons to low levels of oxygen (Rothman 1983). When these neurons are first dissociated and put into culture, they have no synaptic connections and survive a one-day exposure to 95% nitrogen/5% carbon dioxide (95% $N_2$/5% $CO_2$) or sodium cyanide (1 mM). However, after two weeks in culture, when excitatory glutamatergic synapses have developed (Rothman & Samaie 1985), both of these treatments leads to extensive neuronal degeneration.

The addition of high concentrations of magnesium, which blocks transmitter release, largely prevents this neuronal loss. The initial interpretation of these results was that blockade of transmitter release protected neurons from hypoxia, although a more modern interpretation recognizes that magnesium can gate the NMDA channel (Nowak et al 1984, Mayer et al 1984) and thereby exert a post-synaptic effect as well. Subsequent in vitro experiments provided additional evidence linking glutamate to hypoxic-ischemic neuronal damage. Pretreating cultures with the nonselective excitatory amino acid antagonist $\gamma$-D-glutamylglycine (Davies et al 1982) was as effective as magnesium in preventing neuronal loss (Rothman 1984). Selective antagonism of NMDA receptors with either competitive or non-competitive antagonists also attenuated the neuronal injury caused by hypoxia (Goldberg et al 1987, Rothman et al 1987b, Weiss et al 1986) or glucose deprivation (Monyer & Choi 1988) in cortical or hippocampal cultures. A neuroprotective effect was also seen when the antagonist was added after completion of a combined oxygen and glucose deprivation insult (Goldberg et al 1988a). In contrast, excitatory amino acid antagonists were not effective against combined oxygen and glucose deprivation in cultures of rat basal ganglia (Goldberg et al 1986), perhaps reflecting the paucity of glutamatergic neurons in these cultures and the severity of the insult delivered.

Parallel experiments examining the pathophysiology of hypoxia in rodent brain slices have also indicated that excitatory transmission is related to neuronal injury. Kass & Lipton (1982) found that slices of rat dentate gyrus irreversibly lost evoked field potentials if they were exposed to 95% $N_2$/5% $CO_2$ for ten minutes. If the hypoxic exposure occurred after calcium was removed from the extracellular perfusate and the magnesium concentration elevated, this field potential reduction was markedly attenuated. As mentioned above, this manipulation should both diminish the release of the excitatory transmitter and block activity elicited by activation of NMDA receptors. The same laboratory has since shown that specific block of NMDA receptors reduces hypoxic damage in their slices (Lobner & Lipton 1987). Similar observations have been made in hippocampal slices (Clark & Rothman 1987, Rothman et al 1987b). Pretreatment with either nonselective excitatory amino acid antagonists, NMDA antagonists, or high extracellular magnesium almost completely prevents the irreversible loss of the evoked CA1 synaptic potential which otherwise occurs after a 40-min exposure to 95% $N_2$/5% $CO_2$. NMDA antagonists also repolarize neurons that have started to depolarize in hypoxic hippocampal slices (Rader et al 1988).

Although there is some agreement among different investigators that excitatory transmitter release plays a role in hypoxic brain slice damage, experiments with excitatory amino acid antagonists have not been uniformly positive. Aitken and colleagues (1988) were not able to show a statistically significant recovery from hypoxia of NMDA antagonist-treated slices. The severity of their model and the low dose of a competitive racemic antagonist that may have some agonist activity, are possible explanations for this negative result.

## In Vivo Hypoxia-Ischemia

The first clear-cut demonstration that glutamate plays a role in hypoxic-ischemic brain damage in vivo came from the experiments of Simon and his associates (1984). They showed that direct intrahippocampal injection of the competitive NMDA antagonist 2-amino-7-phosphonoheptanoate (APH) reduced the loss of CA1 pyramidal neurons produced by transient carotid ligation in the rat. APH was also found to protect the rat caudate from injury induced by hypoglycemia (Wieloch 1985). A number of different laboratories have now confirmed and extended these results with various NMDA antagonists (Gill et al 1988, Kochhar et al 1988, Marcoux et al 1988, Church et al 1988, Natale et al 1988, Boast et al 1988, Steinberg et al 1988, Park et al 1988). However, some negative results have been reported with global ischemia models (Block & Pulsinelli 1987, Jensen & Auer 1988, Wieloch 1988), and as W. A. Pulsinelli (personal communication) has pointed out, not all studies have excluded a neuroprotective effect of brain hypothermia. Other laboratories have found that glutamate antagonists can prevent much of the forebrain damage seen in neonatal rodents after combined hypoxic and ischemic insult (McDonald et al 1987, Prince & Feeser 1988, Andine et al 1988, Olney et al 1989). Several of these laboratories have observed neuroprotective benefits with systemic administration of the antagonist drug after the initiation of ischemia (McDonald et al 1987, Boast et al 1988, Gill et al 1988, Steinberg et al 1988, Park et al 1988).

In addition to the demonstration that antagonists of glutamate, especially those acting at the NMDA receptor-channel complex, limit damage from anoxia, other lines of evidence have implicated glutamate and excitatory synaptic transmission in the etiology of hypoxic-ischemic brain injury. Investigators in two different laboratories (Johansen et al 1986, Onadera et al 1986) selectively lesioned some of the excitatory glutamatergic inputs to the hippocampus and allowed sufficient time for their terminals to degenerate. When these animals were subjected to carotid ligation, they showed dramatically preserved hippocampi. There is also an excellent correlation between the presence of glutamate receptors and vulnerability to brain ischemia. The CA1 region of the hippocampus, which is particularly rich in NMDA receptors (Monaghan et al 1983), is the most susceptible region to this type of injury.

Some of the most convincing evidence linking extracellular glutamate and ischemic injury has been provided by direct measurement of extracellular glutamate concentration with in vivo microdialysis. During a 10-

min period of experimental ischemia, glutamate levels rose from under 5 $\mu$M to approximately 30 $\mu$M (Benveniste et al 1984, Hagberg et al 1985). After 30 min of ischemia, the glutamate concentration exceeded 500 $\mu$M, a concentration which is rapidly toxic to cultured cortical neurons (Choi et al 1987). The origin of the increased extracellular glutamate has not been unequivocally established. Glutamate reuptake is impaired in ischemic synaptosomes (Silverstein et al 1986) and both anoxia and hypoglycemia reduce amino acid uptake into glia in vitro (Drejer et al 1985). Release of glutamate may also increase under ischemic conditions, possibly because collapsed ion gradients lead to the leakage of glutamate from the intracellular space (Kauppinen et al 1988). These results suggest that both neuronal and glial dysfunction could contribute to the excessive "glutamate burden." However, the protective effect of lesioning excitatory afferents implies that excitatory glutamatergic terminals must be present to initiate this process.

## Limitations of the Present Glutamate Hypothesis

The idea that hypoxic-ischemic brain damage can be explained by overstimulation of glutamate receptors is appealing. Unfortunately, glutamate is not the sole cause of ischemic neuronal damage and may have little importance under some conditions. Several variables outside the glutamate system, including ischemia type (focal vs. global), animal species, brain temperature, blood-brain barrier integrity, edema formation, and effects of other neurotransmitters, probably all influence the outcome following hypoxia-ischemia.

Of note, ischemic damage is common in the myelinated tracts of the brain, which consist largely of axons and oligodendrocytes. Myelinated tracts do not show the sensitivity to excitotoxic damage that characterizes brain regions composed largely of neuronal cell bodies and dendrites (e.g. cortex and hippocampus); in fact, excitotoxic lesions were initially described as "axon sparing" (Olney 1978). Therefore, ischemic injury to white matter cannot be attributed directly to glutamate. In addition, even neurons richly endowed with excitatory amino acid receptors are not indefinitely protected from anoxic damage by glutamate antagonists. Cultured cortical neurons maintained in 95% $N_2$/5% $CO_2$ in the presence of high concentrations of NMDA antagonists still die after 16 hours (Goldberg et al 1987). This result is not very surprising, as all mammalian cells will eventually die if deprived of oxygen and nutrients. An excessively severe ischemic insult may account for some of the negative outcomes reported with glutamate antagonists in stroke models.

Along these lines, an important area for future investigation is the relationship between glutamate neurotoxicity and cerebral infarction—

the consolidated zone of pancellular necrosis produced by profound focal ischemia. One might expect glutamate antagonists to lose effectiveness against ischemia sufficiently profound to destroy both neurons and glia, but recent evidence suggests that substantial reductions of infarct size may occur (Simon 1988 and personal communication, Park et al 1988). Why this might be so is unclear at present, but it is potentially of great therapeutic significance. Could glutamate-induced neuronal damage be responsible for other changes—a "bad neighborhood" effect involving a local buildup of lytic enzymes or other toxic factors, or changes in local blood flow— that secondarily contribute to the destruction of non-neuronal elements?

## Therapeutic Issues

Although glutamate neurotoxicity is not the explanation for all ischemia-related brain damage, the positive results discussed above suggest that it may be a highly appropriate target for rational stroke therapy at the present time. A variety of questions, however, will have to be answered before any extensive clinical trials can be considered (Albers et al 1989).

First, we will have to determine the most effective pharmacological strategies for minimizing glutamate neurotoxicity. Since most glutamate damage is probably mediated via NMDA receptors, some type of NMDA antagonist is a logical consideration. Different types of NMDA antagonists will have to be studied to determine differences in efficacy. In addition, there are classes of neurons that are more susceptible to non-NMDA receptor-mediated damage, and it will be important to determine the incremental neuroprotective benefit to be gained by blocking non-NMDA receptors. Alternatively, it may be desirable to try to minimize the presynaptic synthesis or release of glutamate. Hypoxic neuronal injury in cortical culture can be attenuated by removal of the glutamate precursor, glutamine, from the bathing medium (Goldberg et al 1988b). Adenosine agonists, which likely act presynaptically to reduce transmitter release, can improve neuronal survival in both in vitro (Goldberg et al 1988c) and in vivo models of hypoxia-ischemia (Evans et al 1987, von Lubitz et al 1988); adenosine antagonists worsen injury (Wieloch et al 1986, Rudolphi et al 1987).

Second, we will have to determine whether any such interference with glutamate leads to unacceptable side effects. If antagonism is restricted to the NMDA receptor system, non-NMDA receptor-mediated fast excitatory synaptic transmission should be preserved. However, many available non-competitive NMDA antagonists interact with the phencyclidine binding site in the NMDA channel, and may share disturbing psychotomimetic effects with that compound. Of special concern is the fact that certain noncompetitive NMDA antagonists such as ketamine and MK-801 increase the cerebral metabolic rate for glucose, an undesirable effect in ischemic brain (Hammer & Herkenham 1983, Nehls et al 1988). Blocking NMDA currents with either potent competitive antagonists (Boast et al 1988) or antagonists of the glycine receptor (Fletcher & Lodge 1988, Kemp et al 1988) may avoid these problems (Mathisen et al 1988, Nehls et al 1988).

Third, we will have to have more precise information about the time course of the development of irreversible neuronal damage after the onset of hypoxia-ischemia in man. If most patients have already suffered irreversible damage by the time they seek attention, glutamate antagonists will offer them little. However, pathological evidence in both rodents and humans suggests that damage in the CA1 region of the hippocampus develops hours to days after transient ischemia (Pulsinelli et al 1982, Petito et al 1987). If much of this delay reflects ongoing glutamate neurotoxicity, there may be a substantial temporal "therapeutic window"—as seen in experimental stroke models—during which it may prove possible to halt progressive neuronal degeneration and improve neurological outcome.

ACKNOWLEDGMENTS

This work was supported by National Institutes of Health grants NS26907 (D. W. C.) and NS19988 (S. M. R.). We thank Deborah Howard for assistance with the manuscript.

*Literature Cited*

Aitken, P. G., Balestrino, M., Somjen, G. C. 1988. NMDA antagonists: Lack of protective effect against hypoxic damage in CA1 region of hippocampal slices. *Neurosci. Lett.* 89: 187–92

Albers, G. A., Goldberg, M. P., Choi, D. W. 1989. N-methyl-D-aspartate antagonists: Ready for clinical trial in brain ischemia? *Ann. Neurol.* 25: 398–403

Andine, P., Lehmann, A., Ellren, K., Wennberg, E., Kjellmer, L., et al. 1988. The excitatory amino acid antagonist kynurenic acid administered after hypoxia/ischemia in neonatal rats offers neuroprotection. *Neurosci. Lett.* 90: 208–12

Benveniste, H., Drejer, J., Schousboe, A., Diemer, N. H. 1984. Elevation of the extracellular concentrations of glutamate and aspartate in rat hippocampus during transient cerebral ischemia monitored by intracerebral microdialysis. *J. Neurochem.* 43: 1369–74

Block, G. A., Pulsinelli, W. A. 1987. Excitatory amino acid receptor antagonists: Failure to prevent ischemic neuronal damage. *J. Cereb. Blood Flow Metab,* 7(1): S149 (Abst.)

Boast, C. A., Gearhardt, S. C., Pastor, E., Lehmann, J., Etienne, P. E., et al. 1988. The N-methyl-D-aspartate antagonists CGS 19755 and CPP reduce ischemic brain damage in gerbils. *Brain Res.* 442: 345–48

Choi, D. W. 1985. Glutamate neurotoxicity in cortical cell culture is calcium dependent. *Neurosci. Lett.* 58: 293–97

Choi, D. W. 1987. Ionic dependence of glutamate neurotoxicity in cortical cell culture. *J. Neurosci.* 7: 369–79

Choi, D. W. 1988a. Calcium-mediated neurotoxicity: Relationship to specific channel types and role in ischemic damage. *Trends Neurosci.* 11: 465–69

Choi, D. W. 1988b. Glutamate neurotoxicity and diseases of the nervous system. *Neuron* 1: 623–34

Choi, D. W., Koh, J., Peters, S. 1988. Pharmacology of glutamate neurotoxicity in cortical cell culture: Attenuation by NMDA antagonists. *J. Neurosci.* 8: 185–96

Choi, D. W., Maulucci-Gedde, M. A., Kriegstein, A. R. 1987. Glutamate neurotoxicity in cortical cell culture. *J. Neurosci.* 7: 357–68

Church, J., Zeman, S., Lodge, D. 1988. Ketamine and MK-801 as neuroprotective agents in cerebral ischemia/hypoxia. See Domino & Kamenka 1988, pp. 747–56

Clark, G. D., Rothman, S. M. 1987. Blockade of excitatory amino acid receptors protects anoxic hippocampal slices. *Neuroscience* 21: 665–71

Curtis, D. R., Johnston, G. A. R. 1974. Amino acid transmitters in the mammalian central nervous system. *Ergeb. Physiol. Biol. Chem. Exp. Pharmakol.* 69: 98–188

Davies, J., Evans, R. H., Jones, A. W., Smith, D. A. S., Watkins, J. C. 1982. Differential activation and blockade of excitatory amino acid receptors in the mammalian and amphibian central nervous systems. *Comp. Biochem. Physiol. C* 72: 211–24

Domino, E. F., Kamenka, J. M., eds. 1988. *Sigma and Phencyclidine-Like Compounds as Molecular Probes in Biology.* Ann Arbor: NPP Books

Drejer, J., Benveniste, H., Diemer, N. H., Schousboe, A. 1985. Cellular origin of ischemia-induced glutamate release from brain tissue in vivo and in vitro. *J. Neurochem.* 45: 145–51

Evans, M. C., Swan, J. H., Meldrum, B. S. 1987. An adenosine analogue, 2-chloroadenosine, protects against long term development of ischaemic cell loss in the rat hippocampus. *Neurosci. Lett.* 83: 287–92

Ferriero, D. M., Arcavi, L. J., Sagar, S. M., McIntosh, T. K., Simon, R. P. 1988. Selective sparing of NADPH-diaphorase neurons in neonatal hypoxia-ischemia. *Ann. Neurol.* 24: 670–76

Fletcher, E. J., Lodge, D. 1988. Glycine reverses antagonism of N-methyl-D-aspartate (NMDA) by 1-hydroxy-3-aminopyrrolidone-2 (HA-966) but not by D-2-amino-5-phosphonovalerate (D-AP5) on rat cortical slices. *Eur. J. Pharmacol.* 151: 161–62

Foster, A., Gill, R., Woodruff, G. N. 1988. Neuroprotective effects of MK-801 in vivo: Selectivity and evidence for delayed degeneration mediated by NMDA receptor activation. *J. Neurosci.* 8: 4745–54

Garthwaite, G., Garthwaite, J. 1986. Neurotoxicity of excitatory amino acid receptor agonists in rat cerebellar slices: Dependence on calcium concentration. *Neurosci. Lett.* 66: 193–98

Gill, R., Foster, A. C., Woodruff, G. M. 1988. MK-801 is neuroprotective in gerbils when administered during the postischemic period. *Neuroscience* 25: 847–55

Goldberg, M. P., Monyer, H., Choi, D. W. 1988a. Cortical neuronal injury in vitro following combined glucose and oxygen deprivation: Ionic dependence and delayed protection by NMDA antagonists. *Soc. Neurosci. Abstr.* 14: 745

Goldberg, M. P., Monyer, H., Choi, D. W. 1988b. Hypoxic neuronal injury in vitro depends on extracellular glutamine. *Neurosci. Lett.* 94: 52–57

Goldberg, M. P., Monyer, H., Weiss, J. H., Choi, D. W. 1988c. Adenosine reduces cortical neuronal injury induced by oxygen or glucose deprivation in vitro *Neurosci. Lett.* 89: 323–27

Goldberg, M. P., Weiss, J. H., Pham, P. C., Choi, D. W. 1987. N-methyl-D-aspartate receptors mediate hypoxic neuronal injury in cortical culture. *J. Pharmacol. Exp. Ther.* 243: 784–91

Goldberg, W. J., Kadingo, R. M., Barrett, J. N. 1986. Effects of ischemia-like conditions on cultured neurons: Protection by low $Na^+$, low $Ca^{2+}$ solutions. *J. Neurosci.* 6: 3144–51

Hagberg, H., Lehmann, A., Sandberg, M., Nystrom, B., Jacobson, I., et al. 1985. Ischemia-induced shift of inhibitory and excitatory amino acids from intra- to extracellular compartments. *J. Cereb. Blood Flow Metab.* 5: 413–19

Hammer, R. P., Herkenham, M. 1983. Altered metabolic activity in the cerebral cortex of rats exposed to ketamine. *J. Comp. Neurol.* 220: 396–404

Jensen, M. C., Auer, R. N. 1988. Ketamine fails to protect against ischemic neuronal necrosis in the rat. *Br. J. Anaesth.* 61: 206–10

Johansen, F. F., Jorgensen, M. B., Diemer, N. H. 1986. Ischemic CA1 pyramidal cell loss is prevented by preischemic colchcine destruction of dentate gyrus granule cells. *Brain Res.* 377: 344–47

Kass, I. S., Lipton, P. 1982. Mechanisms involved in irreversible anoxic damage to the in vitro rat hippocampal slice. *J. Physiol.* 332: 459–72

Kauppinen, R. A., McMahon, H. T., Nicholls, D. G. 1988. $Ca^{2+}$-dependent and $Ca^{2+}$-independent glutamate release, energy status and cytosolic free $Ca^{2+}$ concentration in isolated nerve terminals following metabolic inhibition: Possible relevance to hypoglycaemia and anoxia. *Neuroscience* 27: 175–82

Kemp, J. A., Foster, A. C., Leeson, P. D., Priestley, T., Tridgett, R., et al. 1988. 7-Chlorokynurenic acid is a selective antagonist at the glycine modulatory site of the N-methyl-D-aspartate receptor complex. *Proc. Natl. Acad. Sci. USA* 85: 6547–50

Kochhar, A., Zivin, J. A., Lyden, P. D., Mazzarella, V. 1988. Glutamate antagonist therapy reduces neurologic deficits produced by focal central nervous system ischemia. *Arch. Neurol.* 45: 148–53

Koh, J., Choi, D. W. 1988a. Vulnerability of cultured cortical neurons to damage by excitotoxins: Differential susceptibility of neurons containing NADPH-diaphorase. *J. Neurosci.* 8: 2153–63

Koh, J., Choi, D. W. 1988b. Zinc alters excitatory amino acid neurotoxicity on cortical neurons. *J. Neurosci.* 8: 2164–71

Koh, J., Choi, D. W. 1988c. Cultured striatal neurons containing NADPH-diaphorase or acetylcholinesterase are selectively resistant to injury by NMDA receptor agonists. *Brain Res.* 446: 374–78

Lobner, D., Lipton, P. 1987. Glutamate receptors and irreversible anoxic damage in hippocampal slices: Mechanism of interaction. *Soc. Neurosci. Abstr.* 13: 647

Lucas, D. R., Newhouse, J. P. 1957. The toxic effect of sodium L-glutamate on the inner layers of the retina. *Arch. Ophthalmol.* 58: 193–201

MacDermott, A. B., Mayer, M. L., Westbrook, G. L., Smith, S. J., Barker, J. L. 1986. NMDA-receptor activation increases cytoplasmic calcium concentration in cultured spinal cord neurones. *Nature* 321: 519–22

Marcoux, F. W., Goodrich, J. E., Probert, A. W., Dominick, M. A. 1988. Ketamine prevents glutamate-induced calcium influx and ischemic nerve cell injury. See Domino & Kamenka 1988, pp. 735–46

Mathisen, J. E., Rothman, S. M., Contreras, P. C., Deuel, R. K. 1988. Comparison of regional 14C-2-deoxyglucose (2-DG) uptake in rat brain following parenteral administration of either D-2-amino-5-phosphonoheptanoate (APH) or ketamine. *Soc. Neurosci. Abstr.* 14: 480

Mayer, M. L., Westbrook, G. L., Guthrie, P. B. 1984. Voltage-dependent block by Mg$^{2+}$ of NMDA responses in spinal cord neurons. *Nature* 309: 261–63

McDonald, J. W., Silverstein, F. S., Johnston, M. V. 1987. MK-801 protects the neonatal brain from hypoxic-ischemic damage. *Eur. J. Pharmacol.* 140: 359–61

Meldrum, B. 1985. Possible therapeutic applications of antagonists of excitatory amino acid neurotransmitters. *Clin. Sci.* 68: 113–22

Monaghan, D. T., Holets, V. R., Toy, D. W., Cotman, C. W. 1983. Anatomical distributions of four pharmacologically distinct $^3$H-L-glutamate binding sites. *Nature* 306: 176–79

Monyer, H., Choi, D. W. 1988. Morphinans attenuate cortical neuronal injury induced by glucose deprivation in vitro. *Brain Res.* 446: 144–48

Natale, J. E., Schott, R. J., D'Alecy, L. G. 1988. Ketamine reduces neurological deficit following 10 minutes of cardiac arrest and resuscitation in canines. See Domino & Kamenka 1988, pp. 717–26

Nehls, D. G., Kurumaji, A., Park, C. K., McCulloch, J. 1988. Differential effects of competitive and non-competitive N-methyl-D-aspartate antagonists on glucose use in the limbic system. *Neurosci. Lett.* 91: 204–10

Nicoletti, F., Wroblewski, J. T., Novelli, A., Alho, H., Guidotti, A., Costa, E. 1986. The activation of inositol phospholipid metabolism as a signal-transducing system for excitatory amino acids in primary cultures of cerebellar granule cells. *J. Neurosci.* 6: 1905–11

Nowak, L., Bregestovski, P., Ascher, P., Herbet, A., Prochiantz, A. 1984. Magnesium gates glutamate activated channels in mouse central neurons. *Nature* 307: 462–65

Olney, J. W. 1978. Neurotoxicity of excitatory amino acids. In *Kainic Acid as a Tool in Neurobiology*, ed. E. G. McGeer, J. W. Olney, P. L. McGeer, pp. 95–171. New York: Raven

Olney, J. W., Price, M. T., Samson, L., Labruyere, J. 1986. The role of specific ions in glutamate neurotoxicity. *Neurosci. Lett.* 65: 65–71

Olney, J. W., Ikonomidou, C., Mosinger, J. L., Frierdich, G. 1989. MK-801 prevents hypobaric-ischemic neuronal degeneration in infant rat brain. *J. Neurosci.* 9: 1701–4

Olney, J. W., Sharpe, L. G. 1969. Brain lesions in an infant rhesus monkey treated with monosodium glutamate. *Science* 166: 386–88

Onodera, H., Sato, G., Kogure, K. 1986. Lesions to Schaffer collaterals prevent ischemic death of CA1 pyramidal cells. *Neurosci. Lett.* 68: 169–74

Park, C. K., Nehls, D. G., Graham, D. I., Teasdale, G. M., McCulloch, J. 1988. Focal cerebral ischaemia in the cat: Treatment with the glutamate antagonist MK-801 after induction of ischaemia. *J. Cereb. Blood. Flow Metab.* 8: 757–62

Petito, C. K., Feldmann, E., Pulsinelli, W. A., Plum, F. 1987. Delayed hippocampal damage in humans following cardiorespiratory arrest. *Neurology* 37: 1281–86

Prince, D. A., Feeser, H. R. 1988. Dextromethorphan protects against cerebral infarction in a rat model of hypoxia-ischemia. *Neurosci. Lett.* 85: 291–96

Pulsinelli, W. A., Brierley, J. B., Plum, F. 1982. Temporal profile of neuronal damage in a model of transient forebrain ischemia. *Ann. Neurol.* 11: 491–98

Rader, R. K., Watson, G. B., Lanthorn, T. H. 1988. Pharmacological characterization of the persistent depolarization induced by experimental ischemia. *Soc. Neurosci. Abstr.* 14: 189

Rothman, S. M. 1983. Synaptic activity mediates death of hypoxic neurons. *Science* 220: 536–37

Rothman, S. M. 1984. Synaptic release of excitatory amino acid neurotransmitter mediates anoxic neuronal death. *J. Neurosci.* 4: 1884–91

Rothman, S. M. 1985. The neurotoxicity of excitatory amino acids is produced by passive chloride influx. *J. Neurosci.* 5: 1483–89

Rothman, S. M., Olney, J. W. 1986. Glutamate and the pathophysiology of hypoxic-ischemic brain damage. *Ann. Neurol.* 19: 105–11

Rothman, S. M., Samaie, M. 1985. The physiology of excitatory synaptic transmission in cultures of dissociated rat hippocampus. *J. Neurophysiol.* 54: 701–13

Rothman, S. M., Thurston, J. H., Hauhart, R. E. 1987a. Delayed neurotoxicity of excitatory amino acids in vitro. *Neuroscience* 22: 471–80

Rothman, S. M., Thurston, J. H., Hauhart, R. E., Clark, G. D., Solomon, J. S. 1987b. Ketamine protects hippocampal neurons from anoxia in vitro. *Neuroscience* 21: 673–78

Rudolphi, K. A., Keil, M., Hinze, H. J. 1987. Effect of theophylline on ischemically induced hippocampal damage in mongolian gerbils: A behavioral and histopathological study. *J. Cereb. Blood Flow Metab.* 7: 74–81

Schousboe, A. 1981. Transport and metabolism of glutamate and GABA in neurons and glial cells. *Int. Rev. Neurobiol.* 22: 1–45

Siesjo, B. K. 1988. Historical overview. Calcium, ischemia, and death of brain cells. *Ann. NY Acad. Sci.* 522: 638–61

Silverstein, F. S., Buchanan, K., Johnston, M. V. 1986. Perinatal hypoxia-ischemia disrupts high-affinity [$^3$H]-glutamate uptake into synaptosomes. *J. Neurochem.* 47: 1614–19

Simon, R. P. 1988. Focal ischemia, excitatory amino acid antagonists and penumbra. *Neurochem. Int.* 12(Suppl. 1): 23

Simon, R. P., Swan, J. H., Griffiths, T., Meldrum, B. S. 1984. Blockade of N-methyl-D-aspartate receptors may protect against ischemic damage in the brain. *Science* 226: 850–52

Sladeczek, F., Pin, J. P., Recasens, M., Bockaert, J., Weiss, S. 1985. Glutamate stimulates inositol phosphate formation in striatal neurones. *Nature* 317: 717–19

Steinberg, G. K., Saleh, J., Kunis, D. 1988. Delayed treatment with dextromethorphan and dextrorphan reduces cerebral damage after transient focal ischemia. *Neurosci. Lett.* 89: 193–97

Tecoma, E. S., Choi, D. W. 1989. GABAnergic neocortical neurons are resistant to NMDA receptor-mediated injury. *Neurology* 39: 676–82

Van Harreveld, A. 1959. Compounds in brain extracts causing spreading depression of cerebral cortical activity and contraction of crustacean muscle. *J. Neurochem.* 3: 300–15

von Lubitz, D. K. J. E., Dambrosia, J. M., Kempski, O., Redmond, D. J. 1988. Cyclohexyl adenosine protects against neuronal death following ischemia in the CA1 region of the hippocampus. *Stroke* 19: 1133–39

Watkins, J. C., Olverman, H. J. 1987. Agonists and antagonists for excitatory amino acid receptors. *Trends Neurosci.* 10: 265–72

Weiss, J. H., Goldberg, M. P., Choi, D. W. 1986. Ketamine protects cultured neocortical neurons from hypoxic injury. *Brain Res.* 380: 186–90

Wieloch, T. 1985. Hypoglycemia-induced neuronal damage prevented by an N-methyl-D-aspartate antagonist. *Science* 230: 681–83

Wieloch, T. 1988. MK-801 does not protect against brain damage in a rat model of cerebral ischemia. *Neurochem. Int.* 12 (Suppl. 1): 24

Wieloch, T., Koide, T., Westerberg, E. 1986. Inhibitory neurotransmitters and neuromodulators as protective agents against ischemic brain damage. In *Pharmacology of Cerebral Ischemia*, ed. J. Krieglestein, pp. 191–97. New York: Elsevier

Annu. Rev. Neurosci. 1990. 13:183–94

# POSTEMBRYONIC NEURONAL PLASTICITY AND ITS HORMONAL CONTROL DURING INSECT METAMORPHOSIS

*Janis C. Weeks*

Institute of Neuroscience, University of Oregon, Eugene, Oregon 97403

*Richard B. Levine*

Arizona Research Laboratories, Division of Neurobiology, University of Arizona, Tucson, Arizona 85721

## Introduction

Much attention has been focused recently upon postembryonic neuronal plasticity, especially as evidence accumulates for the continued modification of neuronal form in adults (e.g. DeVoogd & Nottebohm 1981, Purves & Hadley 1985, Kurz et al 1986). The behavioral correlates of such changes, as well as the cellular and molecular mechanisms by which they are induced and regulated, are of great interest. A striking natural example of such neuronal plasticity is offered by insect metamorphosis. Metamorphosis entails substantial changes in body form and behavior, and the nervous system must be reorganized to accommodate these changes. Notably, there is a continuation into postembryonic life of many neuro-developmental processes that are normally viewed as being restricted to embryonic life, such as neuronal birth and death, and changes in neuronal structure and synaptic connectivity. The relative simplicity of the insect nervous system offers an excellent opportunity to study these phenomena in individually identified neurons. Furthermore, many aspects of nervous system metamorphosis are controlled by steroid hormones, just as these hormones exert profound effects on the developing vertebrate nervous

183

0147–006X/90/0301–0183$02.00

system (Arnold & Gorski 1984). As we hope to document in this review, studies of the reorganization of the nervous system during metamorphosis and its endocrine control yield insights relevant to numerous areas of contemporary neurobiology.

The moth, *Manduca sexta*, has provided a particularly attractive system in which to study metamorphosis, as indicated by our heavy emphasis in this review on this species. It is large enough to allow standard electrophysiological and neuroanatomical techniques to be used at all post-embryonic stages, and its endocrinology is well characterized. Important contributions have also been provided from recent studies of metamorphosis in the fruitfly *Drosophila*, in which genetic and molecular techniques offer information that complements that obtained in *Manduca*. Our purpose in this review is to provide an overview of recent advances, and to identify promising new approaches to understanding neural events during metamorphosis.

## The Endocrinology of Metamorphosis

Metamorphosis is controlled by the relative blood levels of ecdysteroids (ecdysone and 20-hydroxyecdysone) and juvenile hormone, which influence target cells via changes in gene expression (reviewed in Riddiford 1985). The fluctuations in blood levels of ecdysteroids and juvenile hormone during metamorphosis are known in more detail in *Manduca* than in any other insect. Every molt is triggered by an ecdysteroid surge, with the direction of the molt determined by the juvenile hormone titer. Following embryonic development and hatching from the egg, the caterpillar undergoes four molts over a period of about 2 wk, each triggered by an elevation of blood ecdysteroids in the presence of juvenile hormone.[1] Several key endocrine events occur during the nine days of the fifth (final) larval instar. Early in the instar the juvenile hormone titer drops, followed by a small "commitment pulse" of ecdysteroids in the absence of juvenile hormone, which reprograms tissues for pupal development; e.g. in epidermal cells that secrete the cuticular exoskeleton, larval-specific genes become inactivated. The commitment pulse also triggers wandering behavior, during which the larva burrows underground to pupate. This is followed by a larger, molt-inducing "prepupal peak" of ecdysteroids, that activates pupal-specific genes in epidermal cells and culminates in ecdysis to the pupa. The prepupal peak of ecdysteroids is accompanied by a low level of juvenile hormone, to which most tissues are indifferent due to their previous exposure to the commitment pulse. Development of the adult moth

---

[1] This description is based on insects reared under typical laboratory conditions.

within the pupal case takes 18 days and is triggered by a prolonged rise and fall of ecdysteroids in the absence of juvenile hormone.

Most studies of the relationship between hormonal fluctuations and nervous system metamorphosis involve experimental manipulation of hormone titers (e.g. Weeks & Truman 1986a). Ecdysteroids originate from the prothoracic glands located in the first thoracic segment, whereas juvenile hormone is released from brain-associated endocrine organs. It is therefore particularly convenient to study the effects of hormones on the abdominal nervous system, because ecdysteroids and juvenile hormone can be eliminated from the abdomen by ligating the body at the abdominal-thoracic junction and discarding the anterior fragment. Ligated abdomens typically survive for weeks, and the missing hormones can be replaced by direct infusion into the blood, or by topical application to the body surface.

## Postembryonic Neurogenesis

During embryonic development of the *Manduca* central nervous system (CNS), neurons are produced from a stereotyped array of neuroblasts in a manner similar to that of *Drosophila* and the grasshopper (Thomas et al 1984). In the insect brain, neurogenesis typically continues post-embryonically (Edwards 1969, Nordlander & Edwards 1969a,b, White & Kankel 1978); the new neurons contribute to the extensive elaboration of the visual (Bate 1978) and olfactory (Tolbert et al 1983) neuropils. In addition, it has been established recently that neuroblasts in the thoracic and abdominal ganglia of *Manduca* produce large numbers of neurons, termed *imaginal nest cells*, throughout larval life (Booker & Truman 1987a). These cells differentiate as interneurons during adult development (Booker & Truman 1987b, Witten & Truman 1988). Based on studies in *Drosophila*, Truman & Bate (1988) suggest that the neuroblasts that divide postembryonically may be retained embryonic neuroblasts that delay their final divisions until the larval stage.

The postembryonic mitotic activity of the *Manduca* neuroblasts and the subsequent differentiation of their progeny are controlled by ecdysteroids and juvenile hormone (Booker & Truman 1987b). Interestingly, chemical ablation of specific subpopulations of the imaginal nest cells produces only minor behavioral deficits in adult moths (Truman & Booker 1986). It remains to be determined what functional role the imaginal nest cells play in the nervous system, and how these late-differentiating neurons are integrated into pre-existing neural circuits.

## Programmed Neuron Death

The death of identified neurons during metamorphosis and its hormonal control have been studied extensively in *Manduca* (reviewed by Truman

1987). A wave of programmed neuron death follows each metamorphic molt, to eliminate neurons that are not needed for the next life stage. Most studies have concerned the deaths of motoneurons that innervate abdominal muscles (Taylor & Truman 1974, Truman 1983, Truman & Schwartz 1984, Giebultowicz & Truman 1984, Levine & Truman 1985, Weeks & Truman 1985, 1986b, Weeks 1987).

Motoneurons present in the larval abdomen can have several fates. Some innervate the same target muscle until adulthood. At the larval-pupal transformation, many motoneurons are rendered targetless by the degeneration of larval abdominal muscles. Some of these motoneurons persist and later innervate newly generated adult muscles (see below), but many die. The best-studied of these dying motoneurons is PPR, which innervates a retractor muscle of the larval abdominal proleg (Weeks & Truman 1984). The target muscle of PPR degenerates before pupal ecdysis, and PPR dies two days after pupation. By combining endocrine and surgical manipulations, Weeks & Truman (1985, 1986b) have shown that the rise in blood ecdysteroids during the prepupal peak triggers PPR's death. This response requires previous exposure to ecdysteroids in the absence of juvenile hormone (i.e. the commitment pulse), but PPR is indifferent to juvenile hormone during the prepupal period. Furthermore, interactions with its target muscle do not play a role in PPR's death. The independence of the fates of motoneurons from that of their muscles during metamorphosis has been demonstrated repeatedly (Truman & Schwartz 1984, Bennett & Truman 1985, Weeks 1987, Kent & Levine 1988c and in preparation). After emergence of the adult moth, approximately 50% of the persistent larval motoneurons die in response to the decline in blood ecdysteroids at the end of adult development (Truman 1983, Truman & Schwartz 1984). The elimination of larval motoneurons in the pupal and adult stages reflects the progressive simplification of the abdominal musculature that accompanies metamorphosis. The finding that a rise or fall in ecdysteroid levels can trigger neuronal death (in the pupa and adult, respectively) emphasizes the general finding that the interpretation of a particular hormonal signal is gated by the cell's previous history of endocrine exposure (discussed in Weeks & Truman 1986a).

Although motoneuron death is clearly regulated by ecdysteroid action on the nervous system, it has not yet been possible technically to identify specific motoneurons as direct ecdysteroid targets. All of the data to date are consistent with the dying motoneurons' being direct targets, but this assumption has yet to be demonstrated experimentally. An important step in this direction has been the autoradiographic studies of Fahrbach & Truman (1989), who showed that subsets of *Manduca* motoneurons accumulate radio-labeled ecdysteroid analogs at specific times during metamorphosis. It remains to be established which of the radioactively labeled motoneurons are those known from physiological studies. As has been found in other systems, the programmed death of *Manduca* neurons may involve de novo mRNA and protein synthesis; for instance, cycloheximide treatment can prevent the death of motoneurons (Fahrbach & Truman 1988; J. C. Weeks & B. H. G. Debu, unpublished data). mRNAs associated with the ecdysteroid-triggered degeneration of *Manduca* muscle have been identified (Schwartz & Kay 1987), and probes developed in muscle may also prove useful in the nervous system. Once neurons that are direct targets for ecdysteroids and juvenile hormone have been identified, it should be possible to approach the hormonal regulation of neuron death at the molecular level (see below).

Although ecdysteroids play a key role in controlling neuron death, other factors may intervene. For instance, Fahrbach & Truman (1987) showed that sectioning the nerve cord to interrupt descending neural input prevents the death of an identified *Manduca* motoneuron, MN-11, after adult emergence. Another important influence is segmental location, as illustrated by the finding that a proleg motoneuron, APR, dies in only a subset of abdominal ganglia after pupation (Weeks & Ernst-Utzschneider 1989). Finally, the sexually-dimorphic motoneuron death that occurs in the terminal ganglion of *Manduca* (Giebultowicz & Truman 1984) suggests, since there are no known differences in the hormone titers of male and female larvae, that the genes involved in sex determination can also influence motoneuron death. An important goal for the future is to elucidate how transsynaptic influences, segmental location, and genetic factors may interact with endocrine cues to control motoneuron survival.

## Changes in Neuronal Arbors

One of the more remarkable aspects of metamorphosis is that individual neurons can exhibit markedly different structures and functions in the different life stages (reviewed by Truman et al 1986, Levine 1987, Levine & Weeks 1989). The most detailed studies have involved motoneurons, although some information is also available for persistent sensory neurons and interneurons. A combination of electrophysiological and anatomical techniques are used to demonstrate the persistence of individual neurons. For instance, motoneurons can be re-identified in different stages by backfilling the nerve containing their axons (Taylor & Truman 1974, Truman & Reiss 1976, Thorn & Truman 1989), or by intracellular recording and dye injection at different stages (Levine & Truman 1982, 1985). A more definitive technique, which has been useful for following motoneurons innervating targets that are reorganized dramatically (such as the larval and adult thoracic legs of *Manduca*) involves retrogradely labeling a larval motoneuron by applying a fluorescent dye to its target muscle, and then visualizing the labeled motoneuron in the adult for intracellular recording (Kent & Levine 1988b,c). Another approach to following neuronal fates during metamorphosis is to use antibodies to neurotransmitters or peptides. This technique has been used to follow a persistent serotonin-immunoreactive neuron in the *Manduca* brain (Kent et al 1987), and to document metamorphic changes in populations of neurons staining for a variety of substances in flies (e.g. White et al 1986, Cantera & Nässel 1987, Nässel et al 1988, Vallés & White 1988, Budnik & White 1988).

Unlike motoneurons that innervate the same muscle throughout metamorphosis, persistent *Manduca* motoneurons that change their target muscles typically show morphological changes. Many motoneurons whose larval muscles degenerate exhibit significant regression around the time of pupation. In the case of the motoneurons innervating proleg retractor muscles, dendritic regression is triggered by the prepupal peak of ecdysteroids and is independent of the degeneration of the motoneurons' target muscles (Weeks & Truman 1985, 1986b). Furthermore, although some motoneurons that regress may subsequently die, the two phenomena are not necessarily linked; many motoneurons that regress at pupation survive and later regrow their dendritic fields during adult development (Truman & Reiss 1988, Kent & Levine 1988a–c, Weeks & Ernst-Utzschneider 1989). Definitive proof that regression and death are separate developmental options was provided by showing that the two events can be uncoupled within an individual motoneuron by manipulating ecdysteroid levels (Weeks 1987).

Larval abdominal motoneurons that innervate new muscles in the adult undergo extensive dendritic growth, with a typical pattern being the expansion of a larval unilateral arbor to a bilateral arbor in the adult (Levine & Truman 1985, Thorn & Truman 1989, Weeks & Ernst-Utzschneider 1989). Dendritic outgrowth depends on the elevation of ecdysteroids that normally accompanies adult development, and can be blocked if ecdysteroid release is delayed (such as during pupal diapause), or by treating with juvenile hormone at specific times (Levine & Truman 1985, Truman & Reiss 1988). As is discussed below, the regression or expansion of motoneuron arbors is correlated with developmental changes in synaptic connections.

The best-studied persistent sensory neurons innervate mechanosensory hairs located on a region of the larval abdomen that in the pupa develops into a specialized sensory structure called the *gin trap*. The larval afferents have characteristic axonal arbors within the CNS that expand significantly at pupation in response to the prepupal peak of ecdysteroids (Levine et al 1985, 1986, Levine 1989). Due to the location of their cell bodies in the periphery, these sensory neurons have proved to be ideal for studies of the endocrine dependence of neuronal outgrowth. Local application of juvenile hormone to the presumptive gin trap region during the commitment pulse blocks the sensory neurons' reprogramming, so that at pupation a heterochronic mosaic pupa is formed that bears a small patch of larval epidermis and hairs. The arbors of the treated sensory neurons fail to expand despite their location in a pupal CNS, thus indicating that the endocrine environment of the cell body is of paramount importance in directing changes in neuronal shape (Levine et al 1986). Similarly, local application of 20-hydroxyecdysone to presumptive gin trap sensory neurons in abdomens that were ligated to prevent the prepupal peak causes pupal growth of the sensory neurons arbors within an otherwise larval CNS (Levine 1989). These data provide the strongest evidence to date of a direct action of ecdysteroids and juvenile hormone in controling neuronal structure. The hormonally directed changes in sensory neuron arbors have demonstrable behavioral consequences (see below).

Developmental studies of persistent interneurons are in their infancy, but at least some larval interneurons can be re-identified in pupae (Sandstrom & Weeks 1988, Waldrop & Levine 1988, Levine & Weeks 1989). To ensure reliable re-identification, these studies have concentrated on interneurons that do not undergo major structural modifications, but it is likely that other interneurons do change morphologically.

Although hormonally triggered changes in neuronal structure are now well documented in *Manduca*, a challenge for the future is to elucidate the mechanisms involved. For instance, it is desirable to determine whether the neurons whose structures change are direct hormone targets or whether transsynaptic interactions with neurons or muscles are involved (e.g. Murphey et al 1975, Schneiderman et al 1982, Murphey 1986). A direct hormone action is consistent with the observed inability to prevent morphological changes in motoneurons by manipulating presynaptic afferents or target muscles (Weeks & Truman 1985, Kent & Levine 1988c, Jacobs & Weeks 1990). Although the evidence is good that hormones act directly on sensory neurons to trigger morphological changes, and the circumstantial evidence suggests that the same is true of motoneurons, more direct approaches are desirable. For instance, it would be of great interest to identify gene products involved in the hormonal responses of *Manduca* neurons (see below).

## Changes in Synaptic Connections and Behavior

The relationship between structural and functional changes in *Manduca* neurons during metamorphosis has been reviewed recently by Levine (1987), Weeks et al (1989), and Levine & Weeks (1989). In several cases it has been possible to correlate the hormonally mediated regression or growth of neuronal arbors with changes in synaptic connections and behavior. For instance, the contralateral growth of abdominal moto-neuron arbors during adult development is associated with the acquisition of new, monosynaptic excitatory inputs from stretch receptor sensory neurons that are important for postural reflexes of the moth (Levine & Truman 1982). During the larval-pupal transformation, a proleg with-drawal reflex mediated by monosynaptic excitatory connections between mechanosensory proleg sensory neurons and the proleg retractor moto-neurons (Weeks & Jacobs 1987, Peterson & Weeks 1988) is lost as the motoneurons regress (Weeks et al 1989). In heterochronic mosaics that have larval sensory neurons and regressed pupal motoneurons, the syn-aptic coupling between the sensory and motoneurons is reduced to the same extent as in normal pupae (Weeks & Jacobs 1988, Jacobs & Weeks 1990). Thus, the loss of the reflex behavior during metamorphosis appears to be due to regression of the motoneuron dendrites. Heterochronic mosaics have also been used to demonstrate that expansion of the axonal arbors of gin trap sensory neuron arbors within the CNS is necessary but not sufficient for the sensory neurons' ability to evoke the pupal-specific gin trap reflex (Levine et al 1986, 1989, Levine 1989). Thus, during meta-morphosis, the remodeling of neuronal arbors is involved in the acquisition of new behaviors needed for the next life stage, and the elimination of outmoded behaviors leftover from the previous stage.

These studies have all concerned relatively simple reflexes involving sensory and motoneurons, but most of the behavioral changes during metamorphosis undoubtedly involve changes in interneuron connections (e.g. Levine & Truman 1982, 1983, Mesce & Truman 1988, Miles & Weeks 1988 and in preparation, Waldrop & Levine 1989, Weeks et al 1989). Therefore, it will be especially illuminating to follow structural and func-tional changes in interneurons during metamorphosis. Initial studies indi-cate that the responses of abdominal interneurons to sensory input may differ in the larval and pupal stages (B. Waldrop and R. B. Levine, unpub-lished observations), and that the size of interneuron-evoked synaptic potentials in motoneurons may change markedly (D. J. Sandstrom and J. C. Weeks, unpublished observations).

## Future Directions

We have attempted to convey the substantial progress that has taken place in understanding the cellular changes that accompany insect meta-morphosis. Several approaches will be important for future progress. Con-tinued electrophysiological and anatomical studies of identified neurons and their synaptic interactions will provide further insights into how nervous systems can be modified to produce new behaviors, and how such modifications are regulated hormonally. Insect metamorphosis is also an ideal situation in which to explore the molecular mechanisms underlying neuronal development and differentiation, an area that ought to yield particularly exciting results in the coming years. The molecular biology of ecdysteroid and juvenile hormone action in *Manduca* epidermis is understood in considerable detail (e.g. Riddiford 1985, 1987). By utilizing the complementary advantages afforded by *Drosophila* and *Manduca*, it should be possible to identify neuronal genes that respond to ecdysteroids and juvenile hormone (e.g. Restifo & White 1988), and to determine how their products affect neuronal structure and function. Relevant to this approach are recent demonstrations that many genes involved in initial pattern formation within *Drosophila* embryos are expressed later during embryonic or postembryonic neural development (e.g. Brower 1987, Doe & Scott 1988, Doe et al 1988). These genes, or others known to be involved in early neuronal differentiation in *Drosophila* (e.g. Campos-Ortega 1988, Thomas et al 1988, Caudy et al 1988), might also be involved in the differentiation of neurons derived from postembryonic neurogenesis or the re-differentiation of persistent neurons. By continued multidisciplinary approaches to the study of insect metamorphosis, we hope to better understand how hormones influence structural and functional plasticity in all nervous systems.

ACKNOWLEDGMENTS

Research carried out in the authors' laboratories was supported by National Institute of Health (NIH) grant NS23208, a National Science Foundation (NSF) Presidential Young Investigator Award, and an Alfred P. Sloan Fellowship to J. C. Weeks; and NIH grant NS24822, NIH training grant NS07309, and NSF grant BNS8607066 to R. B. Levine. We thank Dr. K. S. Kent and Dr. B. A. Trimmer for their comments on the manuscript.

*Literature Cited*

Arnold, A. P., Gorski, R. A. 1984. Gonadal steroid induction of structural sex differences in the CNS. *Annu. Rev. Neurosci.* 7: 413–42

Bate, C. M. 1978. Development of sensory systems in arthropods. In *Handbook of Sensory Physiology, Development of Sensory Systems*, ed. M. Jacobson, 9: 1–53. Berlin: Springer-Verlag. 469 pp.

Bennett, K., Truman, J. W. 1985. Steroid-dependent survival of identifiable neurons in cultured ganglia of the moth *Manduca sexta*. *Science* 229: 58–60

Booker, R., Truman, J. W. 1987a. Postembryonic neurogenesis in the CNS of the tobacco hornworm, *Manduca sexta*. I. Neuroblast arrays and the fate of their progeny during metamorphosis. *J. Comp. Neurol.* 255: 548–59

Booker, R., Truman, J. W. 1987b. Postembryonic neurogenesis in the CNS of the tobacco hornworm, *Manduca sexta*. II: Hormonal control of imaginal nest cell degeneration and differentiation during metamorphosis. *J. Neurosci.* 7: 4107–14

Brower, D. L. 1987. Ultrabithorax gene expression in *Drosophila* imaginal discs and larval nervous system. *Development* 101: 83–92

Budnik, V., White, K. 1988. Catecholamine-containing neurons in *Drosophila melanogaster*: Distribution and development. *J. Comp. Neurol.* 268: 400–13

Campos-Ortega, J. A. 1988. Cellular interactions during early neurogenesis of *Drosophila melanogaster*. *Trends Neurosci.* 11: 400–4

Cantera, R., Nässel, D. R. 1987. Postembryonic development of serotonin-immunoreactive neurons in the blowfly CNS. II. The thoracico-abdominal ganglia. *Cell Tissue Res.* 250: 449–59

Caudy, M., Vässin, H., Brand, M., Tuma, R., Jan, L. Y., Jan, Y. N. 1988. *Daughterless*, a *Drosophila* gene essential for both neurogenesis and sex determination, has sequence similarities to *myc* and the *achaete-scute* complex. *Cell* 55: 1061–67

DeVoogd, T., Nottebohm, F. 1981. Gonadal hormones induce dendritic growth in the adult avian brain. *Science* 214: 202–4

Doe, C. Q., Scott, M. P. 1988. Segmentation and homeotic gene function in the developing nervous system of *Drosophila*. *Trends Neurosci.* 11: 101–6

Doe, C. Q., Smouse, D., Goodman, C. S. 1988. Control of neuronal fate by the *Drosophila* segmentation gene *even-skipped*. *Nature* 333: 376–78

Edwards, J. S. 1969. Postembryonic development and regeneration of the insect nervous system. *Adv. Insect Physiol.* 6: 97–137

Fahrbach, S. E., Truman, J. W. 1987. Possible interactions of a steroid hormone and neural inputs in controlling the death of an identified neuron in the moth *Manduca sexta*. *J. Neurobiol.* 18: 497–508

Fahrbach, S. E., Truman, J. W. 1988. Cycloheximide inhibits ecdysteroid-regulated neuronal death in the moth *Manduca sexta*. *Soc. Neurosci. Abstr.* 14: 368

Fahrbach, S. E., Truman, J. W. 1989. Autoradiographic identification of ecdysteroid binding cells in the nervous system of the moth *Manduca sexta*. *J. Neurobiol.* In press

Giebultowicz, J. M., Truman, J. W. 1984. Sexual differentiation in the terminal ganglion of the moth *Manduca sexta*: Role of sex-specific neuronal death. *J. Comp. Neurol.* 226: 87–95

Jacobs, G. A., Weeks, J. C. 1990. Post-

synaptic changes at a sensory-to-motor synapse contribute to the developmental loss of a reflex behavior during insect metamorphosis. *J. Neurosci.* In press

Kent, K. S., Hoskins, S. G., Hildebrand, J. G. 1987. A novel serotonin-immunoreactive neuron in the antennal lobe of the Sphinx moth *Manduca sexta* persists throughout postembryonic life. *J. Neurobiol.* 18: 451–65

Kent, K. S., Levine, R. B. 1988a. Neural control of leg movements in a metamorphic insect: Sensory and motor elements of the larval thoracic legs in *Manduca sexta*. *J. Comp. Neurol.* 271: 559–76

Kent, K. S., Levine, R. B. 1988b. Neural control of leg movements in a metamorphic insect: Persistence of the larval leg motor neurons to innervate the adult legs of *Manduca sexta*. *J. Comp. Neurol.* 276: 30–43

Kent, K. S., Levine, R. B. 1988c. Reorganization of an identified leg motor neuron during metamorphosis of the moth *Manduca sexta*. *Soc. Neurosci. Abstr.* 14: 1004

Kurz, E. M., Sengelaub, D. R., Arnold, A. P. 1986. Androgens regulate the dendritic length of mammalian motoneurons in adulthood. *Science* 232: 345–98

Levine, R. B. 1987. Neural reorganization and its endocrine control during metamorphosis. *Curr. Top. Dev. Biol.* 21: 341–65

Levine, R. B. 1989. Expansion of the axonal arborizations of persistent sensory neurons during insect metamorphosis: The role of 20-hydroxyecdysone. *J. Neurosci.* 9: 1045–54

Levine, R. B., Pak, C., Linn, D. 1985. The structure, function, and metamorphic reorganization of somatotopically projecting sensory neurons in *Manduca sexta* larvae. *J. Comp. Physiol.* 157: 1–13

Levine, R. B., Truman, J. W. 1982. Metamorphosis of the nervous system: Changes in the morphology and synaptic interactions of identified cells. *Nature* 299: 250–52

Levine, R. B., Truman, J. W. 1983. Peptide activation of a simple neural circuit. *Brain Res.* 279: 335–38

Levine, R. B., Truman, J. W. 1985. Dendritic reorganization of abdominal motoneurons during metamorphosis of the moth, *Manduca sexta*. *J. Neurosci.* 5: 2424–31

Levine, R. B., Truman, J. W., Linn, D., Bate, C. M. 1986. Endocrine regulation of the form and function of axonal arbors during insect metamorphosis. *J. Neurosci.* 6: 293–99

Levine, R. B., Waldrop, B., Tamarkin, D.

1989. The use of hormonally-induced mosaics to study alterations in the synaptic connections made by persistent sensory neurons during insect metamorphosis. *J. Neurobiol.* 20: 362–38

Levine, R. B., Weeks, J. C. 1989. Reorganization of neural circuits and behavior during insect metamorphosis. In *Perspectives on Neural Systems and Behavior*, ed. T. J. Carew, D. Kelley, pp. 195–228. New York: Liss

Mesce, K. A., Truman, J. W. 1988. Metamorphosis of the ecdysis motor pattern in the hawkmoth *Manduca sexta*. *J. Comp. Physiol.* 163: 287–99

Miles, C. I., Weeks, J. C. 1988. Developmental changes in pre-ecdysis motor patterns of the moth, *Manduca sexta*. *Soc. Neurosci. Abstr.* 14: 998

Murphey, R. K. 1986. Competition and dynamics of axon arbor growth in the cricket. *J. Comp. Neurol.* 251: 100–10

Murphey, R. K., Mendenhall, B., Palka, J., Edwards, J. S. 1975. Deafferentation slows the growth of specific dendrites of identified giant interneurons. *J. Comp. Neurol.* 159: 407–18

Nässel, D. R., Ohlsson, L., Cantera, R. 1988. Metamorphosis of identified neurons innervating thoracic neurohemal organs in the blowfly: Transformation of cholecystokininlike immunoreactive neurons. *J. Comp. Neurol.* 267: 343–56

Nordlander, R. H., Edwards, J. S. 1969a. Postembryonic brain development in the monarch butterfly *Danaus plexippus plexippus* L. I. Cellular events during brain morphogenesis. *Wilhelm Roux Arch. Entwicklungsmech. Org.* 162: 197–217

Nordlander, R. H., Edwards, J. S. 1969b. Postembryonic brain development in the monarch butterfly *Danaus plexippus plexippus* L. II. The optic lobes. *Wilhelm Roux Arch. Entwicklungsmech. Org.* 163: 197–220

Peterson, B. A., Weeks, J. C. 1988. Somatotopic mapping of sensory neurons innervating mechanosensory hairs on the larval prolegs of *Manduca sexta*. *J. Comp. Neurol.* 275: 128–44

Purves, D., Hadley, R. D. 1985. Changes in the dendritic branching of adult mammalian neurons revealed by repeated imaging *in situ*. *Nature* 315: 404–6

Restifo, L. L., White, K. 1988. Expression of a steroid hormone regulated gene in the CNS of *Drosophila melanogaster*. *Soc. Neurosci. Abstr.* 14: 733

Riddiford, L. M. 1985. Hormone action at the cellular level. In *Comprehensive Insect Physiology, Biochemistry, and Pharmacology*, ed. G. A. Kerkut, L. I. Gilbert, 8: 37–84. New York: Pergamon. 595 pp.

Riddiford, L. M. 1987. Hormonal control of sequential gene expression in insect epidermis. In *Molecular Entomology*, ed. J. H. Law, pp. 211–22. New York: Liss. 500 pp.

Sandstrom, D. J., Weeks, J. C. 1988. Identified interneurons in larval and pupal abdominal ganglia of the tobacco hornworm, *Manduca sexta*. *Soc. Neurosci. Abstr.* 14: 1003

Schneiderman, A. M., Matsumoto, S. G., Hildebrand, J. G. 1982. Trans-sexually grafted antennae influence the development of sexually dimorphic neurons in the brain of *Manduca sexta*. *Nature* 298: 844–46

Schwartz, L. M., Kay, B. K. 1987. *De novo* transcription and translation of new genes is required for the programmed death of the intersegmental muscles of the tobacco hawkmoth, *Manduca sexta*. *Soc. Neurosci. Abstr.* 13: 8

Taylor, H. M., Truman, J. W. 1974. Metamorphosis of the abdominal ganglia of the tobacco hornworm, *Manduca sexta*. *J. Comp. Physiol.* 90: 367–88

Thomas, J. B., Bastiani, M. J., Bate, N., Goodman, C. S. 1984. From grasshopper to *Drosophila*: A common plan for neuronal development. *Nature* 310: 203–7

Thomas, J. B., Crews, S. T., Goodman, C. S. 1988. Molecular genetics of the single-minded locus: A gene involved in the development of the *Drosophila* nervous system. *Cell* 52: 133–41

Thorn, R., Truman, J. W. 1989. Sex-specific neuronal respectification during the metamorphosis of the genital segments of the tobacco hornworm moth, *Manduca sexta*. *J. Comp. Neurol.* 284: 489–503

Tolbert, L. P., Matsumoto, S. M., Hildebrand, J. G. 1983. Development of synapses in the antennal lobe of the moth, *Manduca sexta*, during metamorphosis. *J. Neurosci.* 3: 1158–75

Truman, J. W. 1983. Programmed cell death in the nervous system of an adult insect. *J. Comp. Neurol.* 216: 445–52

Truman, J. W. 1987. The insect nervous system as a model system for the study of neuronal death. *Curr. Top. Dev. Biol.*, 21: 99–116

Truman, J. W., Bate, M. 1988. Spatial and temporal patterns of neurogenesis in the central nervous system of *Drosophila melanogaster*. *Dev. Biol.* 125: 145–57

Truman, J. W., Booker, R. 1986. Adult-specific neurons in the nervous system of *Manduca sexta*: Selective chemical ablation using hydroxyurea. *J. Neurobiol.* 17: 613–25

Truman, J. W., Reiss, S. E. 1976. Dendritic reorganization of an identified motoneuron during metamorphosis of the tobacco hornworm, *Manduca sexta. Science* 192: 477–79

Truman, J. W., Reiss, S. E. 1988. Hormonal regulation of the shape of identified motoneurons in *Manduca sexta. J. Neurosci.* 8: 765–75

Truman, J. W., Schwartz, L. M. 1984. Steroid regulation of neuronal death in the moth nervous system. *J. Neurosci.* 4: 274–80

Truman, J. W., Weeks, J. C., Levine, R. B. 1986. Developmental plasticity during metamorphosis of an insect nervous system. In *Comparative Neurobiology: Modes of Communication in the Nervous System*, ed. M. J. Cohen, F. Strumwasser, pp. 25–44. New York: Wiley. 397 pp.

Vallés, A. M., White, K. 1988. Serotonin-containing neurons in *Drosophila melanogaster*: Development and distribution. *J. Comp. Neurol.* 268: 414–28

Waldrop, B., Levine, R. B. 1988. Interneurons involved in multisegmental reflexes in larvae and pupae of the moth, *Manduca sexta. Soc. Neurosci. Abstr.* 14: 1003

Waldrop, B., Levine, R. B. 1989. Development of the gin trap reflex in *Manduca sexta* a comparison of larval and pupal motor responses. *J. Comp. Physiol.* In press

Weeks, J. C. 1987. Time course of hormonal independence for developmental events in neurons and other cell types during insect metamorphosis. *Dev. Biol.* 124: 163–76

Weeks, J. C., Ernst-Utzschneider, K. 1989. Respecification of larval proleg motoneurons during metamorphosis of the tobacco hornworm, *Manduca sexta*: segmental dependence and hormonal regulation. *J. Neurobiol.* 20: 569–92

Weeks, J. C., Jacobs, G. A. 1987. A reflex behavior mediated by monosynaptic connections between hair afferents and motoneurons in the larval tobacco hornworm, *Manduca sexta. J. Comp. Physiol.* 160: 315–29

Weeks, J. C., Jacobs, G. A. 1988. Developmental changes in postsynaptic neurons cause loss of a monosynaptic reflex during metamorphosis in *Manduca sexta. Soc. Neurosci. Abstr.* 14: 998

Weeks, J. C., Truman, J. W. 1984. Neural organization of peptide-activated ecdysis during the metamorphosis of *Manduca sexta*. II. Retention of the proleg motor pattern despite loss of the prolegs at pupation. *J. Comp. Physiol.* 155: 423–33

Weeks, J. C., Truman, J. W. 1985. Independent steroid control of the fates of motoneurons and their muscles during insect metamorphosis. *J. Neurosci.* 5: 2290–2300

Weeks, J. C., Truman, J. W. 1986a. Steroid control of neuron and muscle development during the metamorphosis of an insect. *J. Neurobiol.* 17: 249–67

Weeks, J. C., Truman, J. W. 1986b. Hormonally mediated reprogramming of muscles and motoneurons during the larval-pupal transformation of the tobacco hornworm, *Manduca sexta. J. Exp. Biol.* 125: 1–13

Weeks, J. C., Jacobs, G. A., Miles, C. I. 1989. Hormonally-mediated modifications of neuronal structure, synaptic connectivity, and behavior during metamorphosis of the tobacco hornworm, *Manduca sexta. Am. Zool.* In press

White, K., Hurteau, T., Punsal, P. 1986. Neuropeptide-FRMFamide-like immunoreactivity in *Drosophila*: Development and distribution. *J. Comp. Neurol.* 247: 430–38

White, K., Kankel, D. R. 1978. Patterns of cell division and movement in the formation of the imaginal nervous system in *Drosophila melanogaster. Dev. Biol.* 65: 296–321

Witten, J. L., Truman, J. W. 1988. Regulation of transmitter choice in an identified lineage in *Manduca sexta. Soc. Neurosci. Abstr.* 14: 28

*Annu. Rev. Neurosci. 1990. 13:195–225*

# SEGMENTATION AND SEGMENTAL DIFFERENTIATION IN THE DEVELOPMENT OF THE CENTRAL NERVOUS SYSTEMS OF LEECHES AND FLIES

*Michael Levine and Eduardo Macagno*

Department of Biological Sciences, Columbia University, New York, New York 10027

## INTRODUCTION

There are two current views regarding the phyletic organization of the arthropods. In one view, all arthropods [chelicerates, crustaceans, and uniramians (onychophorans, myriapods, and hexapods)] belong to a single phylum, arising through a complex evolutionary radiation that occurred some 500 to 600 million years ago; the second view holds that the arthropods are polyphyletic, since embryological features suggest separate evolutionary relationships between polychaetes and crustaceans, and between oligochaetes and uniramians (Anderson 1973, Gupta 1979). Regardless which of these is correct, the metamerism of annelids and arthropods is generally thought to have arisen in a common ancestor of both groups, and both structural and embryological analyses provide evidence for a close relationship between the two groups. One such analysis of several shared features has led to the proposal of a particular phylogenetic affinity between leeches and insects (Sawyer 1984). The relative closeness of these relationships has been somewhat compromised, however, by recent cross-species comparisons of nucleic acid sequences. Metamery is also a feature of the chordates, and, interestingly, sequence analysis of 18s rRNA does not indicate that, among the eucoelomates, annelids and arthropods are

195

0147–006X/90/0301–0195$02.00

any more closely related to each other than either is to the chordates or echinoderms (Field et al 1988). Further analysis of conservation and divergence at the molecular level is clearly necessary before solid conclusions about evolutionary relations can be made from these results, but they do indicate that our present understanding is incomplete.

The metameric organization of ectodermal and mesodermal derivatives along the anteroposterior body axis is a property common to annelids and arthropods. Both groups display reiterated features that permit the definition of segments throughout most of the length of the body. Arthropods have fixed, species-specific numbers of segments; among annelids, only the leeches have a fixed number, while oligochaetes and polychaetes do not. Similarity among segments is generally greater in annelids than it is in arthropods, the latter showing much greater segmental differentiation. Accompanying the specialization of segments is an increased functional interdependence of groups of segments, culminating in the organization of the body into a relatively small number of tagmata, as in the dipterans. In some cases, this leads to the loss in the adult of overt signs of segmentation at some positions, between segments that undergo fairly complete fusion. These general properties apply not only to external features, but to internal organs as well, and in particular to the nervous system.

We concentrate in this review on results of recent studies of the development of the central nervous systems of the leech and the fruit fly, particularly on those concerned with the processes of segmentation and segmental differentiation. The genesis of segmentation in leeches and flies has received intense scrutiny in recent years, and it has become clear that this process has parallels in the two groups but also differs in some instructive and interesting ways. Comparison of the developmental strategies evolved by these two groups may thus lead to further insight into basic mechanisms of morphogenesis. Furthermore, the increasing availability of specific markers has fostered significant advances in our knowledge of the development of their nervous systems. The study of segmental differences and how they arise during neurogenesis, moreover, can provide information on the relative roles of intrinsic and extrinsic determinants of neuronal phenotype. Each group offers particular advantages; the larger size of precursor cells and neurons has allowed a more complete analysis of cell lineage and cell properties in the leech, while genetic analysis has led to the isolation and characterization of many important developmental genes in the fruit fly. Advances in molecular techniques and the realization that genes with similar function may be isolated on the basis of sequence homology, along with the development of cellular techniques that permit work with smaller cells, suggest an expansion of approaches in both groups and a concomitant convergence of the kinds of information that will become available.

Space constraints have dictated the exclusion of several important areas from this review, particularly current studies of the peripheral nervous system.

## SEGMENT FORMATION IN LEECHES AND FLIES

The great majority of leeches have 32 segments, of which seven contribute to the tail region and four to the head, which also contains structures of non-metameric origin (Sawyer 1986). The 21 middle body segments are quite similar to each other, containing many iterated structures, such as ganglia, peripheral sensory structures, and body wall muscles. Many features, however, are heterogeneously distributed. Nephridia, for example, are found in body segments 2–18 in some leech families and in body segments 2–4 and 8–18 in others (Sawyer 1986); the sexual organs are generally found in body segments 5 (male) and 6 (female). In the CNS, some neurons have homologues in every segmental ganglion, some in several ganglia, and some in a single one. Some ganglia have many more neurons than others (Macagno 1980). Flies have fewer segments, divided into three groups that form the head, thorax, and abdomen. The head comprises an anterior region of non-metameric origin plus six segments (Jurgens et al 1986, Regulski et al 1987), while the thorax comprises three and the abdomen nine segments. Segments within the head are highly differentiated with respect to each other and to segments in the other regions; those in the thorax are more similar to each other, although dorsal structures are quite different, while those in the abdomen resemble one another to a great extent. The embryonic ventral nerve cord consists of a chain of three thoracic and nine abdominal ganglia that are very similar to each other. Three subesophageal ganglia connect the ventral cord to the supraesophageal ganglion (Poulson 1950, Hartenstein & Campos-Ortega 1984, Campos-Ortega & Hartenstein 1985). Late larval and pupal development lead to a great expansion of the cephalic neural centers and the thoracic ganglia, and a lesser growth of those in the abdomen. In the adult fly, the ventral nerve cord condenses into a single mass dominated by the thoracic components.

## Drosophila

Larval and adult body segments are composed of mesodermal and ectodermal derivatives; endodermal tissues do not display an overt metameric organization at any time during the *Drosophila* life cycle. During gastrulation, the presumptive mesoderm is formed by the invagination of the ventral furrow (Poulson 1950, Campos-Ortega & Hartenstein 1985). The ventral mesoderm and overlying ventral and lateral ectoderm correspond to the germ band, which will form several of the posterior head segments as well as each of the three thoracic and nine abdominal segments. The dorsal-most regions of the ectoderm do not contribute to metameric structures. The germ band first elongates by expanding posteriorly (to a maximum length 75% greater than the length of the egg) and folding back on itself along the dorsal surface. Transient metameric processes are observed soon after the completion of germ band elongation, approximately 5 to 6 hr following fertilization (Martinez-Arias & Lawrence 1985). However, final segment boundaries are not established until the onset, at about 7 to 8 hr, of germ band shortening, when the germ band retracts to its original length and position along the ventral surface. Such borders are morphologically defined by intersegmental furrows within the germ band, and by sites of attachment for segmentally repeated muscles of the body wall (Campos-Ortega & Hartenstein 1985). Each of the segments that are now defined have essentially identical organizations and are composed of three segmentally repeated tissues: (a) epidermis, or dermomere; (b) neural, or neuromere; and (c) mesoderm, or myomere.

Embryonic cells come to express distinct sets of genes and follow different pathways of morphogenesis based on their spatial coordinates within the early embryo. Many of the genes that specify this positional information have been defined in past genetic screens, based on disruptions in the cuticle pattern of advanced-stage embryos (Lewis 1978, Nüsslein-Volhard & Wieschaus 1980, Nüsslein-Volhard et al 1984, Jurgens et al 1984, Wieschaus et al 1984, Wakimoto et al 1984). At least 40 zygotically-active genes have been identified, which represent the vast majority of regulatory genes that control the embryonic pattern (reviewed in Levine & Harding 1987, Akam 1987, Ingham 1988). These genes fall into two groups: (a) those that differentiate the anteroposterior (A-P) pattern; and (b) those that control the dorsoventral (D-V) pattern. Most of the genes (32 of 40) control the A-P pattern and participate in the process of segmentation during early development. Nearly two thirds of these genes have been cloned and characterized. The majority (17 of 20) encode sequence-specific transcription factors and contain a known DNA-binding motif: 14 contain a copy of the evolutionarily conserved homeo box (McGinnis et al 1984, Laughon & Scott 1984, Levine & Hoey 1988), whereas the other three contain the zinc finger motif (Miller et al 1985, Rosenberg et al 1986, Tautz et al 1987, Nauber et al 1988, Oro et al 1988, Rhodes & Klug 1988). These studies strongly suggest that the specification of positional information in the early *Drosophila* embryo involves the modulation of gene expression at the level of transcription.

Nearly every one of the A-P regulatory genes that has been examined displays a unique pattern of expression and is expressed in a specific subset of cells in the early embryo (Carroll & Scott 1985, DiNardo et al 1985, Harding et al 1986, Macdonald et al 1986, Ingham et al 1986, Frasch et al 1987). The patterns are consistent with the hypothesis that many or all of the ~6000 cells that comprise the early embryo contain unique combinations of regulatory gene products. It is thought that these different permutations of gene expression play an important role in selecting cell fate and specifying diverse patterns of morphogenesis (Garcia-Bellido 1975, 1977). The misexpression of one or more of these genes in the early embryo can result in transformations of cell fate, and subsequent disruptions in the spatial organization of advanced-stage embryos (Struhl 1985, Schneuwly et al 1987, Kuziora & McGinnis 1988). For example, the expression of the homeotic gene *Antp* is normally restricted to the presumptive thorax in developing embryos, where it is important for morphogenesis of thoracic structures such as wings and legs (Kaufman et al 1980, Struhl 1981, Levine et al 1983). Misexpression of normal *Antp* products in head regions can cause the "classical" homeotic trans-formation of antennae into legs (Schneuwly et al 1987). Thus, a key problem regarding the specification of positional information in *Dro-sophila* is a question of regulation: How does a regulatory gene come to be expressed in exactly the right subset of cells at the right times?

The A-P genes can be divided into two groups: (*a*) segmentation genes and (*b*) homeotic genes. Segmentation genes divide the embryo into a repeating series of homologous body segments (Nüsslein-Volhard & Wie-schaus 1980). Homeotic genes act afterwards to make these segments morphologically and functionally distinct (Lewis 1978, Kaufman et al 1980). The segmentation genes fall into three classes: gap genes, pair-rule genes, and segment polarity genes (Nüsslein-Volhard & Wieschaus 1980). The precise expression of each of these segmentation genes depends upon a regulatory hierarchy that occurs in the early embryo (reviewed in Ingham 1988). The first step in the segmentation process involves the establishment of localized expression of the five gap genes within broad, overlapping domains along the anteroposterior axis (Jackle et al 1986, Nauber et al 1988, Oro et al 1988). The initiation of these localized patterns depends on maternal factors present in the unfertilized egg (i.e. Driever & Nüsslein-Volhard 1988a,b). The gap genes influence the segmentation process indirectly, by regulating the expression of pair-rule and homeotic genes (Meinhardt 1986, Carroll & Scott 1986, Howard & Ingham 1986, Frasch & Levine 1987). At least six of the eight pair-rule genes show a periodic pattern of expression along the length of early embryos, such that the products encoded by these genes accumulate in alternating segmental primordia (Hafen et al 1984, Macdonald et al 1986, Ingham et al 1986, Frigerio et al 1986, Frasch et al 1987, Gergen & Butler 1988). One of the central problems of the segmentation process is how relatively few gap genes provide sufficient information to define highly localized "stripes" of pair-rule gene expression. The pair-rule genes, in turn, implement segmentation by modulating the expression of segment polarity genes (Harding et al 1986, Macdonald et al 1986, DiNardo & O'Farrell 1987, Ingham et al 1988). For example, by the onset of gastrulation, the segment polarity gene *en* comes to be expressed in a series of 14 evenly-spaced stripes in the middle body region of gastrulating embryos (DiNardo et al 1985, Kornberg et al 1985). Each segmental primordium is composed of a strip of about four cells in width that encompasses the circumference of the embryo. *engrailed* (*en*) is expressed in the posterior-most cells of each primordium, where it is required for the establishment and maintenance of the posterior compartment and segment borders. In summary, the segmentation process depends upon a regulatory hierarchy that culminates with the precise expression of about ten different segment polarity genes in unique sets of cells in each segmental primordium. This hierarchy primarily involves a cascade of transcription factors and occurs during a brief interval of embryogenesis, beginning with the onset of cellular blastoderm formation and terminating with the completion of germ band elongation.

## Leeches

Recent studies of the development of segmentation in leeches have been greatly facilitated by the injection into single cells of dyes that serve as lineage tracers, a technique pioneered by Stent, Weisblat and collaborators (Weisblat et al 1980; reviewed in Stent & Weisblat 1985). Analyses of this type have been carried out in glossiphoniid leeches because they have large eggs and hence large teloblasts, which are more easily injected with the tracer dyes. Early cell divisions give rise to five bilateral pairs of teloblasts, known as M, N, O, P, and Q. The M teloblast gives rise largely to mesodermal structures, whereas the others are responsible mostly for ectodermal derivatives. Each teloblast undergoes a series of unequal divisions that generate linear arrays, or bandlets, of primary blast cells. The ten bandlets of blast cells come to lie in parallel along the anteroposterior axis of the developing embryo, forming a sheet of cells known as the germinal plate. Segmental tissues arise from this sheet, the clones derived from older blast cells contributing to the more anterior segments. The 32 segments are formed, therefore, in a temporal sequence that reflects the order of divisions that form the bandlets (Fernández 1980a,b, Fernández & Stent 1982).

The number of blast cells that contribute progeny to the 32 segments is not the same for all bandlets: the M, O, and P teloblasts provide 32 or 33 blast cells and the N and Q teloblasts 64. More than these numbers of blast cells are generated, but the extra ones degenerate. Although these numbers suggest that an adult segment (defined operationally as a morphological unit comprising a central ganglion and the peripheral territory it directly innervates) might arise from clones derived from single blast cells from the M, O, and P bandlets and pairs of blast cells from the N and Q bandlets, studies of the segmental distribution of the progeny of individual blast cells show that this is not entirely the case (Weisblat & Shankland 1985, Bissen & Weisblat 1987). In summary:

1. Two consecutive blast cells provide the segmental complement from each N bandlet. One of them generates anteriorly located cells, the other posterior ones.
2. Two consecutive blast cells provide the segmental complement from each Q bandlet, but their progenies are not simply segregated into anterior and posterior groups of cells.
3. For both the O and P bandlets, a single blast cell contributes its progeny to two adjacent segments; conversely, each segment contains contributions from two consecutive blast cells from these bandlets.
4. Each blast cell from an M bandlet appears to contribute progeny to at least three neighboring segments, and perhaps more.

It is clear, therefore, that some clones derived from individual blast cells do not obey segmental boundaries, but instead intermingle with those derived from neighboring blast cells. Thus, segments defined on the basis of the periodicity of structural components do not correspond to polyclones derived from the primary blast cells. It is not clear, from the studies here summarized, how regional specification beyond segmentation, i.e. into head, middle body, or tail domains, takes place in the leech. This process may require the action of homeotic genes, like those found in the fly, and further elucidation of this aspect of leech development may require the isolation of such genes in the leech.

The lineal identity of each cell in the leech appears to be invariant, to the extent that this has been tested (Kramer & Weisblat 1985). This invariance could be construed to mean that cell lineage and cell fate are causally related or, in other words, that intrinsic constraints determine cell fate. Thus, when an N or a Q teloblast is deleted, its progeny is found to be absent from the germinal plate. In contrast, the deletion of one of the teloblasts that will give rise to the O and P lineages always results in the progeny of the remaining cell following a P lineage, independent of which of the two teloblasts is killed (Shankland & Weisblat 1984; reviewed in Shankland 1987a). Thus the two cells are called the O/P teloblasts, since either of them can give rise to O or P pattern elements. Commitment of an O/P blast cell to the O pathway does not take place when this cell is born, nor for several hours afterwards, during which time it does not divide. However, once the blast cell begins to divide, its descendant clone gradually loses its potential to produce P pattern elements in response to the ablation of the P bandlet. The P bandlet always lies closer to the animal pole of the embryo; the interactions that define this positional property appear to involve not only the O bandlet, but also the overlying epithelium (Ho & Weisblat 1987). Further evidence that some blast cells have the potential to generate cells with other than their normal fates follows from other ablation experiments, in which blast cells are longitudinally translocated to other segments (Shankland 1984, Martindale & Shankland 1988). In such cases, the translocated blast cells take part in the formation of the host segment in a manner appropriate to the host but not to the donor segment.

Comparison of the lineages of cells in the subesophageal segments versus more posterior body segments reveals an interesting dichotomy. Early O/P blast cell divisions generate similar sublineages in both regions, and these eventually produce the same set of descendants. However, the sublineages themselves arise through different lineages, thus demonstrating that comparable complements of cells can result from alternative lineage paths during normal development (Shankland 1987b). Whereas similar fates from different lineages could result from the utilization of different combinations of intrinsic factors (e.g. activation of the same genes in different sequences; Shankland 1987a), this observation also underscores the need to consider possible roles for factors other than lineage in the generation of the leech embryo. Positional information, as we discussed above, plays a critical role in fly development; its role in the leech needs to be investigated further experimentally.

A major difference between the segmentation process in the fly and the leech is the degree to which regulatory determinants are pre-localized in the egg. Thus, in the fly, early development entails the sequential operation of a set of transcription factors resulting in the specification of the segmental features of the embryo. The first step in this process involves pre-localized maternal factors, such as *bicoid*, which are essential for the correct initiation of early zygotic gene expression. In the leech, segmentation follows from the sequence of asymmetric divisions that give rise to segmental founder cells. Determination of the functions of the leech equivalents of the fly gap, pair-rule, and segment polarity genes should prove very interesting. For example, is it conceivable that leech "pair-rule" genes perform roles in the leech analogous to their functions in flies? We presently lack morphological evidence for the activity of such genes in the annelid.

# EARLY NEUROGENESIS

## Early Neurogenesis and Segmentation in Drosophila

One of the first indications of metamery during *Drosophila* development is seen within the developing neuromeres that comprise the ventral cord of the CNS (Campos-Ortega & Hartenstein 1985). Here we consider the segmental origins of the ventral nerve cord but do not discuss the genesis of the procephalon, since this is not well known.

The ventral nerve cord develops from stem cells (known as neuroblasts) that segregate from the medial or neurogenic region of the ventral ectoderm (Poulson 1950). Segregation of cells commences immediately after gastrulation and occurs over a period of 3–4 hr. During this time, about 20% of the cells in the neurogenic region enlarge and migrate out of the ectoderm, forming bilateral monolayers of neuroblasts on the interior surface of the ventral ectoderm.

Neuroblast segregation occurs in three separate waves, with the first wave taking place during a brief span of only 10 min (Campos-Ortega & Hartenstein 1985). The first wave of neuroblasts forms two longitudinal columns on either side of the germ band. By the completion of this process, each presumptive neuromere is composed of an average of 24 neuroblasts, with 12 neuroblasts located bilaterally on either side of the ventral midline (Poulson 1950, Hartenstein & Campos-Ortega 1984, Campos-Ortega & Hartenstein 1985). It is not known whether the four neuroblasts that lie along a longitudinal column within a presumptive hemi-neuromere reflect the compartmentalized cell identities that have been established for the epidermis. The epidermal regions of each segment (a dermomere) consists of at least two "compartments," a posterior compartment and an anterior compartment (Morata & Lawrence 1975, Crick & Lawrence 1975, Kornberg 1981). Compartments do not intermingle during development, and they are composed of cells that derive from several precursor cells (that is, they are polyclonal). The segment polarity gene *en* plays a particularly crucial role in defining the posterior vs. anterior compartment, and it has been shown that *en* RNAs and proteins are normally expressed in the cells that comprise the posterior compartment, but not in cells of the anterior compartment (DiNardo et al 1985, Kornberg et al 1985). More recent studies on the pair-rule class of segmentation genes suggest the occurrence of at least three, and possibly as many as four, compartments per segment (Meinhardt 1986). Just following cellularization (at about $2\frac{1}{2}$–3 hr after fertilization), each segment primordium consists of a transversal strip of about four cells along the anterior-posterior axis. Only one of the cells within each of these primordia (the one located most posteriorly) expresses *en* and gives rise to the posterior compartment. Each of the other three cells also comes to express a distinctive set of segmentation genes (see below), and each appears to possess a unique developmental potential (reviewed in Levine & Harding 1987, Akam 1987, Ingham 1988). There is morphologic evidence for the occurrence of these four distinct cell states based on examining cuticular structures derived from the epidermal portions of each segment. It is possible, but currently not known, that similar processes define multiple cell identities for the neuroblast progenitors of a given presumptive neuromere.

Neuroblasts undergo a series of asymmetric divisions, which ultimately yield a total of about 250 neurons per hemi-neuromere. On average, each neuroblast produces eight to ten ganglion mother cells, the exact number depending on the location of the neuroblast (Poulson 1950, Seecof 1977). A ganglion mother cell undergoes one terminal, symmetric division that results in two cells that differentiate into neurons. Thus, on average, each neuroblast contributes 16 to 20 neurons to a segmental ganglion of the ventral nerve cord.

There are several transient manifestations of metamery prior to the formation of segmental or parasegmental borders during germ band shortening. The anatomical relationship betweeen these early metameric processes and the final segmentation pattern is not clear. One of the earliest metameric repeats that is observed is the arrangement of the neuroblasts within the presumptive abdominal neuromeres. As indicated above, following the second wave of segregation, neuroblasts are organized within a bilaterally symmetric "3 by 4" lattice composed of three longitudinal and four transverse rows (Campos-Ortega & Hartenstein 1985). However, each of the future abdominal neuromeres shows a peculiar periodicity in the arrangement of the neuroblasts within the intermediate longitudinal row, whereby there is a periodic lack of neuroblasts. Thus, each presumptive abdominal neuromere contains only two, not four, neuroblasts within the intermediate row, and shows a "neuroblast-gap-neuroblast-gap" pattern. Interestingly, this periodic gap of neuroblasts is closely coupled with a transient metamery of the overlying epidermis. Just beneath the intermediate row of neuroblasts there are clusters of epidermoblasts that divide several minutes earlier than their neighbors. It appears that these precocious epidermal clusters are located at the positions of the gaps within the presumptive abdominal neuromeres. At a slightly later stage of neurogenesis, new clusters of neuroblasts appear on either side of the ventral midline as a result of a third wave of neuroblast segregation. These clusters of neuroblasts are evenly spaced along the germ band and are separated by "bridges" of epidermoblasts.

Recent morphological studies indicate the occurrence of transient, metamerically repeated furrows at the completion of germ band elongation that do not appear to coincide with the final segment borders of larvae and adults. These furrows are thought to define "parasegments," which correspond to the posterior compartment of one segment and the anterior compartment of the adjacent segment (Martinez-Arias & Lawrence 1985). The parasegment appears to be the fundamental domain of function for many of the homeotic selector genes (Struhl 1984, Lawrence et al 1987; see below). It is difficult to assess whether presumptive neuromeres and/or ventral ganglia are of segmental or parasegmental origin. The mature neurons that comprise each ganglion project their axons into a neuropile that is divided by two commissures. Histologic studies suggest that each ganglion is segmental, with both commissures being located within the posterior half of the segment and most of the associated cell bodies in the anterior half. However, the patterns of *en* and homeotic gene expression in the CNS is consistent with the possibility that ganglia are parasegmental. *en* transcripts accumulate in the cell bodies located just anterior to the commissures of a given ganglion, a finding that suggests that they occur within the posterior compartment, not the anterior compartment as indicated by the histologic studies (Wedeen et al 1986, Carroll et al 1988). Moreover, the expression patterns of several homeotic genes, including *Antp* and *Ubx*, are consistent with the possibility of a parasegmental origin for ventral ganglia (Levine et al 1983, Akam & Martinez-Arias 1985). *Antp* products are largely confined to parasegment 4 of early embryos, which corresponds to the posterior compartment of T1 plus the anterior compartment of T2. Later in development *Antp* transcripts are primarily detected in the cell bodies of the "T1" ganglion and the cells associated with the commissures in "T2." A similar pattern of *Ubx* expression has been observed for the T3/A1 ganglia. The morphology of the intersegmental nerve is also consistent with a parasegmental organization of the CNS. This nerve is formed by the joining of two nerve roots from adjacent ganglia: the anterior root from the more posterior ganglion and the posterior root from the more anterior ganglion (Campos-Ortega & Hartenstein 1985).

## The Roles of Fly "Neurogenic" Genes in Cell Lineage Decisions

The embryonic nerve cord arises from a group of neuroblasts that segregate from the ventral ectoderm and form clusters that become the segmental ganglia. The initial event, the determination of which ectodermal cells become neuroblasts and which epidermoblasts, appears to be under the control of the so-called "neurogenic" genes (reviewed in Campos-Ortega 1988, Artavanis-Tsakonas 1988). [It should be noted that genes of the *achaete-scute* complex are likely to play crucial roles in this process as well (see Romani et al 1987, Villares & Cabrera 1987, Cabrera et al 1987).] A variety of experimental approaches, including genetic, molecular, laser ablation, and cell transplantation methods, have yielded data implicating these genes in cell interactions, mediated by an intricate signal transduction system, that function as a switch between the two cell lineages.

Considered in cross-section, approximately 72 cells comprise the circumference of the early *Drosophila* embryo. The 15–16 cells at the ventral midline invaginate to form the ventral furrow, and will give rise to mesodermal derivatives, including muscles. The 13 ventrolateral cells on either side of the ventral furrow correspond to the neurogenic ectoderm. Just after cellularization (about 3 hr after fertilization), the neurogenic region includes a total of 1600 cells that nearly span the length of the embryo, with 800 cells on either side of the ventral furrow. Two distinct cell populations originate from this region during gastrulation and germ band elongation (from about 3 to 5 hr after fertilization): epidermoblasts and neuroblasts. As development proceeds, about 25% of the ectodermal cells, those destined to become neuroblasts, leave the surface and migrate toward the interior of the embryo; the future epidermoblasts remain at the surface. This spatial segregation of neuroblast and epidermoblast lineages does not involve intervening mitoses, but occurs in the context of a fixed number of blastoderm cells established at cellularization.

Laser ablation studies in grasshoppers first suggested the importance of cell-cell interactions in the differentiation of the ectodermal lineages (Doe & Goodman 1985). When individual neuroblasts were ablated, they were replaced by ectodermal cells that would normally remain at the surface and give rise to epidermoblasts. These and other observations suggest that all ectodermal cells have the potential of becoming neuroblasts, and that commitment of a cell to the neuroblast lineage results in the inhibition of neighboring cells from also following the neuroblast pathway. As a consequence, the neuroblast lineage would correspond to the developmental "ground state." Consistent with this notion is the observation that "knock-out" (or null) mutations in any one of the "neurogenic" genes in *Drosophila* cause all of the cells within the neurogenic region to develop as neuroblasts at the expense of the epidermoblast lineage (see below). In contrast, no known null mutations cause the reciprocal transformation, i.e. all cells following the epidermoblast lineage.

The ten known "neurogenic" genes of *Drosophila* fall into two classes: six zygotically active genes that are expressed during embryogenesis, and four maternally expressed genes that encode products deposited in the unfertilized egg (Campos-Ortega 1988, Artavanis-Tsakonas 1988). Several of the zygotic genes are also expressed prior to fertilization. The zygotic genes have been studied in the most detail, and include *Enhancer of split*

[*E(spl)*], *Notch* (*N*), *Delta* (*Dl*), *mastermind* (*mam*), *big brain* (*bib*), and *neuralized* (*neu*). As mentioned above, null mutations in any one of these genes result in the same embryonic lethal phenotype, all cells within the neurogenic region becoming neuroblasts. As expected, such mutants show a gross hypertrophy of the embryonic CNS and a corresponding reduction of ventral epidermal structures. Several lines of evidence suggest that the neurogenic genes encode both the "signal" and the "receptor" for an intercellular regulatory signal, generated by cells committed to the neuroblast pathway, which inhibits neighboring cells from also entering the neuroblast pathway and forces them to follow the epidermoblast lineage. Mutations that disrupt either the signal or the receptor should give the same neuralizing phenotype, whereby all cells follow the developmental ground state, the neuroblast lineage, by default.

The results of cell transplantation studies suggest that five of the six zygotic "neurogenic" genes are involved in the specification of the signal and that the *E(spl)* gene is involved in the reception and/or processing of this signal (Technau & Campos-Ortega 1986). In these experiments, small numbers of ectodermal cells from the neurogenic region of a donor embryo carrying a null mutation in one of these genes are implanted into the neurogenic region of a wild-type recipient. Mutant donor cells that give rise to epidermoblasts must have a normal receptor, which allows them to respond to the normal signal in the wild type environment. Among mutants in the six genes, only those in *E(spl)* fail to give rise to epidermoblasts, thus indicating that only mutations in this gene affect the receptor and/or signal processing.

Other observations are also compatible with the notion of this function for *E(spl)*. For example, although *E(spl)*⁻ embryos show the same neuralized phenotype as null mutations in any of the other neurogenic genes, a special class of *E(spl)* mutant alleles produce the opposite phenotype. Embryos derived from females that are heterozygous for such dominant gain-of-function *E(spl)* alleles show hypotrophy of the CNS and a corresponding increase in the ventral epidermis. These mutations probably encode a constitutively active "receptor." The same phenotype is seen also in embryos that possess extra copies of the wild-type *E(spl)*⁺ allele. The conclusion from these genetic experiments is that raising the level of *E(spl)*⁺ activity (either by increasing the dose or by making the product constitutively active) increases the likelihood of a cell becoming an epidermoblast.

If *E(spl)* is the last step in the epidermoblast/neuroblast switch, then which of the remaining neurogenic genes specify the signal that activates *E(spl)*? As discussed above, transplantation experiments suggest that all five of the remaining neurogenic genes are reasonable candidates for encoding such a signal. However, studies of double mutant combinations suggest that two of these genes, *Notch* and *Dl*, may be particularly important. A striking interaction is observed between $E(spl)^-$ and $Notch^-$ null mutations. Although these two genes are completely unlinked, such mutations fail to complement (De la Concha et al 1986). That is, $E(spl)^-/+$; $Notch^-/+$ double heterozygotes show an embryonic lethal, neuralizing phenotype, similar to that observed for $E(spl)^-$ or $Notch^-$ homozygotes. Such interactions are also observed between $E(spl)^-$ and $Dl^-$ alleles. These results suggest that the simultaneous reduction of both $E(spl)^+$ and $Notch^+$ (or $Dl^+$) activity results in the breakdown of the signal transduction system. Perhaps a single copy of $Notch^+$ or $Dl^+$ reduces the level of signal, and together with a decrease in the level of receptor [resulting from only one copy of $E(spl)^+$], the threshold of activated receptor is inadequate to trigger the epidermoblast pathway of development.

*Notch, Dl,* and *E(spl)* have been cloned and characterized (Wharton et al 1985, Kidd et al 1986, Vassin et al 1987, Hartley et al 1988). Based on nucleotide sequence analysis, it appears that both *Notch* and *Dl* encode transmembrane proteins. The putative extracytoplasmic domains of the two proteins contain regions that are homologous to epidermal growth factor (EGF). The *Notch* protein appears to be composed of 2703 amino acid residues, and the bulk of the extracytoplasmic region is composed of EGF-like repeats. There are 36 of these cysteine-rich repeats, each of which contains 38 amino acid residues (Wharton et al 1985, Kidd et al 1986). In overall structure, the *Notch* protein is most homologous to the EGF precursor protein. The putative *Delta* protein appears to be smaller than *Notch* and is composed of only 880 amino acid residues. The extracytoplasmic domain contains nine EGF-like repeats (Vassin et al 1987). The significance of the homology to EGF, however, remains to be elucidated.

The deduced structures of the *Notch* and *Delta* proteins pose a possible paradox: How do integral membrane proteins provide the source of a signal that can act on neighboring cells? One possibility is that the EGF-like repeats are cleaved from the *Notch* and/or *Delta* proteins, and can serve as a diffusable ligand at a distance. There is currently no direct biochemical evidence that supports or rejects this possibility. A preferred model is that cell-cell contact is required for the presentation of the *Notch* and/or *Delta* signals. It is possible that the EGF-like repeats remain on the extracellular surface of a given cell and interact with the *E(spl)* receptor on neighboring cells, thereby activating *E(spl)* and triggering the epidermoblast pathway of development. Previous genetic studies are consistent with such a mechanism. Genetic mosaics have been used to examine the development of patches of *Notch^-* cells in a wild-type background (Hoppe

& Greenspan 1986). Mutant cells in relatively large clones or patches become neuroblasts for the most part, although they are surrounded by wild-type cells. In contrast, small $Notch^-$ clones become almost entirely epidermoblasts, which is consistent with the cell transplantation studies described above. The failure of most cells in large $Notch^-$ clones to follow an epidermoblast pathway suggests that the signal does not diffuse over a large distance, whereas the finding that most cells in small clones do indicates that the signal acts locally, perhaps through cell-cell contacts. The recent demonstration that a putative protein product of $E(spl)$ contains homology with the beta subunit of mammalian G proteins is compatible with its role in processing the signal generated by the other neurogenic genes (Hartley et al 1988).

## Cell Lineages and Cell Fates in the Leech

Cells derived from all five teloblasts contribute to the segmental ganglia of the leech nerve cord; the great majority of the central neurons are N-derived (Weisblat et al 1984). Some peripheral neurons have also been assigned a lineage, but most of the PNS cannot be identified with lineage tracers for technical reasons. A clone of cells derived from a single teloblast (M, N, O, P, or Q) is referred to as a *kinship group*. There is little or no crossing of the midline, the left and right kinship groups remaining within their own sides and showing mirror-symmetric cell distributions. In the description that follows, therefore, we consider only the lineages on one side of the animal.

A central hemiganglion arises from two primary blast cells from each of the N, O, P, and Q kinship groups. The central nervous system arises in the territory of the germinal plate corresponding to the N and O bandlets. Hence, neurons and glia from these bandlets do not need to migrate extensively to get to the CNS, in contrast to those arising from the more lateral P and Q bandlets (Torrence & Stuart 1986). The progenies of the two N blast cells that contribute to each hemisegment are largely segregated into rostral and caudal domains within a hemiganglion, inter-mixing slightly with each other but not at all with those of adjacent ganglia. Almost all of the N-derived neurons are found in the CNS, but two or three neurons that arise from the more caudal clones migrate laterally away from the ganglionic rudiment and into positions along the nerve roots (Weisblat et al 1984).

The more anterior of the two O blast cells that contributes to a hemi-segment generates an anterior dorsal cluster of neurons and at least one giant glial cell in the CNS, as well as one peripheral neuron. Its progeny also extends into the next, more rostral segment. The other O blast cell gives rise to middle and posterior central neurons and to two peripheral neurons in this hemisegment, as well as cells in the caudal neighbor. The clones of the two O blast cells interdigitate somewhat, and hence do not define a very clear anteroposterior margin between them. Two P primary blast cells also contribute to one hemisegment, their clones extending into the adjacent neighbors. The more anterior P blast cell gives rise to most of the P-derived central neurons, as well as six peripheral neurons, while the more posterior one gives rise to three central and one peripheral neuron. As with the O blast cell clones, the P clones do not segregate from one another along an anteroposterior boundary.

Of the two types of Q blast cells, one generates a few central neurons and glia, plus several peripheral neurons, while the other generates only a few peripheral neurons (Weisblat et al 1984). Since the Q bandlet is the most laterally located in the germinal plate, Q neuroblasts and glioblasts destined for the CNS need to migrate significant distances along stereotypic routes in the germinal plate in order to reach their definitive locations (Torrence & Stuart 1986). At least some of these cells appear to proliferate when they reach the ganglionic rudiment, but some of the migrating cells may be postmitotic.

The M kinship group provides the smallest contribution to the nervous system, only three or four central neurons per hemiganglion. These are derived from a single M primary blast cell, which also gives rise to other structures in the same segment as well as the two more rostral segments (Weisblat & Shankland 1985).

Details of the relationship between kinship group and cell identity in the CNS has been explored in *Haementaria* (Kramer & Weisblat 1985). Of the approximately 400 neurons in a segmental ganglion, 260 to 320 belong to the N, 40 to 100 to the O, 16 to 24 to the P, 12 to 18 to the Q, and 6 to 12 to the M kinship groups. In addition, any particular identified neuron always originates from the same teloblast, suggesting that the cell lineage of each neuron is determinate. However, neurons of the same functional type do not necessarily belong to the same kinship group. For example, one pair of pressure mechanosensory cells, those innervating dorsal skin, belongs to the O kinship group, but the pair that innervates ventral skin belongs to the P kinship group. Similarly, the glial cells arise from several teloblasts (N, O, P, and Q). The kinship groups of the monoamine-containing neurons have been determined (Stuart et al 1987) in several species. All the serotonin-containing neurons belong to the N kinship group, whereas the dopamine-containing neurons belong to the O, P, and Q kinship groups.

There is an overproduction of cells in the segmental ganglia during early embryonic development, and ~20% of the cells degenerate (Stewart et al 1986). Some of this cell death represents the loss in every segment of homologous cells (Stewart et al 1987), some of it corresponds to the loss of certain cells in specific segments (Macagno & Stewart 1987). A possible model for the genesis of segmentally heterogeneous distributions of neurons is that the initial production of cells gives rise to a full complement of possible neuronal phenotypes in every ganglion, followed by the selective loss of different subsets at each location. Our present knowledge suggests that this model might not apply to all cell phenotypes; a few observations suggest that some cell lineages may in fact yield different adult neuronal phenotypes (E. Macagno, unpublished observations). However, this notion is very preliminary and further work is needed to resolve it.

It would appear, therefore, that cell lineages may be more determinate in the leech than they are in the fly, though compensation for ablated precursors has been demonstrated in both. Hence, if the early cell fates in the leech are essentially determined by lineage, what then might be the roles of the leech equivalents of the fly neurogenic genes? Surely, if present in leeches, they will be involved in some aspect of cell-cell interactions, although it is conceivable that flies have co-opted a rather general signal transduction mechanism for the more specialized task of establishing epidermal vs. neuronal lineages. Or, perhaps, they may act later in leech development, and affect decisions different from the epidermal/neural choice.

## DIFFERENTIATION

The nervous systems of leeches and flies show segmental differences at various levels, in numbers of neurons, in the distribution and branching patterns of homologues, and in the fates of cells that are derived from iterated lineages. A question of particular relevance in the context of this review is the extent to which such segmental differences can be assigned to endogenous determinants, such as segment-specific gene expression controlled by neurons themselves, and to what extent these arise from interactions with other tissues or other epigenetic factors. Segment-specific neuronal gene expression under the control of homeotic and other segmentation genes has been documented in *Drosophila* (e.g. Doe et al 1988a,b), suggesting that at least some aspects of neuronal phenotypes may be determined endogenously. However, many segment-specific phenotypes appear to result from interactions, as discussed below.

### Segmental Differences in Cell Number

Ganglia in adult leeches have on the order of 400 neurons in each of the 21 body segments, excepting those that contain the male (segment 5) and female (segment 6) sexual organs, which contain more (Macagno 1980).

The head ganglion has two parts, a supraesophageal region, which is not metamerically derived (Weisblat et al 1984), and a subesophageal region, which arises from the fusion of four metameric ganglia early in development and contains about 1600 neurons. The tail ganglion is formed from the fusion of seven metameric ganglia and presumably contains on the order of 3000 neurons, though this number has never been measured directly.

The manner in which the adult complement of neurons is attained has been studied in the hirudinid leeches *Haemopis marmorata* and *Hirudo medicinalis* (Stewart et al 1986, Baptista & Macagno 1988). In these leeches, the ganglia associated with the sexual organs (hereafter called the sex ganglia) contain a few hundred extra neurons that are not found in other segments (Macagno 1980). During the first third of embryonic development (about ten days), blast cells from each of the five kinship groups (described in a previous section) proliferate and generate a population of cells in each metameric ganglion whose number exceeds the adult complement by about 10–20%. A period of cell death during the second third of embryogenesis brings this number down to about 400 in all segments, including the sex segments (Stewart et al 1986). Near the end of embryogenesis, but particularly throughout postembryonic development, the number of cells increases in the sex ganglia, with the full complement achieved in sexually mature animals (Stewart et al 1986, Baptista & Macagno 1988). Thus, this segmental difference in cell number appears to be due to differential cell proliferation rather than to differential cell death, as appears to be the case in some other systems.

A recent series of experiments shows that the generation of extra cells in the sex ganglia depends upon an interaction between these ganglia and the male organ (Baptista & Macagno 1988). When the organ is removed before the sixteenth day of embryogenesis, the extra cells fail to appear and the sex ganglia end up with the same number of cells as other segmental ganglia. A similar result is obtained if the nerves connecting the male organ to the sex ganglia is ablated. If these operations are carried out after the sixteenth day but before the extra cells begin to appear, however, the extra cells are generated, thus suggesting that the interaction with the male organ serves a triggering function rather than a continuous one. A reasonable interpretation of these observations is that neuroblasts, which might be present in all segmental ganglia, receive a signal in the sex ganglia that causes them to continue dividing and thus generate the extra cells. Countering this possibility, however, is the additional observation that an ectopic male organ, transplanted at ten days, fails to induce the appearance of extra cells in the ganglion that now innervates it (Baptista & Macagno 1988), although other phenotypic changes do occur (e.g. in the branching pattern of certain neurons; Loer et al 1987; see below). However, it is possible that transplantation at ten days is too late to elicit extra cells. An alternative hypothesis that also fits these results is that the sources of the extra cells are segment-specific neuroblasts that are found only in the sex ganglia. Segmental differences of this kind may well be specified by homeotic genes such as have been characterized in the fly (see below). Recent observations (Wysocka-Diller et al 1989) of the embryonic patterns of expression of leech homeobox genes suggest that cognates of the fly homeotic genes are present in the leech.

Segmental differences in cell number in the ventral CNS of the adult fly appear to arise from segmental differences in the numbers of larval neuroblasts that produce the majority of the adult neurons (Truman & Bate 1988). Although the embryonic ventral ganglia have similar numbers of neurons ($\sim 250$), postembryonic addition of cells is considerably greater in the thoracic than in the abdominal ganglia; over 4000 neurons are added to each of the former, less than a hundred to each of the latter. These neurons arise from neuroblasts, which are more numerous in thoracic (at least 47) than in abdominal (6 in most segments) ganglia. In addition, thoracic neuroblasts produce on the average 80–100 neurons each, while those in the abdomen yield only at most 12 neurons. One hypothesis regarding the source of the larval neuroblasts is that they are persistent embryonic neuroblasts that, after undergoing a set number of cell divisions in the embryo, had stopped dividing and decreased in size, becoming indistinguishable from other cells. Another possibility is that the larval neuroblasts are cells that were put aside in the embryo, to begin generating cells at a later time. In either case, how segmental differences occur in their number and in the number of divisions they undergo is not known.

## Segment-specific Distribution of Homologous CNS Neurons

Leech segmental ganglia arise from identical complements of primary blast cells and, to a large extent, have very similar complements of neurons. For example, all the body ganglia have six pairs of touch sensory neurons and a pair of Retzius cells (Muller et al 1981). However, there are many examples of identified neurons, homologues of which are heterogeneously distributed in the nerve cord. Homologues in the leech are defined by several properties, such as cell body size and position, branching pattern, function, transmitter phenotype, and the kinship group they belong to. Homologues of some neurons are found in most body ganglia (e.g. pairs of annulus erector or AE motor neurons in body ganglia 1 to 18; Gillon & Wallace 1984), some in a few (e.g. the unpaired posteriomedial serotonergic or PMS neurons in ganglia 2 to 7 in *Hirudo*; Macagno & Stewart

1987), and some in only one ganglion (e.g. the paired rostral penile evertor or RPE motor neurons in ganglion 6; Zipser 1979).

Observations in early embryogenesis of several different species suggest that there are three alternative ways to attain heterogeneity in homologue distribution: (*a*) initially homogeneous distribution, followed by segment-specific cell death; (*b*) initially homogeneous distribution, followed by segment-specific differentiation of phenotypes; (*c*) heterogeneous distribution from the beginning. For example, in *Hirudo*, the PMS neurons mentioned above first appear in pairs in all body ganglia and only later achieve their adult distribution through cell death (Macagno & Stewart 1987). In *Haementaria*, two pairs of P cells appear in all segmental ganglia, but one pair is lost in the fifth and sixth body segments, although whether this loss occurs by cell death or by these cells differentiating into other cell types is unclear (Loer et al 1986). In contrast, in glossiphoniid embryos, pairs of PMS neurons never appear in body segments 8–21, as in adults, though cell death reduces the pairs to single cells in segments 1–7 (Stuart et al 1987). Data supporting the second alternative, that the final distributions are attained by segment-specific cell differentiation, are less convincing but strongly suggestive (Stewart 1985). Cell death, as mentioned above, trims the initial population of cells in a ganglion by 10–20%, in some cases deleting all cells of a particular type (Stewart et al 1987), in others one of a pair in each ganglion (Macagno & Stewart 1987), in still others probably homologues in specific segments. The mechanisms responsible for these segmental specificities are unknown at present, and could comprise both endogenous instructions, such as programmed cell death, and interactions with tissues outside the CNS whose distribution is segmentally heterogeneous.

## Segment-specific Branching Patterns of Homologues

A particular branching pattern is a feature shared, to a great extent, by neurons of the same type, such as segmentally iterated homologues. Homologous neurons, however, frequently show segment-specific variations, for example, by having or lacking specific processes as a function of position along the nerve cord. Differences in both central arborization and pattern of extraganglionic axonal projections have been reported in many different species (e.g. Bate et al 1981, Gao & Macagno 1987a,b, Gillon & Wallace 1984, Glover & Mason 1986, Goodman et al 1981, Mittenthal & Wine 1978, Shafer & Calabrese 1981, Wilson 1979).

The question of how segmental differences in patterns of axonal projections arise during embryonic development has been recently investigated for several identified leech neurons. In all the cases studied, no segmental differences are seen early in development. All cells, irrespective of ganglion identity, appear to generate all the axonal projections that are seen in the adult. Segmental differences are manifested later by some of the homologues, when certain of their projections cease growing and retract (Wallace 1984, Glover & Mason 1986, Gao & Macagno 1987a,b). The retraction of processes does not seem to be a property controlled by the neuron itself but is the result of an interaction with other cells. For one group of neurons, the HA, AP, and AE neurons (Muller et al 1981), the interaction is among homologues. These cells extend axonal projections to the contralateral periphery and longitudinal projections to neighboring ganglia. The longitudinal projections grow to the next ganglia, where, if they encounter ipsilateral homologues, they stop growing and retract. If they do not encounter a homologue, they continue growing and branch to and innervate the periphery in the adjacent segments. A longitudinal projection does not retract, moreover, if the homologue is ablated early, before retraction begins (Gao & Macagno 1987a,b). Whether this apparent interaction between homologues is direct, contact-mediated inhibition or a competition for some required factor is not known at present. More recent experiments indicate that if a competition does take place, it is not for peripheral targets but for some factor within the CNS. Moreover, the normally retracted longitudinal projections can be rescued and maintained if the lateral projections are disconnected from peripheral targets before retraction begins (Gao & Macagno 1988).

Another leech neuron, the Retzius cell, also retracts longitudinal projections, but only in the two segments containing the sexual organs, and independently of its neighboring homologues (Jellies et al 1987, Loer et al 1987). Instead, this process retraction appears to be triggered by the innervation of the sexual organ. No retraction takes place if the organs are removed early (Loer et al 1987), whereas Retzius cells at other locations retract longitudinal projections if they innervate a naturally occurring (Macagno et al 1986) or transplanted ectopic organ (Loer et al 1987).

## The Roles of Fly Segmentation and Homeotic Genes

SEGMENTATION GENES    Nearly every one of the segmentation and homeotic genes that has been characterized is expressed in specific subsets of embryonic neurons, both central and peripheral (Carroll & Scott 1985, White & Wilcox 1985, Carroll et al 1986, Frasch et al 1987, Doe et al 1988a). Since most of these genes encode proteins that are likely to act as transcription factors, it is possible that they might also participate in key aspects of neurogenesis. Perhaps one or more of these regulatory proteins modulate the expression of "target" genes that are responsible for specific neuronal properties. In several instances the expression observed in neural tissues is quite different from that observed in the early embryo. For example, the pair-rule gene *fushi tarazu* (*ftz*) is expressed in a series of seven periodic stripes along the anteroposterior axis. This pattern is transient and disappears prior to the onset of neurogenesis, which begins at about 7 to 8 hr following fertilization. There is no obvious lineage relationship between the neural pattern of *ftz* expression seen in older embryos and this striped pattern (Carroll & Scott 1985, Doe et al 1988b). In fact, whereas the early pattern is periodic and involves only alternating segmental primordia, neural expression occurs in about 30 of the 250 neurons that comprise each hemisegment at 8 to 9 hr of development. By 12 hr, after the completion of ventral cord formation and the onset of condensation, *ftz* is no longer expressed in the CNS.

The neural expression of *ftz* is not fortuitous, but instead involves *cis* regulatory sequences within the *ftz* promoter that are distinct from the sequences responsible for the early periodic pattern (Hiromi et al 1985, Hiromi & Gehring 1987, Doe et al 1988b). It is possible, in principle, that the neural pattern of *ftz* expression does not involve selective transcription, but instead depends upon a post-transcriptional process. For example, perhaps all neurons efficiently transcribe *ftz*, but its mRNA is subject to differential degradation, thereby permitting stable accumulation of high steady-state levels of the transcript in certain neurons. The *ftz* promoter has been studied by attaching different 5′ sequences to the reporter gene *lacZ* and introducing these promoter fusions into flies by P-element mediated germ line transformation. The most obvious and important conclusion of these studies is that both the early striped pattern and the later neural expression are regulated at the level of transcription. These studies also show that CNS expression depends on regulatory sequences (called the neural element) located within a region of the *ftz* promoter that is distinct and separable from the sequences responsible for the early pattern (Hiromi et al 1985, Hiromi & Gehring 1987). When attached to a foreign promoter, the *ftz* neural element can induce the expression of a reporter gene in the usual "*ftz*" neurons of the CNS (Y. Hiromi, personal communication).

Another pair-rule gene, *even-skipped* (*eve*), also shows a later, different pattern of expression in the CNS following its participation in the segmentation process during early development (Frasch et al 1987). *eve* is expressed in 16 of the ~250 neurons per hemiganglion but, unlike *ftz*, its pattern of expression persists in the CNS for the duration of embryonic and larval development. Preliminary promotor fusion studies have identified discrete *eve* neural regulatory sequences (R. Warrior and M. Levine, unpublished results). The demonstration of a "purposeful" mechanism for the CNS expression of *ftz* (and possibly *eve*) suggests that one or both of these genes play important roles in some aspect of neurogenesis, such as determining cell fate or cell differentiation. Stronger support for this possibility is based on examining the morphology of identified neurons in embryos that lack the normal activity of these genes.

An assessment of *ftz* and *eve* function in the CNS has been obtained by examining conditional mutants. In the case of *ftz*, a synthetic mutant was obtained by attaching the normal *ftz* gene to a defective promoter lacking neural regulatory sequences. This defective *ftz* gene was introduced into *ftz⁻* embryos by P-element mediated germ line transformation, and was shown to "rescue" the early segmentation defect while failing to produce detectable levels of *ftz* protein in the CNS (Doe et al 1988b). However, examination of the phenotypes of seven (aCC, pCC, MP1, dMP2, vMP2, RP1, and RP2) of the ~30 neurons known to express *ftz* normally revealed only one discernible abnormality. The axon of the RP2 neuron failed to grow anteriorly and ipsilaterally, as it normally does, instead growing sometimes contralaterally, along with the axon of its sibling RP1, other times sending axons in two directions. This result was interpreted as suggesting a partial switch in cell fate, RP2 becoming more like RP1 in the absence of *ftz* activity. The lack of an effect in the other neurons examined may mean that *ftz* has no function in these cells, or that its function is redundant and can be supplied by other regulatory products.

The uncoupling of segmentation and neural activity was more easily achieved for *eve* because of the availability of a temperature-sensitive mutation in this locus. Embryos homozygous for this mutation were grown at permissive (low) temperatures during the early period when *eve* function is critical for the segmentation process, then switched to high temperature during the time that neural expression occurs, thereby inactivating the *eve* protein. Two of the identified neurons that normally express *eve*, aCC and RP2, were specifically disrupted due to the loss of *eve⁺* gene activity, whereas the pCC neuron appeared unaffected (Doe et al 1988a). These three identified neurons are both *eve⁺* and *ftz⁺*, yet each responds differently to the loss of *eve* and/or *ftz* neural function. Loss of function by either gene fails to affect axonal morphology of the pCC neuron, while only loss of *eve* function affects its sibling, the aCC neuron; in the absence of *ftz* expression, the aCC neuron expresses *eve* and has normal axonal projections. The transformation of the RP2 neuron seen in the *eve* mutants is similar to that seen when *ftz⁺* activity is lost. In fact, the loss of *ftz* activity causes a failure to activate *eve* expression in the RP2 neuron, thus suggesting that the abnormal RP2 morphology observed is due to the loss of *eve* activity in both cases. Perhaps *ftz* normally activates *eve* expression in RP2, which in turn modulates the activities of genes more directly responsible for axonogenesis. It is likely that this regulation occurs at the level of transcription, since the temperature-sensitive mutation that completely abolishes *eve* neural (and segmentation) function corresponds to a single amino acid substitution within the homeo box, the domain that has been shown to mediate the sequence-specific DNA binding activity of the *eve* protein (Frasch et al 1988).

In addition to *ftz* and *eve*, three of the other four segmentation proteins that have been examined show selective patterns of expression in the embryonic CNS. Two of these additional genes, the gap genes *hb* and *Kr*, display completely novel CNS patterns that are virtually unrelated to the expression patterns that are associated with their early segmentation activities (Rosenberg et al 1986, Tautz et al 1987, Gaul & Jackle 1987). By analogy to *ftz* and *eve*, it is reasonable to expect that the *hb* and *Kr* promoters will contain specific neural sequences, separable from the regulatory elements required for segmentation. Both *hb* and *Kr* contain multiple copies of the zinc finger DNA binding motif, and hence might have regulatory functions in the CNS (Rosenberg et al 1986, Tautz et al 1987). However, there is no information at present that supports or rejects this view. The demonstration that all three classes of segmentation genes, gap, pair-rule, and segment polarity, are expressed in the CNS suggests that the localized expression of these genes might involve the same regulatory hierarchy that operates in the early embryo (see section above). This seems unlikely, however, since the regulatory relationships established among at least some of the genes in the early embryo are different from the interactions that occur in the CNS. For example, during early development *eve* exerts (either directly or indirectly) a negative effect on *ftz* expression, whereas *ftz* has no apparent effect on *eve* expression (Harding et al 1986, Macdonald et al 1986, Frasch et al 1988). Just the opposite situation is seen in the CNS, as discussed above (Doe et al 1988a,b).

HOMEOTIC GENES    There is a close correlation between the sites of function for each of the seven homeotic selector genes in epidermal/cuticular tissues and their sites of expression in the embryonic and larval CNS. For example, the homeotic gene *Antp* is required for the morphogenesis of epidermal derivatives of the thorax (Kaufman et al 1980, Struhl 1981). Based on both RNA localization studies and antibody staining, the site of peak *Antp* expression corresponds to the meso- and metathoracic ganglia of the embryonic ventral cord (Levine et al 1983, Carroll et al 1986). Such colinear expression of homeotic genes in epidermal and neural tissues is quite different from the uncoupling of the early vs. neural expression seen for many of the segmentation genes (see above). Another important difference between the patterns of homeotic and segmentation genes in the CNS is that a given segmentation gene is expressed in the same subset of neurons in every ganglion of the cord, whereas a particular homeotic gene is expressed at peak levels in many or all of the neurons of just one or several adjacent ganglia (Canal & Ferrús 1987; reviewed in Doe & Scott 1988). This fundamental difference in expression might be expected to reflect distinct roles for segmentation and homeotic genes in neural development. For example, segmentation genes might control reiterated features common to segmentally homologous neurons, such as axonogenesis, while homeotic genes might control the segment-specific properties that sometimes distinguish homologous cells or structures in different regions of the CNS.

Most of the homeotic genes in *Drosophila* are located in one of two clusters or complexes, called the Antennapedia complex (ANT-C; Kaufman et al 1980) and the Bithorax complex (BX-C; Lewis 1978, Sanchez-Herrero et al 1985). Several studies have implicated some of these genes in specific aspects of neural differentiation. The first indication came from studies on the projection of an interneuron, the giant fiber, in the adult nervous system (Thomas & Wyman 1984). Normally, this axon extends along the ventral midline from the brain to the mesothoracic neuromere, where, apparently in response to a specific cue present in this region, the axon turns laterally and synapses onto particular motoneurons. BX-C mutations that cause the transformation of the metathorax into a mesothorax also affect the morphology of the giant fiber. Specifically, the fiber grows further, into the duplicated mesothoracic neuromere, and makes turns and synapses in both mesothoracic regions of the CNS. One interpretation of these observations is that the *Antp* gene, which is responsible for the morphogenesis of mesothoracic structures, either directly or indirectly specifies a cue extrinsic to the giant fiber, possibly involving the synaptic targets of the fiber. In BX-C mutations there is a posterior expansion of *Antp* activity into the metathoracic domain of the CNS, perhaps reflecting the homeotic transformation to a mesothoracic phenotype. This misexpression of *Antp* might be the cause of the transformation of metathoracic neurons to their mesothoracic homologues and consequently the duplication of the cues that affect the growth of the giant fiber.

Evidence that BX-C genes participate in neuronal morphogenesis is based on studies of the specialized structures in each of the three thoracic ganglia of larvae and pupae that are known as the leg neuromeres (Teugels & Ghysen 1983, 1985). Certain mutations in the BX-C cause the transformation of epidermal tissues in the first abdominal segment into corresponding structures of the metathorax. In such mutants there is a corresponding duplication of the leg neuromere in the first abdominal ganglion. This result alone does not prove that BX-C$^+$ products act autonomously in neural tissues; an alternative explanation is that homeotic transformations in epidermal tissues induce changes in associated neural tissues. Evidence that such an inductive process is not responsible for the duplication of the leg neuromeres stems from the demonstration that certain "weak" BX-C mutations can uncouple transformations of the epidermis and duplications of the leg neuromeres.

## CONCLUDING REMARKS

In summary, we believe that the leech and fly systems offer complementary strengths and weaknesses in examining contemporary problems in neurogenesis. The primary advantage of *Drosophila* is the remarkable progress made in recent years in both the identification and molecular characterization of the key regulatory genes that control early development. An unexpected finding is that about half of these regulatory genes encode proteins that are related by a 60 amino acid residue DNA binding motif called the homeo box. Mounting evidence indicates that the homeo box proteins function as sequence-specific transcription factors and directly modulate gene expression. Preliminary studies suggest that at least some homeo box proteins might play critical roles in neuronal morphogenesis, perhaps affecting processes such as cell migration, cell death, cell adhesion, or axonal guidance. Homeo box genes are present in virtually every eukaryote that has been investigated, including the leech (McGinnis 1985, Wysocka-Diller et al 1989), and some are expressed in discrete regions of the mammalian CNS during development (reviewed in Holland & Hogan 1988). Thus, a critical evaluation of homeo box gene function might be central to an understanding of evolutionarily conserved mechanisms of neurogenesis. However, the fly and mammalian systems pose several serious drawbacks toward addressing this issue. Most importantly, there is only limited information concerning the details of specific interactions among identified neurons, and small cell size frequently restricts electrophysiological studies of cell properties and synaptic circuitry. Leeches might offer a more experimentally tractable system, due to considerable progress in obtaining lineage information and the relative ease of injecting identified neurons and stem cells with tracers and macromolecules. Disruption of homeo box gene function in specific leech neurons might be achieved through various "reverse genetic" techniques such as the injection of antibodies or oligonucleotides.

Acknowledgments

We thank Marty Shankland and Barry Yedvobnick for their useful comments on the manuscript.

*Literature Cited*

Akam, M. 1987. The molecular basis for metameric pattern in the *Drosophila* embryo. *Development* 101: 1–22

Akam, M., Martinez-Arias, A. 1985. The distribution of *Ultrabithorax* transcripts in *Drosophila* embryos. *EMBO J.* 4: 1690–1700

Anderson, D. T. 1973. *Embryology and Phylogeny in Annelids and Arthropods.* Oxford: Pergamon

Artavanis-Tsakonas, S. 1988. The molecular biology of the *Notch* locus and the fine tuning of differentiation in *Drosophila. Trends Genet.* 4: 95–100

Baptista, C. A., Macagno, E. R. 1988. The role of the sexual organs in the generation of postembryonic neurons in the leech *Hirudo medicinalis. J. Neurobiol.* 19: 707–26

Bate, C. M., Goodman, C. S., Spitzer, N. C. 1981. Embryonic development of identified neurons: Segment-specific differences in the H cell homologues. *J. Neurosci.* 1: 103–6

Bissen, S. T., Weisblat, D. A. 1987. Early differences between alternate n blast cells in leech embryos. *J. Neurobiol.* 18: 251–69

Cabrera, C. V., Martinez-Arias, A., Bate, M. 1987. The expression of three members of the *achaete-scute* gene complex correlates with neuroblast segregation in *Drosophila. Cell* 50: 425–33

Campos-Ortega, J. A. 1988. Cellular interactions during early neurogenesis in *Drosophila melanogaster. Trends Neurosci.* 11: 400–5

Campos-Ortega, J. A., Hartenstein, V. 1985. *The Embryonic Development of Drosophila Melanogaster.* Berlin: Springer-Verlag

Canal, I., Ferrús, A. 1987. The expression of *Ultrabithorax* (*Ubx*) during development of the nervous system of *Drosophila. J. Neurogenet.* 4: 161–77

Carroll, S. B., DiNardo, S., O'Farrell, P. H., White, R., Scott, M. P. 1988. Temporal and spatial relationships between homeotic and segmentation gene expression in *Drosophila* embryos: Distribution of the *fushi tarzu, engrailed, Sex combs reduced, Antennapedia* and *Ultrabithorax* proteins. *Genes Dev.* 2: 350–60

Carroll, S. B., Laymon, R. A., McCutcheon, M. A., Ryley, P. D., Scott, M. P. 1986. The localization and regulation of *Antennapedia* protein expression in *Drosophila* embryos. *Cell* 47: 113–22

Carroll, S. B., Scott, M. P. 1985. Localization of the *fushi tarazu* protein during *Drosophila* embryogenesis. *Cell* 43: 47–57

Carroll, S. B., Scott, M. P. 1986. Zygotically active genes that affect the spatial expression of the *fushi tarazu* segmen-tation gene during early *Drosophila* embryogenesis. *Cell* 45: 113–26

Crick, F. H. C., Lawrence, P. A. 1975. Compartments and polyclones in insect development. *Science* 189: 340–45

De la Concha, A., Dietrich, U., Weigel, D., Campos-Ortega, J. A. 1986. Functional characterization of neurogenic genes of *Drosophila melanogaster. Genetics* 118: 499–508

DiNardo, S., Kuner, J. M., Theis, J., O'Farrell, P. H. 1985. Development of embryonic pattern in *D. melanogaster* as revealed by accumulation of the nuclear *engrailed* protein. *Cell* 43: 59–69

DiNardo, S., O'Farrell, P. H. 1987. Establishment and refinement of segmental pattern in the *Drosophila* embryo: Spatial control of *engrailed* expression by pair-rule genes. *Genes Dev.* 1: 1212–25

Doe, C. Q., Goodman, C. S. 1985. Early events in insect neurogenesis. *Dev. Biol.* 111: 206–19

Doe, C. Q., Hiromi, Y., Gehring, W. J., Goodman, C. S. 1988b. Expression and function of the segmentation gene *fushi tarazu* during *Drosophila* neurogenesis. *Science* 239: 170–75

Doe, C. Q., Scott, M. P. 1988. Segmentation and homeotic gene function in the developing nervous system of *Drosophila. Trends Neurosci.* 11: 101–6

Doe, C. Q., Smouse, D., Goodman, C. S. 1988a. Control of neuronal fate by the *Drosophila* segmentation gene *even-skipped. Nature* 333: 376–78

Driever, W., Nüsslein-Volhard, C. 1988a. A gradient of *bicoid* protein in *Drosophila* embryos. *Cell* 54: 83–93

Driever, W., Nüsslein-Volhard, C. 1988b. The *bicoid* protein determines position in the *Drosophila* embryo in a concentration-dependent manner. *Cell* 54: 95–104

Fernández, J. 1980a. Embryonic development of the Glossiphoniid leech *Theromyzon rude*: Characterization of developmental stages. *Dev. Biol.* 76: 245–62

Fernández, J. 1980b. Embryonic development of the Glossiphoniid leech *Theromyzon rude*: Structure and development of the germinal bands. *Dev. Biol.* 78: 407–34

Fernández, J., Stent, G. 1982. Embryonic development of the hirudinid leech *Hirudo medicinalis*: Structure, development and segmentation of the germinal plate. *J. Embryol. Exp. Morphol.* 72: 71–96

Field, K. G., Olsen, G. J., Lane, D. J., Giovannoni, S. J., Ghiselin, M. T., Raff, E. C., Pace, N. R., Raff, R. A. 1988. Molecular phylogeny of the animal kingdom. *Science* 239: 748–53

Frasch, M., Hoey, T., Rushlow, C., Doyle, H., Levine, M. 1987. Characterization and localization of the *even-skipped* protein of *Drosophila*. *EMBO J.* 6: 749–59

Frasch, M., Levine, M. 1987. Complementary patterns of *even-skipped* expression involve their differential regulation by a common set of segmentation genes in *Drosophila*. *Genes Dev.* 1: 981–95

Frasch, M., Warrior, R., Tugwood, J. D., Levine, M. 1988. Molecular analysis of *even-skipped* mutants in *Drosophila* development. *Genes Dev.* 2: 1824–38

Frigerio, G., Burri, M., Bopp, D., Baumgartner, S., Noll, M. 1986. Structure of the segmentation gene *paired* and the *Drosophila* PRD gene set as part of a gene network. *Cell* 47: 735–46

Gao, W.-Q., Macagno, E. R. 1987a. Extension and retraction of axonal projections by some developing neurons in the leech depends upon the existence of neighboring homologues. I. The HA cells. *J. Neurobiol.* 18: 43–59

Gao, W.-Q., Macagno, E. R. 1987b. Extension and retraction of axonal projections by some developing neurons in the leech depends upon the existence of neighboring homologues. II. The AP and AE neurons. *J. Neurobiol.* 18: 295–313

Gao, W.-Q., Macagno, E. R. 1988. Axon extension and retraction by leech neurons: Severing early projections to peripheral targets prevents normal retraction of other projections. *Neuron* 1: 269–77

Garcia-Bellido, A. 1975. Genetic control of wing disc development in *Drosophila*. In *Cell Patterning*, pp. 161–82. Amsterdam: Assoc. Sci. Publ.

Garcia-Bellido, A. 1977. Homeotic and atavic mutations in insects. *Am. Zool.* 17: 613–20

Gaul, U., Jackle, H. 1987. How to fill a gap in the *Drosophila* embryo. *Trends Genet.* 3: 127–32

Gergen, J. P., Butler, B. A. 1988. Isolation of the *Drosophila* segmentation gene *runt* and analysis of its expression during embryogenesis. *Genes Dev.* 2: 1179–93

Gillon, J. W., Wallace, B. G. 1984. Segmental variation in the arborization of identified neurons in the leech central nervous system. *J. Comp. Neurol.* 228: 142–48

Glover, J. C., Mason, A. 1986. Morphogenesis of an identified leech neuron: Segmental specification of axonal outgrowth. *Dev. Biol.* 115: 256–60

Goodman, C. S., Bate, C. M., Spitzer, N. C. 1981. Embryonic development of identified neurons: Origin and transformation of the H cell. *J. Neurosci.* 1: 94–102

Gupta, A. P., ed. 1979. *Arthropod Phylogeny.* New York: Van Nostrand Reinhold

Hafen, E., Kuroiwa, A., Gehring, W. J. 1984. Spatial distribution of transcripts from the segmentation gene *fushi tarazu* of *Drosophila*. *Cell* 27: 825–31

Harding, K., Rushlow, C., Doyle, H. J., Hoey, T., Levine, M. 1986. Cross-regulatory interactions among pair-rule genes in *Drosophila*. *Science* 233: 953–59

Hartenstein, V., Campos-Ortega, J. A. 1984. Early neurogenesis in wild-type *Drosophila melanogaster*. *Wilhelm Roux's Arch. Dev. Biol.* 193: 308–25

Hartley, D. A., Preiss, A., Artavanis-Tsakonas, S. 1988. A deduced gene product from the *Drosophila* neurogenic locus, *Enhancer of Split*, shows homology to mammalian G-protein beta subunit. *Cell* 55: 785–95

Hiromi, J., Gehring, W. J. 1987. Regulation and function of the *Drosophila* segmentation gene *fushi tarazu*. *Cell* 50: 963–74

Hiromi, J., Kuroiwa, A., Gehring, W. J. 1985. Control elements of the *Drosophila* segmentation gene *fushi tarazu*. *Cell* 43: 603–13

Ho, R. K., Weisblat, D. A. 1987. A provisional epithelium in leech embryo: Cellular origins and influence on a developmental equivalence group. *Dev. Biol.* 120: 520–34

Holland, P. W. H., Hogan, B. L. M. 1988. Expression of homeo box genes during mouse development: A review. *Genes Dev.* 2: 773–82

Hoppe, P. E., Greenspan, R. J. 1986. Local function of the *Notch* gene for embryonic ectodermal pathway choice in *Drosophila*. *Cell* 46: 773–83

Howard, K., Ingham, P. W. 1986. Regulatory interactions between the segmentation genes *fushi tarazu*, *hairy* and *engrailed* in the *Drosophila* blastoderm. *Cell* 44: 949–59

Ingham, P. W. 1988. The molecular genetics of embryonic pattern formation in *Drosophila*. *Nature* 335: 25–34

Ingham, P. W., Baker, N. E., Martinez-Arias, A. 1988. Regulation of segment polarity genes in the *Drosophila* blastoderm by *fushi tarazu* and *even-skipped*. *Nature* 331: 73–75

Ingham, P. W., Ish-Horowicz, D., Howard, K. R. 1986. Correlative changes in homeotic and segmentation gene expression in *Kruppel* mutant embryos of *Drosophila*. *EMBO J.* 5: 1659–65

Jackle, H., Tautz, D., Schuh, R., Seifert, E., Lehmann, R. 1986. Cross-regulatory interactions among the gap genes of *Drosophila*. *Nature* 324: 668–70

Jellies, J., Loer, C. M., Kristan, W. B. 1987. Morphological changes in leech Retzius neurons after target contact during embryogenesis. *J. Neurosci.* 7: 2618–29

Jurgens, G., Lehmann, R., Schardin, M., Nüsslein-Volhard, C. 1986. Sequential organization of the head in the embryo of *Drosophila melanogaster. Wilhelm Roux's Arch. Dev. Biol.* 195: 359–77

Jurgens, G., Weischaus, E., Nüsslein-Volhard, C., Kluding, H. 1984. Mutations affecting the larval cuticle in *Drosophila melanogaster.* II: Zygotic loci on the third chromosome. *Wilhelm Roux's Arch. Dev. Biol.* 193: 283–95

Kaufman, T. C., Lewis, R., Wakimoto, B. 1980. Cytogenetic analysis of chromosome-3 in *Drosophila melanogaster:* The homeotic gene complex in polytene chromosome interval 84A-B. *Genetics* 94: 115–33

Kidd, S., Kelley, M. R., Young, M. W. 1986. Sequence of the *Notch* locus of *Drosophila melanogaster:* Relationship of the encoded protein to mammalian clotting and growth factors. *Mol. Cell Biol.* 6: 3094–3108

Kornberg, T. 1981. *engrailed:* A gene controlling compartment and segment formation in *Drosophila. Proc. Natl. Acad. Sci. USA* 78: 1095–99

Kornberg, T., Siden, I., O'Farrell, P., Simon, M. 1985. The *engrailed* locus of *Drosophila: In situ* localization of transcripts reveals compartment-specific expression. *Cell* 40: 45–53

Kramer, A. P., Weisblat, D. A. 1985. Developmental neural kinship groups in the leech. *J. Neurosci.* 5: 388–407

Kuziora, M. A., McGinnis, W. 1988. Autoregulation of a homeotic selector gene. *Cell* 55: 477–85

Laughon, A., Scott, M. P. 1984. Sequence of a *Drosophila* segmentation gene: Protein structure homology with DNA-binding proteins. *Nature* 310: 25–31

Lawrence, P. A., Johnston, P., Macdonald, P., Struhl, G. 1987. The *fushi tarazu* and *even-skipped* genes delimit the borders of parasegments in *Drosophila* embryos. *Nature* 328: 440–42

Levine, M., Hafen, E., Garber, R. L., Gehring, W. J. 1983. Spatial distribution of *Antennapedia* transcripts during *Drosophila* development. *EMBO J.* 2: 2037–46

Levine, M., Harding, K. 1987. Spatial regulation of homeo box gene expression in *Drosophila. Oxford Surv. Eukary. Genes* 5: 116–42

Levine, M., Hoey, T. 1988. Homeobox proteins as sequence-specific transcription factors. *Cell* 55: 537–40

Lewis, E. B. 1978. A gene complex controlling segmentation in *Drosophila. Nature* 276: 565–70

Loer, C. M., Jellies, J., Kristan, W. B. 1987. Segment-specific morphogenesis of leech Retzius neurons requires particular peripheral targets. *J. Neurosci.* 7: 2630–38

Loer, C. M., Schley, C., Zipser, B., Kristan, W. B. 1986. Development of segmental differences in the pressure mechanosensory neurons of the leech *Haementaria ghilianii. J. Comp. Neurol.* 254: 403–9

Macagno, E. R. 1980. Number and distribution of neurons in leech segmental ganglia. *J. Comp. Neurol.* 190: 283–302

Macagno, E. R., Peinado, A., Stewart, R. R. 1986. Segmental differentiation in the leech nervous system: Specific phenotypic changes associated with ectopic targets. *Proc. Natl. Acad. Sci. USA* 83: 2746–50

Macagno, E. R., Stewart, R. R. 1987. Cell death during gangliogenesis in the leech: Competition leading to the death of PMS neurons has both random and nonrandom components. *J. Neurosci.* 7: 1911–18

Macdonald, P. M., Ingham, P., Struhl, G. 1986. Isolation, structure, and expression of *even-skipped:* A second pair-rule gene of *Drosophila* containing a homeo box. *Cell* 47: 721–34

Martindale, M. Q., Shankland, M. 1988. Developmental origin of segmental differences in the leech ectoderm: Survival and differentiation of the distal tubule cells is determined by the host segment. *Dev. Biol.* 125: 290–300

Martinez-Arias, A., Lawrence, P. A. 1985. Parasegments and compartments in the *Drosophila* embryo. *Nature* 313: 639–42

McGinnis, W. 1985. Homeo box sequences of the Antennapedia class are conserved only in higher animal genomes. *Cold Spring Harbor Symp. Quant. Biol.* 50: 263–70

McGinnis, W., Levine, M. S., Hafen, E., Kuroiwa, A., Gehring, W. J. 1984. A conserved DNA sequence in homeotic genes of the *Drosophila* Antennapedia and bithorax complexes. *Nature* 308: 428–33

Meinhardt, H. 1986. Hierarchical inductions of cell states: A model for segmentation in *Drosophila. J. Cell Sci. Suppl.* 4: 357–81

Miller, J., McLachlan, A. D., Klug, A. 1985. Repetitive zinc-binding domains in the protein transcription factor IIIA from *Xenopus* oocytes. *EMBO J.* 4: 1609–14

Mittenthal, J. E., Wine, J. J. 1978. Segmental homology and variation in flexor motoneurons in the crayfish abdomen. *J. Comp. Neurol.* 177: 311–34

Morata, G., Lawrence, P. A. 1975. Control of compartment development by the engrailed gene in Drosophila. Nature 255: 614–18

Muller, K. J., Nicholls, J. G., Stent, G. S. 1981. Neurobiology of the Leech. Cold Spring Harbor, NY: Cold Spring Harbor Lab.

Nauber, U., Pankratz, M. J., Kienlin, A., Seifert, E., Klemm, U., Jackle, H. 1988. Abdominal segmentation of the Drosophila embryo requires a hormone receptor-like protein encoded by the gap gene knirps. Nature 336: 489–92

Nüsslein-Volhard, C., Wieschaus, E. 1980. Mutations affecting segment number and polarity in Drosophila. Nature 287: 795–801

Nüsslein-Volhard, C., Wieschaus, E., Kluding, H. 1984. Mutations affecting the larval cuticle in Drosophila melanogaster. I: Zygotic loci on the second chromosome. Wilhelm Roux's Arch. Dev. Biol. 193: 267–82

Oro, A. E., Ong, E. S., Margolis, J. S., Posakony, J. W., McKeown, M., Evans, R. M. 1988. The Drosophila gene knirps-related is a member of the steroid-receptor gene superfamily. Nature 336: 493–96

Poulson, D. F. 1950. Histogenesis, organogenesis, and differentiation in the embryo of Drosophila melanogaster (Meigen). In Biology of Drosophila, ed. M. Demerec, pp. 168–274. New York: Wiley

Regulski, M., McGinnis, N., Chadwick, R., McGinnis, W. 1987. Developmental and molecular analysis of Deformed, a homeotic gene controlling Drosophila head development. EMBO J. 6: 767–77

Rhodes, D., Klug, A. 1988. Zinc fingers: A novel motif for nucleic acid binding. In Nucleic Acids and Molecular Biology, ed. F. Eckstein, D. M. J. Lilley, Vol. 2. Berlin: Springer-Verlag

Romani, S., Campuzano, S., Modolell, J. 1987. The achaete-scute complex is expressed in neurogenic regions of Drosophila embryos. EMBO J. 6: 2085–92

Rosenberg, U. B., Schroder, C., Preiss, A., Kienlin, A., Cote, S., Riede, I., Jackle, H. 1986. Structural homology of the product of the Drosophila Kruppel gene with Xenopus transcription factor-IIIA. Nature 319: 336–39

Sanchez-Herrero, E., Vernos, I., Marco, R., Morata, G. 1985. Genetic organization of the Drosophila bithorax complex. Nature 313: 108–13

Sawyer, R. T. 1984. Arthropodization in the Hirudinea: Evidence for a phylogenetic link with insects and other Uniramia? Zool. J. Linn. Soc. London 80: 303–22

Sawyer, R. T. 1986. Leech Biology and Behaviour. Oxford: Oxford Univ. Press

Schneuwly, S., Klemenz, R., Gehring, W. J.

1987. Redesigning the body plan of Drosophila by ectopic expression of the homeotic gene Antennapedia. Nature 325: 816–18

Seecof, R. L. 1977. A genetic approach to the study of neurogenesis and myogenesis. Am. Zool. 17: 577–84

Shafer, M. R., Calabrese, R. L. 1981. Similarities and differences in the structure of segmentally homologous neurons that control the heart in the leech, Hirudo medicinalis. Cell Tissue Res. 214: 137–53

Shankland, M. 1984. Positional determination of supernumerary blast cell death in the leech embryo. Nature 307: 541–43

Shankland, M. 1987a. Position-dependent cell interactions and commitments in the formation of the leech nervous system. Curr. Top. Dev. Biol. 21: 31–63

Shankland, M. 1987b. Cell lineage in leech embryogenesis. Trends Genet. 3: 314–19

Shankland, M., Weisblat, D. A. 1984. Stepwise commitment of blast cell fates during the positional specification of the O and P cell lines in the leech embryo. Dev. Biol. 106: 326–42

Stent, G. S., Weisblat, D. A. 1985. Cell lineage in the development of invertebrate nervous systems. Annu. Rev. Neurosci. 8: 45–70

Stewart, R. R. 1985. The genesis of segmental differences in two species of leeches. PhD dissertation. Columbia Univ., New York

Stewart, R. R., Gao, W.-Q., Peinado, A., Zipser, B., Macagno, E. R. 1987. Cell death during gangliogenesis in the leech: Bipolar cells appear and then degenerate in all ganglia. J. Neurosci. 7: 1919–27

Stewart, R. R., Spergel, D., Macagno, E. R. 1986. Segmental differentiation in the leech nervous system: The genesis of cell number in the segmental ganglia of Haemopis marmorata. J. Comp. Neurol. 253: 253–59

Struhl, G. 1981. A homeotic mutation transforming leg to antenna in Drosophila. Nature 292: 635–38

Struhl, G. 1984. Splitting the bithorax complex of Drosophila. Nature 308: 454–57

Struhl, G. 1985. Near reciprocal genotypes caused by inactivation or indiscriminate expression of the Drosophila segmentation gene ftz. Nature 318: 677–80

Stuart, D. K., Blair, S. S., Weisblat, D. A. 1987. Cell lineage, cell death, and the developmental origin of identified serotonin- and dopamine-containing neurons in the leech. J. Neurosci. 7: 1107–22

Tautz, D., Lehmann, R., Schnurch, H., Schuh, R., Seifert, E., Kienlin, A., Jones, K., Jackle, H. 1987. Finger protein of novel structure encoded by *hunchback*, a second member of the gap class of *Drosophila* segmentation genes. *Nature* 327: 383–89

Technau, G. M., Campos-Ortega, J. A. 1986. Lineage analysis of transplanted individual cells in embryos of *Drosophila melanogaster*. *Wilhelm Roux's Arch. Dev. Biol.* 195: 445–54

Teugels, E., Ghysen, A. 1983. Independence of the numbers of legs and leg ganglia in *Drosophila bithorax* mutants. *Nature* 304: 440–42

Teugels, E., Ghysen, A. 1985. Domains of action of bithorax genes in *Drosophila* central nervous system. *Nature* 314: 558–61

Thomas, J. B., Wyman, R. J. 1984. Duplicated neural structure in Bithorax mutant *Drosophila*. *Dev. Biol.* 102: 531–33

Torrence, S. A., Stuart, D. K. 1986. Gangliogenesis in leech embryos: Migration of neural precursor cells. *J. Neurosci.* 6: 2736–46

Truman, J. W., Bate, M. 1988. Spatial and temporal patterns of neurogenesis in the central nervous system of *Drosophila melanogaster*. *Dev. Biol.* 125: 145–57

Vassin, H., Bremer, K. A., Knust, E., Campos-Ortega, J. A. 1987. The neurogenic gene *Delta* of *Drosophila melanogaster* is expressed in neurogenic territories and encodes a putative transmembrane protein with EGF-like repeats. *EMBO J.* 6: 3431–40

Villares, R., Cabrera, C. V. 1987. The *achaete-scute* gene complex of *D. melanogaster*: Conserved domains in a subset of genes required for neurogenesis and their homology to *myc*. *Cell* 50: 415–24

Wakimoto, B. T., Turner, F. R., Kaufman, T. C. 1984. Defects in embryogenesis in mutants associated with the Antennapedia gene complex of *Drosophila melanogaster*. *Dev. Biol.* 102: 147–72

Wallace, B. G. 1984. Selective loss of neurites during differentiation of cells in leech central nervous system. *J. Comp. Neurol.* 228: 149–53

Wedeen, C., Harding, K., Levine, M. 1986. Spatial regulation of Antennapedia and Bithorax gene expression by the *Polycomb* locus in *Drosophila*. *Cell* 44: 739–48

Weisblat, D. A., Harper, G., Stent, G. S., Sawyer, R. T. 1980. Embryonic cell lineages in the nervous system of the glossiphoniid leech *Helobdella triserialis*. *Dev. Biol.* 76: 58–78

Weisblat, D. A., Kim, S. Y., Stent, G. S. 1984. Embryonic origins of cells in the leech *Helobdella triserialis*. *Dev. Biol.* 104: 65–85

Weisblat, D. A., Shankland, M. 1985. Cell lineage and segmentation in the leech. *Philos. Trans. R. Soc. London Ser. B* 312: 39–56

Wharton, K. A., Johansen, K. M., Xu, T., Artavanis-Tsakonas, S. 1985. Nucleotide sequence from the neurogenic locus *Notch* implies a gene product that shares homology with proteins containing EGF-like repeats. *Cell* 43: 567–81

White, R. A. H., Wilcox, M. 1985. Distribution of *Ultrabithorax* proteins in *Drosophila*. *EMBO J.* 4: 2035–43

Wieschaus, E., Nüsslein-Volhard, C., Jurgens, G. 1984. Mutations affecting larval cuticle in *Drosophila melanogaster*. III: Zygotic loci on the X-chromosome and fourth chromosome. *Wilhelm Roux's Arch. Dev. Biol.* 193: 296–307

Wilson, J. A. 1979. The structure and function of serially homologous leg motor neurons in locust. II. Physiology. *J. Neurobiol.* 10: 153–67

Wysocka-Diller, J. W., Aisemberg, G., Baumgarten, M., Levine, M., Macagno, E. R. 1989. Characterization of a homolog of Bithorax Complex genes in the leech *Hirudo medicinalis*. *Nature*. In press

Zipser, B. 1979. Identifiable neurons controlling penile eversion in the leech. *J. Neurophysiol.* 42: 455–64

*Annu. Rev. Neurosci. 1990. 13:227–55*
Copyright © 1990 by Annual Reviews Inc. All rights reserved

# CARBOHYDRATES AND CARBOHYDRATE–BINDING PROTEINS IN THE NERVOUS SYSTEM

*T. M. Jessell,\* M. A. Hynes,\* and J. Dodd†*

Center for Neurobiology and Behavior, Departments of \*Biochemistry and Molecular Biophysics, Howard Hughes Medical Institute, and †Physiology and Cellular Biophysics, Columbia University, New York, New York 10032

## INTRODUCTION

The possibility that cellular interactions within the vertebrate nervous system are mediated by cell surface carbohydrates has been considered on numerous occasions. The initial suggestions that carbohydrate structures might mediate neural cell adhesion were based on the expression of complex oligosaccharides, in particular gangliosides, by neural cells, and the detection of cell surface glycosyltransferases that were proposed to function nonenzymatically as receptors for surface oligosaccharides (Roseman 1970, Roth et al 1971, Marchase 1977, Shur & Roth 1975). Subsequent progress in elucidating the function of cell surface carbohydrates in the nervous system has, however, been slow, in part because of the difficulties (*a*) in purifying and characterizing complex oligosaccharides that are expressed on small subsets of neural cells and (*b*) in generating these structures synthetically. In addition, the identification of cell surface molecules such as neural cell adhesion molecules (NCAM), N-cadherin, and integrins (Edelman 1986, Takeichi 1988, Ruoslahti & Pierschbacher 1987, Rutishauser & Jessell 1988) has focused attention on mechanisms of neural cell adhesion that involve direct protein-protein interactions.

Increasing evidence indicates that interactions between surface oligosaccharides and carbohydrate-binding proteins mediate cell adhesion and recognition between nonneural cells. Thus, the stage- and species-specific

227

0147–006X/90/0301–0227$02.00

binding of a sperm to the zona pellucida coat that surrounds mammalian oocytes has been shown to result from the interaction of a sperm receptor with an O-linked oligosaccharide that is present on the zona pellucida glycoprotein, ZP-3 (Wassarman 1987). Evidence also indicates that the polylactosamine oligosaccharides present on the blastomeres of pre-implantation mouse embryos are involved in adhesive interactions that occur during compaction (Fenderson et al 1984, Rastan et al 1985, Bayna et al 1988). In addition, the specific homing of recirculating lymphocytes to peripheral lymphoid targets (Rosen et al 1985, Rosen & Yednock 1986, Gallatin et al 1986, Brandley et al 1987) appears to depend, at least in part, on receptors that recognize specific carbohydrate structures. Finally, the hepatic asialoglycoprotein receptor that is responsible for the clearance of circulating serum glycoproteins (Ashwell & Harford 1982) represents the best-characterized surface protein with a defined physiological role in the recognition of carbohydrate structures.

With the availability of monoclonal antibodies (MAbs), the complex expression patterns of oligosaccharides on neural cells has become more readily apparent. Some of the defined carbohydrate antigens are highly restricted to subsets of vertebrate neurons and reveal molecular gradients in developing neural tissues. In addition, several carbohydrate-binding proteins with specificity for neural cell surface oligosaccharides have recently been detected in the vertebrate nervous system. In this review we discuss briefly the evidence emerging from both neural and nonneural systems that indicates that cell surface carbohydrate structures may indeed play important roles in mediating neural cell recognition and adhesion.

## DIVERSITY OF CARBOHYDRATE STRUCTURES ON VERTEBRATE CELLS

The complex oligosaccharides expressed by vertebrate cells are associated with ceramides in glycolipids or attached via N- or O-linkages to protein backbones. Several classes of carbohydrate structures can be defined on the basis of their polysaccharide backbone sequences (Table 1). Lactoseries

**Table 1** Oligosaccharide classification by polysaccharide backbone sequence

| Lactoseries | |
|---|---|
| (Type 1) | Gal($\beta$1-3)GlcNAc($\beta$1-3)Gal$\beta$1-4Glc-R |
| (Type 2) | Gal($\beta$1-4)GlcNAc($\beta$1-3)Gal$\beta$1-4Glc-R |
| Globoseries | GalNAc$\beta$1-3Gal$\alpha$1-4Gal$\beta$1-4Glc-R |
| Ganglioseries | Gal$\beta$1-3GalNAc$\beta$1-4Gal$\beta$1-4Glc-R |

carbohydrates contain the [Gal$\beta$1-3(4)GlcNAc-R] structure, globoseries carbohydrates contain the [GalNAc$\beta$1-3Gal$\alpha$1-4Gal-R] backbone and ganglioseries structures contain the [Gal$\beta$1-3GalNAc$\beta$1-R] sequence (Hakomori 1981, Feizi 1985). These backbone sequences can be modified extensively by the addition of branched or terminal saccharides, thus generating many structurally distinct members of each class. In addition, the attachment of the same saccharide via multiple linkages, the existence of branched carbohydrate chains of the same or differing structure, and extensive variation in sialic acid content (Schauer 1982) provide the potential for an enormous diversity of complex oligosaccharide structures.

The assembly of complex oligosaccharides is achieved by the coordinated and sequential activity of glycosyltransferase enzymes (Beyer & Hill 1982). Each enzyme is capable of adding specific saccharides via defined linkages to a highly restricted set of oligosaccharide substrates. The structural diversity in cell surface oligosaccharides must therefore be defined in large part by the cellular expression and substrate specificity of these glycosyltransferases.

## CARBOHYDRATE-BINDING PROTEINS

A large number of endogenous proteins have been characterized that bind to distinct surface oligosaccharides on vertebrate cells. These carbohydrate-binding proteins can be subdivided into several categories on the basis of their primary structure and biochemical properties (see Drickamer 1988, Barondes 1988):

1. Calcium-dependent carbohydrate-binding proteins (C-type lectins): Lectins of this class require the presence of calcium for carbohydrate-binding activity and are defined on the basis of the homology with the carbohydrate-recognition domain of the rat hepatic asialoglycoprotein receptor (Drickamer 1988). The common structural organization of the binding domain results from a conserved set of 18 amino acids that includes cysteine residues involved in disulphide bond formation, which is essential for carbohydrate-binding activity (Drickamer 1988). Different members of this class exhibit distinct sugar-binding specificities that may result from nonconserved amino acids within the binding domain. This class of lectins also includes the chicken hepatic *N*-acetylglycosamine receptor, the soluble rat mannose-binding proteins (Drickamer et al 1986, Ezekowitz et al 1988), the pulmonary surfactant apoprotein (Haagsman et al 1987), cartilage proteoglycan (Krusius et al 1987, Halberg et al 1988), and a lymphocyte homing receptor (Siegelman et al 1989, Lasky et al 1989). Several other proteins have been identified that exhibit this conserved carbohydrate-binding domain but that have not yet been demonstrated to function as lectins.

2. Calcium-independent, soluble carbohydrate-binding proteins (S-lac type lectins): Several low-molecular-weight soluble carbohydrate-binding proteins have been isolated from a wide range of tissues and constitute a separate class of lectins (Barondes 1984, 1988). S-lac lectins form a structurally homologous class, sharing a number of conserved amino acids that may be critical for carbohydrate-binding function (see Drickamer 1988). In contrast to C-type lectins, S-lac lectins are inhibited by oxidation and do not contain invariant cysteine residues. They exhibit a degree of conservation similar to that of the C-type lectins, and most of them share binding specificity for $\beta$-galactosides. Some of these lectins, however, have markedly different binding affinities for more complex lactosamine-based oligosaccharides (Leffler & Barondes 1986). Included in this class are two proteins termed RL-14.5 and RL-29 that are expressed with striking selectivity within the developing nervous system. Soluble lectins with mannose-binding properties have also been identified in neural tissues (Zanetta et al 1987), although the structure of these proteins has not yet been established.

3. Membrane-bound or soluble glycosyltransferases: Glycosyltransferases have been proposed to function as carbohydrate-binding proteins when their appropriate nucleotide sugar donors are not available (Bayna et al 1986). The cell surface expression of glycosyltransferases, though still controversial, appears likely, and represents a situation in which the enzyme may bind carbohydrate substrates without enzymatic transfer of additional saccharides. Comparison of the primary structure of galactosyltransferases (Shaper et al 1986, 1988, Narimatsu et al 1986) and sialyltransferase (Weinstein et al 1987), predicted by cDNA clones, does not indicate any striking sequence similarities between glycosyltransferases with distinct saccharide-binding properties.

# CARBOHYDRATE-MEDIATED CELL ADHESION AND RECOGNITION

In several nonneural systems there is now quite compelling evidence that interactions between carbohydrates and carbohydrate-binding proteins mediate cell adhesion and recognition. In this section we discuss briefly the evidence for carbohydrate recognition in nonneural systems with the intention of providing a framework for assessing the role of similar or identical molecules expressed within the nervous system.

## Fertilization

The binding of sperm to mammalian eggs exhibits a striking species- and stage-specificity. This simple example of selective cell recognition has provided an accessible system with which to examine the molecular basis of intercellular recognition in vertebrates. The initial specificity in mammalian sperm-egg interactions appears to result from the binding of receptors on the sperm surface to a ligand present on the zona pellucida membrane that surrounds the egg (Wassarman 1987). In the mouse, the zona pellucida consists of only three major glycoproteins, termed ZP-1, ZP-2 and ZP-3. It has been established that the ZP-3 glycoprotein mediates the binding of sperm (Bleil & Wassarman 1980, Wassarman 1987). The ZP-3 protein contains both N-linked and O-linked oligosaccharides. By comparing the effect of proteolytic degradation of ZP-3 with the selective cleavage of N- and O-linked oligosaccharides from the protein, Wassarman and colleagues have shown that the O-linked oligosaccharides on ZP-3 serve as the ligand recognized by sperm receptors (see Wassarman 1987). The precise structure of the ZP-3 oligosaccharides that function as the sperm receptor is not yet known. The addition of fucans and other fucose polymers, however, inhibits the binding of sperm in a wide variety of species, suggesting that sperm binding may be dependent on a fucose group (Huang & Yanagimachi 1984). The use of selective glycosidases has also implicated an $\alpha$-linked terminal galactose as a crucial structural component of the ZP-3 O-linked oligosaccharide receptor (Bleil & Wassarman 1988).

The nature of receptors on the sperm surface that bind to the O-linked oligosaccharide on ZP-3 is not clear. One candidate is a galactosyltransferase (GalTase) (Lopez et al 1985). Several lines of evidence are consistent with this possibility. Inhibitors of GalTase activity, in particular $\alpha$-lactalbumin and UDP-dialdehyde, inhibit sperm binding when assayed in vitro. The addition of soluble GalTase or anti-GalTase antibodies also inhibits the binding of sperm. In addition, UDP-galactose (the nucleotide sugar donor for the enzyme), but not other nucleotide sugars, is able to dissociate sperm from the egg zona pellucida (Lopez et al 1985). Since this GalTase catalyzes the transfer of galactose to terminal $N$-acetylglucosamine, the ZP-3 oligosaccharide might be expected to exhibit a terminal $N$-acetylglucosamine residue. The demonstration that a pre-existing terminal $\alpha$-galactose on ZP-3 is required for sperm receptor activity (Bleil & Wassarman 1988), together with the evidence that the receptor involves a fucose residue, suggests that multiple saccharide determinants are involved in sperm recognition of ZP-3. In some invertebrates, the sperm receptor appears to be a lectin termed bindin (Glabe et al 1982). It is still possible, therefore, that lectins on the surface of mammalian sperm contribute to interactions with the ZP-3 O-linked oligosaccharides. Recent studies have identified the asialoglycoprotein receptor (ASGP-R) or a closely related protein on the surface of mammalian sperm (Abdullah &

Kierszenbaum 1989). Further characterization of the O-linked oligo-saccharides on ZP-3 should provide information on functionally relevant carbohydrate-binding proteins on the sperm surface.

## Adhesion of Blastomeres in Cleavage-stage Mouse Embryos

Studies of preimplantation mouse embryos have provided evidence for roles of both calcium-dependent adhesion molecules and oligosaccharides in blastomere adhesion. At the early 8-cell stage, individual mouse blasto-meres exhibit a low degree of cell-cell contact and retain defined cell boundaries (Calarco-Gillam 1985). A striking increase in blastomere adhesion and in cell-cell contact occurs at this stage of development, a process termed compaction. The earliest phase of compaction appears to be $Ca^{2+}$-dependent, whereas at later times in the compaction process, adhesion becomes progressively less dependent on the presence of $Ca^{2+}$ (Ducibella & Anderson 1975). The $Ca^{2+}$-dependent cell adhesion molecule, E-cadherin, is expressed on blastomeres at the 8-cell stage and mediates the early, $Ca^{2+}$-dependent phase of compaction (Hyafil et al 1980, Takeichi 1986). Several studies have suggested that oligosaccharides play a role in the later, $Ca^{2+}$-independent adhesive events that underlie compaction.

The possibility that oligosaccharides are involved in blastomere adhesion derived initially from analysis of a set of lactoseries oligo-saccharides, originally defined as stage-specific embryonic antigens (SSEAs) (Solter & Knowles 1978). These lactoseries structures, in par-ticular the fucosylated SSEA-1 structure, are expressed on mouse blasto-meres immediately preceding compaction. The addition of soluble, multi-valent neo-glycoproteins that express the SSEA-1-reactive trisaccharide inhibits the compaction of early mouse embryos (Fenderson et al 1984). Inhibition of compaction is specific to the SSEA-1 determinant, since structural analogues do not inhibit compaction. In addition, when embry-onal cell surface polylactosaminoglycans are cleaved by the bacterial enzyme endo-beta-galactosidase, the recompaction of previously dis-sociated blastomeres is significantly delayed (Rastan et al 1985). Similar enzyme treatment does not appear to affect earlier stages of compaction, a finding that suggests that the cell surface oligosaccharides sensitive to endo-beta-galactosidase are involved only in the later stages of com-paction.

Mouse blastomeres and teratocarcinoma cells express lectin-like mol-ecules on their surfaces. A 56 kDa lectin purified from mouse terato-carcinoma cells (Grabel 1984, Grabel et al 1985) has been reported to react with fucosylated oligosaccharides similar to those present on 8-cell stage embryos. There is also evidence for the involvement of cell surface Gal-

Tases in embryonic compaction (Bayna et al 1988). GalTase activity on the surface of mouse blastomeres increases markedly during the later phases of compaction. Moreover, addition of α-lactalbumin or antibodies that eliminate GalTase activity result in the decompaction of mouse morulae. The compaction of preimplantation mouse embryos may therefore prove useful as a system with which to analyze the respective contributions of cadherin- and carbohydrate-mediated adhesion on a single mammalian cell.

## Lymphocyte Homing

One of the initial events in the recirculation of lymphocytes is their migration from the bloodstream to lymphatic ducts in peripheral lymphoid tissues. This process has been termed lymphocyte homing and involves the adhesion of lymphocytes to a specialized set of high endothelial venules (HEV) that are located in the post-capillary beds (Gallatin et al 1986). Different subsets of recirculating lymphocytes interact in an organ-specific manner with HEV in peripheral lymph nodes and in mucosa-associated lymphoid tissues such as Peyer's Patches.

There is now considerable evidence that lymphocyte homing to HEV is mediated by a set of lymphocyte cell surface proteins that interact with oligosaccharide structures located on the HEV cell surface. The binding of lymphocytes to HEV cells in vitro is inhibited selectively by mannose-6-phosphate and by yeast phosphomannan polysaccharides (Stoolman et al 1984, Yednock et al 1987a,b). Sialidase treatment of HEV cells also perturbs lymphocyte adhesion (Rosen et al 1985), a result that suggests that circulating lymphocytes recognize a complex set of oligosaccharides on HEV cells. Polyacrylamide surfaces derivatized with carbohydrates have been used to define, more precisely, the carbohydrate-binding specificity of homing receptors on the lymphocyte surface (Brandley et al 1987). Lymphocytes adhere selectively to polyacrylamide gels derivatized with either phosphomannans or fucose sulfate polymers such as fucoidan (Brandley et al 1987). The binding of lymphocytes to both saccharides suggests that distinct classes of homing receptors with differing saccharide specificities may exist on a single population of lymphocytes. Phosphomannans have been detected in a variety of mammalian glycoconjugates, and HEV cells are known to express high levels of sulphated glycoconjugates (Andrews et al 1983), although their identity is unclear.

The structure of a murine lymphocyte homing receptor with peripheral lymph node specificity has recently been determined (Siegelman et al 1989, Lasky et al 1989). This receptor contains a carbohydrate-binding domain with structural homology to the hepatic asialoglycoprotein receptor (see below). In addition, the receptor contains multiple EGF-like domains. Collectively, these findings provide strong evidence for the involvement of cell surface oligosaccharides and lectins in selective cell recognition by lymphocytes.

## Glycoprotein Recognition by Hepatic Lectins

The endocytosis of circulating serum glycoproteins by hepatocytes revealed one of the first physiological roles for carbohydrate recognition in vertebrate species. In rat, the removal of terminal sialic acid residues from native serum glycoproteins exposes penultimate galactose residues and results in the accelerated clearance of these glycoproteins from the circulation (Ashwell & Harford 1982). This process results from the recognition of desialylated glycoproteins by an integral membrane receptor on the hepatocyte surface, the ASGP-R. The specificity of the rat ASGP-R for galactose residues can be demonstrated by blocking the endocytosis of glycoproteins by enzymatic alteration of terminal saccharide residues (Ashwell & Harford 1982).

Studies on the interactions between hepatic lectins and their carbohydrate ligands have provided important insights into the factors that may regulate the affinity of carbohydrate-mediated interactions in vertebrate cells. A direct correlation between binding affinity and the degree of saccharide substitution has been demonstrated with model neoglycoproteins that vary in the number of saccharides substituted on the protein (Kawaguchi et al 1980, Lee & Lee 1982, Kuhlenschmidt et al 1984, Lehrman et al 1986). The extent of oligosaccharide branching also significantly affects lectin binding affinity. For example, triantennary oligosaccharides exhibit a markedly greater inhibitory potency at the ASGP-R than biantennary oligosaccharides of the same linear structure (Lee 1989). The degree of oligosaccharide branching may therefore be critical in determining the affinity of binding to protein receptors.

Although both the chick and rat hepatic ASGP-Rs are single subunit transmembrane proteins, they exist in the membrane as multimers. The active species of the chicken hepatic lectin appears to be a hexamer (Drickamer 1988). Thus, although the receptor monomer is capable of binding to its carbohydrate ligand, the multimeric nature of these receptors in the plasma membrane may contribute to their enhanced affinity for ligands that contain clusters of terminal sugars (Kuhlenschmidt et al 1984). The multimeric composition of these carbohydrate-binding receptors may also extend the range of their carbohydrate-binding capabilities, either by producing a shift in sugar binding specificity and/or by increasing the affinity of binding of these proteins to branched oligosaccharides with different backbone structures (Lee 1989).

# DIVERSITY OF CARBOHYDRATE EXPRESSION IN THE NERVOUS SYSTEM

The analysis of the comparatively simple interactions between cells in early embryos and in nonneural tissues has provided evidence that surface carbohydrates are involved in cell recognition and adhesion. Within the nervous system there is no direct evidence that carbohydrates subserve similar adhesive or recognition functions. Indirect support for this idea has, however, derived from the restricted expression patterns of cell surface carbohydrates, the identification of carbohydrate-binding proteins in subsets of neurons, and preliminary functional studies on gangliosides and glycosyltransferases. The complexity of cell interactions in the nervous system has made determination of the precise function of these carbohydrates and binding proteins substantially more difficult than, for example, in sperm-egg interactions or blastomere compaction. Despite this, progress has been made in ascribing potential functions to these molecules in several regions of the nervous system.

## Carbohydrate-Mediated Adhesion in the Retino-Tectal System

The topographic projection of retinal ganglion cell axons onto the tectal surface represents one of the best-studied neural systems with which to examine the formation of specific connections in the nervous system. Biochemical support for the existence of carbohydrate gradients in the retino-tectal system was originally obtained by measuring the adhesion of dissociated dorsal or ventral retinal cells to topographically appropriate tectal regions in vitro (Barbera et al 1973, Barbera 1975, Marchase 1977). Marchase (1977) established that treatment of ventral retinal cells or ventral tectum with protease blocked the preferential adhesion of the retinal cells to their matching tectal halves, whereas similar protease treatment of dorsal retinal cells or dorsal tectum did not alter adhesion. In contrast, treatment of dorsal retinal or dorsal tectal cells with $N$-acetylhexosaminidase or sialidase resulted in a decrease in specific retino-tectal adhesion (Marchase 1977). Marchase (1977) proposed the existence of two opposing gradients of complementary molecules: a protease-insensitive molecule containing a terminal $\beta$-$N$-acetylgalactosamine residue that is more concentrated in dorsal retina and tectum, and a protease-sensitive molecule that is more concentrated in the ventral retina and tectum.

Support for this model was obtained with the demonstration that the binding of $G_{M2}$ ganglioside to retina and tectum exhibits a ventral-to-dorsal gradient, thus suggesting that $G_{M2}$ may be the substrate for the ventrally located protease-sensitive molecule (Marchase 1977). One candidate for the protease-sensitive molecule is therefore a galactosyl transferase (GalTase), $G_{M1}$ synthetase, which recognizes $G_{M2}$ and converts it to $G_{M1}$ by the addition of a terminal galactose residue. Enzymatic activity similar to that of $G_{M1}$ synthetase was detected in a ventral to dorsal gradient in the retina (Marchase 1977). Functional evidence that gangliosides may be involved in the specification of retino-tectal projections has been provided by independent experiments that revealed selective adhesion of neural retinal cells to immobilized gangliosides, including $G_{M2}$ (Blackburn et al 1986). These adhesive interactions appear to be specific in that they are not detected between neural retinal cells and other charged lipids (e.g. sulphatides, phospholipids) or between gangliosides and other cell types, such as hepatocytes. In addition, of several gangliosides tested, $G_{M2}$, $G_{D3}$, and $G_{D1a}$ supported a greater strength and extent of adhesion than $G_{T1b}$, $G_{M1}$, and $G_{D1b}$, a result that suggested the presence of a receptor on retinal cells that can distinguish between gangliosides (Blackburn et al 1986). Additional evidence that $G_{M2}$ and $G_{M1}$ synthetase contribute to specific retino-tectal connections has not emerged. Furthermore, although the experiments described above establish the presence within the retino-tectal system of complementary patterns of potentially interactive molecules, a problem in any functional interpretation lies in the fact that cell-to-cell or cell-to-substrate adhesion was examined and may not reflect the properties of retinal growth cones. Nonetheless, it remains a viable idea that a graded distribution of oligosaccharides and carbohydrate-binding molecules on the surface of retinal axons interacts with a reverse gradient on the tectum to produce specific connections.

In further support, immunocytochemical studies have revealed selective expression patterns of several gangliosides in the retino-tectal system. The monoclonal antibody 18B8 detects a number of developmentally regulated ganglioside species in chicken retina and brain (Dubois et al 1986). The major ganglioside species recognized by antibody 18B8 in retina, $G_{T3}$, is associated with the cell bodies of most neurons in the retina in early development but becomes progressively restricted to synaptic layers during later development (Grunwald et al 1985). Other gangliosides, distinct from those detected with monoclonal antibody 18B8, are recognized by MAb JONES (Constantine-Paton et al 1986, Blum & Barnstable 1987). The JONES antigen is a 9-O-acetylated derivative of the $G_{D3}$ ganglioside and is similar or identical to the D1.1-reactive ganglioside that is expressed in the developing rat neuroectoderm (Levine et al 1984). The pattern of expression of the 9-O-acetylated form of the $G_{D3}$ ganglioside is independent of that of the nonacetylated $G_{D3}$ ganglioside recognized with MAb R24 (see Table 2). In retina and tectum, the JONES antigen is distributed on neurons and glia in a dorso-ventral gradient (Constantine-Paton et al

1986), whereas $G_{D3}$ staining appears to be uniformly distributed, thus suggesting that the selectivity of expression of the acetylated form is dependent on the spatial restriction of specific biosynthetic or degradation enzymes (Blum & Barnstable 1987).

The Jones antigen, $G_{D3}$, and two other gangliosides with a Jones-reactive epitope are found in a number of other developing neural tissues. The distribution correlates with the presence of migrating neuroblasts (Mendez-Otero et al 1988), which suggests that the function of the D1.1/Jones antigen in the retina and other regions of the embryonic central nervous system is in the regulation of early neuroepithelial cell migration. The distribution of the D1.1 antigen has been shown to overlap with that of fibronectin and the fibronectin receptor (Stallcup 1988, Stallcup et al 1989). The adhesion of post-natal rat cerebellar cells to fibronectin is inhibited by antibodies to the fibronectin receptor, by Arg-Gly-Asp peptides that block the recognition of fibronectin by its receptor, and also by antibodies to D1.1 itself. The D1.1 ganglioside may therefore be required to enhance the attachment to fibronectin of cells expressing the fibronectin receptor. The termination of expression of D1.1 on neural epithelial cells at the cessation of mitosis may permit post-mitotic neuroblasts to detach from fibronectin and migrate away from germinal zones (Stallcup 1988, Stallcup et al 1989).

Recent biochemical and functional studies have extended the original observation (Roth et al 1971) that $N$-acetylgalactosaminyl-transferase (GalNacTase) is found on the surface of embryonic chick neural retinal cells (Balsamo et al 1986). GalNacTase from embryonic chick neural retina can be isolated both as a soluble protein and as a particulate complex associated with its endogenous acceptor (Balsamo et al 1986). The two GalNacTase forms are immunologically cross-reactive but have different molecular masses. Under most conditions, the enzyme remains associated with its endogenous carbohydrate acceptor (Balsamo et al 1986), thus suggesting the possibility of a lectin-like function for this enzyme in the retina. Immunochemical analysis, using antibodies that do not distinguish the two forms of the enzyme, reveals that at least one form of the enzyme is associated with cells throughout the retina in the early embryo but becomes restricted to synaptic layers and to the outer segment of the photoreceptor in the retina of adult animals. The isolation of both soluble and particulate forms of the enzyme from the retina, together with the developmental change in its anatomical localization, raises the possibility that one form of the enzyme contributes to retinal cell adhesion. Oligosaccharides, therefore, remain putative mediators of cell adhesion or recognition along the dorso-ventral axis of the retina and tectum.

In contrast, the recent studies of Bonhoeffer and colleagues suggest that topography along the anterior-posterior axis involves tectal proteins that

**Table 2**  Complex oligosaccharide expression within the nervous system

| Structural class | MAbs | Distribution | Refs. |
|---|---|---|---|
| Lactosyl | | | |
| Galβ1-4GlcNAcβ1-3Galβ1-R | A5, 1B2/1B7 | Primary sensory neurons, rodents | Dodd & Jessell 1985 |
| Galα1-3Galβ1-4GlcNAcβ1-R | 2C5, αB | Primary sensory neurons, rodents | Coackham et al 1982, Oriol et al 1984 |
| Galα1-3Galβ1-4GlcNAcβ1-R<br>  &#124;<br>  Fucα2-3 | LA4, LD2<br>TC6, KH10 | Primary sensory neurons rodents, olfactory neurons | Dodd et al 1984<br>Dodd & Jessell 1985 |
| Galβ1-3GlcNAcβ1-3Galβ1-R | FC10.2 | Rodent primary sensory neurons | Dodd & Jessell 1985 |
| Galβ1-4GlcNAcβ1-R<br>  &#124;<br>  Fucα1-3 | αSSEA-1<br>αX, α7A<br><br>AC4 | Early CNS neurons<br><br>Subsets peripheral sensory, olfactory neurons, tastebuds, rodents, primary sensory neurons (avian) | Yamamoto et al 1985<br><br>J. Dodd et al, unpublished |
| Fucα1-2Galβ1-4GlcNAcβ1-R | αH | Tastebuds, olfactory neurons | Oriol et al 1984 |
| Fucα1-2Galβ1-4GlcNAcβ1-R<br>  &#124;<br>  Fucα1-4 | αLeb | Tastebuds | Akabas et al 1988 |

| Structure | Antibody | Expression | Reference |
|---|---|---|---|
| $SO_4$-3GlcUA$\beta$1-3Gal$\beta$1-4GlcNAc$\beta$1-3Gal$\beta$1-4Glc$\beta$1-R | HNK-1 | Differentiating CNS neurons (avian & rodent) | Schwarting 1987 |
| | NC1<br>Leu7<br>L2<br>4F4 | Neural crest cells (avian)<br>Subsets sensory neurons | see Schachner 1989<br>Dodd & Jessell 1986 |
| **Globoseries** | | | |
| R-3GalNAc$\beta$1-3Gal$\alpha$1-4R | $\alpha$SSEA-3 | Subsets of primary sensory neurons, rodent | Dodd et al 1984 |
| NeuAc$\alpha$2-3Gal$\beta$1-3GalNAc$\beta$1-R | $\alpha$SSEA-4 | Subsets of primary sensory neurons, rodent | Dodd et al 1984 |
| **Ganglioseries** | | | |
| Gal$\alpha$1-3Gal$\beta$1-3GalNAc$\beta$1-4Gal$\beta$1-4Glc$\beta$1-R<br>$\quad\mid\quad\quad\quad\quad\quad\mid$<br>$\quad$Fuc$\alpha$1-2$\quad\quad$NeuAc$\alpha$2-3 | TC6, LA4, LD2, KH10 | Subsets primary sensory neurons | Dodd & Jessell 1985 |
| Neu5,9Ac$\alpha$2-8NeuAc$\alpha$2-3Gal$\beta$1-4Glc$\beta$1-R | D1.1, Jones | Differentiating neurons retinal neurons & glia | Levine et al 1984<br>Constantine-Paton et al 1986 |
| NeuAc$\alpha$2-8NeuAc$\alpha$2-3Gal$\beta$1-R | R24 | Differentiating neurons | Blum & Barnstable 1987 |
| NeuAc$\alpha$2-8NeuAc$\alpha$2-8NeuAc$\alpha$2-3Gal$\beta$1-4Glc$\beta$1-R | 18B8 | Retinal neurons | Grunwald et al 1985 |
| (NeuAc$\alpha$2-8NeuAc$\alpha$2-8)$_n$ = PSA | 5A5, MenB<br>$\alpha$-K1 | Expressed in association with NCAM in many cell types | J. Dodd et al, unpublished<br>Rougon et al 1986<br>Frosch et al 1985 |

inhibit the migration of incoming temporal retinal axons (Walter et al 1987).

# CARBOHYDRATES AND CARBOHYDRATE-BINDING PROTEINS IN SENSORY SYSTEMS

## Oligosaccharide Expression on Sensory Neurons

Considerable evidence now indicates that primary sensory neurons and peripheral sensory receptor cells in the olfactory, gustatory, auditory, and somatosensory systems express specific complex oligosaccharide antigens (Dodd et al 1984, Dodd & Jessell 1985, 1986, Jessell & Dodd 1985, Akabas et al 1988, Oriol 1983, Oriol et al 1984). In this review we focus on the primary sensory (dorsal root ganglion) neurons of the somatosensory system. More than a dozen functional classes of dorsal root ganglion (DRG) neurons have been defined on anatomical and functional grounds. Each distinct class of sensory afferents project to specific domains in the spinal cord that coincide with the laminar divisions originally defined on the basis of spinal cord neuronal cytoarchitecture (see Jessell & Dodd 1986).

One approach in the analysis of mechanisms underlying the development and extension of sensory axons has been to identify cell surface molecules that are expressed on functionally and anatomically distinct subsets of DRG neurons. MAbs directed against defined carbohydrate epitopes have identified three classes of oligosaccharides—globoseries, lactoseries and ganglioseries (see Table 1)—that are expressed on subsets of DRG neurons (Table 2) (Dodd et al 1984, Dodd & Jessell 1985, Chou et al 1989). The expression of these oligosaccharides on subsets of DRG neurons correlates with the central projection sites of physiologically characterized subclasses of DRG neurons (Dodd & Jessell 1985) and defines subsets of DRG neurons during development (Dodd & Jessell 1986).

MAbs anti-SSEA-3 and anti-SSEA-4 recognize globoseries oligosaccharides that are expressed on DRG neurons of both intermediate and large diameter (Dodd et al 1984). The central terminals of SSEA-3$^+$ and SSEA-4$^+$ DRG neurons are located in lamina III and the medial part of lamina IV, with a sparse projection to lamina I. This subset of DRG neurons does not contain neuropeptides or other cytochemical markers that define subpopulations of small-diameter DRG neurons (Dodd et al 1984, Jessell & Dodd 1986). The location of afferent terminals that express globoseries oligosaccharides suggests that they define myelinated primary afferents involved in the transmission of low-threshold cutaneous information to deeper laminae and possibly high-threshold mechanoreceptive or thermoreceptive information to lamina I.

A distinct subset, constituting approximately 50% of adult DRG neurons, expresses the type 2 lactoseries structure detected by MAbs A5 and 1B2/1B7 (Dodd & Jessell 1985). This antigen appears on the surface of DRG neurons during embryonic development and is present on about 10% of neurons in embryonic day (E) 18 DRG. During the first postnatal week, the proportion of neurons expressing this oligosaccharide structure increases to its adult value. In the adult, this $N$-acetyllactosamine structure is restricted to small- and intermediate-diameter DRG neurons that have central terminals in laminae I and II of the dorsal horn. Neurons that express type 2 lactoseries structures also contain neuropeptides and other selective markers of nociceptive sensory neurons (Dodd & Jessell 1985).

The expression of a distinct set of α-galactose-extended ganglioseries or lactoseries structures is associated with a more restricted subset of small-diameter DRG neurons (Dodd & Jessell 1985, Chou et al 1989). The α-galactose-extended structure identified by MAb LD2 is expressed by a subpopulation of small DRG neurons that project to the dorsal region of lamina II. The LD2-reactive sensory neurons include all those that express somatostatin but not those that contain substance P (Dodd & Jessell 1985). A closely related α-galactose-extended structure, recognized by MAb LA4, is found on a separate population of small- and intermediate-diameter DRG neurons (constituting 40–50% total DRG neurons) that project predominantly to the ventral part of lamina II (Dodd & Jessell 1985). Other complex lactoseries oligosaccharide structures delineate additional subsets of small-diameter DRG neurons (Jessell & Dodd 1986). The distribution of afferent terminals expressing these oligosaccharides indicates that they define C fibers and also some Aδ fibers. These observations provide evidence that there is a high degree of specificity in the expression of complex oligosaccharides on functional subsets of DRG neurons. The expression of distinct oligosaccharide structures by DRG neurons is detectable soon after sensory neuron differentiation and is clearly evident during late embryonic and early post-natal stages. This correlates with the time during which sensory neurons extend axons into the spinal cord and toward targets in the periphery and establish specific synaptic connections (see Dodd & Jessell 1986). Since similar or identical molecules, in particular those based on the $N$-acetyllactosamine backbone sequence, have been implicated in cell adhesion and recognition of other embryonic cells, it is possible that the oligosaccharides on DRG neurons serve the same general functions.

## Carbohydrate-Binding Proteins in Sensory Systems

Several proteins that are capable of binding to the β-galactoside structures expressed by primary sensory neurons have been described (Sarkar et al

1975, Leffler & Barondes 1986) and many of these are present in neural tissues (Kobiler & Barondes 1977, Eisenbarth et al 1978, Joubert et al 1987, 1988). These lectins appear to exist in soluble form and belong to the S-lac class. As such, they represent one class of proteins with the potential to interact with lactoseries oligosaccharides found on the surface of DRG neurons. Antibodies have been used to localize lactose-binding lectins in DRG and spinal cord (Regan et al 1986). Two lectins, termed RL14.5 and RL29, which have molecular weights of 14,500 and 29,000 (Cerra et al 1985), can be detected in DRG and spinal cord (Regan et al 1986). The lectins can be detected in DRG neurons soon after the differentiation of sensory neuroblasts and are present in central sensory axons that terminate in the dorsal horn of the spinal cord. RL-14.5 is detected in DRG by E14 and continues to be expressed at later stages of development; the highest levels of immunoreactivity are detected from E20 to postnatal day (P)5. DRG neurons that express highest levels of RL-14.5 have small or intermediate diameters, although larger-diameter neurons do appear to express low levels of the lectin. In contrast, RL-29 is not detectable by western blot or immunocytochemistry in DRG neurons until E16. Afferent fibers containing RL-14.5 are distributed at highest density in the superficial dorsal horn; a lower intensity of labeling is associated with afferents in laminae III and IV and in the dorsal columns. The distribution of RL-29 in the DRG and spinal cord is more restricted than that of RL-14.5 and is absent from deeper laminae of the dorsal horn. These observations indicate a high degree of specificity in the expression of these two soluble lactose-binding lectins in subsets of developing mammalian neurons.

cDNAs encoding RL-14.5 have been isolated (Hynes et al 1989, Clerch et al 1988). High levels of RL-14.5 mRNA are present in embryonic and postnatal DRG and spinal cord, with much lower levels in other regions of the central nervous system (Hynes et al 1989). Both DRG neurons and motoneurons express RL-14.5 mRNA soon after their differentiation, and this selective expression pattern persists in the adult nervous system. All embryonic and adult DRG neurons appear to express RL-14.5 mRNA, but the level of expression varies about five-fold between individual neurons. In the adult spinal cord, RL-14.5 is detected at high levels in motoneurons, and there is no detectable expression in other cells. Within the adult brain, hybridization is largely restricted to motoneuron nuclei in the brainstem (Hynes et al 1989). These observations confirm the selective expression pattern of RL-14.5 detected by immunocytochemistry and Western blot analysis.

The predominant expression of RL-14.5 and RL-29 in sensory neurons and motoneurons suggests that one possible function of these lectins may be to cross-link carbohydrate structures on the surface of axons with carbohydrates expressed on target cells within the spinal cord or periphery (Dodd & Jessell 1986) or expressed by adjacent cells or on matrix molecules in the peripheral environment through which these axons migrate. Lactoseries oligosaccharides that are potential ligands for RL-14.5 and RL-29 have been demonstrated on a variety of peripheral cell surface and extracellular matrix components, including the axonal adhesion molecule NILE (Margolis et al 1986), the glycoprotein laminin (Fujiwara et al 1988), and the glycosaminoglycan chains on chondroitin sulfate proteoglycans (Krusius et al 1986). In addition, the HNK-1 carbohydrate epitope that has been identified on a variety of cell and matrix adhesion molecules is a sulfated glucuronic acid derivative of a lactoseries oligosaccharide (Chou et al 1986) (see Table 2 and below), a finding that suggests that this structure may be recognized by soluble lactose-binding lectins.

It is also possible that RL-14.5 and RL-29 have intracellular functions in sensory neurons and motoneurons. RL-29, in particular, can be detected in the cytoplasm and nucleus of DRG neurons as well as other cells (Barondes 1984, Regan et al 1986). Although no direct evidence indicates that such lectins function within the nucleus, the recent observations that many nuclear proteins (Hart et al 1987), including some transcription factors (Jackson & Tjian 1988), are glycosylated raises the possibility that this class of lectins may be involved in regulating the function of some nuclear proteins.

RL-29 appears to be identical to a rat IgE-binding protein clone from rat basophilic leukemia cells (Albrandt et al 1987). It is the rat homologue of CBP-35, a lactose-binding lectin isolated from mouse 3T3 fibroblasts (Jia & Wang 1988). The structure of CBP-35, deduced from cDNA clones, has revealed that the protein consists of two domains: the carboxy terminal portion resembles the other soluble lactose-binding proteins, whereas the amino terminus exhibits homology with the heterogeneous ribonucleoprotein complex (Jia & Wang 1988). CBP-35 does appear to be a component of ribonucleoprotein complexes in mouse 3T3 cells (Laing & Wang 1988). The function of lectins within the nuclei of neurons and other cells, however, remains to be determined.

# CARBOHYDRATE-BINDING PROTEINS ON OTHER NEURAL CELLS

## Galactosyltransferases

Some progress has been made in the characterization of ganglioside-binding proteins in neural tissues. Using radiolabeled $G_{T1b}$-coupled neoglycoprotein as ligand, Tiemeyer et al (1989) have demonstrated that rat brain membranes contain a high affinity ($K_D = 2$–4 nM) ganglioside-binding protein. This protein exhibits specificity for the structurally related gangliosides $G_{Q1b}$, $G_{T1b}$, and $G_{D1b}$ and is absent, or expressed at low levels, in nonneural tissues. Early studies on the expression and function of glycosyltransferases on neural retinal cells (Marchase 1977, Bayna et al 1986) have been extended by reports that neural crest cells express surface GalTase activity (Runyan et al 1986). Perturbation of enzyme function on neural crest cells by the GalTase modifier protein, alpha lactalbumin, inhibits neural crest cell migration on complex extracellular matrix substrates but not on fibronectin substrates. A similar inhibition of neural crest cell migration was observed after the addition of competitive GalTase substrates, whereas addition of the sugar nucleotide catalytic substrate UDP-galactose, enhances the rate of migration. These observations suggest that cell surface GalTase may contribute to neural cell migration on extracellular matrix substrates.

Laminin appears to be a major extracellular matrix substrate for GalTase. Inhibition of GalTase activity on melanoma cells with antibodies to alpha-lactalbumin blocks cell spreading on laminin substrates although initial cell attachment is not affected (Runyan et al 1988). Conversely, UDP-galactose enhances melanoma cell spreading. The substrates for GalTase have been reported to be N-linked oligosaccharides, primarily on the A chain of laminin, since pretreatment of laminin with $N$-glycanase blocks melanoma cell spreading. It is possible, therefore, that some of the adhesive function of laminin reside in the molecule's oligosaccharide side chains. The oligosaccharide-mediated interactions of laminin may augment or complement cell adhesion mediated via protein domains that interact with the integrin family of receptors on neural cells.

## Soluble Lectins in the Cerebellum

Two soluble lectins that exhibit mannose-binding specificity have been characterized in the developing rat cerebellum. These lectins have molecular weights of 31,000–33,000 and appear to be expressed by oligodendrocytes in rat cerebellum (Zanetta et al 1985, 1987). Immunocytochemical studies have localized one of these lectins, termed CSL, to the regions of contact between oligodendrocytes. Anti-CSL antibodies have been reported to inhibit the compaction of myelin in vitro (Kuchler et al 1988), but the function of these lectins in normal cerebellar development has not been determined.

## Synapse-Specific Carbohydrates and Carbohydrate-Binding Proteins

The potential contribution of carbohydrate structures to synaptic development has been examined in most detail at the neuromuscular junction. A carbohydrate structure specific to the postsynaptic junction of rat skeletal muscle has been identified by labeling with plant lectins, in particular Dolichos biflorus agglutinin (DBA), that recognize terminal $N$-acetylgalactosamine (GalNAc) residues (Sanes & Cheney 1982). Lectins specific for sugars other than GalNAc label both synaptic and extrasynaptic regions of muscle fibers equally (Scott et al 1988). Two different synapse-specific glycoconjugates contain terminal $\beta$-GalNAc residues and represent candidate molecules recognized by DBA lectin. One is the asymmetric form of the enzyme acetylcholinesterase (AchE) and the second is an as yet unidentified glycolipid recognized by anti-SSEA-3 antibodies (Scott et al 1988). The significance of two distinct synapse-specific molecules that contain similar or identical carbohydrate structures is presently unclear. A synapse-specific carbohydrate may contribute to the preferential localization of motoneurons to restricted sites on the muscle. Obata et al (1977) have previously reported that the globoside glycolipid, which is also recognized by anti-SSEA-3 antibodies, perturbs synaptogenesis when added to nerve-muscle co-cultures.

# CARBOHYDRATE STRUCTURES ON NEURAL CELL ADHESION MOLECULES

## HNK-1 Epitope

The analysis of sperm binding during fertilization has established that oligosaccharide structures associated with glycoproteins can function in cell recognition independently of the protein to which they are attached. Several of the glycoproteins that have been implicated in neural cell adhesion have been shown to express a common carbohydrate epitope (McGarry et al 1985, see Kunemund et al 1988). The structure of this epitope has been determined as a 3-sulfated glucuronyl-substituted lacto-series oligosaccharide (Chou et al 1986, Ariga et al 1987) (see Table 2) that is found on both glycolipids and glycoproteins. This epitope is defined by monoclonal antibodies HNK-1, 4F4, Leu-7, NC1, L2, and several others (Schwarting et al 1987, see Schachner 1989). Neural cell surface glycoproteins that express the HNK-1 epitope include NCAM (Edelman 1986, Rutishauser & Goridis 1986), L1 (NILE) (Kruse et al 1984), TAG-1 (Dodd et al 1988), cytotactin (tenascin) (see Kunemund et al 1988), JI (Kruse et al 1985), myelin associated glycoprotein (McGarry et al 1983), the fibronectin receptor $\alpha$-subunit (a member of the integrin class of receptors) (Pesheva et al 1987), and the myelin protein, $P_O$ (Bollenson & Schachner 1987). This epitope is probably also expressed by many additional uncharacterized glycoproteins.

The functional role of this carbohydrate structure is still unclear. Antibodies to the carbohydrate have been reported to perturb neuron-astrocyte and astrocyte-astrocyte adhesion (Keilhauer et al 1985) and neuron-oligodendrocyte and oligodendrocyte-oligodendrocyte adhesion (Poltorak et al 1987) in vitro and to inhibit process outgrowth of vertebrate neurons on conditioned medium or basement membrane glycoprotein substrates (Riopelle et al 1986). It is, however, difficult to exclude steric inhibition of functional protein domains as a consequence of antibody binding to the carbohydrate epitope. The absence of N-linked oligosaccharides expressing the L2 epitope on NCAM does not appear to affect the adhesive properties of the molecules, a finding that suggests, at least in this case, that binding function is not dependent on the presence of this carbohydrate structure (Cole & Schachner 1987). Glycolipids that express this determinant appear to decrease the migration of cells from neural tissue explants and to inhibit neurite extension on poly-D-lysine or laminin substrates (Kunemund et al 1988). These observations suggest that in some systems the HNK-1/L2 epitope may play a role in neural cell adhesion and neurite extension.

## Polysialylation of Cell Adhesion Molecules

A second, and more selective, carbohydrate modification of glycoproteins involved in neural cell adhesion is the existence of alpha 2-8-linked polysialyl chains on NCAM. Differential splicing of NCAM RNA generates at least three protein backbone forms of this general adhesion molecule (Cunningham et al 1987). Each of these protein forms can be modified by the addition of long chain alpha 2-8 N-acetyl neuraminic acid polymers (Finne et al 1983). On human neuroblastoma cells these polymers consist of extended chains of at least 55 sialyl residues (Livingston et al 1988) and appear to be attached to the core protein by conventional high-mannose linkages (Finne et al 1983, Crossin et al 1984). It is striking that with the possible exception of the electroplax $Na^+$ channel (James & Agnew 1987), NCAM is the only neural protein that exhibits this type of polysialyl chain modification. There must therefore be a high degree of specificity in the structure of the N-linked core saccharide and/or adjoining protein sequence to account for the selectivity of polysialyl chain elongation. Biochemical studies with Golgi-enriched fractions of embryonic rat brain have demonstrated the presence of an endogenous cytidine 5'-monophospho-N-acetylneuraminic acid: poly alpha 2-8 sialosyl sialyltransferase (McCoy et al 1985, McCoy & Troy 1987).

The expression of the highly sialylated forms of NCAM is developmentally regulated. Antibodies specific for the polysialyl chain (Rougon et al 1986, Sunshine et al 1987) have revealed that the highly sialylated form of NCAM first appears at the time of neural differentiation and persists in most regions of the nervous system until neural circuitry is established, at which time, in most cases, NCAM expression reverts to a form with a lower degree of sialylation. The developmental transition in the degree of sialylation of NCAM appears to reflect the de novo synthesis of protein forms with a low sialic acid content rather than the cleavage of polysialyl chains by an endogenous polysialyl endoneuraminidase (Friedlander et al 1985), although it is possible that such an enzyme is present in neural tissue.

Endoglycosidase treatment of NCAM affects the binding properties of the protein. The rate of aggregation of vesicles containing the low sialic acid form of NCAM is considerably greater than that of vesicles containing an equal amount of the high sialic acid form (Hoffman & Edelman 1983). That the polysialyl chains act directly as ligands in cell adhesion is unlikely. Instead, several lines of evidence suggest that the presence of a hydrated polysialyl domain may exert a steric constraint on the protein backbone that regulates the affinity of NCAM homophilic binding. A phage endoneuraminidase has been identified that selectively cleaves alpha 2-8 linked polysialic acid. This enzyme has been useful in assessing the contribution of the polysialyl chain to NCAM functions (Rutishauser et al 1985). Enzymatic removal of the polysialic acid chain from NCAM expressed on the neurites of primary sensory neurons in vitro has been shown to enhance the fasciculation of neurites on collagen substrates. This enhancement of fasciculation can be reversed by anti-NCAM antibodies. In contrast, on laminin substrates, identical enzyme treatment promotes neurite-substrate interactions and decreases fasciculation (Rutishauser et al 1988). This observation has been interpreted to indicate that the interaction of laminin receptors on the neurite surface with substrate-bound laminin may be decreased by steric hindrance from the presence of extended polysialyl chains on NCAM. In addition, the injection of phage endoneuraminidase into the eyes of developing chick embryos has been reported to perturb histogenesis in the developing neural epithelium and to result in the formation of an ectopic retinal axon fiber layer in the optic nerve (Rutishauser et al 1985).

One interpretation of these findings is that in addition to regulation of the affinity of NCAM homophilic binding, extended polysialic acid polymers on NCAM may exert a more general regulation of cell membrane apposition (Rutishauser et al 1988). For example, the high expression of polysialic acid on embryonic axons at the time of their extension may permit axons to avoid certain adhesive interactions, such as the establishment of stable junctions, while retaining the capacity to recognize relevant guidance and target cues.

## Regulation of NCAM Function by Heparan Sulfate Proteoglycan

The homophilic binding properties of NCAM also appear to be enhanced or stabilized by noncovalent association with a heparan sulfate proteoglycan (Cole & Glaser 1986). The involvement of heparan sulfate in the binding function of NCAM can be demonstrated by the competitive inhibition of neural cell adhesion to NCAM substrates by addition of heparin or heparan sulfate. The heparin-binding domain has been localized to the N-terminal 25 kDa fragment of NCAM within the second immunoglobulin loop of the protein (Cole & Akeson 1989), near the homophilic NCAM binding site (Cole et al 1986).

## PROSPECTS

Over the past decade, a role for cell surface oligosaccharides in cell-cell recognition and adhesion in vertebrates has received strong experimental support in several diverse systems. The clearest functional evidence for carbohydrate recognition has emerged from simple cell interactions, in particular sperm-egg recognition and lymphocyte homing. The complexity of cellular interactions in the nervous system ensures that a complete understanding of the role of carbohydrate recognition in neural function is still in the future. Several features emerge from a comparison of neural and nonneural systems in which carbohydrate-mediated cell interactions are implicated, however, that provide an encouraging prognosis of future progress in this field.

The structural and immunological characterization of oligosaccharides that has been achieved in recent years has revealed a high degree of conservation in basic oligosaccharide structures on functionally distinct neural and nonneural cell types. Thus, the developmental profile of cell surface oligosaccharides on preimplantation mouse embryos is recapitulated, in part, on developing primary sensory neurons and at newly formed nerve-muscle synapses. The detection of cell surface glycosyltransferases on developing neural and nonneural cells and the availability of probes with which to modify or perturb glycosyltransferase function has revived interest in the role of these enzymes as mediators of cell recognition. In the case of galactosyltransferases, in particular, the diversity of probes with which to perturb binding function provides strong support for their role in recognition or adhesion. A clearer understanding of the contribution of glycosyltransferases to the adhesive properties of cells that coincidentally express other adhesive glycoproteins may provide insights into the combined function of these distinct molecular systems.

Although the structural classes of oligosaccharides have been at least partially delineated, the extent of diversity in carbohydrate-binding proteins that interact with cell surface oligosaccharides is less well resolved. Biochemical analysis and molecular cloning has begun to define the existence of several structurally distinct classes of carbohydrate-binding proteins. There is now extensive documentation of the developmental expression and neuronal specificity of individual glycosyltransferases and soluble $\beta$-galactoside-binding proteins. In nonneural systems, however, the $Ca^{2+}$-dependent carbohydrate-binding proteins of the hepatic ASGP-R class have been shown to represent the most diverse molecular and functional class. It seems probable that members of this family of carbohydrate-binding proteins will be present within the nervous system. Identification of these and other carbohydrate-binding proteins should permit a more powerful genetic and biochemical dissection of the function of carbohydrate-mediated cell interaction in the vertebrate nervous system.

ACKNOWLEDGMENTS

Research in the authors' laboratories is supported by the Howard Hughes Medical Institute (M.A.H., T.M.J.), the National Institutes of Health (J.D., T.M.J.), the Irma T. Hirschl Foundation (J.D.), The McKnight Foundation (J.D., T.M.J.), and The Klingenstein Foundation (J.D.). T.M.J. is an investigator of the Howard Hughes Medical Institute. We thank Rita Lenertz for her assistance in preparing the manuscript.

*Literature Cited*

Abdullah, M., Kierszenbaum, A. L. 1989. Identification of rat testis galactosyl receptor using antibodies to liver asialoglycoprotein receptor: Purification and localization on surfaces of spermatogenic cells and sperm. *J. Cell Biol.* 108: 367–75

Akabas, M. H., Dodd, J., Al-Aqwati, A. 1988. A bitter substance induces a rise in intracellular calcium in a subpopulation of rat taste cells. *Science* 242: 1047–50

Albrandt, K., Orida, N. K., Liu, F.-T. 1987. An IgE-binding protein with a distinctive repetitive sequence and homology with an IgG receptor. *Proc. Natl. Acad. Sci. USA* 84: 6859–63

Andrews, P., Milsom, D. W., Stoddart, R. W. 1983. Glycoconjugates from high endothelial cells. I. Partial characterization of a sulphated glycoconjugate from the high endothelial cells of rat lymph nodes. *J. Cell Sci.* 59: 231–44

Ariga, T., Kohriyama, T., Freddo, L., Latov, N., Saito, M., Kon, K., Ando, S., Suzuki, M., Hemling, M. E., Rinehart, K. L., Kusunoki, S., Yu, R. K. 1987. Characterization of sulfated glucuronic acid containing glycolipids reacting with IgM M proteins in patients with neuropathy. *J. Biol. Chem.* 262: 848–54

Ashwell, G., Harford, J. 1982. Carbohydrate-specific receptors of the liver. *Annu. Rev. Biochem.* 51: 531–54

Balsamo, J., Pratt, R. S., Lilien, J. 1986. Chick neural retina N-acetylgalactosaminyltransferase/acceptor complex: Catalysis involves transfer of N-acetylgalactosamine phosphate to endogenous acceptors. *Biochemistry* 25: 5402–7

Barbera, A. J. 1975. Adhesive recognition between developing retinal cells and the optic-tecta of the chick embryo. *Dev. Biol.* 46(1): 167–91

Barbera, A. J., Marchase, R. B., Roth, S. 1973. Adhesive recognition and retinotectal specificity. *Proc. Natl. Acad. Sci. USA* 70(9): 2482–86

Barondes, S. H. 1988. Bifunctional properties of lectins: Lectins redefined. *Trends Biochem.* 13: 480–82

Barondes, S. H. 1984. Soluble lectins: A new class of extracellular proteins. *Science* 223: 1259–64

Bayna, E. M., Runyan, R. B., Scully, N. F., Reichner, J., Lopez, L. C., Shur, B. D. 1986. Cell surface galactosyltransferase as a recognition molecule during development. *Mol. Cell Biochem.* 72: 141–51

Bayna, E. M., Shaper, J. H., Shur, B. D. 1988. Temporally specific involvement of cell surface $\beta$-1,4 galactosyltransferase during mouse embryo morula compaction. *Cell* 53: 145–57

Beyer, T. A., Hill, R. L. 1982. Glycosylation pathways in the biosynthesis of non-reducing terminal sequences in oligosaccharides of glycoproteins. In *The Glycoconjugates*, ed. M. I. Horowitz, 3: 25–45. New York: Academic

Blackburn, C. C., Swank-Hill, P., Schnarr, R. L. 1986. Gangliosides support neural retina cell adhesion. *J. Biol. Chem.* 261: 2873–81

Bleil, J. D., Wassarman, P. M. 1980. Mammalian sperm-egg interaction: Identification of a glycoprotein in mouse egg zonae pellucidae possessing receptor activity for sperm. *Cell* 20: 873–82

Bleil, J. D., Wassarman, P. M. 1988. Galactose at the nonreducing terminus of O-linked oligosaccharides of mouse egg ZP3 is essential for the glycoprotein's sperm receptor activity. *Proc. Natl. Acad. Sci. USA* 85: 6778–82

Blum, A. S., Barnstable, C. J. 1987. O-acetylation of a cell-surface carbohydrate creates discrete molecular patterns during neural development. *Proc. Natl. Acad. Sci. USA* 84: 8716–20

Bollensen, E., Schachner, M. 1987. The peripheral myelin glycoprotein $P_O$ expresses the L2/HNK-1 and L3 carbohydrate structures shared by neural adhesion molecules. *Neurosci. Lett.* 82: 77–82

Brandley, B. K., Ross, T. S., Schnarr, R. L. 1987. Multiple carbohydrate receptors on lymphocytes revealed by adhesion to immobilized polysaccharides. *J. Cell Biol.* 105: 991–97

Calarco-Gillam, P. 1985. Cell-cell interactions in mammalian preimplantation development. In *Developmental Biology: A Comprehensive Synthesis*, ed. L. Browder, 2: 329–72. New York: Plenum

Cerra, R. F., Gitt, M. A., Barondes, S. H. 1985. Three soluble rat $\beta$-galactoside-binding lectins. *J. Biol. Chem.* 260: 10474–77

Chou, D. K., Dodd, J., Jessell, T. M., Costello, C. E., Jungalwala, F. B. 1989. Identification of $\alpha$-galactose ($\alpha$-Fucose)-asialo $G_{M1}$ glycolipid expressed by subsets of rat dorsal root ganglion neurons. *J. Biol. Chem.* 264: 3409–15

Chou, D. K., Ilyas, A. A., Evans, J. E., Costello, C., Quarles, R. H., Jungalwala, F. B. 1986. Structure of sulfated glucuronyl glycolipids in the nervous system reacting with HNK-1 antibody and some IgM paraproteins in neuropathy. *J. Biol. Chem.* 261(25): 11717–25

Clerch, L. B., Whitney, P., Hass, M., Brew, K., Miller, T., Werner, R., Massaro, D. 1988. Sequence of a full-length cDNA for rat lung $\beta$-galactoside-binding protein: Primary and secondary structure of the lectin. *Biochemistry* 27: 692–99

Coackham, H. B., Garson, J. A., Harper, A. A., Harper, E., Lawson, S. N., Randle, B. J. 1982. Monoclonal antibody 205: A new marker for a subset of small neurons in the rat dorsal root ganglion. *J. Physiol.* 332: 60

Cole, G. J., Akeson, R. 1989. Identification of a heparin binding domain of the neural cell adhesion molecule NCAM using synthetic peptides. *Neuron* 2: 1157–65

Cole, G. J., Glaser, L. 1986. A heparin-binding domain from NCAM is involved in neural cell-substratum adhesion. *J. Cell Biol.* 102: 403–12

Cole, G. J., Loewy, A., Cross, N. V., Akeson, R., Glaser, L. 1986. Topographic localization of the heparin-binding domain of the neural cell adhesion molecule NCAM. *J. Cell Biol.* 103: 1739–44

Cole, G. J., Schachner, M. 1987. Localization of the L2 monoclonal antibody binding site on N-CAM and evidence for its role in N-CAM mediated cell adhesion. *Neurosci. Lett.* 78: 227–32

Constantine-Paton, M., Blum, A. S., Mendez-Otero, R., Barnstable, C. J. 1986. A cell surface molecule distributed in a dorsoventral gradient in the perinatal rat retina. *Nature* 324: 459–61

Crossin, K. L., Edelman, G. M., Cunningham, B. A. 1984. Mapping of three carbohydrate attachment sites in embryonic and adult forms of the neural cell adhesion molecule. *J. Cell Biol.* 99: 1848–55

Cunningham, B. A., Hemperly, J. J., Murray, B. A., Prediger, E. A., Brackenbury, R., Edelman, G. M. 1987. Neural cell adhesion molecule: Structure, immunoglobulin-like domains, cell surface modulation, and alternative RNA splicing. *Science* 236: 799–806

Dodd, J., Jessell, T. M. 1985. Lactoseries carbohydrates specify subsets of dorsal root ganglion neurons projecting to superficial dorsal horn of rat spinal cord. *J. Neurosci.* 5: 3278–94

Dodd, J., Jessell, T. M. 1986. Cell surface glycoconjugates and carbohydrate-binding proteins: Possible recognition signals in sensory neurone development. *J. Exp. Biol.* 129: 225–38

Dodd, J., Morton, S. B., Karagogeos, D., Yamamoto, M., Jessell, T. M. 1988. Spatial regulation of axonal glycoprotein expression on subsets of embryonic spinal neurons. *Neuron* 1(2): 105–16

Dodd, J., Solter, D., Jessell, T. M. 1984. Monoclonal antibodies against carbohydrate differentiation antigens identify subsets of primary sensory neurons. *Nature* 311: 469–72

Drickamer, K. 1988. Two distinct classes of carbohydrate-recognition domains in animal lectins. *J. Biol. Chem.* 263: 9557–60

Drickamer, K., Dordal, M. S., Reynolds, L. 1986. Mannose-binding proteins isolated from rat liver contain carbohydrate-recognition domains linked to collagenous tails. *J. Biol. Chem.* 261: 6878–87

Dubois, C., Magnani, J. L., Grunwald, G. B., Spitalnik, S. L., Trisler, G. D., Nirenberg, M., Ginsburg, V. 1986. Monoclonal antibody 18B8, which detects synapse-associated antigens, binds to ganglioside $G_{T3}(II^3(NeuAc)_3 LacCer)$. *J. Biol. Chem.* 261(8): 3826–30

Ducibella, A. T., Anderson, E. 1975. Cell shape and membrane changes in the eight-cell mouse embryos: Prerequisites for morphogenesis of the blastocyst. *Dev. Biol.* 47: 45–58

Edelman, G. M. 1986. Cell adhesion molecules in the regulation of animal form and tissue pattern. *Annu. Rev. Cell Biol.* 2: 81–116

Eisenbarth, G. S., Ruffolo, R. R., Walsh, F. S., Nirenberg, M. 1978. Lactose sensitive lectin of chich retina and spinal cord. *Biochem. Biophys. Res. Commun.* 83: 1246–52

Ezekowitz, R. A. B., Day, L. E., Herman, G. A. 1988. A human mannose-binding protein is an acute phase reactant that shares sequence homology with other vertebrate lectins. *J. Exp. Med.* 167: 1034–46

Feizi, T. 1985. Demonstration by monoclonal antibodies that carbohydrate structures of glycoproteins and glycolipids are onco-developmental antigens. *Nature* 314: 53–57

Fenderson, B. A., Zehavi, U., Hakomori, S. I. 1984. A multivalent lacto-*N*-fucopentose III-lysyllysine conjugate decompacts preimplantation mouse embryos, while the free oligosaccharide is ineffective. *J. Exp. Med.* 160: 1591–96

Finne, J., Finne, V., Deagostini-Bazin, H., Goridinis, C. 1983. Occurrence of α2–8 linked polysialosyl units in a neural cell adhesion molecule. *Biochem. Biophys. Res. Commun.* 112: 482–87

Friedlander, D. R., Brackenbury, R., Edelman, G. M. 1985. Conversion of embryonic form to adult forms of N-CAM in vitro: Results from de novo synthesis of adult forms. *J. Cell Biol.* 101: 412–19

Frosch, M., Görgen, I., Boulnois, G. J., Timmis, K. N., Bitter-Suermann, D. 1985. NZB mouse system for production of monoclonal antibodies to weak bacterial antigens: Isolation of an IgG antibody to the polysaccharide capsules of *Eschericia coli* K1 and group B meningococci. *Proc. Natl. Acad. Sci. USA* 82: 1194–98

Fujiwara, K., Shinkai, H., Deutzman, R., Paulsson, M., Timpl, R. 1988. Structure and distribution of N-linked oligosaccharide chains on various domains of mouse tumour laminin. *Biochem. J.* 252: 453–61

Gallatin, M., St. John, T. P., Siegelman, M., Reichert, R., Butcher, E. C., Weissman, I. L. 1986. Lymphocyte homing receptors. *Cell* 44: 673–80

Glabe, C. G., Grabel, L. B., Vacquier, V. D., Rosen, S. D. 1982. Carbohydrate specificity of sea urchin sperm bindin: A cell surface lectin mediating sperm-egg adhesion. *J. Cell Biol.* 94: 123–28

Grabel, L. B. 1984. Isolation of a putative cell adhesion-mediating lectin from teratocarcinoma stem cells and its possible role in differentiation. *Cell Differ.* 15: 121–24

Grabel, L. B., Singer, M. S., Martin, G. R., Rosen, S. D. 1985. Isolation of a teratocarcinoma stem cell lectin implicated in intercellar adhesion. *FEBS Lett.* 183(2): 228–31

Grunwald, G. B., Fredman, P., Magnani, J. L., Trisler, D., Ginsburg, V., Nirenberg, M. 1985. Monoclonal antibody 18B8 detects gangliosides associated with neuronal differentiation and synapse formation. *Proc. Natl. Acad. Sci. USA* 82: 4008–12

Haagsman, H. P., Haegood, S., Sargeant, I., Buckley, D., White, R. T., Drickamer, K., Benson, B. J. 1987. The major lung surfactant protein, SP28-36, is a calcium-dependent, carbohydrate-binding protein. *J. Biol. Chem.* 262: 13877–80

Hakomori, S. 1981. Glycosphingolipids in cellular interaction, differentiation, and oncogenesis. *Annu. Rev. Biochem.* 50: 733–64

Halberg, D. F., Proulx, G., Doege, K., Yamada, Y., Drickamer, K. 1988. A segment of the cartilage proteoglycan core protein has lectin-like properties. *J. Biol. Chem.* 263: 9486–90

Hart, G. D., Snow, C. M., Senior, A., Haltiwanger, R. S., Gerace, L., Hart, G. W. 1987. Nuclear pore complex glycoproteins contain cytoplasmically disposed O-linked N-acetylglucosamine. J. Cell Biol. 104: 1157–64

Hoffman, S., Edelman, G. M. 1983. Kinetics of homophilic binding by embryonic and adult forms of the neural cell adhesion molecule. Proc. Natl. Acad. Sci. USA 80: 5762–66

Huang, T. T. F., Yanagimachi, R. 1984. Fucoidin inhibits attachment of guinea pig spermatazoa to the zona pellicida through binding to the inner acrisomal membrane and equitorial domains. Exp. Cell Res. 153: 363–73

Hyafil, F., Morello, D., Babinet, C., Jacob, F. 1980. A cell surface glycoprotein involved in the compaction of embryonal carcinoma cells and cleavage stage embryos. Cell 21: 927–34

Hynes, M. A., Gitt, M. A., Barondes, S. H., Jessell, T. M., Buck, L. B. 1989. Selective expression of a lactose-binding lectin gene in subsets of central and peripheral neurons. Submitted

Jackson, S. P., Tjian, R. 1988. O-glycosylation of eukaryotic transcription factors: Implications for mechanisms of transcriptional regulation. Cell 55: 125–33

James, L. W. M., Agnew, W. S. 1987. Multiple oligosaccharide chains in the voltage-sensitive Na channel from electrophorus electricus: Evidence for alpha-2,8-linked polysialic acid. Biochem. Biophys. Res. Commun. 148: 817–26

Jessell, T. M., Dodd, J. 1985. Structure and expression of differentiation antigens on functional subclasses of primary sensory neurones. Philos. Trans. R. Soc. London Ser. B 308: 271–81

Jessell, T. M., Dodd, J. 1986. Neurotransmitters and differentiation antigens in subsets of sensory neurons projecting to the spinal dorsal horn. In Neuropeptides in Neurologic and Psychiatric Diseases, ed. J. B. Martin, J. D. Barchas, pp. 111–33. New York: Raven

Jia, S., Wang, J. L. 1988. Carbohydrate binding protein 35. Complementary DNA sequence reveals homology with proteins of the heterogeneous nuclear RNP. J. Biol. Chem. 263: 6009–11

Joubert, R., Caron, M., Bladier, D. 1988. Distribution of beta-galactoside specific lectin activities during pre- and post-natal mouse brain development. Cell. Mol. Biol. 34: 79–87

Joubert, R., Caron, M., Bladier, D. 1987. Brain lectin-mediated agglutinability of dissociated cells from embryonic and post-natal mouse brain. Brain Res. 433: 146–50

Kawaguchi, K., Kuhlenschmidt, M., Rose-man, S., Lee, Y. C. 1980. Synthesis of some cluster galactosides and their effect on the hepatic galactose-binding system. Arch. Biochem. Biophys. 205: 388–95

Keilhauer, G., Faissner, A., Schachner, M. 1985. Differential inhibition of neurone-neurone, neurone-astrocyte and astrocyte-astrocyte adhesion by L1, L2 and NCAM antibodies. Nature 316: 728–30

Kobiler, D., Barondes, S. H. 1977. Lectin activity from embryonic chick brain, heart and liver: Changes with development. Dev. Biol. 60: 326–30

Kruse, J., Keilhauer, G., Faissner, A., Timpl, R., Schachner, M. 1985. The J1-glyco-protein—a novel nervous system cell adhesion molecule of the L2/HNK-1 family. Nature 316: 146–48

Kruse, J., Milhammer, R., Wernecke, H., Faissner, A., Sommer, I., Goridis, C., Schachner, M. 1984. Neural cell adhesion molecules and myelin-associated glyco-protein share a common carbohydrate moiety recognized by monoclonal anti-body-L2 and antibody-HNK-1. Nature 311: 153–55

Krusius, T., Finne, J., Margolis, R. K., Margolis, R. U. 1986. Identification of an O-glycosidic mannose-linked sialylated tetrasaccharide and keratan sulfate oligo-saccharides in the chondroitin sulfate pro-teoglycan of brain. J. Biol. Chem. 261: 8237–42

Krusius, T., Gehlsen, K. R., Ruoslahti, E. 1987. A fibroblast chondroitin sulfate pro-teoglycan core protein contains lectin-like and growth factor-like sequences. J. Biol. Chem. 262: 13120–25

Kuchler, S., Fressinaud, C., Sarlieve, L. L., Vincendon, G., Zanetta, J. P. 1988. Cerebellar soluble lectin is responsible for cell adhesion and participates in myelin com-paction in cultures at rat oligodendro-cytes. Dev. Neurosci. 10: 199–212

Kuhlenschmidt, T. B., Kuhlenschmidt, M. S., Roseman, S., Lee, Y. C. 1984. Binding and endocytosis of glycoproteins by iso-lated chicken hepatocytes. Biochemistry 23: 6437–44

Kunemund, V., Jungalwala, F. B., Fischer, G., Chou, D. K. H., Keilhauer, G., Schachner, M. 1988. The L2/HNK-1 carbohydrate of neural cell adhesion mol-ecules is involved in cell interactions. J. Cell Biol. 106: 213–23

Laing, J. G., Wang, J. L. 1988. Identification of carbohydrate binding protein 35 in heterogeneous nuclear ribonucleoprotein complex. Biochemistry 27: 5329–34

Lasky, L. A., Singer, M. S., Yednock, T. A., Dowbendo, D., Fennie, C., Rodriguez, H., Nguyen, T., Stachel, S., Rosen, S. D. 1989. Cloning of a lymphocyte homing receptor reveals a lectin domain. *Cell* 56: 1045–55

Lee, Y. C. 1989. Binding modes of mammalian hepatic Cal/GalNAc receptors. In *CIBA Found. Symp.* In press

Lee, Y. C., Lee, R. T. 1982. Neoglycoproteins as probes for binding and cellular uptake of glycoconjugates. In *The Glycoconjugates*, ed. M. I. Horowitz, 4: 57–83. New York: Academic

Leffler, H., Barondes, S. H. 1986. Specificity of binding of three soluble rat lung lectins to substituted and unsubstituded mammalian beta-galactosides. *J. Biol. Chem.* 261(22): 10119–26

Lehman, M. A., Haltiwanger, R. S., Hill, R. L. 1986. The binding of fucose-containing glycoproteins by hepatic lectins. The binding specificity of the rat liver fucose lectin. *J. Biol. Chem.* 261: 7426–32

Levine, J. M., Beasley, L., Stallcup, W. B. 1984. The D1.1 antigen: A cell surface marker for germinal cells of the central nervous system. *J. Neurosci.* 4: 820–31

Livingston, B. D., Jacobs, J. L., Glick, M. C., Troy, F. A. 1988. Extended polysialic acid chains (*n* greater than 55) in glycoproteins from human neuroblastoma cells. *J. Biol. Chem.* 263: 9443–48

Lopez, L. C., Bayna, E. M., Litoff, D., Shaper, N. L., Shur, B. D. 1985. Receptor function of mouse sperm surface galactosyltransferase during fertilization. *J. Cell Biol.* 101: 1501–10

Marchase, R. B. 1977. Biochemical investigations of retinotectal adhesive specificity. *J. Cell Biol.* 75: 237–57

Margolis, R. K., Greene, L. A., Margolis, R. U. 1986. Poly (*N*-acetyllactosaminyl) oligosaccharides in glycoproteins of PC12 pheochromocytoma cells and sympathetic neurons. *Biochemistry* 25: 3463–68

McCoy, R. D., Troy, F. A. 1987. CMP-Neu-NAC: poly-alpha-2,8-sialosyl sialosyltransferase in neural cell membranes. *Methods Enzymol.* 138: 627–37

McCoy, R. D., Vimr, E. R., Troy, F. A. 1985. CMP-NeuNAc: poly-alpha-2,8-sialosyl sialyltransferase and the biosynthesis of polysialosyl units in neural cell adhesion molecules. *J. Biol. Chem.* 260: 12695–99

McGarry, R. C., Helfand, S. L., Quarles, R. H., Roder, J. C. 1983. Recognition of myelin-associated glycoprotein by the monoclonal antibody HNK-1. *Nature* 306: 373–78

Mendez-Otero, R., Schlosshauer, B., Barnstable, C. J., Constantine-Paton, M. 1988. A developmentally regulated antigen associated with neural cell and process migration. *J. Neurosci.* 8: 564–79

Narimatsu, H., Sinha, S., Brew, K., Okayama, H., Qasba, P. K. 1986. Cloning and sequencing of cDNA of bovine *N*-acetylglucosamine (B, 104) galactosyltransferase. *Proc. Natl. Acad. Sci. USA* 83: 4720–24

Obata, K., Oide, M., Handa, S. 1977. Effects of glycolipids on *in vitro* development of neuromuscular junction. *Nature* 266: 369–71

Oriol, R. 1983. Immunofluorescent localization and genetic control of the synthesis of ABH and Lewis antigens. In *Red Cell Membrane Glycoconjugates and Related Genetic Markers*, ed. J. P. Cartron, P. Rouger, C. Salman, pp. 139–49. Paris: Librairie Arnette

Oriol, R., Cooper, J. E., Davies, D. R., Keeling, P. W. N. 1984. ABO antigens invascular endothelium and some epithelial tissues of baboons. *Lab. Invest.* 50: 514–18

Pesheva, P., Horwitz, A. F., Schachner, M. 1987. Integrin, the cell surface receptor for fibronectin and laminin, expresses the L2/HNK-1 and L3 carbohydrate structures shared by adhesion molecules. *Neurosci. Lett.* 83: 303–6

Poltorak, M., Sadoul, R., Keilhauer, G., Landa, C., Schachner, M. 1987. The myelin-associated glycoprotein (MAG), a member of the L2/HNK-1 family of neural cell adhesion molecules, is involved in neuron-oligodenrocyte and oligodendrocyte-oligodendrocyte interaction. *J. Cell Biol.* 105: 1893–99

Rastan, S., Thorpe, S. J., Scudder, D. P., Brown, S., Gooi, H. C., Feizi, T. 1985. Cell interactions in preimplantation embryos: Evidence for involvement of saccharides of the poly-*N*-acetyllactosamine series. *J. Embryol. Exp. Morphol.* 87: 115–28

Regan, L., Dodd, J., Barondes, S. H., Jessell, T. M. 1986. Selective expression of endogenous lactose-binding lectins and lactoseries glycoconjugates in subsets of rat sensory neurons. *Proc. Natl. Acad. Sci. USA* 83: 2248–52

Riopelle, R. J., McGarry, R. C., Roder, J. C. 1986. Adhesion properties of a neuronal epitope recognized by the monoclonal antibody HNK-1. *Brain Res.* 367: 20–25

Roseman, S. 1970. The synthesis of complex carbohydrates by multiglycosyltransferase systems and their potential function in intercellular adhesion intercellular adhesion. In *Chemistry and Physics of Lipids*, pp. 270–97. Amsterdam: North-Holland

Rosen, S. D., Yednock, T. A. 1986. Lymphocyte attachment to high endothelial venules during recirculation: A possible role for carbohydrates as recognition determinants. *Mol. Cell Biochem.* 72: 153–64

Rosen, S. D., Singer, M. S., Yednock, T. A., Stoolman, L. M. 1985. Involvement of sialic acid on endothelial cells in organ-specific lymphocyte recirculation. *Science* 228: 1005–7

Roth, S., McGuire, E. J., Roseman, J. 1971. Evidence for cell-surface glycosyltransferases: Their potential role in cellular recognition. *J. Cell Biol.* 51(21): 536–47

Rougon, G., Dubois, C., Buckley, N., Magnani, J. L., Zollinger, W. 1986. A monoclonal antibody against meningoccoccus group B polysaccharides distinguishes embryonic from adult NCAM. *J. Cell Biol.* 103: 2429–37

Runyan, R. B., Maxwell, G. D., Shur, B. D. 1986. Evidence for a novel enzymatic mechanism of neural crest cell migration on extracellular glycoconjugate matrices. *J. Cell Biol.* 102: 432–41

Runyan, R. B., Versalovic, J., Shur, B. D. 1988. Functionally distinct laminin receptors mediate cell adhesion and spreading: The requirement for surface galactosyltransferase in cell spreading. *J. Cell Biol.* 107: 1863–71

Ruoslahti, E., Pierschbacher, M. 1987. New perspectives in cell adhesion: RGD and integrins. *Science* 238: 491–97

Rutishauser, U., Acheson, A., Hall, A. K., Mann, D. M., Sunshine, J. 1988. The neural cell adhesion molecule (NCAM) as a regulator of cell-cell interactions. *Science* 240: 55–57

Rutishauser, U., Goridis, C. 1986. N-CAM: The molecule and its genetics. *Trends Genet.* 2: 72–76

Rutishauser, U., Jessell, T. M. 1988. Cell adhesion molecules invertebrate neural development. *Physiol. Rev.* 68: 819–55

Rutishauser, U., Watanabe, M., Silver, J., Troy, F. A., Vimr, E. R. 1985. Specific alteration of NCAM-mediated cell adhesion by an endoneuraminidase. *J. Cell Biol.* 101: 1842–49

Sanes, J. R., Cheney, M. 1982. Lectin binding reveals a synapse specific carbohydrate in skeletal muscle. *Nature* 300: 646–47

Sarkar, M., Liao, J., Kabat, E. A., Tanake, T., Ashwell, G. 1975. The binding site of rabbit hepatic lectin. *J. Biol. Chem.* 254: 3170–74

Schachner, M. 1989. Families of neuron adhesion molecules. *CIBA Found. Symp.* In press

Schauer, R., Cornfield, A. P. 1982. Occurrence of sialic acids. In *Sialic Acids, Chemistry, Metabolism and Function*, ed. R. Schauer, pp. 5–50. New York: Springer-Verlag

Schwarting, G. A., Jungawalla, F. B., Chou, D. K. H., Boyer, A. M., Yamamoto, M. 1987. Sulfated glucuronic acid-containing glycoconjugates are temporally and spatially regulated antigens in the developing mammalian nervous system. *Dev. Biol.* 120: 65–76

Scott, L. J. C., Bacou, F., Sanes, J. R. 1988. A synapse-specific carbohydrate of the neuromuscular junction association with both acetylcholinesterase and a glycolipid. *J. Neurosci.* 8(3): 932–44

Shaper, N. L., Hollis, G. F., Douglas, J. G., Kirsch, I. R., Shaper, J. H. 1988. Characterization of the full length cDNA for murine B-1-4-galactosyltransferase. *J. Biol. Chem.* 263: 10420–28

Shaper, N. L., Shaper, J. H., Meuth, J. L., Fox, J. L., Chang, H., Kirsch, I. R., Hollis, G. F. 1986. Bovine galactosyltransferase: Identification of a clone by direct immunological screening of a cDNA expression library. *Proc. Natl. Acad. Sci. USA* 83: 1573–77

Shur, B. D., Roth, S. 1975. Cell surface glycosyltransferases. *Biochim. Biophys. Acta* 282: 18–30

Siegelman, M. H., Van deRijn, M., Weissman, I. L. 1989. Mouse lymph node homing receptor cDNA clone encodes a glycoprotein revealing tandem interaction domains. *Science* 243: 1165–71

Solter, D., Knowles, B. B. 1978. Monoclonal antibody defining a stage-specific mouse embryonic antigen (SSEA-1). *Proc. Natl. Acad. Sci. USA* 75: 5565

Stallcup, W. B. 1988. Involvement of gangliosides and glycoprotein fibronectin receptors in cellular adhesion to fibronectin. *Exp. Cell Res.* 177: 90–102

Stallcup, W. B., Pytela, R., Ruoslahti, E. 1989. A neuroectoderm associated ganglioside participates in fibronectin receptor-mediated adhesion of germinal cells to fibronectin. *Dev. Biol.* 132: 212–29

Stoolman, L. M., Tenforde, T. S., Rosen, S. D. 1984. Phosphomannosyl receptors may participate in the adhesive interaction between lymphocytes and high endothelial venules. *J. Cell Biol.* 99: 1535–40

Sunshine, J., Balak, K., Rutishauser, U., Jacobson, M. 1987. Changes in neural cell adhesion molecule (NCAM) structure during vertebrate neural development. *Proc. Natl. Acad. Sci. USA* 84: 5986–90

Takeichi, M. 1986. Molecular basis for teratocarcinoma cell-cell adhesion. In *Developmental Biology*, ed. L. M. Browder, 2: 373–86. New York: Plenum

Takeichi, M. 1988. The cadherins: Cell-cell adhesion molecules controlling animal morphogenesis. *Development* 102: 639–55

Tiemeyer, M., Tasuda, Y., Schnaar, R. L. 1989. Ganglioside-specific binding protein on rat brain membranes. *J. Biol. Chem.* 264: 1671–81

Walter, J., Henke-Fahle, S., Bonhoeffer, F. 1987. Avoidance of posterior tectal membranes by temporal retina axons. *Development* 101: 909–14

Wassarman, P. M. 1987. Early events in mammalian fertilization. *Annu. Rev. Cell Biol.* 3: 109–42

Weinstein, J., Lee, E. U., McEntee, K., Lai, P. H., Paulson, J. C. 1987. Primary structure of β-galactoside alpha 2,6-sialyl-transferase. Conversion of membrane-bound by cleavage of the NH2-terminal signal anchor. *J. Biol. Chem.* 262: 17735–43

Yamamoto, M., Boyer, A. M., Schwarting, G. S. 1985. Fucose-containing glycolipids are stage- and region-specific antigens in the developing embryonic brain of rodents. *Proc. Natl. Acad. Sci. USA* 82: 3045–49

Yednock, T. A., Butcher, E. C., Stoolman, L. M., Rosen, S. D. 1987a. Receptors involved in lymphocyte homing: Relationship between a carbohydrate-binding receptor the MEL-14 antigen. *J. Cell Biol.* 104: 725–31

Yednock, T. A., Stoolman, L. M., Rosen, S. D. 1987b. Phosphomannosyl-derivatized beads detect a receptor involved in lymphocyte homing. *J. Cell Biol.* 104: 713–23

Zanetta, J. P., Dontenwill, M., Meyer, A., Roussel, G. 1985. Isolation and immuno-histochemical localization of a lectin-like molecule from the rat cerebellum. *Brain Res.* 349: 233–43

Zanetta, J. P., Meyer, A., Kuchler, S., Vincedon, G. 1987. Isolation and immuno-chemical study of a soluble cerebellar lectin delineating its structure and function. *J. Neurochem.* 49: 1250–57

*Annu. Rev. Neurosci. 1990. 13:257–81*

# PERCEPTUAL NEURAL ORGANIZATION: SOME APPROACHES BASED ON NETWORK MODELS AND INFORMATION THEORY

*Ralph Linsker*

IBM Research Division, T. J. Watson Research Center,
Yorktown Heights, New York 10598

## INTRODUCTION

To understand neural processing we need to study structure and function at many levels of organization, from subcellular to systemic. We also need to understand the linkages between levels. First, what are the mechanisms at a lower level that generate structures at a higher one? Second, of all the possible structures that could be formed from the given constituents, only some are in fact generated by the lower-level mechanisms. Are the generated structures optimal, or favored over the other structures, with respect to some property? If so, we may be able to describe the lower-level mechanisms as implementing an organizing, or optimization, principle. Third, can we account for such putative organizing principles in terms of their adaptive value to the animal?

This review explores linkages between lower-level mechanisms and functional architecture in the processing of sensory information. It brings together two lines of study. The first of these is the investigation of how lower-level mechanisms can generate the types of neural structures that are found in the early processing stages of perceptual systems. This approach involves modeling the formation and modification of neuronal connections by simple rules (for example of Hebb type), expressing these rules as a mathematical procedure or algorithm, and using computer simulations or mathematical analysis to determine what structures the rules generate. The

257

0147–006X/90/0301–0257$02.00

types of structures or patterns whose formation has been studied include topographic maps, orientation-selective receptive fields, and ocular dominance and orientation columns. The second line of study consists of a set of general ideas about how data may be encoded and transformed in a perceptual system, and the informational purposes that these transformations may serve. I review both approaches, then discuss recent results that suggest how the approaches may be unified—that is, how lower-level mechanisms may be used to create perceptual processing stages that implement certain types of optimal encoding principles.

## FROM SALIENT EXPERIMENTAL FEATURES TO MODELS OF SELF-ORGANIZATION

How do specific patterns of neural connectivity and functional architecture develop, how may they be plastically altered, and how do these patterns subserve perceptual functions? A great deal of experimental progress has been made concerning these issues. Theoretical work in this area has several purposes: to seek common rules and principles that may account for a range of observations, to predict new features of neural organization and cell response, and to provide a view of biological information processing that integrates several levels of organization.

Experimental evidence indeed suggests that common rules and principles may underlie important aspects of sensory processing. First, cortical regions subserving different processing functions share similar intrinsic structure (Mountcastle 1978). Second, by altering the character of the input to a sensory processing region, one can induce patterns of organization and response properties that differ from those normally found in that region, but in an apparently lawful way (e.g. Constantine-Paton & Law 1978, Kaas et al 1983, Merzenich et al 1984, Métin & Frost 1989, Rauschecker 1987, Sur et al 1988).

The biological systems of interest exhibit immense complexity. The models to be discussed are by comparison extremely simple. The purpose of this simplicity is to allow us to gain understanding of how underlying rules can generate structure and function, so that essential complexities can then be added in an insightful way.

### Types of Pattern-Generating Models

When a structure or organized pattern is found in nature, various types of patterning models may account for it. Two extreme types are patterning by "explicit specification" and patterning by a process of so-called "self-organization." In the first type of model, the pattern is directly determined or strongly influenced by a pre-existing pattern in an underlying substrate. An example of this type is Sperry's (1943, 1963) chemoaffinity model of topographic map formation, in which each cell has a specific chemical "address." When such a model is used, the problem of explaining the original pattern's emergence is replaced by the problem of explaining how the underlying pattern of "addresses" itself arises from lower-level rules. In some models, an explicit specification may determine the pattern in detail; in others, it may determine only certain features of the pattern, such as its overall orientation or its coarse-grained arrangement.

In a model of self-organization, on the other hand, the pattern develops from an initially homogeneous structure as a result of processes that incrementally change each element of the system according to a relatively simple set of rules (Turing 1952). Typically such rules are local; that is, each incremental change in one element depends only upon the state of a few other elements. In the generation of certain properties such as topographic maps, both self-organization and a partial form of explicit specification appear to play a role (Udin & Fawcett 1988). This review focuses mainly on the self-organizing aspects of patterning models.

## Basic Structure of the Models

The models discussed here can all be understood with reference to a common basic structure and set of patterning rules, although the details vary and not all of the components are present in each model. The structure consists of a "source" and a "target" layer of cells, with feedforward connections from source to target and lateral connections between target cells. Each connection is characterized by a number called its "strength." The patterning process consists of repeatedly modifying the connection strengths (and in some cases creating or destroying connections) according to the rules until a final configuration develops.

Each connection strength is modified in a way that depends upon the "state" of the connected cells. In "marker-based" models the state of a cell is defined as the amount of a marker substance of some type (there may be more than one type of marker). In "activity-dependent" models the state is defined as a measure of neuronal signaling activity such as a firing rate.

The patterning rules consist of three parts: (a) a "transmission" rule that determines how each target cell's state depends upon the states of the source cells connected to it (the target cell's state is typically related to an average of the source cells' states weighted by their connection strengths); (b) a "lateral interaction" rule that describes how the states of nearby target cells are modified by interactions within the target layer; (c) an "update" rule that modifies each feedforward (and in some cases lateral) connection according to the degree to which the states of the connected cells are similar.

In activity-dependent models, the update rule typically changes each connection strength in a way that depends upon the presynaptic signaling activity and the postsynaptic firing rate or depolarization potential. In this case, the change usually depends upon the degree of correlation between the pre- and postsynaptic quantities, with stronger correlation causing an increase in strength. I refer to this type of neural activity-dependent rule as "Hebb-like" (see Brown et al 1990 for review), while recognizing that Hebb's original proposal (Hebb 1949) referred to cell firing, not depolarization, and that it provided no statement of the conditions under which strengths could decrease (cf. Stent 1973).

Note that both Hebb-like rules and marker-based rules play similar roles in the patterning process, although the mechanisms to which they refer are very different. In each case a connection is strengthened when the pre- and postsynaptic cells are correlated with each other—either in their signaling activity or in the possession of similar amounts of a marker.

It is striking that many salient features of perceptual neural organization emerge in models containing the basic elements described above: a *positive feedback* process (whether marker-based or Hebb-like) in which large connection strength causes the states of the connected source and target cells to be more similar, and greater similarity tends further to increase the connection strength; and a *lateral interaction* process that causes the states of nearby target cells to tend to become either more or less similar to each other, depending upon the interaction.

Models of neural self-organization developed since the early 1970s have emphasized different aspects of these rules and invoked different constraints or assumptions concerning connectivity and input and output signaling activity. They have been directed toward various goals such as elucidating basic principles of pattern formation, showing how specific features emerge, or modeling specific underlying mechanisms in greater detail. The remainder of this section reviews some key ways in which these models have increased our understanding of how simple rules lead to complex structures that resemble those found biologically.

## Self-Organizing Models of Topographic Map Formation

In a topographic map, source and target cells are connected in such a way that positional ordering is preserved. Experimental results on topographic map formation have been recently reviewed (Udin & Fawcett 1988). How are the main features of topographic maps generated in self-organizing models?

NEIGHBOR-PRESERVING MAPS    For marker-based models, the essential idea is described by the so-called "tea trade analogy" (von der Malsburg & Willshaw 1977, Willshaw & von der Malsburg 1979). In this model, one assumes that the states of nearby source cells are similar. Each target cell acquires a state that is the average of the source cells' states weighted by their connection strengths. The lateral interaction then "blends" the states of nearby target cells so that these states become more similar to each other. The update rule strengthens connections between source and target cells having similar states. The result of this process is that nearby source cells map to nearby target cells.

How can a topographic map arise in an activity-dependent model (Willshaw & von der Malsburg 1976)? Start with initially random non-negative feedforward connection strengths. (For simplicity, all pairs of source and target cells can be assumed to be connected. A strength that declines to and remains at zero corresponds to an absent connection in a more realistic model.) Many patterns of signaling activity in the source layer are presented to the network, one at a time. Each input presentation consists simply of a localized region, or "spot," of activity against a quiet background. If certain constraints (specific to the model) are met, each such presentation will cause a localized group of target cells to be active. (In particular, the lateral interactions are set up so that the active target cells tend to form a localized region, rather than being dispersed.) The connections between pairs of active source and target cells are then strengthened by a small amount (this is the Hebb-like rule), and compensating reductions are made in the strengths of other connections. After many such spot patterns are presented, the resulting pattern of connection strengths will be topographic; that is, nearby source cells will make their strongest connections to nearby target cells.

Kohonen (1982a,b) has described a related algorithm in which a simple geometric computation is substituted for the more detailed properties of the Hebb-like rule and lateral interactions, and a topographic map emerges. For each presentation of an input activity "spot," this algorithm finds the particular target cell that fires most strongly, and then changes the target cells' response properties so that the maximally responding target cell and its neighbors will respond more strongly to a spot at the same input location in the future. The algorithm also applies more generally to cases in which the input patterns are ordered according to properties other than spatial position.

In both marker-based and activity-dependent models, the degree of selectivity of a target cell—that is, the range of different input patterns to which the cell responds strongly—can depend upon (*a*) the update rule (e.g. Bienenstock et al 1982); (*b*) constraints on total connection strength, or on some other function of the strengths, for each source or target cell

(Willshaw & von der Malsburg 1976); and (c) the particular form of lateral interaction (Grossberg 1976).

Let us consider several important experimental and theoretical questions concerning topographic maps that go beyond the basic issue, of mapping neighbors to neighbors, discussed above.

OVERALL MAP ORIENTATION    The self-organizing model rules do not by themselves favor one map orientation over another. In biological systems, temporal or chemical mechanisms that favor certain regions to be connected first, or preferentially, may induce an orientation bias. This roughly corresponds to weakly biasing either the initial connection strengths (Willshaw & von der Malsburg 1976, 1979) or the patterning rules (Whitelaw & Cowan 1981, Fraser 1985). Also, the shapes of the source and target layers may favor a particular orientation, owing to boundary effects. More generally, source and target layer boundary conditions can influence map formation in significant ways that are not limited to overall orientation (Schwartz 1977). Models having such biases can be thought of as hybrids of self-organizing and explicit-specification models. Experimental manipulations (reviewed by Udin & Fawcett 1988) are important for determining the extent to which the biological mechanisms fit either type of model.

LOCALLY OPTIMAL MAPS THAT ARE NOT GLOBALLY OPTIMAL    If separate regions of the network independently become organized, the resulting maps may be out of register along their boundaries or may have conflicting orientations. To avoid this, one can start with a coarse topographic map that is subsequently refined by a local process. As an example, a two-step process in which the refinement stage is activity-dependent is found during map regeneration in goldfish (Schmidt 1985). Alternatively, the lateral interaction distance in the target layer can be made initially large, then decreased during development (Kohonen 1982a,b). In a biological system, growth of the target layer relative to the lateral interaction distance could achieve a similar result. Yet another approach is to limit early map formation to a single region of nucleation (Willshaw & von der Malsburg 1976).

CONTINUOUS VS. DISCRETE MAPS    Some self-organizing models generate topographic maps in which the receptive fields of a sequence of target cells either (a) shift in a continuous and overlapping manner, as described above, or (b) form clusters with discontinuous jumps as one crosses from one column-like target region to another (Takeuchi & Amari 1979). Discontinuous mappings also can emerge in models with more complex dynamics (Pearson et al 1987).

MAP MAGNIFICATION FACTORS    In certain models (Amari 1980, Kohonen

1982a,b, Pearson et al 1987, Linsker 1988b), if one region of the source layer is stimulated by an activity "spot" more often than another, that source region is mapped onto a larger region of the target layer. From one theoretical standpoint (see A PRINCIPLE OF MAXIMUM PRESERVATION below) it can be desirable for the magnification factor of each source region to be proportional to the frequency of "spot" stimulation, so that each target region of given size is activated an equal fraction of the time. Although the Kohonen algorithm described above was originally thought to have this property (Kohonen 1982a,b), it has since been shown not to (Ritter & Schulten 1986, Kohonen 1988). A model that does generate maps having this proportionality property is described by Linsker (1989b), and has been generalized to cases in which the activity patterns are more complex and the magnification factor does not depend simply upon frequency of stimulation.

## Ocular Dominance

Banded regions of ocular dominance are found in cat and monkey striate cortex (LeVay et al 1975, Hubel & Wiesel 1977, Shatz & Stryker 1978) and in frog tectum following implantation of a third eye (Constantine-Paton & Law 1978). Their formation depends upon correlated electrical activity (Stryker & Harris 1986, Reiter & Stryker 1988).

Two questions addressed by self-organizing models are:

1. How does the process of segregation into ocularity domains interact with the process of topographic map formation?
2. Why do patterns ranging from regular stripes to less regular bands or patches form, and what determines their characteristic dimensions?

The interaction between topographic map and ocularity domain formation is studied in a marker-based model (von der Malsburg 1979) having three classes of markers: one for ocularity and one each for horizontal and vertical position. The multiple-marker model is an extension of topographic map models to a case in which similarities in properties other than position affect the mapping. The overall "similarity" between the marker content of two cells is defined as a combined measure of the similarities of the markers in each class. The overall effect is to map source to target in such a way that the similarity between nearby target cells is maximized on average. This type of mapping criterion is either explicitly introduced or emergent in other self-organizing models, both marker-based and activity-dependent, as well (see for example Kohonen 1982a,b, 1988, Linsker 1988b, Durbin & Mitchison 1988).

Models that incorporate a more detailed interaction between topography and ocularity in map formation include an activity-dependent model (von der Malsburg & Willshaw 1976) in which same-eye input activities are locally correlated and opposite-eye activities are anti-correlated, and a model (Fraser 1985) involving markers that mediate an adhesive interaction, activity-dependent modification, and global positional biasing effects.

Why do ocular dominance stripes and related patterns form? Suppose we ignore visual field position, and simply distinguish between inputs from the two eyes. Each target cell receives connections of initially random strength from each eye. Now let the target cells interact, and their connection strengths change incrementally, according to the following assumed rule (Swindale 1980): Each cell's ocularity preference changes to be more nearly like the average preference of its near neighbors, and to be more nearly opposite the average preference of its midrange neighbors. The result is that regions of each ocularity preference form and segregate into (a) locally parallel stripes (with forks and bends) if the lateral interaction rules are unbiased between the two eyes, or (b) islands of one ocularity in a "sea" of the other if the rules favor the latter ocularity to a sufficient degree. Stripes tend to run into the layer boundary at a right angle, consistent with the observed perpendicularity of stripes at the area 17/18 border in, for example, macaque (Hubel & Wiesel 1977). In addition to these simulation results, mathematical analysis of the onset of segregation shows that as stripes start to form, their width has a particular preferred value that depends upon the scale of the lateral interaction (Swindale 1980). Stripe formation has also been analyzed and simulated in a more detailed activity-dependent model in which each target cell receives inputs from source cells lying in a fixed topographic arrangement (Miller et al 1986, 1989).

## Orientation-Selective Cells and Columns

Cells selectively responsive to edges or bars having a particular orientation, and arranged in columns containing cells of similar orientation preference, were discovered by Hubel & Wiesel in cat (Hubel & Wiesel 1962, 1963, reviewed by Frégnac & Imbert 1984) and macaque (Hubel & Wiesel 1968, 1977). In macaque, they are present at birth (Wiesel & Hubel 1974) prior to structured visual experience. Recent advances in optical imaging of electrical activity (Blasdel & Salama 1986) have provided a more detailed picture of the columnar structure in macaque. The detailed mechanisms that mediate orientation-selective response are still unclear (see review by Ferster & Koch 1987).

The early experimental findings motivated attempts to explain how orientation selectivity could be generated by activity-dependent self-organizing models. To see what has been learned from these models, it is useful to think of the problem in two parts.

1. When the input patterns presented to the network (during the connection modification process) differ from one another with respect to only one property, how does each target cell learn to respond selectively to patterns having a particular value (or range of values) of that property? That is, how does selectivity per se emerge when there is only one pattern selection criterion?

2. When the input patterns differ from one another in many respects, what determines the particular property, or combination of properties, to which each target cell will learn to respond selectively? In particular, how can orientation selectivity emerge in self-organizing models either in the absence of structured visual experience, corresponding to the prenatal development of well-formed selectivity in macaque, or under the influence of realistic postnatal experience?

HOW DOES SELECTIVITY EMERGE?    In the activity-dependent models of topographic map formation discussed above, topographic ordering arises when each input activity pattern consists of a localized spot. Two input activity patterns are similar to each other if their spots are at similar positions. The update rule, and in some models the lateral interactions, cause each target cell to become selective for a subset of similar patterns, and the lateral interactions cause nearby target cells to have similar response properties.

In the first self-organizing models to address orientation selectivity (von der Malsburg 1973, Bienenstock et al 1982), each input pattern consists of an arbitrarily oriented bar of activity against a quiet background. (There can be added random noise that does not affect the results.) The bar's shape and center position are the same for all the input patterns. The only measure of pattern similarity is thus bar orientation similarity.

Von der Malsburg (1973) considers a layer of target cells with fixed center-surround lateral connections. Each source cell is connected to each target cell (topographic mapping is neither present initially nor does it emerge). The result is that most of the target cells become selectively responsive to bars lying at or near some orientation, and nearby target cells tend to become selective for bars of similar orientation.

Bienenstock et al (1982) consider a single target cell whose synaptic inputs can be thought of as forming a circular ring whose center is the same as that of each bar pattern. A version of a Hebb-like rule is used in which the change in strength depends upon the difference in value between the postsynaptic activity and an adjustable threshold. This threshold is important in their model for ensuring that the target cell becomes selective for one bar pattern or a subset of similar patterns.

Because of the simple centered-bar form chosen for the input patterns, the emergence of orientation selectivity in both these models closely parallels the emergence of positional selectivity in the formation of topographic maps. In fact, the activity pattern on a semicircular portion of the ring of synapses in the latter model consists of an interval (or one-dimensional "spot") of activity whose position is given by the bar orientation. The development of orientation selectivity in this model thus corresponds exactly to the development of positional selectivity in a network having a one-dimensional source "layer" (that has been "wrapped around" so that its endpoints meet).

COLUMN FORMATION    We have noted that von der Malsburg's (1973) model generates target cells whose orientation preferences are similar within a region of the target layer; that is, the cells form orientation "columns." In the model studied by Swindale (1982), each target cell is labeled by a vector whose magnitude and direction are modified by lateral interactions with the labels at other sites. The approach is the same as that used for ocular dominance stripe formation (Swindale 1980), in which each cell was labeled by its degree of left-eye and right-eye preference. The same type of center-surround lateral interaction rule is used, so that each label incrementally changes to become more nearly like its near neighbors and unlike its midrange neighbors. The result is that irregularly banded regions of similarly labeled cells emerge. In this model, the questions of why orientation-selective receptive fields arise, and how the particular lateral interaction between labels may arise, are not addressed.

By invoking a similar center-surround type of lateral interaction, von der Malsburg & Cowan (1982) show that if one postulates that different groups of target cells have, for example, horizontal and vertical orientation preference, then the lateral interactions induce the formation of a sequence of cells having the intermediate orientation preferences.

WHY ORIENTATION SELECTIVITY?    Linsker (1986a–c) has analyzed the development of a network in which random uncorrelated signaling activity is preprocessed by a layer of cells having topographically arranged center-surround receptive fields. The output from this "source" layer of cells is in turn provided as input to a target layer via Hebb-modifiable connections. Even in the absence of lateral connections, a layer of orientation-selective target cells emerges under certain conditions. Since the architecture and development rules possess no orientational bias, the emergence of orientation selectivity is an example of a "symmetry-breaking" process (Linsker 1986b, 1988c). Under different conditions a layer of center-surround cells emerges. The two parameters that determine which cell type develops are the radius of the source region that provides input to each target cell, and the degree of correlation needed to cause strength increase via the Hebb-like rule.

If the parameter values lie in a range that leads to orientation-selective cell formation, then adding weak lateral connections of fixed strength to the model causes orientation columns to form. In the model, feedforward connections suffice to generate orientation-selective cells. The simulations discussed do not address the question of the extent to which lateral interactions can influence the formation of orientation-selective cells (as opposed to their columnar organization).

The emergence of orientation-selective cells and columns in this model is explained by analyzing how the Hebb-like rule creates geometric patterns of connection strengths when the pairwise signal-activity correlations in the source layer depend upon the relative positions of the source cells (Linsker 1986b, 1987). These activity correlations arise because of the center-surround preprocessing. Even if one uses a Hebb-like rule different from the particular one used in this model, orientation-selective cells can emerge (see for example Kammen & Yuille 1988). The mathematical reason is that the orientation-selective solutions are formed from combinations of the first few eigenfunctions (those having the largest eigenvalues) of the covariance matrix of activity in the source layer (Linsker 1987).

Experimentally, one finds a series of several layers of cells having substantially center-surround receptive field properties in retina and LGN of cat and macaque monkey, and layer IV of macaque striate cortex, followed by the onset of well-tuned orientation selectivity. Although the feedforward connections in the model can in general be of both excitatory and inhibitory types, orientation selectivity emerges even when the connections from the center-surround layer are constrained to be excitatory (Linsker 1987, and unpublished results). This is of interest since the geniculocortical connections which immediately precede the first well-tuned orientation-selective stage in cat are thought to be exclusively excitatory.

In Linsker's model, the center-surround cells responsible for the preprocessing themselves emerge by a self-organizing process, given only random input and topographically arranged connections to each cell from a neighborhood of cells of the previous layer (Linsker 1986a). The means by which the center-surround layer is generated—whether by an activity-dependent self-organizing process as in the model, or by retinal interactions mediated by anatomically complex connections—does not affect the emergence of orientation selectivity in the later processing stage of the model.

The model as a whole shows that a sequence of feature-analyzing cell types of progressive complexity can emerge via a Hebb-like rule in a multistage feedforward network. Structured rather than random uncorrelated input can also be provided to the first layer, and the developmental results studied (Linsker 1988a).

Models relating to orientation selectivity that are not self-organizing models of cell formation are outside the scope of this review; two examples are (*a*) a nonadaptive model of cell formation (Braitenberg 1985) and (*b*) a demonstration, using a "learning by error correction" algorithm, that orientation-selective elements perform useful intermediate processing functions in a network computation of curvature in a scene (Lehky & Sejnowski 1988).

## NETWORK MODELS, OPTIMIZATION PRINCIPLES, AND INFORMATION THEORY

The results of the preceding section provide a stimulus for investigating why a Hebb-like rule, with or without lateral connections, generates a sequence of feature-analyzing cell types in a layered network. This section explores several senses in which the types of patterning rules studied can create structures that are optimal from an information processing standpoint.

### What Does a Hebb-like Rule Optimize?

First consider a single cell with feedforward but no lateral connections. To understand intuitively the effect of a Hebb-like rule, an analogy is useful: Imagine a committee whose recommendation ("output activity") on any issue is the weighted average of its members' opinions ("input activities"), each opinion being weighted according to the member's voting strength. A member who consistently agrees (disagrees) with the committee's opinion receives an increase (decrease) in voting strength. The operation of this Hebb-like rule transforms a committee of members having random voting strengths, whose averaged output is rarely a strong recommendation (either positive or negative), into a committee whose members of high voting strength tend to agree more often, and whose recommendations tend to be strong (in either direction) more of the time.

What does this result mean for a model cell (Linsker 1988a)? Consider a cell whose output activity is a linear function of its input activities. A histogram of the output values over a range of input presentations will have a spread, or variance, that depends upon the connection strengths. If the Hebb-like rule has a term that changes each strength by an amount proportional to the product of (*a*) the input activity at that connection (minus its mean value) and (*b*) the cell's output activity (minus its mean value), this term will tend to increase the cell's output variance. Depending upon the other terms in the update rule, the net result can be to maximize the variance subject to some constraint, for example on the sum of the strengths (Linsker 1986b, 1988a,c) or the sum of the squares of the strengths (Amari 1977, Oja 1982). A cell with low output variance tends to be nonselective—its responses to different input presentations are similar. High variance tends to correspond to greater selectivity. Even when the cell's output variance is itself constrained (e.g. by a maximum allowed firing rate), or when the cell's response is nonlinear, a suitable Hebb-like rule can act to maximize the cell's responsiveness to statistical structure (or "features") in the input presentations, and to minimize its responsiveness to uncorrelated inputs or random processing noise (Linsker 1988a).

Under certain conditions, a Hebb-like rule generates cells that are optimal in several ways. As we have seen, such a rule produces cells that respond selectively to statistically significant properties of the input presentations. In particular, a suitably constructed rule (Oja 1982) produces a cell that performs a standard computation in statistical feature extraction called "principal component analysis." [Watanabe (1985) discusses in detail the application of this and related statistical methods to pattern recognition.] A cell produced by such a Hebb-like rule also has an "optimal inference" property: The average error incurred when using the cell's output value to estimate its input values is less for such a cell than for any other linear response cell. Finally, under certain conditions, the output from such a cell conveys maximum information about its input activity values, compared with cells having arbitrary connection strengths (Linsker 1988a).

## The Role of Lateral Interactions

If a Hebb-like rule can cause individual target cells to develop so as to optimize certain properties, can the addition of lateral interactions allow an assembly, or an entire layer, of target cells to develop in an optimal or near-optimal way with respect to some property that is important for information processing?

When both feedforward and lateral connections are present, target cells not only can develop feature analyzing properties by means of a Hebb-like rule (as shown above), but can "cooperate" or "compete" with each other in the formation of these properties (see, for example, von der Malsburg 1973, Grossberg 1976, Kohonen 1982a,b, Rumelhart & Zipser 1985, Linsker 1986c, 1988a,b, 1989b, Pearson et al 1987, Hinton 1989). An optimization method (Durbin & Willshaw 1987) based on the "tea trade" model (von der Malsburg & Willshaw 1977) has been developed.

In some of these models the process of modifying the connection strengths corresponds closely to a statistical method for classifying or detecting regularities in data sets (see Lippmann 1987 for review). For example, a network algorithm for forming clusters of related input presentations, developed in connection with the "adaptive resonance theory" (Carpenter & Grossberg 1988), is related to the "sequential leader clustering" algorithm in statistics. Kohonen's (1982a,b) algorithm for "feature map" formation, discussed above, is related to a version of the "k-means" algorithm (MacQueen 1967) for partitioning a set of data points into a number of groups such that the members of each group are similar to each other according to a particular measure. This type of partitioning is useful for extracting common features from raw data, and also for the data compression problem of communicating an item of data, by stating which group it belongs to, in such a way that the original datum can be inferred with minimum error. [Cf. the "optimal inference" result of the previous section, the literature on "vector quantization" (reviewed by Gray 1984), and the work of Watanabe (1969, 1985) on feature extraction.] I discuss the choice of "similarity" measure further in the next section.

Might some type of statistical algorithm be playing an important role in the biological processing of sensory data? If so, what property if any is being optimized, how can the optimization process be realized by a biological network, and of what value is the process to the animal? Although these questions are necessarily speculative, I discuss here a concrete approach to them, in which some basic ideas from information theory play an important role (Linsker 1988a). To place recent connections between information theory and perceptual network algorithms in perspective, it is important to review the history, since the 1950s, of the idea that information theory may relate to the organization of sensory and perceptual processing. The relevant information-theoretic ideas are discussed here in an informal way. For the classic development of information theory see Shannon (1949); a tutorial treatment of some of the ideas as they relate to sensory processing is given in Linsker (1988a).

## Information Theory and Perception

REDUNDANCY REDUCTION    Sensory input, such as a time sequence of visual scenes, is not random and uncorrelated, but contains structure and regularities. Knowing certain aspects of the input allows us to infer other aspects; the input is partially redundant. Attneave (1954, p. 189) proposed that

> A major function of the perceptual machinery is to strip away some of the redundancy of stimulation, to describe or encode incoming information in a form more economical than that in which it impinges on the receptors.

The suggested approach is "equivalent to that of a communications engineer" who exploits regularities in an input scene, in order to encode the input signals so that they may be sent along an essentially noise-free channel of limited capacity and reconstructed by the recipient with high fidelity (low distortion). [For examples of how the computations performed by cells during early visual processing can be analyzed from the standpoints of data compression and reconstruction, see Daugman (1988) and Oğuztöreli & Caelli (1986).]

Attneave's recognition that encoding for redundancy reduction is related to the identification of specific features such as edges or corners in a scene, and that a redundancy-reducing strategy might be important for perceptual processing, is an important insight. The idea is related to Craik's (1943) view of the brain as building a model of the external world that incorporates the world's lawful regularities and constraints. Attneave's idea of sensory processing as data compression, however, leaves some crucial questions unanswered: What role if any is played by the "recipient" in the communications analogy? If the original scene is never "reconstructed" by the brain (and why should it be?), what is the meaning of the fidelity criterion? That is, if not all input information can be preserved, what criterion determines what to discard? Finally, how do biological constraints and costs influence what types of encoding are "economical," and how does a biological system carry out the encoding?

In a series of papers, Barlow has studied several of the physiological correlates of redundancy reduction as a perceptual processing strategy:

1. In a model having discrete (e.g., binary) inputs, with no processing noise to induce coding or transmission errors, one can encode the input "message" by removing correlations (a type of redundancy) and thus produce an output that is more compact than the input, yet is able to regenerate the input in a completely reversible way (Barlow 1959, 1961). The compaction process can be driven by a constraint limiting the average firing rate to a low value in later processing stages (Barlow's principle of "economy of impulses"). Note that as long as the code is fully reversible, the question of what fidelity measure to use does not arise.

2. When the input signals are continuous-valued (or can take on a very large number of values), it is important to treat processing elements having limited resolution (Barlow 1969). This, however, leaves open the question of when two stimuli should be considered "similar" enough not to be resolved by the network, and of how "similarity" with respect to different properties should be compared. The issue is related to the choice of appropriate fidelity measure (above), and recurs in connection with Marr's work (below).

3. Building on Hebb's (1949) discussion of the importance of identifying

correlated inputs, one can regard a Hebb-like rule as forming cells that detect "suspicious coincidences" (Phillips et al 1984, Barlow 1985). These "coincidences" are sets of events that signal the existence of structure (e.g. coherent motion of an object), since they would not jointly occur in the absence of such structure. Legéndy (1970, 1975) discusses a similar idea, that of forming cells that respond selectively to "surprising" combinations of stimuli, in the context of a suggestion that individual cells may possess complex memory and processing capabilities to carry out the requisite computations.

4. The existence of multiple cortical areas with different mappings allows different types of regularities to be detected, if one assumes that a constraint on connection length requires cells to be relatively close together for correlations in their activity to be detected, and consequently for redundancy to be reduced (Barlow 1986).

THE INFORMON MODEL    An information-theoretic idea is explicitly used as the motivation for a synaptic learning rule in the "informon" model of Uttley (1970, 1979). The "mutual information" (Shannon 1949) between two messages is, informally, a measure of the information that either message "conveys about" the other. For example, if two signals tend to co-occur, their mutual information is high. If knowing one signal has no effect on one's expectation of what the other signal is, their mutual information is zero. In Uttley's model both variable and fixed strength synapses provide input to a cell. Each variable strength is incrementally adjusted so that the cell's output becomes as nearly decorrelated with the synaptic input as possible. The update rule is motivated by the idea of minimizing the mutual information between input and output at each synapse, although in practice a function different from the mutual information is used. This learning rule is very similar to the Rescorla-Wagner conditioning rule and to the Widrow-Hoff "LMS" algorithm for supervised learning (Widrow & Stearns 1985). The rule adjusts the strengths so that the net effect of the inputs at the variable-strength synapses is to cancel, as nearly as possible, the net reinforcement or "teacher" signal at the fixed-strength synapses. The output of the "informon" cell is then equal to the residual difference between the two sets of signals. This rule differs from a Hebb-like rule, in which correlations between input and output cause synaptic strength to increase. Nonetheless, the rule causes the cell to detect structure among the inputs to the variable-strength synapses, and to use such structure to match and cancel the "teacher" signal most effectively. I return to the use of a mutual information criterion below (Linsker 1988a) in a different context.

INFORMATION THEORY AND BIOLOGICAL UTILITY    Marr's (1970) discussion of redundancy reduction in sensory processing addresses an issue (introduced above) that is crucial for any connection between information-theoretic or statistical classification schemes and perception. To form generalizations, a system must be able to decide when two events (input presentations) are "similar"—that is, belong in the same class. This "lumping process," however, discards information about the events. By what criterion should this, rather than other, information be discarded? Marr notes that system reliability offers one such criterion: If two events correspond to sufficiently similar sets of signal activities, the events could be confused by the network, and therefore should not be classified differently. He does not, however, consider this a fundamental criterion.

Marr seeks to integrate information-theoretic ideas with biological utility in the following way. His criterion for discarding information is that the loss of that information should not impair the system's ability to diagnose whether the event possesses features of a certain type. The type of feature that is favored is a feature whose presence or absence (in an event) is not sensitive to small changes in the set of other, already-diagnosed features that are present (in the same event). Thus the presence of a feature (or type of redundancy) as such is not sufficient ground for classification; the feature must tend to co-occur with a sufficiently large class of other features (see the "Fundamental Hypothesis," Marr 1970, p. 182).

It is intuitively plausible that the detection of a feature is likely to have greater biological utility if that feature tends to be associated with others. Marr's (1970) theory is developed substantially beyond this point to make neurophysiologic predictions concerning the roles of various cell types. Important questions that are left open include: Can the theory be used to make specific predictions of feature-analyzing properties? If not, would the incorporation of additional biological constraints of some type suffice to allow one to make such predictions? Is the proposed feature classification criterion definite enough that it can be used, even in principle, to construct a perceptual system (or the early stages of one) whose functioning might be compared with what is biologically observed?

A PRINCIPLE OF MAXIMUM INFORMATION PRESERVATION    Motivated by the finding (discussed above) that a Hebb-like rule under certain conditions generates a cell whose output conveys maximum information about its input activities, Linsker (1988a) has proposed an information-theoretic principle for the organization of a biological sensory processing stage having source and target cell layers. This principle of "maximum information preservation" states that the transformation of sensory signals from source to target layer should be chosen such that the target cell activities jointly convey maximum information about the source cell activities. The quantity to be maximized is the average (over time) of the mutual information between the two sets of activities.

The choice of transformation is subject to biological "hardware" constraints. These may include constraints on (a) the type of function each cell can compute (e.g. a linear combination of inputs, or a sigmoid function of the linear combination, or some more complicated function), (b) the spatial extent of lateral connections, and (c) the reliability of signal transmission and processing.

The concept of maximizing average mutual information between input and output can be made more concrete by an analogy to a "guessing game" (cf. Shannon 1949) such as "Twenty Questions." In that game, the questioner's proper strategy is to ask, at each stage, a question whose answer will convey, on average, the maximum possible amount of information (in that case, one bit of information) about the unknown object. In this analogy, the "questioner" is the cell, the "question" is the computation being performed by the cell upon its inputs (e.g. a center-surround cell "asks" the "question": "What degree of contrast is present between the central and peripheral regions of my receptive field?"), and the "answer" is the output activity of the cell. An unreliable "answer" corresponds to the effect of noise associated with the processing stage.

We want to know what set of processing functions, or computations, emerges for an entire layer of cells, according to the principle of maximum information preservation. This is analogous to asking: How should a set of questioners, asking questions *in parallel* about the same object (the input presentation), choose their questions optimally?

If the activity values could take on only discrete values, and if noise could be ignored, the proposed principle would lead to data compression with redundancy reduction along the lines discussed by Attneave and Barlow (above). Processing noise is unavoidable, however, and it has important consequences. It helps to determine which types of information about the input pattern are preserved or discarded, and how redundancy is to be introduced—as well as removed—during processing. The latter point is important if one is to account for the similar response properties of nearby cells in the context of an "optimal encoding" principle.

The optimal choice of cell response properties, according to the principle of maximum information preservation, depends upon the types of statistical regularities present in the set of input presentations, and on the processing constraints. The principle has been applied to simple model networks under various conditions. It generates features that are qualitatively similar to those found in biological systems. These features include topographic maps (Linsker 1988b, 1989b; also see MAP MAGNIFICATION FACTORS above), center-surround cells and cells sensitive to temporal variations in input activity (Linsker 1989a), orientation-selective cells, and column-like assemblies of cells having similar response properties (Linsker 1988a,b, 1989b). In each case the generic quantity being optimized—the average mutual information between source and target layer—is the same.

Note that the quantity being maximized is a function of the processing being carried out by many target cells. This global optimization criterion stands in contrast to other proposed criteria that refer only to an information-theoretic quantity at a single connection (Uttley 1970, 1979) or cell (Pearlmutter & Hinton 1986). A different global information-theoretic quantity is optimized in the "Boltzmann machine" network discussed by Hinton & Sejnowski (1983).

The idea of choosing system parameters so as to maximize an appropriately defined mutual information has been used in several nonbiological contexts, including the choice or placement of sensors in physical systems (e.g. Phua & Dillon 1977 and references therein, Luttrell 1985, Fraser & Swinney 1986) and the choice of acoustic-processing models in speech recognition (Bahl et al 1987). Much remains to be learned about the patterns of information processing that can emerge from this type of information-theoretic principle, particularly in nonlinear systems having multiple processing stages. The potential value of cross-fertilization between neuroscience and nonbiological fields, in advancing our understanding of these issues, is great.

Although the average mutual information between two layers of cells is a global property having a complex mathematical form, a local algorithm that implements the proposed principle has recently been developed for certain types of model networks having feedforward and lateral connections (Linsker 1989b). By a local algorithm we mean one that modifies connection strengths in a way that depends only upon signals available at that connection or cell.

This algorithm, or learning rule, was derived by asking: How should the response properties of each cell be incrementally adjusted, so as to increase the average mutual information between the source and target layer activities by the greatest amount? (It is an example of a "gradient ascent" learning rule.) It is striking that the resulting rule (Linsker 1989b) exhibits properties of Hebb-like modification and cooperative and competitive learning, combined in a particular way, even though no assumptions were made concerning the form of the learning rule or its component properties.

The principle of maximum information preservation has some theoretically attractive features (Linsker 1988a, 1989b), and the existence of local algorithms increases our confidence that biological networks may be capable of implementing it. It does not follow from this that nature in fact makes use of the principle. It will be necessary to see what detailed predictions the principle generates in biologically realistic situations, and to subject these to experimental test.

The proposed principle may be usefully extended in various ways. For example, the principle as stated uses processing noise as the criterion for classifying two patterns as "similar." (That is, "similar" patterns are those which cannot be reliably discriminated from each other.) It can be biologically useful to introduce a different criterion of pattern "similarity." A suitable choice of criterion can facilitate the learning of generalizations or allow resolution to be varied as the focus of attention is shifted. The type of local algorithm mentioned above can still be used to generate near-optimal processing stages provided the "similarity" criterion satisfies certain conditions (Linsker 1989b). [For examples of how the choice of resolution or "similarity" measure is handled in other approaches to learning and perception, see Carpenter & Grossberg's (1988) use of an adjustable "vigilance" parameter, and the "regularization" method for reconstructing aspects of a scene from its image (Poggio & Koch 1985).]

## SUMMARY AND FUTURE DIRECTIONS

This review has focussed on a set of ways in which experimental work, modeling, and theory have interacted in advancing our understanding of the organization of sensory processing systems. These interactions are not of a rigid "bottom-up" or "top-down" type. They involve interplay among—rather than isolated study of—the "computational task," algorithmic, and hardware levels of description (Marr 1982).

The path traced here comprises the following steps:

1. The finding of specific salient experimental features motivates the search for models, especially self-organizing models, that can generate these features.
2. Common elements of architecture and patterning rules are found to underlie a variety of these models.
3. These elements are found, in certain cases, to be associated with optimization principles that appear significant from an information-processing standpoint. In essence, the models are treated as objects of study in their own right, and are found to have properties that might have been difficult or impossible to infer directly from observation of the much more complex biological system. This study leads to links with principles and methods in statistical analysis and information theory.
4. To be able to test whether these putative principles are biologically relevant, it is necessary to develop biologically plausible algorithms that can implement them, and to generate explicit predictions of cell

properties and neural maps that can be compared with experiment. This step has been taken for simple model cases. More work is required to extend the principle and algorithms to networks and sets of input presentations that are of biologically realistic complexity.

If some of the theoretical ideas discussed withstand future experimental tests, it will become important to extend the work in two directions. First, new experimental findings concerning neuronal dynamics (e.g. Gray & Singer 1989) could allow "optimal encoding" strategies to be implemented in new ways. Second, even if the principles discussed are found to describe some of the main effects of feedforward and lateral interactions within a processing stage of a sensory system, it will be important to understand how these principles relate to higher levels of neural organization. The study of attentional mechanisms, other effects of feedback from later processing states, integration across sensory modalities, and sensorimotor integration are examples of such higher-level organizational issues to which the type of approach explored here might be fruitfully extended.

More sophisticated experimental techniques (e.g. for recording and analyzing complex spatial and temporal patterns of activity), increased computational power, and refinement of theoretical ideas are likely to lead to closer and more fruitful contact between experimental and theoretical work. For the theorist, it is important to use experimental findings to (a) constrain the essential features of biological models, (b) inspire and inform the choice of key questions, and (c) enable comparisons with predicted structure and behavior. For the experimentalist, model predictions may suggest appropriate candidate stimuli for studying receptive field properties and neural maps, and may help to guide the analysis and interpretation of complex activity data. Such guidance becomes valuable as increasingly complex sets of response properties are found, and as the characterization of a cell as a detector or analyzer of a well-defined "feature" may become more difficult or untenable.

More generally, if we are to gain an understanding of how and why neural systems are organized in particular ways, it is important to supplement descriptive principles with a developing theoretical framework that can be used to generate testable predictions and to relate structure and function at a variety of organizational levels. Multifaceted interactions among experiment, modeling, and theory will play an important role in achieving these goals.

ACKNOWLEDGMENTS

I thank Drs. Eric Schwartz, Roger Traub, and Christoph von der Malsburg for their comments on the manuscript.

*Literature Cited*

Amari, S.-I. 1977. Neural theory of association and concept-formation. *Biol. Cybern.* 26: 175–85

Amari, S.-I. 1980. Topographic organization of nerve fields. *Bull. Math. Biol.* 42: 339–64

Attneave, F. 1954. Some informational aspects of visual perception. *Psychol. Rev.* 61: 183–93

Bahl, L. R., Brown, P. F., de Souza, P. V., Mercer, R. L. 1987. Speech recognition with continuous-parameter hidden Markov models. *Comput. Speech Lang.* 2: 219–34

Barlow, H. B. 1959. Sensory mechanisms, the reduction of redundancy, and intelligence. In *Mechanisation of Thought Processes, Natl. Phys. Lab. Symp.* 10: Paper 4-1. London: HMSO

Barlow, H. B. 1961. Possible principles underlying the transformations of sensory messages. In *Sensory Communication*, ed. W. A. Rosenblith, pp. 217–34. Cambridge: MIT Press and New York/London: Wiley

Barlow, H. B. 1969. Trigger features, adaptation and economy of impulses. In *Information Processing in the Nervous System*, ed. K. N. Leibovic, pp. 209–26. New York: Springer-Verlag

Barlow, H. B. 1985. Cerebral cortex as model builder. In *Models of the Visual Cortex*, ed. D. Rose, V. G. Dobson, pp. 37–46. New York/London: Wiley

Barlow, H. B. 1986. Why have multiple cortical areas? *Vision Res.* 26: 81–90

Bienenstock, E. L., Cooper, L. N., Munro, P. W. 1982. Theory for the development of neuron selectivity: Orientation specificity and binocular interaction in visual cortex. *J. Neurosci.* 2: 32–48

Blasdel, G. G., Salama, G. 1986. Voltage-sensitive dyes reveal a modular organization in monkey striate cortex. *Nature* 321: 579–85

Braitenberg, V. 1985. Charting the visual cortex. In *Cerebral Cortex*, ed. A. Peters, E. G. Jones, 3: 379–414. New York/London: Plenum

Brown, T. H., Kairiss, E. W., Keenan, C. L. 1990. Hebbian synapses: Biophysical mechanisms and algorithms. *Annu. Rev. Neurosci.* 13: 475–511

Carpenter, G. A., Grossberg, S. 1988. The ART of adaptive pattern recognition by a self-organizing neural network. *Computer* 21(3): 77–88

Constantine-Paton, M., Law, M. I. 1978. Eye-specific termination bands in tecta of three-eyed frogs. *Science* 202: 639–41

Craik, K. J. W. 1943. *The Nature of Explanation.* Cambridge: Cambridge Univ. Press

Daugman, J. G. 1988. Complete discrete 2-d Gabor transforms by neural networks for image analysis and compression. *IEEE Trans. Acoust. Speech Signal Process.* 36: 1169–79

Durbin, R., Mitchison, G. 1988. Dimension reduction models for topographical maps in the cortex. *Neural Networks* 1(Suppl. 1): 491 (Abstr.)

Durbin, R., Willshaw, D. J. 1987. An analogue approach to the travelling salesman problem using an elastic net method. *Nature* 326: 689–91

Ferster, D., Koch, C. 1987. Neuronal connections underlying orientation selectivity in cat visual cortex. *Trends Neurosci.* 10: 487–92

Fraser, A. M., Swinney, H. L. 1986. Independent coordinates for strange attractors from mutual information. *Phys. Rev. A* 33: 1134–40

Fraser, S. E. 1985. Cell interactions involved in neuronal patterning: An experimental and theoretical approach. In *Molecular Bases of Neural Development*, ed. G. M. Edelman, W. E. Gall, W. M. Cowan, pp. 481–507. New York: Wiley

Frégnac, Y., Imbert, M. 1984. Development of neuronal selectivity in primary visual cortex of cat. *Physiol. Rev.* 64: 325–434

Gray, C. M., Singer, W. 1989. Stimulus-specific neuronal oscillations in orientation columns of cat visual cortex. *Proc. Natl. Acad. Sci. USA* 86: 1698–1702

Gray, R. M. 1984. Vector quantization. *IEEE ASSP Mag.* 1(2): 4–29

Grossberg, S. 1976. On the development of feature detectors in the visual cortex with applications to learning and reaction-diffusion systems. *Biol. Cybern.* 21: 145–59

Hebb, D. O. 1949. *The Organization of Behavior.* New York: Wiley

Hinton, G. E. 1989. Connectionist learning procedures. *Artif. Intell.* 40: 185–234

Hinton, G. E., Sejnowski, T. J. 1983. Optimal perceptual inference. *Proc. IEEE Conf. Comput. Vision*, pp. 448–53

Hubel, D. H., Wiesel, T. N. 1962. Receptive fields, binocular interaction and functional architecture in the cat's visual cortex. *J. Physiol.* 160: 106–54

Hubel, D. H., Wiesel, T. N. 1963. Shape and arrangement of columns in cat's striate cortex. *J. Physiol.* 165: 559–68

Hubel, D. H., Wiesel, T. N. 1968. Receptive fields and functional architecture of monkey striate cortex. *J. Physiol.* 195: 215–43

Hubel, D. H., Wiesel, T. N. 1977. Functional

architecture of macaque monkey visual cortex (Ferrier lecture). *Proc. R. Soc. London Ser. B* 198: 1–59

Kaas, J. H., Merzenich, M. M., Killackey, H. P. 1983. The reorganization of somatosensory cortex following peripheral-nerve damage in adult and developing mammals. *Annu. Rev. Neurosci.* 6: 325–56

Kammen, D. M., Yuille, A. L. 1988. Spontaneous symmetry-breaking energy functions and the emergence of orientation selective cortical cells. *Biol. Cybern.* 59: 23–31; (erratum) 59: 430

Kohonen, T. 1982a. Self-organized formation of topologically correct feature maps. *Biol. Cybern.* 43: 59–69

Kohonen, T. 1982b. Analysis of a simple self-organizing process. *Biol. Cybern.* 44: 135–40

Kohonen, T. 1988. *Self-Organization and Associative Memory.* Berlin/Heidelberg/New York: Springer-Verlag. 2nd ed.

Legéndy, C. R. 1970. The brain and its information trapping device. In *Progress of Cybernetics*, ed. J. Rose, 1: 309–38. London/New York/Paris: Gordon & Breach

Legéndy, C. R. 1975. Three principles of brain function and structure. *Int. J. Neurosci.* 6: 237–54

Lehky, S. R., Sejnowski, T. J. 1988. Network model of shape-from-shading: neural function arises from both receptive and projective fields. *Nature* 333: 452–54

LeVay, S., Hubel, D. H., Wiesel, T. N. 1975. The pattern of ocular dominance columns in macaque visual cortex revealed by a reduced silver stain. *J. Comp. Neurol.* 159: 559–76

Linsker, R. 1986a. From basic network principles to neural architecture: Emergence of spatial-opponent cells. *Proc. Natl. Acad. Sci. USA* 83: 7508–12

Linsker, R. 1986b. From basic network principles to neural architecture: Emergence of orientation-selective cells. *Proc. Natl. Acad. Sci. USA* 83: 8390–94

Linsker, R. 1986c. From basic network principles to neural architecture: Emergence of orientation columns. *Proc. Natl. Acad. Sci. USA* 83: 8779–83

Linsker, R. 1987. Towards an organizing principle for perception: Hebbian synapses and the principle of optimal neural encoding. *IBM Res. Rep. RC12830.* IBM Res., Yorktown Heights, NY

Linsker, R. 1988a. Self-organization in a perceptual network. *Computer* 21(3): 105–17

Linsker, R. 1988b. Towards an organizing principle for a layered perceptual network. In *Neural Information Processing Systems, Denver, CO, 1987*, ed. D. Z. Anderson, pp. 485–94. New York: Am. Inst. Phys.

Linsker, R. 1988c. Development of feature-analyzing cells and their columnar organization in a layered self-adaptive network. In *Computer Simulation in Brain Science, Copenhagen, 1986*, ed. R. M. J. Cotterill, pp. 416–31. Cambridge/New York: Cambridge Univ. Press

Linsker, R. 1989a. An application of the principle of maximum information preservation to linear systems. In *Advances in Neural Information Processing Systems, Denver, CO, 1988*, ed. D. S. Touretzky, 1: 186–94. San Mateo, Calif: Morgan Kaufmann

Linsker, R. 1989b. How to generate ordered maps by maximizing the mutual information between input and output signals. *Neural Comput.* 1: 402–11

Lippmann, R. P. 1987. An introduction to computing with neural nets. *IEEE ASSP Mag.* 4(2): 4–22

Luttrell, S. P. 1985. The use of transinformation in the design of data sampling schemes for inverse problems. *Inverse Problems* 1: 199–218

MacQueen, J. 1967. Some methods for classification and analysis of multivariate observations. In *Proc. 5th Berkeley Symp. Math. Stat. Prob.*, ed. L. M. LeCam, J. Neyman, 1: 281–97. Berkeley: Univ. Calif. Press

Marr, D. 1970. A theory for cerebral neocortex. *Proc. R. Soc. London Ser. B* 176: 161–234

Marr, D. 1982. *Vision.* San Francisco: Freeman

Merzenich, M. M., Nelson, R. J., Stryker, M. P., Cynader, M. S., Schoppmann, A., Zook, J. M. 1984. Somatosensory cortical map changes following digit amputation in adult monkeys. *J. Comp. Neurol.* 224: 591–605

Métin, C., Frost, D. O. 1989. Visual responses of neurons in somatosensory cortex of hamsters with experimentally induced retinal projections to somatosensory thalamus. *Proc. Natl. Acad. Sci. USA* 86: 357–61

Miller, K. D., Keller, J. B., Stryker, M. P. 1986. Models for the formation of ocular dominance columns solved by linear stability analysis. *Soc. Neurosci. Abstr.* 12: 1373

Miller, K. D., Keller, J. B., Stryker, M. P. 1989. Ocular dominance column development: Analysis and simulation. *Science* 245: 605–15

Mountcastle, V. B. 1978. An organizing principle for cerebral function: The unit module and the distributed system. In *The Mindful Brain*, ed. G. M. Edelman, V. B. Mountcastle. Cambridge: MIT Press

Oğuztöreli, M. N., Caelli, T. M. 1986. An inverse problem in neural processing. *Biol. Cybern.* 53: 239–45

## 280 LINSKER

Oja, E. 1982. A simplified neuron model as a principal component analyzer. *J. Math. Biol.* 15: 267–73

Pearlmutter, B. A., Hinton, G. E. 1986. G-maximization: An unsupervised learning procedure for discovering regularities. In *Neural Networks for Computing*, ed. J. S. Denker, AIP Conf. Proc. 151. New York: Am. Inst. Phys.

Pearson, J. C., Finkel, L. H., Edelman, G. M. 1987. Plasticity in the organization of adult cerebral cortical maps: A computer simulation based on neuronal group selection. *J. Neurosci.* 7: 4209–23

Phillips, C. G., Zeki, S., Barlow, H. B. 1984. Localization of function in the cerebral cortex. *Brain* 107: 327–61

Phua, K., Dillon, T. S. 1977. Optimal choice of measurements for state estimation. In *1977 Power Industry Computer Applic. Conf., Toronto*, pp. 431–41. New York: IEEE

Poggio, T., Koch, C. 1985. Ill-posed problems in early vision: From computational theory to analogue networks. *Proc. R. Soc. London Ser. B* 226: 303–23

Rauschecker, J. P. 1987. What signals are responsible for synaptic changes in visual cortical plasticity? In *Imprinting and Cortical Plasticity*, ed. J. P. Rauschecker, P. Marler, pp. 193–219. New York: Wiley

Reiter, H. O., Stryker, M. P. 1988. Neural plasticity without postsynaptic action potentials: Less-active inputs become dominant when kitten visual cortical cells are pharmacologically inhibited. *Proc. Natl. Acad. Sci. USA* 85: 3623–27

Ritter, H., Schulten, K. 1986. On the stationary state of Kohonen's self-organizing sensory mapping. *Biol. Cybern.* 54: 99–106

Rumelhart, D. E., Zipser, D. 1985. Feature discovery by competitive learning. *Cogn. Sci.* 9: 75–112

Schmidt, J. T. 1985. Formation of retinotopic connections: Selective stabilization by an activity-dependent mechanism. *Cell. Mol. Neurobiol.* 5: 65–84

Schwartz, E. L. 1977. The development of specific visual connections in the monkey and the goldfish: Outline of a geometric theory of receptotopic structure. *J. Theor. Biol.* 69: 655–83

Shannon, C. E. 1949. In *The Mathematical Theory of Communication*, ed. C. E. Shannon, W. Weaver. Urbana: Univ. Illinois Press

Shatz, C. J., Stryker, M. P. 1978. Ocular dominance in layer IV of the cat's visual cortex and the effects of monocular deprivation. *J. Physiol.* 281: 267–83

Sperry, R. W. 1943. Visuomotor coordination in the newt (Triturus viridescens) after regeneration of the optic nerve. *J. Comp. Neurol.* 79: 33–55

Sperry, R. W. 1963. Chemoaffinity in the orderly growth of nerve fiber patterns and connections. *Proc. Natl. Acad. Sci. USA* 50: 703–10

Stent, G. S. 1973. A physiological mechanism for Hebb's postulate of learning. *Proc. Natl. Acad. Sci. USA* 70: 997–1001

Stryker, M. P., Harris, W. A. 1986. Binocular impulse blockade prevents the formation of ocular dominance columns in cat visual cortex. *J. Neurosci.* 6: 2117–33

Sur, M., Garraghty, P. E., Roe, A. W. 1988. Experimentally induced visual projections into auditory thalamus and cortex. *Science* 242: 1437–41

Swindale, N. V. 1980. A model for the formation of ocular dominance stripes. *Proc. R. Soc. London Ser. B* 208: 243–64

Swindale, N. V. 1982. A model for the formation of orientation columns. *Proc. R. Soc. London Ser. B* 215: 211–30

Takeuchi, A., Amari, S.-I. 1979. Formation of topographic maps and columnar microstructures in nerve fields. *Biol. Cybern.* 35: 63–72

Turing, A. M. 1952. The chemical basis of morphogenesis. *Philos. Trans. R. Soc. London Ser. B* 237: 37–72

Udin, S. B., Fawcett, J. W. 1988. Formation of topographic maps. *Annu. Rev. Neurosci.* 11: 289–327

Uttley, A. M. 1970. The informon: A network for adaptive pattern recognition. *J. Theor. Biol.* 27: 31–67

Uttley, A. M. 1979. *Information Transmission in the Nervous System.* London/ New York/San Francisco: Academic

von der Malsburg, C. 1973. Self-organization of orientation sensitive cells in the striate cortex. *Kybernetik* 14: 85–100

von der Malsburg, C. 1979. Development of ocularity domains and growth behaviour of axon terminals. *Biol. Cybern.* 32: 49–62

von der Malsburg, C., Cowan, J. D. 1982. Outline of a theory for the ontogenesis of iso-orientation domains in visual cortex. *Biol. Cybern.* 45: 49–56

von der Malsburg, C., Willshaw, D. J. 1976. A mechanism for producing continuous neural mappings: Ocularity dominance stripes and ordered retino-tectal projections. *Exp. Brain Res.* 1: 463–69 (Suppl.)

von der Malsburg, C., Willshaw, D. J. 1977. How to label nerve cells so that they can interconnect in an ordered fashion. *Proc. Natl. Acad. Sci. USA* 74: 5176–78

Watanabe, S. 1969. *Knowing and Guessing.* New York: Wiley

Watanabe, S. 1985. *Pattern Recognition: Human and Mechanical.* New York: Wiley

Whitelaw, V. A., Cowan, J. D. 1981. Specificity and plasticity of retinotectal connections: A computational model. *J. Neurosci.* 1: 1369–87

Widrow, B., Stearns, S. D. 1985. *Adaptive Signal Processing.* Englewood Cliffs, NJ: Prentice-Hall

Wiesel, T. N., Hubel, D. H. 1974. Ordered arrangement of orientation columns in monkeys lacking visual experience. *J. Comp. Neurol.* 158: 307–18

Willshaw, D. J., von der Malsburg, C. 1976.

How patterned neural connections can be set up by self-organization. *Proc. R. Soc. London Ser. B* 194: 431–45

Willshaw, D. J., von der Malsburg, C. 1979. A marker induction mechanism for the establishment of ordered neural mappings: Its application to the retinotectal problem. *Proc. R. Soc. London Ser. B* 287: 203–43

*Annu. Rev. Neurosci. 1990. 13:283–307*
*Copyright © 1990 by Annual Reviews Inc. All rights reserved*

# BIOLOGICAL FOUNDATIONS OF LANGUAGE:
## Clues from Sign Language

*Howard Poizner*

Center for Molecular and Behavioral Neuroscience, Rutgers University, Newark, New Jersey 07102

*Ursula Bellugi*

Laboratory for Language and Cognitive Studies, The Salk Institute for Biological Studies, La Jolla, California 92037

*Edward S. Klima*

Department of Linguistics, The University of California at San Diego, La Jolla, California 92093 and Laboratory for Language and Cognitive Studies, The Salk Institute for Biological Studies, La Jolla, California 92037

The general objective of the research reviewed here is to examine the biological foundations of human language. We report on language and its formal architecture, as well as its representation in the brain, through studying languages that have arisen outside of the mainstream of spoken languages; the visual gestural systems developed among deaf people. The study of sign languages offers a unique opportunity for insight into the nature of neural mechanisms for language, since sign languages utilize a transmission modality different from that of spoken languages. To understand the extent to which neural mechanisms for language are shaped by the modality of their implementation, we are naturally led to the study of sign language because we cannot solve the problem for language in general within the spoken-auditory modality. In these studies, one can make use of the difference in transmission modality between sign and speech to investigate the biological foundations of language. Manual communi-

283

0147–006X/90/0301–0283$02.00

cation, with its very different motoric and perceptual substrate, offers a fresh opportunity to unravel the entwined strands of message structure and message performance. Recently developed linguistic analyses have been linked with three-dimensional computergraphic modeling and analysis in a major series of experiments, each bringing to bear a special property of the visual gestural modality on investigating language and brain relationships.

American Sign Language (ASL) displays complex *linguistic* structure, but unlike spoken languages, conveys much of its structure by manipulating *spatial* relations, thus exhibiting properties for which each of the hemispheres of hearing people shows a different predominant functioning. The study of deaf signers with left or right hemisphere damage offers a particularly revealing vantage point for understanding the organization of higher cognitive functions in the brain, and how modifiable that organization may be. We review a series of experimental studies of brain-damaged signers, each focusing on a special property of the visual gestural modality as it bears on the investigation of brain organization for language. By examining the nature of cerebral specialization for a visual-gestural language, we hope to shed light on theoretical questions concerning the determinants of brain organization for language in general.

## STRUCTURE OF SIGNED LANGUAGE

The central issues addressed in this review arise from some new discoveries about the nature of language. Until recently, nearly all we had learned about language had come from the study of spoken languages, but now a lively field of research into signed languages has revealed that there are primary linguistic systems passed down from one generation of deaf people to the next that have become forged into autonomous languages, not derived from spoken languages (Bellugi 1983, 1988, Klima & Bellugi 1979, 1988, Bellugi et al 1983, 1989, Poizner et al 1987, Lane & Grosjean 1980, Stokoe et al 1965, Wilbur 1987). Thus one can examine properties of communication systems that have developed in an alternate transmission system, in the visual gestural channel. The existence of such fully expressive systems arising outside of the mainstream of spoken languages affords a striking new vantage point for investigating the nature of biological constraints on linguistic form.

### Levels of Structure and Layers of Form in Sign Language

ASL shares underlying principles of organization with spoken languages, but the instantiation of those principles occurs in formal devices arising out of the very different possibilities of the visual-gestural mode (Bellugi

1980, Bellugi et al 1988, Poizner et al 1986b). We consider briefly the structure of ASL at different linguistic levels: the layered structure of phonology and morphology and the spatialized syntax of the language.

'PHONOLOGY' WITHOUT SOUND   Research on the structure of lexical signs has shown that, like the words of spoken languages, signs are fractionated into sublexical elements. Recent analyses focus on segmented structure of signed languages, which suggest sequential structure analogous to features and syllables of spoken language (Liddell & Johnson, 1986, Perimutter 1988). Sign languages, however, differ in degree of simultaneity afforded by the articulators such that the elements that distinguish signs (handshapes, movements, places of articulation) are in contrasting spatial arrangements and co-occur throughout the sign.

VERTICALLY ARRAYED MORPHOLOGY   The grammatical mechanisms of ASL exploit elaborately the spatial medium and the possibility of simultaneous and multidimensional articulation. Like spoken languages, ASL has developed grammatical markers that serve as inflectional and derivational morphemes. These markers involve regular changes in form across syntactic classes of lexical items associated with systematic changes in meaning. Some derivationally related forms are shown in Figure 1. In ASL, families of sign forms are related via an underlying stem: The forms share a handshape, a location, and a local movement shape. Inflectional and derivational processes represent the interaction of this stem with other features of movement in space (dynamics of movement, manner of movement, directions of movement, spatial array, and the like) all *layered* with the sign stem.

SPATIALLY ORGANIZED SYNTAX   A major device for the specification of relations among ASL signs is the manipulation of sign forms in space; thus in sign language, space itself bears linguistic meaning. Nominals introduced into ASL discourse may be associated with specific points in a plane of signing space. Spatial indexing to these loci thus allows explicit coreference. The ASL system of verb agreement, like its pronominal system, is also in essence spatialized. Verb signs in one class of verbs may move between the abstract loci in signing space, bearing obligatory markers for person (and number) via spatial indices, thereby specifying subject and object of the verb, as shown in Figure 2a. This spatialized system thus allows explicit reference through pronominals and agreement markers to multiple distinct third person referents. Coreferential nominals are indexed to the same locus point, as is evident in complex embedded structures, such as shown in Figure 2b.

Different spaces may be used to contrast events, to indicate reference to

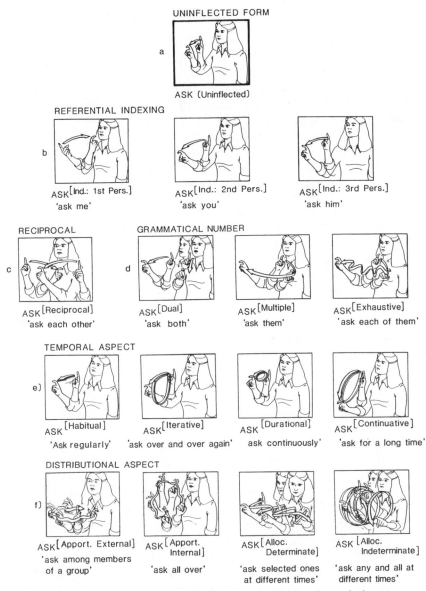

*Figure 1*  A variety of morphological processes in ASL layered on a single root.

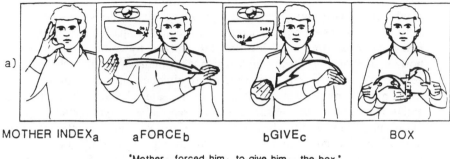

MOTHER INDEX$_a$    $_a$FORCE$_b$    $_b$GIVE$_c$    BOX

'Mother$_i$ forced him$_j$ to give him$_k$ the box.'

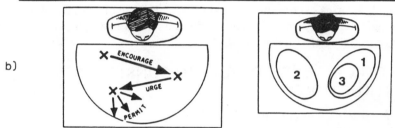

JOHN ENCOURAGE$_a$  $_a$URGE$_b$  $_b$PERMIT$_c$[Exhaustive]  TAKE-UP  CLASS

'John encouraged him$_i$ to urge her$_j$ to permit each of them$_k$ to take up the class.'

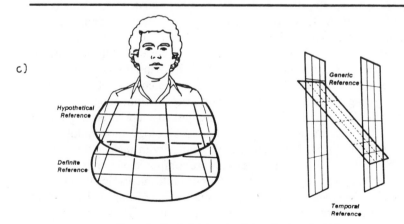

*Figure 2*  Spatialized syntactic mechanisms in ASL.

time preceding the utterance, or to express hypotheticals and counterfactuals (van Hoek 1988). As Figure 2c shows, the grammar of ASL divides sign space along several different axes, and utilizes each of the spatial dimensions for a particular semantic contrast: spatial, temporal, or realis. Contrasts in spatial location are marked by distinctions in the horizontal plane (i.e. left to right), which is the plane of definite reference; generic reference, instead, is marked on an oblique plane. Temporal contrasts may be marked by distinctions in the sagittal plane, thus differentiating events in time. Distinctions between real and hypothetical situations may be marked by the height of the signs, which represent another dimension of semantic contrast; conditionals (e.g. "if it rains") involve a higher plane of signing space than factual statements. Thus ASL structures three-dimensional space according to multidimensional oppositions between basic axes of conceptual organization. The use of spatial loci for referential indexing, verb agreement, and grammatical relations is clearly a unique property of visual gestural systems.

# BIOLOGICAL AND LINGUISTIC CONSTRAINTS ON STRUCTURE

## Sign Language Universals

The study of ASL provides a new perspective on the question of linguistic universals and the role of modality in determining the forms by which universal principles are manifested within a language. Analyses have been performed on the structure and form of an unrelated sign language, Chinese Sign Language (CSL). ASL and CSL are mutually unintelligible sign languages that have arisen independently and differ at all linguistic levels. CSL was selected for study because it has developed in the context of a completely different spoken language, with essentially no inflectional morphology, and because it has developed in the context of a radically different writing system (logographic as opposed to alphabetic).

Despite the differences in the surrounding spoken and written languages, at each level of sign language organization—phonology, morphology and syntax—CSL and ASL have remarkable similarity in *surface* form (Fok et al 1988). Signs in both signed languages are composed of simultaneously layered elements, consisting of a small set of handshapes, locations, and movements. Moreover, CSL, like ASL, makes use of inflected forms to express a range of distinctions: Verbs in CSL inflect for agreement with subject and object and also for distributional distinctions such as dual, reciprocal, multiple, and exhaustive. In CSL, as in ASL, inflectional morphology and stem are fused into one simultaneously articulated unit, and spatial relations are used to convey syntactic relations. As would be expected between two unrelated languages, the grammatical inflections that have arisen in CSL and ASL differ in distinctions marked as well as in formation. There are also systematic differences in phonetic form between the two signed languages, so that, for example, when signers of one sign language learn the other, they sign with the equivalent of a foreign accent (Klima & Bellugi 1979/1988). The two unrelated signed languages use similar principles for grammatical processes, however, and make use of possibilities offered by the visual modality to convey linguistic structure in a highly layered manner, with the active manipulation of space. These studies of CSL and ASL are important both to determining how modality shapes language structure and to cross-modality universals of language organization (Fok et al 1986, 1988).

## The Interplay between Perception of Movement and Perception of Language

Since visual perception and signing differ so radically from auditory perception and speech, one can investigate the interplay between linguistic structure and biological processes quite apart from the particular channel through which the linguistic structure is conveyed. Acquiring American Sign Language can modify perception of some natural movement categories (Poizner 1981, 1983). Deaf ASL signers differ from hearing non-signers in perception of meaningless ASL movements. The data indicates that the modification of natural perceptual categories after language acquisition is not bound to a particular transmission modality, but rather can be a more general consequence of acquiring a formal linguistic system. A recent experiment investigated the psychological representation of visual gestural languages from a cross-linguistic perspective, focusing on the perception of movement, since movement is a key formational building block of sign languages.

Signers from Chinese and American Sign Languages and nonsigning hearing subjects made triadic comparisons of movements that had been isolated from American Sign Language and presented as dynamic point-light displays. Multidimensional scaling of the triadic comparisons revealed marked differences between perception of both groups of signers from that of the hearing nonsigners, replicating and extending previous studies, that showed perceptual modification following acquisition of sign language. Furthermore, American and Chinese signers differed in their perception of one and the same set of movement elements based in part on the differing role of movement in the phonologies of the respective sign languages. Related experiments on the perception of speech make it clear that a speaker's perception of phonemes is determined partly by his phonological knowledge. Thai, for example, partitions the voicing continuum differently from English, and the discrimination functions of Thai and English speakers are correspondingly different (Abramson & Lisker 1970). By comparing perception of linguistic movement across signers from different visual gestural languages, one can begin to uncover—for language in general—how particular phonological knowledge constrains perception and how perception is determined by the psychophysiology of the input-output channels (Poizner et al 1989c, Poizner 1987).

### Three-Dimensional Computergraphic Modeling and Analysis

The study of sign language offers a special opportunity for the investigation of language production, since the movements of the articulators in sign are directly observable. Powerful techniques have been developed for the three-dimensional measurement and analysis of movement. American Sign Language displays all the complex linguistic structure found in spoken language, yet the mechanisms by which the essential grammatical information is conveyed are unique to the modality in which the language developed. This grammatical information is conveyed by changes in the movement and by spatial contouring of the hands and arms. The three-dimensional acquisition, reconstruction, manipulation, and graphic editing of such movement processes is now possible (Poizner et al 1986c, Jennings & Poizner 1988). Two optoelectronic cameras can be used to detect the positions of infrared emitting diodes attached to the major joints of the arms and hands (see Figure 3a). Three-dimensional coordinates are calculated and the movements reconstructed computergraphically. A sequence of positions of the arm in three-dimensional space can be displayed simultaneously, or the actual motion can be recreated in real time. Figure 3b shows stroboscopic reconstructions of three grammatical processes in ASL, and illustrate the fluidity of linguistic processes in the visual modality.

In conjunction with new methods of three-dimensional movement tracking, these computergraphic methods offer new approaches to the analysis of sign language and its underlying neural control. For example, Figure 4 presents reconstructed movements of the hands and arms for a control signer and a signer with Parkinson's disease producing the grammatically inflected sign, LOOK [Exhaustive], meaning "look at each of them." The reconstructions reveal markedly reduced movement amplitudes in the signing of the Parkinsonian signer. These reductions in movement amplitude seem akin to the micrographia, or reduced letter size in the writing of hearing patients with Parkinson's disease. Furthermore, the Parkinsonian signer made few arm movements through space, and utilized distal rather than proximal joints, thus suggesting a difficulty in maintaining over time

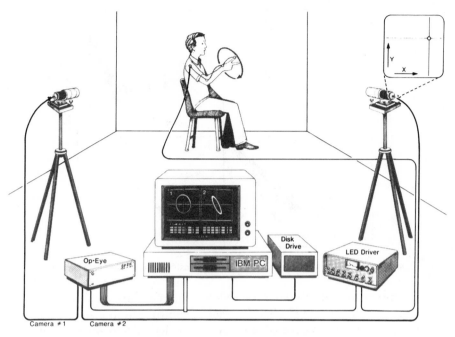

Three-Dimensional Data Acquisition System

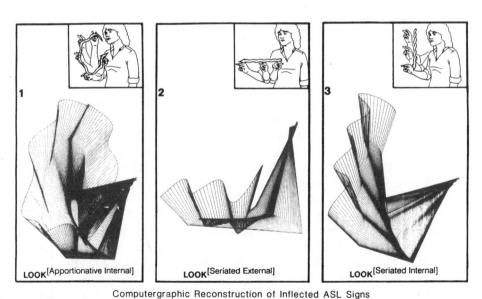

Computergraphic Reconstruction of Inflected ASL Signs

*Figure 3*   Three-dimensional computergraphic analysis of morphological processes in ASL.

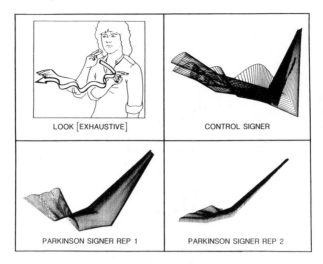

LOOK [EXHAUSTIVE]

CONTROL SIGNER

PARKINSON SIGNER REP 1

PARKINSON SIGNER REP 2

*Figure 4*    Three-dimensional computergraphic reconstructions of Parkinsonian signing.

appropriate force of muscular contraction required of proximal arm movements (Poizner et al 1988). The interactive control and three-dimensional
visualization of reconstructed movements that do or do not serve a linguistic function make possible a powerful analysis of the breakdown of movements that serve *linguistic*, *symbolic*, or strict *motoric* functions.

## BRAIN FUNCTION FOR A VISUOSPATIAL LANGUAGE

The study of sign language breakdown due to brain damage offers a direct
window into cerebral specialization for language, since in sign language
there is interplay between visuospatial and *linguistic* relations within one
and the same system. One broad aim is to investigate the relative contributions of the cerebral hemispheres with special reference to the connection between linguistic functions and the *spatial* mechanisms that convey them. We review a series of studies of the language capacities of deaf
signers having either left or right brain damage. An array of assessment
tools, probes, methods of linguistic analysis, and formal tests have been
administered that constitute a major comprehensive program for investigating the nature of sign language breakdown following unilateral
lesions. The program includes: (*a*) an adaptation, for ASL, of the Boston
Diagnostic Aphasia Examination (BDAE) (Goodglass & Kaplan 1979,
1983); (*b*) a battery of linguistic tests specially designed to assess capacities
of brain-damaged signers vis-á-vis each of the levels of ASL linguistic

structure: "phonology without sound," vertically arrayed morphology, and spatially organized syntax; (*c*) an analysis of production of ASL at all linguistic levels, (*d*) tests of nonlanguage spatial processing and motor control. The battery of language and nonlanguage tasks is administered to left- and right-lesioned signers and to matched deaf control signers. This uniform investigation of left- and right-lesioned signers is important because in this manner performance of these two groups can be directly compared and inferences made regarding the neural substrate underlying a language in a spatial mode (Poizner et al 1987, 1989a, Bellugi et al 1989, Klima et al 1988). These probes permit one to assess patterns of preservation and impairment with right and left hemispheres lesions, and thus gain a deeper understanding of the role of the two hemispheres in the processing of visuospatial language.

We report on on-going studies of brain lesioned signers, with left or right hemisphere lesions. All subjects in these studies were members of deaf communities, had been educated in residential schools for deaf children, and had deaf spouses. All were right handed before their strokes. For each subject, the primary form of communication with family and friends was ASL.

## Nonlanguage Spatial Cognition

Since spatial relations and linguistic relations are so intimately intertwined in ASL, it is important to examine brain organization for nonlanguage spatial functions in order to determine whether the two cerebral hemispheres in deaf signers show differential specialization. Visuospatial deficits were extensive in right-hemisphere-lesioned signers. One subject with right hemisphere damage showed topographic disorientation. Another, who had specialized in building and repairing from blueprints before his stroke, was unable to do simple block designs afterwards. A third, who had been an artist before her stroke, also showed a marked visuospatial impairment, including highly distorted drawings and neglect of left hemispace. Figure 5 presents example performance of eight brain-lesioned signers on a block design test in which subjects must assemble either four or nine three-dimensional blocks to match a two-dimensional model of the top surface. The left-hemisphere-damaged signers (*upper row*) produced correct constructions on the simple block designs and made only featural errors on the more complex designs; in contrast, the right-hemisphere-damaged signers (*lower row*) produced erratic and incorrect constructions and tended to break the overall configurations of the designs. The severe spatial disorganization of the constructions of the right-lesioned signers reflects their severe spatial loss. Across a range of tests, including drawing, block design, attention to visual space, perception of line orientation, facial

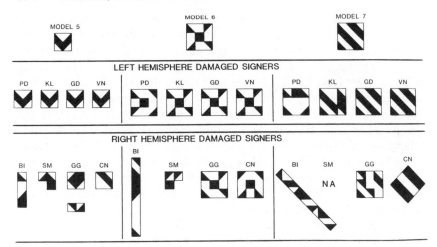

*Figure 5* Performance of left- and right-lesioned signĕrs on the WAIS-R block design task, a nonlanguage visuospatial task. Note the broken configurations and severe spatial disorganization of the right-lesioned signers.

recognition, and visual closure, right-lesioned signers showed many of the classic visuospatial impairments seen in hearing patients with right-hemisphere damage. In contrast, left-lesioned signers showed relatively preserved nonlanguage spatial functioning. In summary, right-lesioned deaf signers show severe impairments in processing of nonlanguage spatial relations. These nonlanguage data show that the right hemisphere in deaf signers develops cerebral specialization for nonlanguage visuospatial functions (Poizner et al 1987).

## GRAMMATICAL IMPAIRMENT IN LEFT-LESIONED SIGNERS

Lesions to the left hemisphere have been found to produce frank sign language aphasias (and relatively preserved nonlanguage spatial functions). Importantly, the studies show that sign language breakdown was not uniform, but rather cleaved along lines of linguistic significance (Poizner et al 1987). Figure 6 shows the rating scale profiles from the ASL adaptation of the Boston Diagnostic Aphasia Examination for three signers with left hemisphere damage (*top*), for matched deaf control subjects (*middle*), and for three right-hemisphere-damaged signers (*bottom*). The middle portion of the figure presents the rating scale profiles of the matched control subjects and shows normal performance. Note that for the left-lesioned signers, the scores are scattered on each scale, spanning virtually the entire range of values. These profiles reflect marked sign language breakdowns. The performance of the right-lesioned signers reflect grammatical (non-aphasic) signing; in fact, their profiles are much like those of the control subjects.

The specific disruption of sign language following left hemisphere lesions is also clearly revealed by performance of brain-damaged signers on tests developed to assess processing various linguistic levels of ASL structure. For example, Figure 7a presents results from a test of the equivalent of phonological processing in sign language. Phonological processing has been considered one of the most strongly left-lateralized aspects of spoken language (Zaidel 1985). In a test for the sign equivalent of "rhyming" left-lesioned, but not right-lesioned signers showed marked impairment. Thus, the specialization of the left hemisphere for phonological processing does not appear to be specific to language in the spoken-auditory mode.

The individual sign language profiles of the left-lesioned signers deviate from normal in different ways, and represent distinct patterns of sign aphasia. We illustrate the nature of the differential breakdown of sign language, following left-hemisphere lesions, through some individual case studies.

## Agrammatic Sign Aphasia

One left-lesioned signer was agrammatic for ASL. After her stroke, she no longer used the indexic pronouns of ASL, and her verbs were without any spatial indices whatever. In fact, her signing was devoid of the syntactic and morphological markings required in ASL; i.e. her signing was what would be classified as agrammatic in hearing aphasics. In addition, her signing was severely impaired, halting and effortful, reduced often to single sign utterances. Her language profile was very much like that of hearing subjects classified as agrammatic; moreover, her lesion was typical of those that produce agrammatic aphasia for spoken language.

## Paragrammatic Sign Aphasia

Another left-lesioned signer was fluent in delivery after his stroke but showed many grammatical paraphasias. He made selection errors and additions within ASL morphology, and erred in the spatialized syntax and discourse processes of ASL. His signing was marked by an overabundance of nominals, a lack of pronominal indices, and failure to mark verb agreement correctly. Figure 7b shows his failure to maintain spatial indices in signed sentences. Such problems may be related to the organizational requirements of spatial planning and spatial memory involved in discourse in ASL, where the formal means for indicating pronominal reference is negotiated on-line and is spatialized. A signer must plan ahead to establish abstract loci for subsequent reference, and must remember where each locus

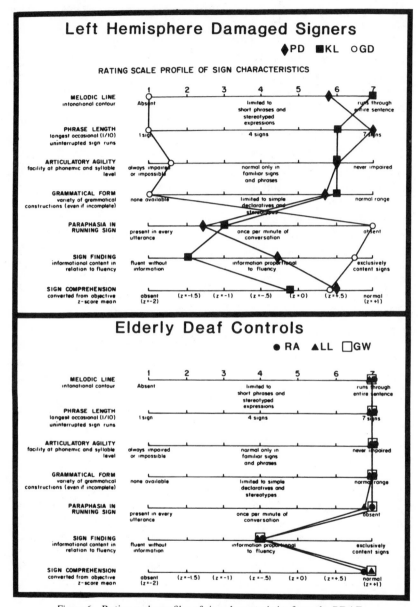

*Figure 6*   Rating scale profiles of sign characteristics from the BDAE.

is placed in the signing plane. Thus, the requirements of a spatially organized framework for syntax and discourse in sign language may be specifically (and differentially) impaired with focal lesions to the left hemisphere.

## Breakdown of Sublexical Sign Structure

A third left-lesioned signer had a fluent sign aphasia after her stroke. She made selection errors in the formational elements of signs which produced the equivalent of phonemic paraphasias in sign language. Her signing, however, was perfectly grammatical, although vague, as she often omitted specifying to whom or what she was referring. She also had a marked sign comprehension loss. Interestingly, this marked and lasting comprehension loss would not be predicted from her lesion if she were hearing; both major language-mediating areas for spoken language (Broca's area and Wernicke's area) were intact. Her lesion was in the inferior parietal lobe, an area known to function for higher order spatial analysis.

## Wernicke-like Sign Aphasia

The case of a fourth left-lesioned signer also had a marked and lasting sign comprehension loss, with a large perisylvian lesion that included the

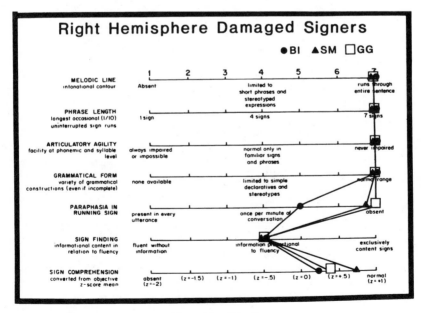

Figure 6 (continued)

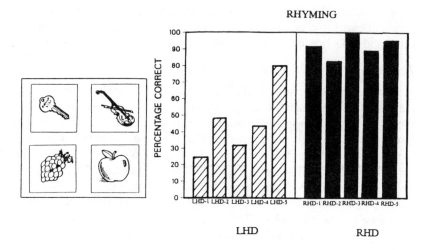

a) Test For Processing Sublexical Structure in ASL: Rhyming

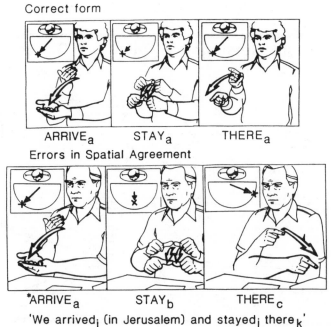

Correct form

ARRIVE$_a$          STAY$_a$          THERE$_a$

Errors in Spatial Agreement

*ARRIVE$_a$          STAY$_b$          THERE$_c$

'We arrived$_i$ (in Jerusalem) and stayed$_j$ there$_k$'

b) Failure to Maintain Spatial Indices in LHD Deaf Signer (P.D.)

inferior parietal lobe. The patient was markedly aphasic in his sign language production, making numerous paraphasias and elaborate semantic blends (see Figure 8a and b). Interestingly, this subject's paraphasias were specific to handshape, a finding that lends support to the view that this parameter may be considered a separate formational tier (Sandler 1987, Corina 1986). Finally, this case presents one of the most striking examples to date of the cleavage between linguistic signs and manual pantomime. In contrast to his severe aphasia for sign language, the patients abilities to communicate in nonlanguage pantomime and his comprehension for pantomime were remarkably spared. Although his comprehension of single signs was impaired, his comprehension of pantomimes depicting the same referents was unimpaired. His production of pantomimes also seemed intact, and he would insert mime into his signing (see Figure 8c). In identifying a flower, he gestured smelling a flower instead of making the sign FLOWER; in identifying a ball, he mimed the act of bouncing a ball instead of signing BALL. This differential brain processing of linguistic (sign) and nonlinguistic gestures (mime) provides an important indication of the left hemisphere's specialization for language-specific functions, and points to separate neuroanatomical pathways that mediate language and gesture (Poizner et al 1989b).

## Converging Evidence from a Hearing Signer

A hearing signer proficient in ASL has been studied during the left intracarotid injection of a barbiturate (the Wada test), and before and after a right temporal lobectomy. The subject, like the deaf signers, was a strong right hander. Neuropsychological and anatomical asymmetries suggested left cerebral dominance for auditory-based language. Emission tomography revealed lateralized activity of left Broca's and Wernicke's regions for spoken language. The Wada test, during which all left language areas were rendered inoperative, caused a marked aphasia in both English and ASL. The patient produced sign paraphasias, preservations, neologisms, and grammatical errors during this temporary inactivation of the left hemisphere. Interestingly, during recovery from the left Wada injection, the patient frequently responded in speech and sign simultaneously—a unique possibility for languages in different modalities—with a mismatch between the two languages and the sign more often in error. The patient

---

*Figure 7* Impaired phonology and syntax in left-lesioned signers. (*a*) A sample item from the Rhyming Test is illustrated in the *top panel*. The correct answers are *key* and *apple*, since signs for these share all major formational parameters but one. Note the impaired performance of left-lesioned signers in processing sublexical structure. (*b*) Failure in spatially organized syntax in a left-lesioned signer.

a)    **ASL Formational Paraphasias**

b)    **Semantic Blend in ASL**

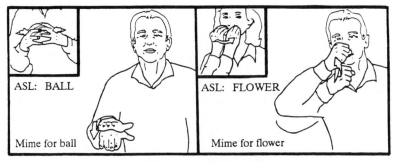

c)    **Mime Substituting for ASL Sign**

*Figure 8*  Language breakdown in a left-lesioned signer. The signer produces numerous sublexical paraphasias (*a*) and semantic blends (*b*) in his signing. The signer substitutes mime for ASL signs as shown in (*c*), thus reflecting a separation between these two manual functions.

subsequently came to brain surgery, and after partial ablation of the right temporal lobe, the abilities to sign and understand signing were unchanged. These data add further support to the notion that anatomical structures of the left cerebral hemisphere subserve language in a visuospatial as well as an auditory mode (Damasio et al 1986).

The data reviewed provide clear evidence for hemispheric differentiation for various cognitive functions in deaf and hearing signers. Specifically, these results show that certain aspects of language processing in individuals whose primary language is a sign language are differentially affected by left versus right hemisphere damage. The findings suggest that the grammaticized aspects of sign language are left hemisphere dominated, and thus that hemispheric specialization for "language" is not specific to *spoken* language. These studies also bear on what may be the basis for this differentiation of hemispheric function. The research into functions of space in ASL and the varying degrees to which formal spatial distinctions are grammaticized or conventionalized in subsystems of the language can provide one important clue to a basis for left versus right hemisphere specialization. It is this issue that we turn to next.

## PRESERVED SIGN LANGUAGE GRAMMAR IN RIGHT-LESIONED SIGNERS

The signers with right-hemisphere lesions present special issues in testing for language impairments; sign language makes linguistic use of space, and these signers show severe nonlanguage spatial deficits. As shown in Figures 6 and 7, the right-hemisphere-damaged signers examined in depth so far are not aphasic for sign language. On the whole, they exhibit error-free signing with a wide range of correct grammatical forms, no agrammatism, and no signing deficits. Furthermore, the right-lesioned signers, but not those with left-hemisphere lesions, were unimpaired on tests of processing ASL at the various linguistic levels. The right-lesioned subjects showed no impairment in the grammatical aspects of their signing, including their spatially organized syntax. They even used the left side of signing space to represent syntactic relations, despite their neglect of left hemispace in nonlanguage tasks. The dissociation between spatial language and nonspatial cognition is brought out in a strong way through an unusual case of a deaf signer who had been an artist prior to her right-hemisphere lesion.

### Dissociations between Spatial Language and Spatial Cognition

As discussed above, signers with lesions to the right hemisphere, but not those with lesions to the left hemisphere, show marked nonlanguage spatial deficits. These marked impairments were found across a broad range of tasks that maximally differentiate performance of hearing subjects with right as opposed to left brain damage (Poizner et al 1987). The case of one right-lesioned signer is particularly dramatic because she had been an accomplished artist before her stroke, with superior nonlanguage visuo-spatial capacities. After her stroke, her visuospatial nonlanguage functioning showed profound impairment. Even when copying simple line drawings, she showed spatial disorganization, massive left-hemispatial neglect, and failure to indicate perspective. Her performance on other constructional tasks was extremely impoverished, even though these were tasks she excelled at before her stroke. And yet her sign language (including spatially expressed syntax) was unimpaired. Her sentences were grammatical and her signs without error. Her lesion was a massive one to the right hemisphere, including areas that would be crucial to language if the lesion had occurred in the left hemisphere of a hearing person. Nonetheless, she showed preserved sign-language functioning, as did the other signers with right-hemisphere lesions. These cases underscore the complete separation in function that can occur between specializations of the right and left cerebral hemispheres in congenitally deaf signers. In light of their severe visuospatial deficits for nonlanguage tasks, the correct use of the spatial mechanisms for signed syntax in right-lesioned signers underscores the abstract nature of these mechanisms in ASL.

## Specialization for Facial Signals with Linguistic Function

The study of the effects of brain damage on deaf ASL signers affords a special opportunity to extend our understanding of brain-language relations and brain function underlying facial expression. In ASL, specific facial and body signals have arisen as a part of the grammar. They co-occur with manual signs and add an additional layer to the grammatical structure. Facial expressions in ASL serve two distinct purposes: they convey emotional information, as with hearing nonsigners; and, more interestingly, they serve to signal grammatical structures (Corina 1989, Reilly et al 1989a, b). Specifically, certain constellations of facial signals serve to mark syntactic structures (relative clauses, topics, and conditionals) and to mark adverbial constructions. Therefore, facial signals in ASL pose an interesting challenge to strict right-hemisphere processing of facial expression. As certain facial signals in ASL function as part of a tightly constrained linguistic system, they may be unaffected by right-hemisphere damage, despite the known mediation of the right hemisphere for facial signals that function effectively (Ley & Bryden 1979, Strauss & Moscovitch 1981).

AFFECTIVE AND LINGUISTIC FACIAL SIGNALS IN RIGHT-LESIONED SIGNERS   The functional dissociation between control of affective as opposed to linguistic facial expressions is being investigated in right- and left-lesioned signers. A right-lesioned signer showed a sparing of linguistic facial expressions in the face of severely attenuated emotional expression. Instances of clear sentential contexts in which either affective facial expression would be expected or specifically linguistic facial signals would be required were examined. Figure 9 shows the dissociation between affective and linguistic facial signals in this right-lesioned signer. This is an important finding, since presumably one and the same muscular system is involved. Thus one cannot account for the finding in terms of weakness of facial musculature, but rather this must be accounted for in terms of a dissociation between linguistic and affective facial expression. Importantly, in contrast, a left-lesioned agrammatic signer showed the opposite dissociation, with preserved affective facial expression, but without required linguistic facial expressions. Converging evidence comes from studies with neurologically intact deaf signers compared with hearing nonsigners that demonstrates that the left hemisphere in deaf signers plays a dominant role in the processing of *linguistic* facial expressions (Corina, 1989). Thus, as for manual signs, these data on processing facial signals suggest that the linguistic function of the signal, rather than its physical form, is the basis for the specialization of the left hemisphere for language.

## NEURAL MECHANISMS FOR SIGN LANGUAGE

### Dissociations between Apraxia and Aphasia

In order to understand the principles of neural organization of language, some investigators have attempted to root speech in the general psychophysiology of motor function. In this view, neural disorders of purposive movement, the apraxias, have been linked to the aphasias, and the left hemisphere has been considered to be primarily specialized for the control of changes in the position of both oral and manual articulators (Kimura, 1982). Other investigators have considered the left hemisphere to be specialized for general *symbolic* functions, including language (Goldstein 1948). The study of the breakdown of sign language and the breakdown of nonlinguistic gesture offers new ways of investigating apraxia and its relation to aphasia. In sign language, one can directly evaluate whether both aphasia and apraxia result from the same underlying substrate, since the expression of language and gesture does not cross transmission modalities.

In order to investigate the relationship between apraxia and aphasia for a gestural language, a series of tests for apraxia were administered to the patients, including tests of production and imitation of both repre-

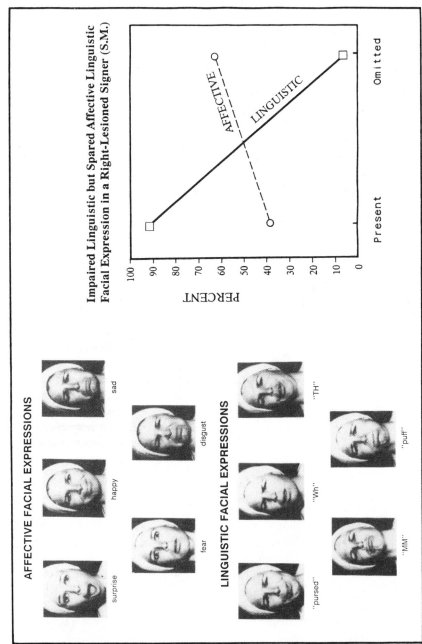

*Figure 9* Dissociation between linguistic and affective facial signals in a right-lesioned signer. The *left panel* contrasts affective facial expressions with specifically linguistic facial expressions in ASL. The *right panel* shows impaired linguistic but spared affective facial expression in a right-lesioned signer.

sentational and nonrepresentational movements. The right-lesioned signers showed no evidence of either aphasia or apraxia. However, for the left-lesioned signers, all of whom were aphasic for sign language, some strong dissociations emerged between their capacities for sign language and their nonlanguage gesture and motor capacities. The language deficits of these patients were on the whole related to specific linguistic components of sign language rather than to an underlying motor disorder. Nor were these deficits related to an underlying disorder in the capacity to express and comprehend symbols of any kind (Poizner et al 1987).

## Brain Systems for Signed and Spoken Language

The study of sign language can serve as a unique vehicle not only for uncovering the basic principles of hemispheric specialization, but also for investigating the nature of those anatomical structures within the left hemisphere that mediate language. Sign languages utilize channels for transmission and reception radically different from those of spoken language, and the anatomic structures important for language have certainly evolved at least in part in connection with the speech channel. Two major language-mediating neural structures, Broca's area and Wernicke's area, are linked to the speech channel. Broca's area is a portion of premotor cortex thought to program movements of the vocal tract. More importantly, Wernicke's area is part of *auditory* association cortex. Although it is clear that lesions in the left but not in the right hemisphere produce sign language aphasias, different areas within the left hemisphere may mediate language in the two modes. There are increasing reports of lesions to the parietal lobe of the left hemisphere that produce aphasias for sign language that would not be predicted for a hearing individual with that lesion (Poizner et al 1987, Chiarello et al 1982). It may well turn out that areas of the left hemisphere more intimately involved with gestural control and higher order spatial analysis emerge as language-mediating areas for sign language.

This research has broad implications for the theoretical understanding of the neural mechanisms underlying the human capacity for language. Patterns of breakdown of a visuospatial language in deaf signers allow new perspectives on the nature of cerebral specialization for language, because sign language entails interplay between visuospatial and linguistic relations within the same system. First, the data show that hearing and speech are not necessary for the development of hemispheric specialization. Sound is **not** crucial. Second, the data show that the two cerebral hemispheres of congenitally deaf signers can develop separate functional specializations for nonlanguage spatial processing and for language processing, even though sign language is conveyed in large part via spatial manipulation. Furthermore, it is the left cerebral hemisphere that is dominant for sign language. It thus appears that linguistic functions and the processing operations required, rather than the form of the signal, promote left hemisphere specialization for language. Finally, by uncovering the neural circuitry for sign language, one can uncover how the neural circuitry for spoken language (as well as sign language) operate for linguistic processing independently of language modality and how such circuitry is modality bound.

## ACKNOWLEDGMENTS

We gratefully acknowledge support by National Institute of Health grants NS 19096, NS 15175, and HD 13249 to U. Bellugi, and National Science Foundation grant BNS86-09085 and National Institute of Health grant NS 25149 to H. Poizner.

*Literature Cited*

Abramson, A. S. Lisker, L. 1970. Discriminability along the voicing continuum: Cross-language tests. *Proc. 6th Int. Cong. Phonet. Sci.*, pp. 569–73. Prague: Acadenuam

Bellugi, U. 1980. The structuring of language: Clues from the similarities between signed and spoken language. In *Signed and Spoken Language: Biological Constraints on Linguistic Form*, ed. U. Bellugi, M. Studdert-Kennedy, pp. 115–40. Dahlem Konferenzen. Weinheim/Deerfield Beach, Fla: Verlag Chemie

Bellugi, U. 1988. The acquisition of a spatial language. In *The Development of Language and Language Researchers, Essays in Honor of Roger Brown*, ed. F. Kessel, pp. 153–85. Hillsdale, NJ: Erlbaum

Bellugi, U. 1983. Language structure and language breakdown in American Sign Language. In *Psychobiology of Language*, ed. M. Studdert-Kennedy, pp. 152–76. Cambridge: MIT Press

Bellugi, U., Klima, E. S., Poizner, H. 1988. Sign language and the brain. In *Language, Communication, and the Brain*, ed. F. Plum, pp. 39–56. New York: Raven

Bellugi, U., Poizner, H., Klima, E. S. 1989. Mapping brain functions for language: Evidence from sign language. In *Signal and Sense: Local and Global Order in Perceptual Maps*, ed. G. Edelman, W. E. Gall, M. Cowan. New York: Wiley. In press

Bellugi, U., Poizner, H., Klima, E. S. 1983. Brain organization for language: Clues from sign aphasia. *Hum. Neurobiol.* 2: 155–70

Chiarello, C., Knight, R., Mandel, M. 1982. Aphasia in a prelingually deaf woman. *Brain* 105: 29–51

Corina, D. 1986. *ASL phonology: A CV perspective*. Presented at Ling. Soc. Am., New York

Corina, D. 1989. Recognition of affective and noncanonical linguistic facial expressions in hearing and deaf subjects. *Brain Cognit.* 9: 227–37

Damasio, A., Bellugi, U., Damasio, H., Poizner, H., Van Gilder, J. 1986. Sign language aphasia during left hemisphere amytal injection. *Nature* 322: 363–65

Fok, Y. Y. A., Bellugi, U., Lillo-Martin, D. 1986. Remembering in Chinese signs and characters. In *Linguistics, Psychology, and the Chinese Languages*, ed. H. Kao, R. Hoosain, pp. 177–202. Hong Kong: Univ. Hong Kong

Fok, Y. Y. A., Bellugi, U., van Hoek, K., Klima, E. S. 1988. The formal properties of chinese languages in space. *Cognitive Aspects of the Chinese Language*, ed. I. M. Liu, H. Chen, M. J. Chen, pp. 187–205. Hong Kong: Asian Res. Ser.

Goldstein, K. 1948. *Language and Language Disturbance*. New York: Grune & Stratton

Goodglass, H., Kaplan, E. 1972/1983. *Assessment of Aphasia and Related Disorders*. Philadelphia: Lea & Febiger

Jennings, P., Poizner, H. 1988. Computergraphic modeling and analysis II: Three dimensional reconstruction and interactive analysis. *J. Neurosci. Methods* 24: 45–55

Kimura, D. 1982. Left-hemisphere control of oral and brachial movements and their relation to communication. *Philos. Trans. R. Soc. London Ser. B* 298: 135–49

Klima, E. S., Bellugi, U. 1979/1988. *The Signs of Language.* Cambridge: Harvard Univ. Press

Klima, E. S., Bellugi, U., Poizner, H. 1988. Grammar and space in sign aphasiology. *Aphasiology* 2(3/4): 319–28

Lane, H., Grosjean, F., eds. 1980. *Recent Perspectives on American Sign Language,* pp. 79–101. Hillsdale, NJ: Erlbaum

Ley, R. G., Bryden, M. P. 1979. Hemispheric difference in processing emotions and faces. *Brain Lang.* 7: 127–38

Liddell, S. K., Johnson, R. E. 1986. American Sign Language compound formation. *Nat. Lang. Ling. Theory* 4: 445–513

Perlmutter, D. 1988. *A moraic theory of ASL syllable structure.* Presented at Theoretical Issues in Sign Lang. Res. Gallaudet Univ. Washington, DC

Poizner, H. 1981. Visual and phonetic coding of movement: Evidence from American Sign Language. *Science* 212: 691–93

Poizner, H. 1983. Perception of movement in American Sign Language: Effects of linguistic structure and linguistic experience. *Percept. Psychophys.* 33(3): 215–31

Poizner, H. 1987. Sign language processing. In *Gallaudet Encyclopedia of Deaf People and Deafness,* ed. J. Van Cleve, pp. 327–32. New York: McGraw-Hill

Poizner, H., Bellugi, U., Klima, E. S. 1989a. Sign language aphasia. *Handbook of Neuropsychology,* ed. F. Boller, J. Grafman, 2: 157–72. Amsterdam: Elsevier

Poizner, H., Bellugi, U., Klima, E. S., Kritchevsky, M. 1986a. *Competition between gesture and language in a brain damaged signer.* Presented at Acad. Aphasia, Nashville, Tenn.

Poizner, H., Corina, D., Bellugi, U., O'Grady, L., Feinberg, T., Dowd, D. 1989b. Sign aphasia with spared pantomime. *J. Clin. Exp. Neuropsychol.* 11: 42

Poizner, H., Fok, Y. Y. A., Bellugi, U. 1989c. Perception of language and perception of motion: Evidence from comparisons across sign languages. *Lang. Sci.* In press

Poizner, H., Klima, E. S., Bellugi, U. 1987. *What the Hands Reveal About the Brain.*
Cambridge: MIT Press/Bradford Books

Poizner, H., Klima, E. S., Bellugi, U., Livingston, R. 1986b. Motion analysis of grammatical processes in a visual-gestural language. In *Event Cognition,* ed. V. McCabe, G. Balzano, pp. 155–74. Hillsdale, NJ: Erlbaum

Poizner, H., Kritchevsky, M., O'Grady, L., Bellugi, U. 1988. *Disturbed prosody in a Parkinsonian signer.* Presented at Acad. Aphasia, Montreal

Poizner, H., Mack, L., Verfaillie, M., Rothi, L., Heilman, K. 1989d. Three-dimensional computergraphic analysis of apraxia: Neural representations of learned movement. *Brain.* In press

Poizner, H., Wooten, E., Salot, D. 1986c. Computergraphic modeling and analysis: A Portable system for tracking arm movements in three-dimensional space. *Behav. Res. Methods Instrum. Comput.* 18: 427–33

Reilly, J., McIntire, M., Bellugi, U. 1989a. Faces: The relationship between language and affect. In *From Gesture to Language in Hearing and Deaf Children,* ed. V. Volterra, C. Erting. New York: Springer-Verlag. In press

Reilly, J., McIntire, M., Bellugi, U. 1989b. Baby face: A new perspective on universals in language acquisition. In *Theoretical Issues in Sign Language Research: Psychology,* ed. P. Siple. New York: Springer-Verlag

Sandler, W. 1987. *Sequentiality and simultaneity in American Sign Language phonology.* PhD thesis. Univ. Texas, Austin

Stokoe, W., Croneberg, C., Casterline, D. 1965. *A Dictionary of ASL on Linguistic Principles.* Washington, DC: Gallaudet College Press

Strauss, E., Moscovitch, M. 1981. Perception of facial expression. *Brain Lang.* 13: 308–32

van Hoek, K. 1988. *Mental space and sign space.* Presented at Linguistic Soc. Am. New Orleans

Wilbur, R. 1987. *American Sign Language.* San Diego: Little, Brown

Zaidel, E. 1985. Language in the right hemisphere. In *The Dual Brain,* ed. D. F. Benson, E. Zaidel. New York: Guilford

*Annu. Rev. Neurosci. 1990. 13:309–36*

# SIGNAL TRANSFORMATIONS REQUIRED FOR THE GENERATION OF SACCADIC EYE MOVEMENTS

*David L. Sparks*

Department of Psychology, University of Pennsylvania, Philadelphia, Pennsylvania 19104

*Lawrence E. Mays*

Department of Physiological Optics and Neurobiology Research Center, University of Alabama at Birmingham, Birmingham, Alabama 35294

## INTRODUCTION

Saccades (high velocity, conjugate rotations of the eyes) function to bring the image of selected visual targets onto regions of the retina containing a high density of photoreceptors. Recent reviews (Robinson 1981a, Fuchs et al 1985) have documented the progress made in describing the neural signals controlling these rapid, precise movements. These papers considered the literature from the perspective of the motoneuron, emphasizing the generation of command signals by brainstem neurons having relatively direct synaptic connections with the motoneuron pools innervating the extraocular muscles. In this review, we summarize what is known about saccadic command signals found at a more central point—the superior colliculus (SC). Using this information as well as kinematic arguments, we discuss the transformations required to convert collicular signals into those needed by premotor pools of neurons. Also, we consider the functional properties of other brainstem neurons in the context of the required transformations.

309

0147–006X/90/0301–0309$02.00

# SACCADIC COMMAND SIGNALS OBSERVED
# IN THE SUPERIOR COLLICULUS

There is now compelling evidence that the intermediate and deep layers of
the SC contain neurons that are critical components of the neural circuitry
initiating and controlling saccadic eye movements (see Sparks 1986, Sparks
& Jay 1986, Sparks & Hartwich-Young 1988 for recent reviews). The
intermediate and deep layers receive inputs from brain areas involved in
the analysis of sensory (visual, auditory, and somatosensory) signals used
to guide saccadic eye movements (see Huerta & Harting 1984, Sparks 1986
for references). Neurons in the intermediate and deep layers project to
brainstem areas known to be important in the generation of saccadic
eye movements and to nuclei containing neurons having monosynaptic
connections with motoneurons (Highstein et al 1974, 1976, Harting 1977,
Huerta & Harting 1984).

Electrophysiological data support the assertion that the SC is involved
in saccade initiation. Many collicular neurons generate a high-frequency
pulse of spike activity that precedes saccade onset by approximately 20
msec (Sparks 1978). These neurons have movement fields (i.e. they discharge before saccades with a particular range of directions and amplitudes)
and, since they are topographically organized according to their movement
fields (Schiller & Koerner 1971, Wurtz & Goldberg 1972a, Sparks et al
1976), form a map of saccadic motor error (defined below). Additionally,
microstimulation of a discrete point in the SC reliably produces a saccade
with a particular direction and amplitude that is comparable in velocity
and trajectory to a visually guided saccade (Robinson 1972a, Schiller &
Stryker 1972). The current required to elicit a short-latency (20–30 msec)
saccade is low (5–20 $\mu$amp for stimulus trains of 40 msec duration, 500
pulses/sec).

In contrast to interpretations of early lesion studies, recent experiments
attest to the importance of the SC in the generation of saccadic eye
movements (see Sparks 1986 for a review). For example, reversible inactivation of collicular neurons severely impairs the direction, amplitude,
and velocity of visually guided saccades (Hikosaka & Wurtz 1985, 1986,
Lee et al 1988); and monkeys with combined lesions of the SC and frontal
eye fields are unable to initiate saccades to visual targets (Schiller et al
1980).

Given the critical role of collicular neurons in the control of saccadic
movements, how is information about the direction, amplitude, and velocity of saccades encoded by these cells? First, the discharge of collicular
neurons is related to certain changes in eye position, not to movement of
the eye to a particular position in the orbit. Each collicular neuron dis-

charges before saccades in its movement field regardless of the original position of the eye in the orbit (Schiller & Koerner 1971, Wurtz & Goldberg 1972a, Sparks et al 1976).

Second, it is the location of the active population of collicular neurons within the topographical map of movement fields, not their frequency of firing, that codes information about saccadic motor error (the direction and amplitude of the saccade required to direct gaze to the target). Saccade-related burst neurons are arranged topographically within the SC corresponding to a motor map first described in detail by microstimulation studies (Robinson 1972a). Although there is a spatial and a temporal gradient of activity within the population of cells discharging before a saccade (Sparks et al 1976), identical bursts of a single neuron may precede many saccades with different directions and amplitudes (Sparks & Mays 1980). Unlike medium-lead burst neurons in the pons (see below), an individual saccade-related burst neuron in the SC does not encode saccadic direction or amplitude by its firing pattern.

Third, the location of the active population of collicular neurons encodes initial saccadic motor error, not dynamic saccadic motor error. When a visual target appears in the peripheral visual field, initially (before the onset of a saccade) saccadic motor error is large. During the saccade, the motor error is rapidly reduced until, if the saccade is accurate, the error is zero. However, during the saccade the site of activity within the SC does not progressively shift toward the rostral pole of the SC, as it would if the site of activity were also encoding the dynamic changes in saccadic motor error occurring during a saccade (see Fuchs et al 1985, Tweed & Vilis 1985 for additional discussion of this point).[1]

Fourth, saccadic velocity may be related to the level of activity within the active population, since reversible deactivation of collicular neurons produces dramatic reductions in saccadic velocity (Hikosaka & Wurtz 1985, Lee et al 1988). Although earlier experiments (Sparks & Mays 1980)

---

[1] In a recent research note, Waitzman and colleagues (Waitzman et al 1988) present findings that they interpret as supporting the hypothesis that the discharge frequency of individual neurons in the SC encodes dynamic motor error. Their major observation is that the burst frequency of many collicular neurons gradually decays during a saccade. The time course of the decay resembles the time course of the decay in motor error. This hypothesis needs to be tested in more detail. Seemingly, the hypothesis would have difficulty explaining why electrical stimulation at a constant frequency for varying time intervals (presumably disrupting the frequency code) produces saccades with a relatively constant amplitude. Moreover, it has been reported (Sparks & Mays 1980) that the duration of the burst of neurons discharging maximally before small saccades (1–3° in amplitude) outlasts, significantly, the duration of the saccade. Finally, it is possible that the observed decay in spike frequency is related to a decrease in saccadic velocity rather than to a decrease in motor error, a possibility not considered in the report.

did not reveal a relationship between the discharge of saccade-related burst neurons and saccadic velocity, this was probably due to the small range of velocities observed when movements of the same direction and amplitude were made to continuously present visual targets. Recently, under conditions in which saccadic velocity had greater variability (saccades to remembered targets or saccades to auditory targets), a positive correlation between the average firing rate of collicular cells and saccadic velocity was observed (Rohrer et al 1987). A similar relationship between the instantaneous firing frequency of collicular cells and saccadic velocity has been described also in cats (Berthoz et al 1986, Munoz & Guitton 1986), animals that show large variations in saccadic velocity, even for visually guided saccades. Since the activity of a given population of SC neurons is associated with a saccade having a particular direction and a particular amplitude, the level of activity within this population is related to the velocity of the overall movement (vectorial velocity) and not directly related to the velocity of either the horizontal or vertical component of the rotation (component velocity).

Fifth, the collicular saccadic command signal may not fully compensate for the original position of the eye in the orbit. Saccades to visual targets compensate for the initial position of the eye in the orbit since systematic errors in accuracy are not observed as a function of initial eye position (Ritchie 1976, Pelisson & Prablanc 1988). Saccades produced by collicular stimulation do not fully compensate for the initial position. For example, stimulation of a collicular site that produces a purely horizontal saccade 10° in amplitude when the eye is in the center of the orbit will produce a saccade with a small downward component if the eye is elevated and a small upward component if, before stimulation, the eye is depressed. Similarly, centrifugal stimulation-induced saccades are smaller than centripetal ones (Segraves & Goldberg 1984, D. L. Sparks, unpublished observations). Noticeable differences in burst parameters as a function of original eye position have not been reported for neurons in the SC.

To summarize, each visually guided or spontaneously occurring saccade is preceded by the discharge of a relatively large population of neurons in the intermediate and deep layers of the SC. The location of this active population encodes initial motor error, the direction and amplitude of the change in eye position required for target acquisition. The rate of firing of neurons in the active population may be a determinant of the vectorial velocity of the ensuing saccade. Apparently, the saccadic command signal originating in the SC does not fully compensate for the differences in innervation required to produce comparable rotations from different starting positions (see below).

# THE OCULOMOTOR PLANT AND ITS INNERVATION

The neural signals observed in the SC are quite different from the inner-vation signals required for saccadic rotations of the eye. These differences become apparent when the properties of the oculomotor plant (the globe, extraocular muscles, orbital suspensory tissues or any other passive orbital tissues influencing rotation of the eye) and kinematic constraints upon the rotations of the eye are considered.

## The Oculomotor Plant

Mathematical models of the oculomotor plant and the innervation required to produce saccadic eye movements have been developed by several investigators (Cook & Stark 1967, Clark & Stark 1974, Collins 1975, Bahill et al 1980, Robinson 1981b). These models are based upon available measurements of muscle and globe parameters but also must employ a number of approximations and simplifying assumptions. Because of a lack of experimental data on the behavior of the globe and muscles in situ, the models are seriously underdetermined (Robinson 1981b). Thus, if parameters are chosen carefully, all the models generate normally appearing saccades, although some require unrealistic neural inputs. Non-etheless, there is considerable agreement on two points.

First, all of the models recognize the highly overdamped characteristics of the plant and require a pulse and step of innervation to produce a saccade. The pulse of innervation produces a phasic increase in muscle tension that overcomes the viscous drag of the orbital tissues and moves the eye at a high velocity. The step of innervation causes the sustained change in muscle tension required to hold the eye in the new position and to overcome the elastic restoring innervations of the orbital tisue. In addition, Robinson's model of the plant (1964) has a third time constant so that the innervation required is a pulse of high-frequency activity fol-lowed by an exponential slide in instantaneous frequency to a final step level of innervation.

Second, the total innervation needed to hold the eyes in a fixed position is strongly dependent upon eye position (Collins 1975, Robinson 1975a); the innervation required to produce a saccade of a given amplitude depends upon both initial position and saccade direction (Robinson 1975a, Optican & Robinson 1980). This indicates that information about changes in eye position present in the SC is not sufficient to produce accurate saccades; information about absolute eye position must be used as well.

## Kinematic Considerations

Lawful relationships observed in kinematic studies must be accounted for by the mechanical properties of the plant or by the innervation signals supplied to the plant. If the small translations associated with ocular rotations are ignored, the eye can be modeled as a perfect ball-and-socket joint requiring three independent variables to describe eye position unambiguously. According to Donders' law, however, saccadic eye movements are accomplished using only two degrees of freedom of rotational movements. This law states that the relationship of the vertical meridian of the eye with respect to a vertical reference (e.g. a plumb line) is invariant for a given eye position, regardless of how the eye got to that position. Thus, the torsion, or twist of the eye about the line of sight, when the eye is directed 10° above and 10° to the left of the primary position, is the same whether that position is achieved by a 10° upward and then a 10° leftward movement, or whether the movements are executed in opposite order. Listing's law specifies the amount of this cyclorotation, relative to the vertical, at any given position. Donders' law is not a universal property of rigid bodies free to rotate in three dimensions. Such rotations are not commutative; the order of the rotations about the horizontal and vertical axes is important in determining the amount of torsion at the final position. Examples of this effect are shown by Nakayama (1975) and Tweed & Vilis (1987). Moreover, the eye is not mechanically constrained to observe Donders' law. That voluntary saccades are accomplished using only two degrees of freedom is a consequence of the pattern of innervation. Donders' law does not hold for movements occurring, for example, during sleep (Nakayama 1975), during pursuit eye movements (Westheimer & McKee 1973, Ferman et al 1987b), or during vergence movements (Nakayama 1983).

The implications of Donders' and Listing's laws for models of the neural mechanisms generating saccades have been discussed by Westheimer (1954, 1957, 1981) and Nakayama (1975, 1983). They state that the neural circuitry must specify an absolute, as opposed to a relative, eye position. If the saccade generator simply issued instructions for relative rotations (i.e. move 3° down and 6° to the right), torsional positions violating Donders' law would often be observed, since the amount of torsion observed would depend upon the presaccadic position of the eye in the orbit (Nakayama 1975, Westheimer 1981).

## Orbital Geometry and Muscle Interactions

The implications of kinematics for neural control depend on orbital geometry and muscle interactions as well. Robinson (1975a) developed a computer model of the plant that analyzed the interactions among the six extraocular muscles. This muscle interaction model provides values for the length, force, and the innervation signal required to generate that force, for each muscle when the eye is in any orbital position. He emphasized that the rotation produced by contraction of any single extraocular muscle depends upon the state of contraction of all the remaining muscles. Moreover, muscles interfere with each other, and there is considerable cross-coupling between muscles (Robinson 1975a). Thus, the innervation patterns required to obey Listing's and Donders' laws cannot be deduced by considering the planes of action of each of the muscles individually. An important outcome of Robinson's model is that iso-innervation curves predicted for each of the six extraocular muscles when constrained by Listing's law are considerably straighter and more nearly parallel to the horizontal and vertical meridians than would have been expected by observing the pulling actions of individual muscles at various eye positions. From the point of view of generating saccadic commands, Robinson (1975a) noted that the innervational participation of a muscle is different from its mechanical participation.

A recent study by Miller (1989) also indicates that orbital geometry may profoundly affect the signal transformations required in brainstem circuits. He used magnetic resonance imaging to generate pictures of the recti muscles and optic nerve of human observers while the eye was in a number of different orbital positions. He reported that there is almost no side-slip of the recti muscles and that the muscle planes, but not necessarily the axes of rotation, are essentially fixed in the orbit. Two models consistent with his observations were considered. In both, side-slip of the muscles is prevented by a "harness" composed of fascia bands connecting the muscles with each other and with the bony orbit. In one model, the muscle sheaths act as a pulley fixed to the orbit such that the axis of rotation moves with the eye. Ignoring the problem of recruitment of motoneurons, this model would permit saccadic commands to be computed in retinotopic or global coordinates, independent of the position of the eye in the orbit. For example, contraction of the lateral rectus would cause rotation along an axis perpendicular to the plane of the horizontal meridian of the eye over a wide range of elevations. In contrast, the second model assumes no significant linkage between the muscle "harness" and the orbit; neural commands would need to be organized in orbital rather than retinotopic coordinates. Models with intermediate characteristics can also be envisioned.

In summary, the most realistic models of the oculomotor plant suggest that either a pulse/step or a pulse/slide/step innervation signal should be observed. Furthermore, the amplitude of the total innervation of both the pulse and the step should vary for identical saccades with different starting positions. This implies that, ultimately, saccade commands must be commands to move the eye to a particular orbital position, not commands to produce relative rotations. Finally, cross-coupling between muscles and orbital geometry are important determinants of the coordinates in which innervation signals must be organized.

# MOTONEURON ACTIVITY

## The Pulse/Slide/Step

The pattern of activity observed at the single motoneuron level is a pulse/slide/step waveform, the pattern of innervation suggested by Robinson's early model (1964) of the plant. A high frequency burst of motoneuron activity precedes, by about 8 msec, the onset of saccades in the on-direction (Fuchs & Luschei 1970, 1971, Schiller 1970, Robinson 1970). The pulse of activity is not rectangular, as some models assume, but decays with both a short and a long time constant to the steady-state level of activity associated with the new position (Fuchs & Luschei 1970, Robinson 1970, Goldstein 1983).

## Nonlinear Relationship Between Muscle Innervation and Eye Position

It is clear from an analysis of the plant (Collins 1975, Robinson 1975a) that the total innervation to a given muscle is a nonlinear increasing function of the rotation in the on-direction. In order to move the eyes in a series of equal amplitude saccades to more eccentric positions, successively greater innervation increments are required. The firing patterns of individual motoneurons show a simpler relationship to eye position than that suggested by analyses of total innervation. For steady fixation, the activity of any motoneuron is linearly proportional to eye position along its on-direction. The static firing rate of a motoneuron can be described by two parameters: a threshold position and the slope of the line of best fit for the firing rate vs. eye position plot (Robinson 1970, Goldstein & Robinson 1986). The major mechanism for generating a nonlinear relationship between muscle innervation and eye position appears to be recruitment of more and more motoneurons. For static positions, the recruitment order of extraocular motoneurons is relatively fixed (Robinson 1970, Keller 1981, Fuchs et al 1988). Motoneurons with larger rate-position slopes tend to be recruited later as the eye moves in the on-direction and, according to Robinson (1970), all motoneurons for a given muscle are recruited by the point at which the eye has moved approximately 20° beyond the primary position in the on-direction. If this observation is true, increased muscle innervation beyond 20° must be generated by increased activity rather than recruitment of additional motoneurons. This implies that the total innervation will be a linear function of eye position once all motoneurons are recruited. Van Gisbergen (1988) plotted total neural innervation as a function of eye position for a hypothetical motoneuron pool. He assumed that the recruitment thresholds of the motoneuron pool were distributed homogeneously over the observed range of recruitment thresholds and that the threshold and the slope of the rate/position curve were correlated (Robinson 1970, Keller 1981). The resulting curve resembled, somewhat, the isometric muscle-tension relation described by Collins (1975) in that it was nonlinear up to the point at which all motoneurons were recruited.

The nonlinear relationship between innervation and eye position during the dynamic phase of saccades is poorly understood. For small saccades, the burst frequency of motoneurons is related to saccadic velocity and amplitude. But for saccades larger than $10°$, burst frequency saturates and becomes unrelated to saccadic amplitude. Because of this saturation, the major determinant of saccadic amplitude for large saccades is burst duration. With respect to recruitment, Robinson (1970) and Goldstein (1983) reported that, during saccades, burst rate depends upon the starting position of the saccade relative to the unit's threshold position; the velocity signals reaching individual neurons in the abducens nucleus may not add linearly to position signals (Goldstein 1983). Whether or not the pulse innervation of the entire motoneuron pool is dependent upon saccadic starting position is unknown (van Gisbergen 1988). Nonetheless, it should be noted that the effective saturation of the pulse in generating force occurs at a lower firing rate than that exhibited by motoneurons. Motoneurons may burst at instantaneous rates of 600 spikes/sec (Robinson 1970), yet artificial stimulation suggests that little additional tension is generated above 200–300 Hz (Robinson 1981b, Nelson et al 1986).

## Organization of Commands in Horizontal and Vertical Coordinates

Hepp & Henn (1985) measured the frequency of firing of mononeurons innervating the six extraocular muscles during periods of fixation within $30°$ of the primary position. They constructed plots of the horizontal and vertical eye positions associated with particular rates of firing of single motoneurons. As expected, the iso-frequency plots of individual motoneurons innervating the horizontal recti formed a family of curves that were approximately parallel with the vertical meridian. Iso-frequency plots of motoneurons innervating the inferior rectus and most motoneurons with upward on-directions were approximately parallel with the horizontal meridian. (For technical reasons, it was not possible to distinguish inferior oblique from superior rectus motoneurons). The iso-frequency gradients of motoneurons innervating the superior oblique muscle depended upon horizontal eye position only when the innervated eye was abducted. The discharge frequency of motoneurons innervating the oblique muscles and the vertical recti was more highly correlated with the horizontal or vertical position of the eye in the orbit than would be expected based upon knowledge of the axis of rotation produced by these muscle pairs acting independently.

The results of the Hepp & Henn (1985) study were, in general, consistent with Robinson's (1975a) analysis of muscle interactions. According to this model, the static innervation signals required by each extraocular muscle are approximately parallel to either the horizontal or the vertical meridian. This suggests that the characteristics of the oculomotor plant itself may simplify the neural computations needed to satisfy Listing's law. Based upon the findings of Hepp & Henn (1985) and Robinson's (1975a) analysis, it is possible that Listing's law would be obeyed approximately, without explicitly computing a torsional command, if signals were simply organized in horizontal and vertical coordinates. The plausibility of this idea depends upon how well Listing's law is actually obeyed. Ferman and colleagues (1987a,b) tested the validity of Listing's law directly. They concluded that the control of torsion by the saccadic system is usually not precise, and that Listing's and Donders' laws are only obeyed approximately. Long-term fluctuations of up to $5°$ of torsion were observed with the eye in the primary position, and systematic deviations from values predicted by Listing's law were seen in secondary and tertiary positions.

In summary, the signals needed by the oculomotor plant and supplied by motoneurons are quite different from those found in the SC. Although collicular cells generate a presaccadic pulse of activity, they do not provide a step signal related to steady eye position. Collicular neurons specify relative rotations of the eye, not commands to move the eye to an absolute orbital position. There is no evidence that the magnitude of the saccade-related burst associated with saccades of a particular direction and amplitude depends upon initial eye position. Finally, the map of collicular activity is not organized along either the planes of action of the extraocular muscles or in horizontal and vertical coordinates.

# FORMATION OF INNERVATION SIGNALS BY BRAINSTEM CIRCUITS AND THE REQUIRED TRANSFORMATIONS OF COLLICULAR SIGNALS

## Formation of the Pulse/Slide/Step Signal

BRAINSTEM SIGNALS    Four major functional classes of neurons are observed in the paramedian pontine reticular formation (PPRF), a brain region critical for the generation of horizontal saccades. Medium-lead burst neurons (MLBs) generate a high-frequency burst before ipsilateral saccades. Omnipause neurons fire at a relatively constant rate during fixation and pause during saccades in all directions. Tonic neurons discharge at a frequency proportional to horizontal eye position, and long-lead burst neurons (LLBNs) generate a low-level increase in activity and a vigorous burst before saccades (see Raphan & Cohen 1978, Fuchs et al 1985 for reviews).

Robinson used three of these neural elements in a local feedback model (1975b) designed to simulate horizontal saccades. The model has two inputs: a signal of the desired horizontal position of the eyes (DHP) and a trigger signal. An important feature of the model is that the input specifying saccade amplitude does so by providing a signal of final eye position in the orbit, not the required displacement of the eye. Saccades are initiated by a trigger signal that briefly inhibits the pause cells, permitting the MLBs to discharge at a rate proportional to horizontal motor error (the difference between DHP and an internal estimate of current horizontal eye position, CHP). In the model, the pulse of activity generated by MLBs is transmitted directly to motoneurons and to a neural integrator. The neural integrator converts the pulse into a step of activity (observed in the activity of tonic cells) that is sent to motoneurons and used as the estimate of CHP. Once activated, this circuit drives the eye at a high velocity until the representation of CHP matches the DHP signal. At that point, the eye stops on target and the pause cells are allowed to resume firing, thereby inactivating the saccadic generator until a new trigger signal arrives.

Robinson's original model produces a pulse (output of MLBs) and step (output of tonic neurons) of innervation as observed in motoneuron activity. A modification of the model proposed by Optican & Miles (1985) also generates the slide component of motoneuron activity. Given such a circuit, the task of more central components of the saccadic system is to provide the input signals required by the pontine network: a trigger signal, and a signal of the DHP. Provided with these signals, the PPRF circuit automatically generates the pulse/slide/step signal that is transmitted to the motoneurons. The computations required to extract these signals from those observed in the SC are discussed below.

Medium-lead burst neurons identified in the rostral midbrain [the interstitial nucleus of Cajal (INC) and the rostral interstitial nucleus of the medial longitudinal fasciculus (riMLF)] have functional properties similar to those observed in the PPRF except that during saccades their discharge seems to be related to the vertical, instead of the horizontal, component of eye movements (Buttner et al 1977a,b, King & Fuchs 1977, 1979, King et al 1981).[2] Robinson's model, as modified by Optican & Miles, could be extended to this population of vertical bursters to generate a pulse/slide/step signal for motoneurons providing innervation signals for vertical rotations of the eye. Since pause cells in the PPRF stop firing before both horizontal and vertical saccades, and microstimulation of the pause region inhibits the occurrence of both horizontal and vertical saccades (Keller 1977, Evinger et al 1977), the pause cells control both the horizontal and vertical pulse generator circuits.

The vertical pulse generator circuit is more complex than the horizontal circuit. Cells in the riMLF and INC project to motoneurons innervating the vertical recti and oblique muscles (Buttner-Ennever & Buttner 1978) and two subsets of vertical MLBs organized in planes similar to those of the anterior and posterior semicircular canals have been functionally identified (Hepp et al 1986). During visually elicited saccades with the head stationary, MLBs with upward or downward on-directions are found bilaterally in the riMLF (Hepp et al 1986). However, these neurons can be subdivided into two subpopulations by observing their activity during sinusoidal head rotation in the dark. For example, during head rotation in the plane of the right-anterior/left-posterior semicircular canals, MLBs in the right riMLF with downward on-directions and MLBs in the left riMLF with upward on-directions are most vigorously activated. The firing rates observed during rotations in this plane are significantly higher than firing rates associated with saccadic eye movements in the light. In a companion study, Vilis and colleagues (Vilis et al 1986) reported that unilateral microinjections of muscimol, a GABA agonist, into the riMLF disrupted saccades along specific planes. Injections in the right riMLF impaired visually elicited saccades to targets in the upper left and lower right quadrants of the visual field. Conversely, injections into the left riMLF impaired saccades to targets in the upper right and lower left quadrants.

As suggested by Robinson (1972b), saccades are controlled, at least in part, by neural circuits subserving the phylogenetically older vestibulo-ocular-reflex (VOR). Accordingly, rapid vertical rotations of the eyes can be executed in two modes (Villis et al 1986). During vestibular nystagmus, quick phases in the plane of head rotation (movements that may not obey Listing's law) are produced by preferentially activating subsets of riMLF

---

[2] Robinson & Zee (1981) hypothesized that the on-directions of vertical burst neurons were organized in the canal planes. They noted that if so, the on-directions of vertical burst cells could not be determined by observing their activity only during visually guided saccades, since the saccadic system operates under the constraints of Listing's law and does not generate rotations in the planes of the anterior and posterior canals.

burst neurons. Visually elicited saccades are produced by bilateral activation of burst neurons in the riMLF. Presumably, variations in the
symmetry of bilateral activation are associated with differences in the
torsional position of the eye. Appropriate innervation ratios for these burst
neurons are required if the saccadic system is to approximate Listing's
law.

REQUIRED TRANSFORMATIONS OF COLLICULAR SIGNALS    Robinson's model
of the saccadic system (1975b) as revised by Optican & Miles (1985) will
automatically produce the pulse/slide/step of innervation if the horizontal
and vertical burst generator circuits are provided with a trigger signal, a
signal of the desired horizontal position of the eye in the orbit, and a signal
of the desired vertical position of the eye in the orbit.

Saccade-related burst neurons (SRBNs) observed in the SC could serve
as the trigger input required by Robinson's model (Robinson 1975b) of
the saccadic system (Sparks & Mays 1980). The axons of SRBNs comprise
a major efferent pathway from the SC to subsequent premotor neurons,
particularly the PPRF. SRBNs meet the requirements for a trigger input;
they generate a high-frequency burst of activity tightly linked to saccade
onset. The burst precedes saccade onset by an appropriate time. Keller's
(1980) observations that 10 out of 10 SRBNs recorded in the SC are
antidromically activated by stimulation of the region of the PPRF containing pause units, supports the suggestion that the SC provides a trigger
input to the PPRF. Only 1 of 11 other collicular neurons with saccade-
related discharges, but lacking the high-frequency burst, was antidromically activated by PPRF stimulation.

Robinson's model (1975b) does not specify how a signal of DHP is
generated, but the formation of such a signal requires additional processing. Neurons in the SC generate commands for changes in eye position or
desired displacement, rather than commands to move the eye to a particular position in the orbit. The motor error signals recorded in these areas
must be decomposed into appropriate horizontal and vertical components
[desired horizontal displacement (DHD), and desired vertical displacement
(DVD)], and an estimate of CHP must be added to the DHD signal to
form a signal of DHP. Modified versions of Robinson's model assume
that the horizontal and vertical pulse/step generators are driven by signals
of desired displacement instead of desired position. The extraction of
desired displacement signals is discussed below.

## Recruitment and Compensation for Presaccadic
## Eye Position

Little experimental attention has been given to the question of how signals
reaching the motoneuron pools control recruitment. Tonic units in the

PPRF (thought to project to motoneurons) are recruited in a relatively continuous manner out to approximately 25° in the on-direction; some of these cells have nonlinear rate position curves, displaying increased slopes at more lateral positions (Luschei & Fuchs 1972, Keller 1974). King and colleagues (1981) did not find a significant correlation between the slope of the rate/position curve and the recruitment threshold for vertical burst-tonic neurons near the INC. In a recent study, Fuchs and colleagues (1988) studied the recruitment order of identified motoneurons and internuclear neurons in the monkey abducens nucleus. Motoneurons in the abducens nucleus innervate the ipsilateral lateral rectus muscle and display a linear increase in firing rate associated with progressively larger temporal rotations of the eye. The axons of internuclear neurons travel in the medial longitudinal fasciculus to the subdivision of the oculomotor nucleus containing motoneurons that innervate the medial rectus of the contralateral eye. Abducens internuclear neurons are involved in the coordination of yoked, conjugate movements of the two eyes and show increases in firing rate when the ipsilateral eye moves temporally and the contralateral eye moves nasally. Fuchs and collaborators (1988) showed that, in contrast to motoneurons, there is only a weak relationship between the rate-position slope and recruitment threshold for internuclear neurons. This suggests that the recruitment order of motoneurons may be established at the level of the motoneuron pool. A single recruitment mechanism shared by all conjugate subsystems might explain the lack of strong eye position effects found in the saccadic subsystem.

In addition to a shared recruitment mechanism for generating the nonlinear innervation signal, separate neural circuits may form supplementary signals to compensate for the presaccadic eye position. Ritchie (1976) reported that rhesus monkeys with lesions of the posterior vermis made hypermetric centripetal saccades and hypometric centrifugal saccades. Optican & Robinson (1980) also studied the effects of cerebellar lesions upon saccades in different parts of the oculomotor range. Their lesions, larger than those made by Ritchie, included the posterior vermis, paravermis, and fastigial nuclei. Monkeys with these lesions made hypermetric saccades, but centripetal saccades were significantly less hypermetric than centrifugal ones. Optican & Robinson (1980) concluded that the cerebellum acts as an interface between visual commands and motor performance by providing (directly or indirectly) different innervation signals to the motoneurons for the same retinal error signal, depending upon the orbital position and direction of the impending saccade. The route by which such signals are conveyed to motoneurons is unknown.

In the same study, Optican & Robinson (1980) found that animals with an intact cerebellum can adjust the gain of the saccadic pulse and step to compensate for surgically induced alterations in the mechanical properties of the muscle. Cerebellar lesions impair the ability of animals to make these gain changes. Since cerebellar lesions also disrupt the ability of the animal to compensate for initial eye position (Ritchie 1976, Optican & Robinson 1980), the mechanism responsible for the adaptive adjustment of saccadic parameters could also be responsible for compensating for initial eye position.

The saccadic command signals found in the SC may need to be supplemented by signals compensating for orbital position. Although not studied in detail, neither the saccade-related discharge of collicular neurons (Wurtz & Goldberg 1972a, Sparks et al 1976) nor the discharge of MLBs in the PPRF (Keller 1974, van Gisbergen et al 1981) appears to depend upon initial eye position. As noted above, stimulation-induced saccades do not fully compensate for initial eye position (Segraves & Goldberg 1984, D. L. Sparks, unpublished observations). The effects of original eye position upon the trajectory of the stimulation-induced movement, however, are small. Moreover, the saccades produced by collicular stimulation are comparable to natural saccades in their amplitude-velocity characteristics, and the eyes maintain their new position after a stimulation-induced saccade. This suggests that the basic motoneuron recruitment mechanism is intact, but there is a failure to provide a small adjustment for initial position.

A neural circuit involving neurons in the nucleus reticularis tegmenti pontis (NRTP) and cerebellar vermis may be associated with the generation of a compensatory signal. The region of the NRTP containing cells with saccade-related activity receives extensive projections from the intermediate layers of the SC (Harting 1977) and sends most, if not all, of its efferent projections to the cerebellum (Brodal & Jansen 1946, Gerrits & Voogd 1986, 1987). Cells in the NRTP with saccade-related activity have movement fields similar to those described for cells in the SC (Crandall & Keller 1985), but, more importantly, the vigor of the burst of NRTP cells differs depending upon the origin of the movement (Crandall & Keller 1985). Also, neurons in the cerebellar vermis display eye-position-dependent variations in saccade-related bursts (McElligott & Gochin 1986), and the direction and size of saccades evoked by cerebellar stimulation are dependent upon initial eye position (McElligott & Keller 1984).

## Coordinates of Premotor Signals

BRAINSTEM SIGNALS    A large body of evidence based upon recording, microstimulation, anatomical, and lesion data (see Raphan & Cohen 1978, Fuchs et al 1985 for reviews) indicates that signals responsible for the horizontal component of saccadic eye movements are generated in the

PPRF. Similarly, neurons in two regions of the rostral mesencephalic reticular formation (the INC and the riMLF) are involved in coordinating the vertical recti and oblique muscles during vertical saccades (Buttner et al 1977a,b, Buttner-Ennever 1977, Buttner-Ennever & Buttner 1978, King & Fuchs 1979, King et al 1981).

Not only must separate horizontal and vertical saccadic signals be formed, these signals must be coordinated during oblique saccades. Oblique saccades usually have relatively straight (rather than curved) trajectories (Figure 1A) because the onsets of the horizontal and vertical pulses are synchronized and the durations of the horizontal and vertical displacements are approximately equal (Guitton & Mandl 1980, Evinger et al 1981, King et al 1986). When the amplitudes of the horizontal and vertical components of an oblique saccade are unequal, the duration of the smaller component is greater than the duration of a pure horizontal or vertical saccade of the same amplitude (Figure 1B). The increased duration of the minor component of oblique saccades is neurally mediated; abducens neurons (King et al 1986) and midbrain MLBs (King & Fuchs 1979) display a decrease in average firing rate and an increase in burst duration under these conditions. This implies that signals controlling the velocity of the horizontal and vertical components of oblique saccades are based upon the amplitude of the total movement.

A direct extension of Robinson's model assuming independent pulse/step controllers for the horizontal and vertical components of saccades cannot account for the fact that visually guided oblique saccades are usually straight rather than curved. Even if the onsets of the pulses gen-

---

*Figure 1*   Examples of nine saccades all having 8° rightward horizontal components but differing in the magnitude and/or direction of the vertical component. B: The velocity profiles associated with three of the saccades illustrated in A. Note the reduction in peak velocity and the increase in duration of the horizontal component of movements having larger vertical components. C: The locations of collicular neurons discharging maximally before each of the nine saccades illustrated in A. D: It is hypothesized that by appropriate synaptic weighting, some tectorecipient neurons in the central mesencephalic reticular formation (cMRF) would display a frequency of firing proportional to the velocity of the horizontal (or vertical) component of oblique saccades. Neurons discharging maximally before saccades with small vertical components (e.g. E) provide large excitatory drive, whereas those discharging maximally before saccades with larger vertical components would provide proportionally smaller excitatory drive. E: A plot (*filled squares*) of the peak velocity of the horizontal component of the nine saccades shown in A as a function of the direction of the saccade (0 = straight right). The firing rate (*solid line*) of a hypothetical neuron receiving inputs from all collicular neurons discharging maximally before saccades with an 8° horizontal component is shown. The strength of synaptic input is weighted according to the cosine of the angle of the movement. (From Sparks 1990).

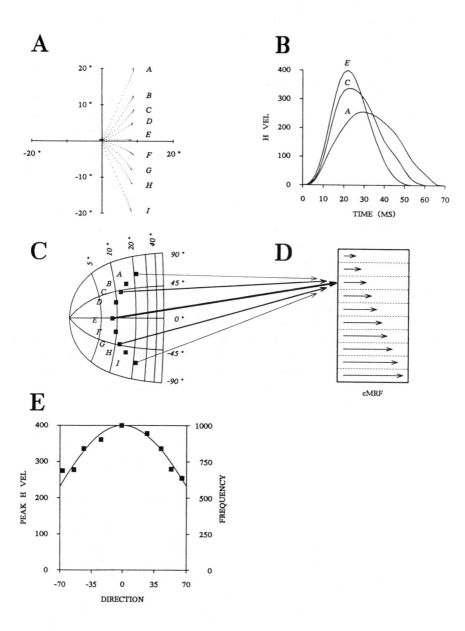

erated by the horizontal and vertical controllers were synchronized via pause cells, the offsets could be asynchronous and result in curved saccades. To prevent this, the velocity of the minor component must be reduced by lowering the discharge frequency of its MLBs so that the durations of the two components are also equal (King et al 1986). Several workers (Tweed & Vilis 1985, van Gisbergen et al 1985, van Gisbergen 1988, Grossman & Robinson 1988, Scudder 1988) have proposed modifications of Robinson's model to account for the coordination of the horizontal and vertical components of oblique saccades. These are considered in more detail below.

REQUIRED TRANSFORMATIONS OF COLLICULAR SIGNALS   The discharge of collicular neurons is related to the direction and amplitude of eye displacement, and is not uniquely related to the amplitude of either the horizontal or the vertical component of the movement. Thus, separate signals of horizontal and vertical motor error must be extracted from the anatomical map of motor error found in the SC. How and where are the separate signals needed by the burst generators formed? Several possibilities exist. Three are discussed in more detail below.

First, each collicular neuron could synapse (with appropriate weighting) directly upon the horizontal and vertical MLBs in the pons and midbrain. In this case, an intermediate stage in which signals of horizontal and vertical motor error are explicitly extracted would not exist. Since MLBs in the pons are not activated with monosynaptic latencies by collicular stimulation (Raybourn & Keller 1977), the available evidence does not support this possibility.

A second possibility is that separate signals of the horizontal and vertical motor error are present in populations of cells intermediate between the SC and horizontal and vertical MLBs. Most recent revisions of Robinson's model (Keller 1981, Tweed & Vilis 1985, Scudder 1988) assume that a signal of horizontal motor error is formed by selective convergence of axons of collicular neurons upon LLBNs, the cells assumed to provide excitatory drive to horizontal MLBs. Direct projections from the SC to the contralateral PPRF have been demonstrated anatomically (Harting 1977) and electrophysiologically (Raybourn & Keller 1977, Keller 1979). Raybourn & Keller (1977) reported that all LLBNs in their sample received short-latency excitatory input from the SC and that the latency of the response was in the monosynaptic range for approximately one third of the sampled neurons. Moreover, each LLBN was driven by stimulating electrodes placed in both rostral and caudal areas of the SC, evidence for convergence of signals thus originating from widespread regions of the SC.

Although anatomical evidence is consistent with the hypothesis that collicular signals converge upon LLBNs, there is no anatomical evidence that the LLBNs provide an input to MLBs. Moreover, it is not clear that LLBNs provide a signal of horizontal motor error. LLBNs display a relatively long period of irregular low-frequency activity as a "prelude" to an intense burst of firing that precedes saccade onset by 10–11 msec (Luschei & Fuchs 1972, Keller 1974, Henn & Cohen 1979, Hepp & Henn 1983). Because most investigators have not plotted the movement fields of LLBNs, there is no unequivocal evidence that the discharge of LLBNs is related only to the horizontal component of oblique saccades. Van Gisbergen and colleagues (1981) reported that information about saccade size, saccade duration, or saccade velocity is not coded in any obvious way by the discharge patterns of individual LLBNs. Their sample of LLBNs often fired well before saccade onset and long after saccade offset. During the saccade, the burst rate varied little with saccade size or eye velocity. They concluded that the signal recorded from single LLBNs would need to be combined with signals from other sources to provide an input that was the basis for the strict relationship between burst parameters and eye velocity observed in MLBs. Clearly, additional studies are needed, designed explicitly to examine the functional properties of LLBNs, as these neurons play a prominent role in current models of the saccadic system.

Recent experiments suggest that neurons in the central midbrain reticular formation (cMRF) could be involved in extracting information about the horizontal component of oblique saccades. Neurons in the intermediate and deeper layers of the SC project to the cMRF (Cohen et al 1986) and, in turn, neurons in the cMRF project to regions of the pontine reticular formation involved in the control of horizontal saccades (Cohen & Buttner-Ennever 1984). Cells in the cMRF burst before contralateral saccades having horizontal components, and microstimulation in this area produces only horizontal saccades (Waitzman 1982). The amplitude of the stimulation-induced horizontal movement depends upon the depth of the stimulating electrode. Small amplitude movements are evoked by stimulation dorsally, and progressively larger movements are evoked as the electrode is moved ventrally (Waitzman 1982, Cohen et al 1985).

Based upon the findings of Cohen and colleagues (Cohen et al 1985), Sparks (1986) proposed a mechanism for extracting a signal of horizontal motor error from the anatomically coded signal observed in the SC. He suggested that the axons of all the collicular neurons residing along an isoΔH curve (same change in horizontal position) converged upon neurons at a specific depth in the cMRF (see Figure 1D). Neurons at that depth would discharge maximally before any saccade having a particular horizontal component, regardless of the vertical component of the movement. Cohen & Buttner-Ennever (1984) observed differential retrograde labeling in rostral and caudal areas of the SC following injection of anatomical tracer material into dorsal and ventral cMRF, but their data cannot be used to support this hypothesis because they looked for patterns of label in the SC related to the overall amplitude of the saccade, not the amplitude of the horizontal component. Other tectorecipient neurons are hypothesized to extract information about the vertical component of oblique saccades.

Van Gisbergen and colleagues (1985) suggested a third possibility for how signals of horizontal and vertical motor error are extracted from the collicular signals. They argue that a signal of the velocity of the combined movement (vectorial velocity) is computed first and later decomposed into separate velocity signals for the horizontal and vertical components (component velocity). These authors hypothesize that the vectorial pulse generator is composed of an array of LLBNs, each of which codes eye velocity in a certain direction. In Scudder's model of the saccadic system (Scudder 1988), LLBNs act as integrators of topographically coded inputs from the SC. This model predicts that the rate of increase of activity of LLBNs will be proportional to the size of the component in the on-direction of the cell. Unfortunately, available data do not permit an assessment of these hypotheses, but earlier studies failed to identify LLBNs displaying a relationship between the firing rates and saccadic velocity (van Gisbergen et al 1981).

The activity of some MLBs could be related to the vectorial velocity of oblique saccades. There are several reports (Luschei & Fuchs 1972, Henn & Cohen 1976, Keller 1977, Strassman et al 1986a,b, Nelson et al 1988) that some MLBs discharge maximally before oblique saccades.

The possibility that the activity of saccade-related burst neurons in the SC encodes vectorial velocity must also be considered. The SC would be a good candidate to coordinate oblique saccades, since information about both the horizontal and the vertical components is present and there appears to be some control of vectorial velocity as well. The velocity of saccades is reduced after collicular ablations (Wurtz & Goldberg 1972b, Schiller et al 1980) or reversible deactivation of collicular neurons (Hikosaka & Wurtz 1985, Lee et al 1988). Preliminary data (see above) are consistent with the hypothesis that the frequency of firing of neurons in the active population is related to the velocity of the saccade along a particular trajectory (vectorial velocity). What is needed is a mechanism by which component (horizontal or vertical) velocity can be scaled according to saccadic direction. A possible mechanism is illustrated in Figure 1C,D. To extend the earlier suggestion of Sparks (1986), neurons encoding the velocity of the horizontal (or vertical) component of oblique saccades by their frequency of firing could be formed by allowing these cells to receive excitatory synaptic inputs from all members of an iso$\Delta$H curve, but scaling the strength of the excitation by the location of the cell along the curve. For example, those members of the $8°$ iso$\Delta$H curve that are located rostrally and near the midline (cells associated with $8°$ horizontal movements having small vertical components) would provide greater excitatory drive than members located caudally (i.e. cells associated with $8°$ horizontal movements and large vertical components).

## Kinematic Constraints

BRAINSTEM SIGNALS    The implications of the rules of rotational kinematics were considered by Tweed & Vilis (1987) for models of the VOR and saccade generation. Current models of the VOR require a neural integrator to transform signals corresponding to eye velocity into signals of eye position. Models of the saccadic system use the same integrator to construct an eye position signal from the output of saccade-related burst neurons in the pons and midbrain. Tweed & Vilis (1987) argued that one-dimensional and two-dimensional saccadic models that subtract current eye position from desired eye position to generate signals cannot be easily extended to handle three-dimensional rotations. They proposed an alternative model implemented with quaternions in which angular velocity was multiplied by position feedback before integration. In their computer simulations, however, both the subtractive and multiplicative models performed well for saccades in the oculomotor range. Moreover, the major failure of the subtractive model discussed in their paper occurred when gaze shifts of $180°$ were attempted, a highly unusual circumstance.

REQUIRED TRANSFORMATIONS OF COLLICULAR SIGNALS    Robinson's model (1975b) of the saccadic system assumes that saccade targets are specified in terms of the final position of the eye in the orbit but does not indicate how these signals are generated. Revisions of Robinson's model, designed to accept collicular commands for changes in eye position, do not address the question of how Donders' law is implemented. Recently, Tweed & Vilis (1990) outlined a model of the saccadic system based upon a revival of Westheimer's (1957) use of quaternions for describing the rotations of the eye. In this model, a signal of the desired gaze vector is coded in head coordinates upstream from the SC. The gaze vector signal is passed through a Listing's law operator and the output is combined with a signal of current eye position to determine the site of collicular activation. The site of activation in the SC controls the rotation of the eye in three dimensions rather than, as usually assumed, in two. Listing's law operator, a constraint on craniotopic eye position, is placed upstream from the SC so that the operation occurs at a level at which saccades are still coded in craniotopic coordinates. Clearly, this model is shaped by the concern of devising a scheme by which Listing's law can be obeyed but, as noted above, it is possible that Listing's law could be approximated by simpler operations in which the torsional component is not computed explicitly.

This issue needs to be resolved. The computations to be performed by the saccadic system are quite different, depending upon which model is correct. According to a simple scheme in which commands are organized in horizontal and vertical coordinates without an explicit computation of the torsional component of a saccade, the topographic map of the SC specifies the amplitude and direction of the saccade. The collicular signal is decomposed into horizontal and vertical commands and routed to the appropriate prenuclear bursters. Just as the SC produces a "standard" command for a saccade of a given direction and amplitude, the prenuclear bursters would have the same output for a given saccade, regardless of initial eye position. Information about eye position would be needed only for the motoneuron recruitment mechanism, which is used by all conjugate subsystems.

According to the Tweed & Vilis model (1990), or other models in which there is an explicit torsion signal, identical saccades must result in the activation of different subsets of prenuclear bursters, depending on initial eye position. For example, the behavior of bursters during a 10° upward saccade started with the eyes 20° to the left must differ from their behavior when the 10° upward saccade begins with the eyes 20° to the right. Otherwise, the post-saccadic eye position would display torsion that would violate Listing's law. A Listing's law transformation predicts that the output of the prenuclear burst neurons should vary for identical saccades as the eyes assume different tertiary positions. Although no obvious differences in burst output have been noted (Keller 1974, van Gisbergen et al 1981), this has not been systematically studied. Furthermore, the magnitude of the expected differences is unknown and would be difficult to compute.

In order to conform to Listing's law, the transformation from the SC coding scheme to that of the prenuclear bursters would be much more complex than the simple two-dimensional scheme which ignored Listing's law. Consider two alternatives. The Listing's law operation might occur downstream of the SC. This would require that the output of the SC be re-routed to the prenuclear bursters, depending upon absolute eye position at the time of the onset of the saccade. Alternatively, the Listing's law operation might be upstream of the SC. This view, favored by Tweed & Vilis (1990) suggests that the SC encodes torsion as well as the horizontal and vertical components of the saccade. Further investigation might uncover a third (torsion) dimension of the collicular map. The Tweed &

Vilis model predicts that microstimulation of a single site in the SC will generate the same three-dimensional eye rotation regardless of the original position of the eye. Thus, eye rotations evoked by collicular stimulation should usually end up in final positions violating Donders' law. If saccades are specified by using only horizontal and vertical coordinates, however, the ocular torsion observed after stimulation-induced changes in eye position should fall within normal limits, regardless of the original eye position.

A more fundamental concern is whether Listing's law indeed imposes constraints on the signal transformations involving the SC. As we have noted, Listing's law is obeyed only approximately (Ferman et al 1987a,b). Furthermore, the analysis of extraocular muscle interaction by Robinson (1975a) implies that Listing's law may be approximated if the commands to the extraocular muscles were simply organized in horizontal and vertical coordinates. Finally, the recent report by Miller (1989) emphasizes that the geometry of the eyeball and muscles is complex and can have profound and unexpected effects on the signal transformations. Thus, it may be premature to speculate on the neuronal Listing's law operator before it is established, experimentally, that one is needed.

## SUMMARY AND CONCLUSIONS

Chronic unit recording experiments conducted over the past two decades have identified many functional classes of neurons with saccade-related activity that reside in a host of brainstem nuclei. Older models of the saccadic system were based upon the properties of only a few of these functional types of neurons. They described the putative flow of signals through the brainstem circuitry and specified some, but not all, of the signal transformations to be performed. How the necessary computations were performed by neurons was not always explicit.

Recent experiments investigating the neural control of saccadic eye movements and modifications of the original models are designed to fill in the details of the broad sketch of saccadic circuitry originally available. This review suggests one strategy for proceeding with this effort. Saccadic command signals observed in the SC require transformations to interface with the burst generators and motoneuron pools innervating the extraocular muscles. Specifying the signal transformations required for this interface should facilitate the design of experiments directed toward an understanding of the functional properties of cells located in nuclei intervening between the SC and the pulse/step circuitry, subsets of neurons that often have no role in models of the saccadic system. In this review, we hypothesize that neurons residing in various tectorecipient brainstem nuclei participate in one or more of the required signal transformations. The pathway from SC to cMRF and PPRF may be involved in the extraction of information about the amplitude and/or velocity of the horizontal component of oblique saccades. The pathway from SC to NRTP and cerebellar vermis may act selectively to generate signals compensating for the presaccadic orbital position. Finally, the activity of LLBNs and MLBs discharging maximally before oblique saccades may form the basis of computations required to match component velocity with overall saccade direction and amplitude. Although the data supporting these speculations are meager at present, such conjectures do form the basis of working hypotheses that can be tested experimentally.

We also considered the implications of kinematic constraints, especially Donders' and Listing's laws, for future investigations. Tweed & Vilis (1987, 1990) proposed models specifically designed to handle these constraints. In their models, eye position is represented on four oculomotor channels: three coding the vector components of eye position, and one carrying a signal inversely related to gaze eccentricity and torsion. Yet, other evidence suggest that simpler computations may suffice for the implementation of laws that are only approximately obeyed. Indeed, the question is whether the laws reflect "a special effort and programming by the nervous system, or are just an adventitious consequence of the mechanics of the peripheral oculomotor plant" (Ferman et al 1987b) when movements are specified in a two-dimensional coordinate system.

ACKNOWLEDGMENTS

The laboratories of the authors are supported by National Institutes of Health grants R01-EY01189, P30-EY03039, and R01-EY03463 and Alfred P. Sloan Foundation grant 86-10-7.

*Literature Cited*

Bahill, A. T., Latimer, J. R., Troost, B. T. 1980. Linear homeomorphic model for human movement. *IEEE Trans. Biomed. Eng.* BME27: 631–39

Baker, R., Berthoz, A., eds. 1977. *Control of Gaze by Brain Stem Neurons. Dev. Neurosci.* Vol. 1. New York: Elsevier/ North-Holland. 514 pp.

Berthoz, A., Grantyn, A., Droulez, J. 1986. Some collicular efferent neurons code saccadic eye velocity. *Neurosci. Lett.* 72: 289–94

Brodal, A., Jansen, J. 1946. The ponto-cerebellar projection in the rabbit and cat. Experimental investigations. *J. Comp. Neurol.* 84: 31–118

Buttner, U., Buttner-Ennever, J. A., Henn, V. 1977a. Vertical eye movement related unit activity in the rostral mesencephalic reticular formation of the alert monkey. *Brain Res.* 130: 239–52

Buttner, U., Hepp, K., Henn, V. 1977b. Neurons in the rostral mesencephalic and paramedian pontine reticular formation generating fast eye movements. See Baker & Berthoz 1977, pp. 309–18

Buttner-Ennever, J. A. 1977. Pathways from the pontine reticular formation to structures controlling horizontal and vertical eye movements in the monkey. See Baker & Berthoz 1977, pp. 89–98

Buttner-Ennever, J. A., Buttner, U. 1978.

A cell group associated with vertical eye movements in the rostral mesencephalic reticular formation of the monkey. *Brain Res.* 151: 31–47

Clark, M. R., Stark, L. 1974. Control of human eye movements: II. A model for the extraocular plant mechanism. *Math. Biosci.* 20: 213–38

Cohen, B., Buttner-Ennever, J. A. 1984. Projections from the superior colliculus to a region of the central mesencephalic reticular formation (cMRF) associated with horizontal saccadic eye movements. *Exp. Brain Res.* 57: 167–76

Cohen, B., Matsuo, V., Raphan, T., Waitzman, D., Fradin, J. 1985. Horizontal saccades induced by stimulation of the central mesencephalic reticular formation. *Exp. Brain Res.* 57: 605–16

Cohen, B., Waitzman, D. M., Buttner-Ennever, J. A., Matsuo, V. 1986. Horizontal saccades and the central mesencephalic reticular formation. *Prog. Brain Res.* 64: 243–56

Collins, C. C. 1975. The human oculomotor control system. See Lennerstrand & Bachy-Rita 1975, pp. 145–80

Cook, G., Stark, L. 1967. Derivation of a model for the human eye-positioning mechanisms. *Bull Math. Biophys.* 29: 153–74

Crandall, W. F., Keller, E. L. 1985. Visual and oculomotor signals in nucleus reticularis tegmenti pontis in alert monkey. *J. Neurophysiol.* 54: 1326–45

Evinger, C., Kaneko, C. R. S., Fuchs, A. F. 1981. Oblique saccadic eye movements of the cat. *Exp. Brain Res.* 41: 370–79

Evinger, C., Kaneko, C. R. S., Johanson, G. W., Fuchs, A. F. 1977. Omnipauser cells in the cat. See Baker & Berthoz 1977, pp. 337–40

Ferman, L., Collewijn, H., Van den Berg, A. V. 1987a. A direct test of Listing's law. I. Human ocular torsion measured in static tertiary positions. *Vision Res.* 27: 929–38

Ferman, L., Collewijn, H., Van den Berg, A. V. 1987b. A direct test of Listing's law. II. Human ocular torsion measured under dynamic conditions. *Vision Res.* 27: 939–51

Fuchs, A. F., Kaneko, C. R. S., Scudder, C. A. 1985. Brainstem control of saccadic eye movements. *Annu. Rev. Neurosci.* 8: 307–37

Fuchs, A. F., Luschei, E. S. 1970. Firing patterns of abducens neurons of alert monkeys in relationship to horizontal eye movement. *J Neurophysiol.* 33: 382–92

Fuchs, A. F., Luschei, E. S. 1971. The activity of single trochlear nerve fibers during eye movements in the alert monkey. *Exp. Brain Res.* 13: 78–89

Fuchs, A. F., Scudder, C. A., Kaneko, C. R. S. 1988. Discharge patterns and recruitment order of identified motoneurons and internuclear neurons in the monkey abducens nucleus. *J. Neurophysiol.* 60: 1874–95

Gerrits, N. M., Voogd, J. 1986. The nucleus reticularis tegmenti pontis and the adjacent rostral paramedian reticular formation: Differential projections to the cerebellum and the caudal brain stem. *Exp. Brain Res.* 62: 29–45

Gerrits, N. M., Voogd, J. 1987. The projection of the nucleus reticularis tegmenti pontis and adjacent regions of the pontine nuclei to the central cerebellar nuclei in the cat. *J. Comp. Neurol.* 258: 52–69

Goldstein, H. P. 1983. *The neural encoding of saccades in the rhesus monkey.* PhD thesis. Johns Hopkins Univ., Baltimore, Md

Goldstein, H. P., Robinson, D. A. 1986. Hysteresis and slow drift in abducens unit activity. *J. Neurophysiol.* 55: 1044–56

Grossman, G. E., Robinson, D. A. 1988. Ambivalence in modelling oblique saccades. *Biol. Cybern.* 58: 13–18

Guitton, D., Mandl, G. 1980. Oblique saccades of the cat: A comparison between the durations of horizontal and vertical components. *Vision Res.* 20: 875–81

Harting, J. K. 1977. Descending pathways from the superior colliculus: An autoradiographic analysis in the rhesus monkey (*Macaca mulatta*). *J. Comp. Neurol.* 173: 583–612

Henn, V., Cohen, B. 1976. Coding of information about rapid eye movements in the pontine reticular formation of alert monkeys. *Brain Res.* 108: 307–25

Hepp, K., Henn, V. 1983. Spatio-temporal recoding of rapid eye movement signals in the monkey paramedian pontine reticular formation (PPRF). *Exp. Brain Res.* 52: 105–20

Hepp, K., Henn, V. 1985. Iso-frequency curves of oculomotor neurons in the rhesus monkey. *Vision Res.* 25: 493–99

Hepp, K., Vilis, T., Henn, V. 1986. Vertical and torsional rapid eye movement generation in the riMLF. *Soc. Neurosci. Abstr.* 12: 1187

Highstein, S. M., Cohen, B., Matsunami, K. 1974. Monosynaptic projections from the pontine reticular formation to the IIIrd nucleus in the cat. *Brain Res.* 75: 340–44

Highstein, S. M., Maekawa, K., Steinacker, A., Cohen, B. 1976. Synaptic input from the pontine reticular nuclei to abducens motoneurons and internuclear neurons in the cat. *Brain Res.* 112: 162–67

Hikosaka, O., Wurtz, R. H. 1985. Modification of saccadic eye movements by

GABA-related substances. I. Effect of muscimol and bicuculline in monkey superior colliculus. *J. Neurophysiol.* 53: 266–91

Hikosaka, O., Wurtz, R. H. 1986. Saccadic eye movements following injection of lidocaine into the superior colliculus. *Exp. Brain Res.* 61: 531–39

Huerta, M. F., Harting, J. K. 1984. The mammalian superior colliculus: Studies of its morphology and connections. In *Comparative Neurology of the Optic Tectum*, ed. H. Vanegas, pp. 687–773. New York: Plenum

Keller, E. L. 1974. Participation of medial pontine reticular formation in eye movement generation in monkey. *J. Neurophysiol.* 37: 316–31

Keller, E. L. 1977. Control of saccadic eye movements by midline brain stem neurons. See Baker & Berthoz 1977, pp. 327–36

Keller, E. L. 1979. Colliculoreticular organization in the oculomotor system. *Progr. Brain Res.* 50: 725–34

Keller, E. L. 1980. Oculomotor specificity within subdivisions of the brain stem reticular formation. In *The Reticular Formation Revisited*, ed. M. A. B. Brazier, pp. 227–40. New York: Raven

Keller, E. L. 1981. Oculomotor neuron behavior. See Zuber 1981, pp. 1–20

King, W. M., Fuchs, A. F. 1977. Neuronal activity in the mesencephalon related to vertical eye movements. See Baker & Berthoz 1977, pp. 319–26

King, W. M., Fuchs, A. F. 1979. Reticular control of vertical saccadic eye movements by mesencephalic burst neurons. *J. Neurophysiol.* 42: 861–76

King, W. M., Fuchs, A. F., Magnin, M. 1981. Vertical eye movement-related responses of neurons in midbrain near interstitial nucleus of Cajal. *J. Neurophysiol.* 46: 549–62

King, W. M., Lisberger, S. G., Fuchs, A. F. 1986. Oblique saccadic eye movements of primates. *J. Neurophysiol.* 56: 769–84

Lee, C., Rohrer, W. H., Sparks, D. L. 1988. Population coding of saccadic eye movements by neurons in the superior colliculus. *Nature* 332: 357–60

Lennerstrand, G., Bach-y-Rita, P., eds. 1975. *Basic Mechanisms of Ocular Motility and Their Clinical Implications*. Oxford: Pergamon. 584 pp.

Luschei, E. S., Fuchs, A. F. 1972. Activity of brain stem neurons during eye movements of alert monkeys. *J. Neurophysiol.* 35: 445–61

McElligott, J. G., Gochin, P. M. 1986. Coding properties of cat cerebellar vermal Purkinje cells during saccadic eye movements. *Int. Congr. Physiol. Sci.*, Satellite Meet. (Abstr.)

McElligott, J. G., Keller, E. L. 1984. Cerebellar vermis involvement in monkey saccadic eye movements: Microstimulation. *Exp. Neurol.* 86: 543–58

Miller, J. M. 1989. Functional anatomy of normal human rectus muscles. *Vision Res.* 29: 223–40

Munoz, D., Guitton, D. 1986. Presaccadic burst discharges of tecto-reticulo-spinal neurons in the alert head-free and -fixed cat. *Brain Res.* 398: 185–90

Nakayama, K. 1975. Coordination of extraocular muscles. See Lennerstrand & Bach-y-Rita 1975, pp. 193–208

Nakayama, K. 1983. Kinematics of normal and strabismic eyes. In *Vergence Eye Movements: Basic and Clinical Aspects*, ed. C. M. Schor, K. J. Ciuffreda, pp. 543–64. Boston: Butterworths

Nelson, J. S., Goldberg, S. J., McClung, J. R. 1986. Motoneuron electrophysiological and muscle contractile properties of superior oblique motor units in cat. *J. Neurophysiol.* 55: 715–26

Nelson, J. S., Hartwich-Young, R., Sparks, D. L. 1988. Medium-lead burst neuron activity during oblique saccades. *Soc. Neurosci. Abstr.* 14: 955

Optican, L. M., Miles, F. A. 1985. Visually induced adaptive changes in primate saccadic oculomotor control signals. *J. Neurophysiol.* 54: 940–58

Optican, L. M., Robinson, D. A. 1980. Cerebellar-dependent adaptive control of primate saccadic system. *J. Neurophysiol.* 44: 1058–76

Pelisson, D., Prablanc, C. 1988. Kinematics of centrifugal and centripetal saccadic eye movements in man. *Vision Res.* 28: 87–94

Raphan, T., Cohen, B. 1978. Brainstem mechanisms for rapid and slow eye movements. *Annu. Rev. Physiol.* 40: 527–52

Raybourn, M. S., Keller, E. L. 1977. Colliculoreticular organization in primate oculomotor system. *J. Neurophysiol.* 40: 861–78

Ritchie, L. 1976. Effects of cerebellar lesions on saccadic eye movements. *J. Neurophysiol.* 39: 1246–56

Robinson, D. A. 1964. The mechanics of human saccadic eye movement. *J. Physiol.* 174: 245–64

Robinson, D. A. 1970. Oculomotor unit behavior in the monkey. *J. Neurophysiol.* 33: 393–404

Robinson, D. A. 1972a. Eye movements evoked by collicular stimulation in the alert monkey. *Vision Res.* 12: 1795–1808

Robinson, D. A. 1972b. On the nature of visual-oculomotor connections. *Invest. Ophthalmol.* 11: 497–503

Robinson, D. A. 1975a. A quantitative analysis of extraocular muscle cooperation and squint. *Invest. Ophthalmol.* 14: 801–25

Robinson, D. A. 1975b. Oculomotor control signals. See Lennerstrand & Bach-y-Rita 1975, pp. 337–74

Robinson, D. A. 1981a. Control of eye movements. In *Handbook of Physiology. The Nervous System. Motor Control*, ed. J. M. Brookhart, V. B. Mountcastle, V. B. Brooks, S. R. Geiger. 2: 1275–1320. Bethesda: Am. Physiol. Soc.

Robinson, D. A. 1981b. Models of the mechanics of eye movements. See Zuber 1981, pp. 21–42

Robinson, D. A., Zee, D. S. 1981. Theoretical considerations of the function and circuitry of various rapid eye movements. *Prog. Oculomotor Res. Dev. Neurosci.* 12: 3–12

Rohrer, W. H., White, J. M., Sparks, D. L. 1987. Saccade-related burst cells in the superior colliculus: Relationship of activity with saccadic velocity. *Soc. Neurosci. Abstr.* 13: 1092

Schiller, P. H. 1970. The discharge characteristics of single units in the oculomotor and abducens nuclei of the unanesthetized monkey. *Exp. Brain Res.* 10: 347–62

Schiller, P. H., Koerner, F. 1971. Discharge characteristics of single units in superior colliculus of the alert rhesus monkey. *J. Neurophysiol.* 34: 920–36

Schiller, P. H., Stryker, M. 1972. Single-unit recording and stimulation in superior colliculus of the alert rhesus monkey. *J. Neurophysiol.* 35: 915–24

Schiller, P. H., True, S. D., Conway, J. L. 1980. Deficits in eye movements following frontal eye-field and superior colliculus ablations. *J. Neurophysiol.* 44: 1175–89

Scudder, C. A. 1988. A new local feedback model of the saccadic burst generator. *J. Neurophysiol.* 59: 1455–75

Segraves, M. A., Goldberg, M. E. 1984. Initial orbital position affects the trajectories of large saccades evoked by electrical stimulation of the monkey superior colliculus. *Soc. Neurosci. Abstr.* 10: 389

Sparks, D. L. 1978. Functional properties of neurons in the monkey superior colliculus: Coupling of neuronal activity and saccade onset. *Brain Res.* 156: 1–16

Sparks, D. L. 1986. Translation of sensory signals into commands for control of saccadic eye movements: Role of primate superior colliculus. *Physiol. Rev.* 66: 118–71

Sparks, D. L. 1990. Neural commands for the control of saccadic eye movements: required transformations of signals observed in the superior colliculus. In *Neural Programming*, ed. M. Ito

Sparks, D. L., Hartwich-Young, R. 1988. The deep layers of the superior colliculus. See Wurtz & Goldberg 1988, pp. 213–55

Sparks, D. L., Holland, R., Guthrie, B. L. 1976. Size and distribution of movement fields in the monkey superior colliculus. *Brain Res.* 113: 21–34

Sparks, D. L., Jay, M. F. 1986. The functional organization of the primate superior colliculus: A motor perspective. *Prog. Brain Res.* 64: 235–42

Sparks, D. L., Mays, L. E. 1980. Movement fields of saccade-related burst neurons in the monkey superior colliculus. *Brain Res.* 190: 39–50

Strassman, A., Highstein, S. M., McCrea, R. A. 1986a. Anatomy and physiology of saccadic burst neurons in the alert squirrel monkey. I. Excitatory burst neurons. *J. Comp. Neurol.* 249: 337–57

Strassman, A., Highstein, S. M., McCrea, R. A. 1986b. Anatomy and physiology of saccadic burst neurons in the alert squirrel monkey. II. Inhibitory burst neurons. *J. Comp. Neurol.* 249: 358–80

Tweed, D., Vilis, T. 1985. A two dimensional model for saccade generation. *Biol. Cybern.* 52: 219–27

Tweed, D., Vilis, T. 1987. Implications of rotational kinematics for the oculomotor system in three dimensions. *J. Neurophysiol.* 58: 832–49

Tweed, D., Vilis, T. 1990. The superior colliculus and spatiotemporal translation in the saccadic system. *Neural Networks.* In press

van Gisbergen, J. A. M. 1988. Models of saccadic control. See Wurtz & Goldberg 1988, pp. 1187

van Gisbergen, J. A. M., Robinson, D. A., Gielsen, S. 1981. A quantitative analysis of generation of saccadic eye movements by burst neurons. *J. Neurophysiol.* 45: 417–22

van Gisbergen, J. A. M., van Opstal, A. J., Schoenmakers, J. J. M. 1985. Experimental test of two models for the generation of oblique saccades. *Exp. Brain Res.* 57: 321–36

Vilis, T., Hepp, K., Schwarz, U., Henn, V., Haas, H. 1986. Unilateral riMLF lesions impair saccade generation along specific vertical planes. *Soc. Neurosci. Abstr.* 12: 1187

Waitzman, D. M. 1982. *Burst neurons in the mesencephalic reticular formation (MRF) associated with saccadic eye movements.* PhD thesis. City Univ. New York, NY

Waitzman, D. M., Ma, T. P., Optican, L. M., Wurtz, R. H. 1988. Superior colliculus neurons provide the saccadic motor error signal. *Exp. Brain Res.* 72: 649–52

Westheimer, G. 1954. Mechanism of saccadic eye movements. *Am. Med. Assoc. Arch. Ophthal.* 52: 710–24

Westheimer, G. 1957. Kinematics of the eye. *J. Opt. Soc. Am.* 47: 967–74

Westheimer, G. 1981. Donders', Listings's and Herings's Laws and their implications. See Zuber 1981, pp. 149–60

Westheimer, G., McKee, S. P. 1973. Failure of Donders' law during smooth pursuit eye movement. *Vision Res.* 13: 2145–53

Wurtz, R. H., Goldberg, M. E. 1972a. Activity of superior colliculus in behaving monkey. III. Cells discharging before eye movements. *J. Neurophysiol.* 35: 575–86

Wurtz, R. H., Goldberg, M. E. 1972b. Activity of superior colliculus in behaving monkey. IV. Effects of lesions on eye movements. *J. Neurophysiol.* 35: 587–96

Wurtz, R. H., Goldberg, M. E., eds. 1988. *Reviews of Oculomotor Research*, Vol. 3, *The Neurobiology of Saccadic Eye Movements.* Amsterdam: Elsevier

Zuber, B. L., ed. 1981. *Models of Oculomotor Behavior and Control.* Boca Raton, Fla.: CRC Press. 289 pp

*Annu. Rev. Neurosci. 1990. 13:337–56*

# CALCIUM CHANNELS IN VERTEBRATE CELLS

*Peter Hess*

Department of Cellular and Molecular Physiology, and Program in Neuroscience, Harvard Medical School, Boston, Massachusetts 02115

## INTRODUCTION

Ca channels are transmembrane proteins that in the open conformation allow the passive flux of Ca ions across the membrane, down the electrochemical gradient. In this review I deal only with voltage-activated, Ca-selective channels in the surface membrane, and restrict the discussion largely to Ca channels in vertebrates. I concentrate on new insights into the functional molecular properties revealed by single channel recording techniques, on structural information gained by biochemical and molecular biological techniques, and on newly recognized pathways of Ca channel modulation.

For a historical account of the discovery of Ca permeability, the reader should consult previous reviews (Reuter 1973, Hagiwara & Byerly 1981, Kostjuk 1981, Tsien 1983). By focusing on the elementary properties of Ca channels, I also fall short of a description of the many important physiological roles of Ca channels. The various aspects of the control of intracellular Ca in neurons, including the role played by neuronal Ca channels, have recently been discussed in a special issue of Trends in Neuroscience (see Tsien et al 1988).

Voltage-sensitive Ca channels are found in most cell types studied today (for review see Bean 1989*a*), and are not restricted to classically excitable cells like neurons, muscle, and heart. One of the significant advances in the understanding of Ca channels in recent years has been the realization that most cell membranes contain several types of voltage-activated Ca channels, which can be classified according to their biophysical properties and pharmacological sensitivities.

337

0147–006X/90/0301–0337$02.00

## The Skeletal Muscle Dihydropyridine Receptor

1,4-dihydropyridine (DHP) Ca channel ligands have been used extensively to study DHP-sensitive Ca channels (for review see Janis et al 1987, Hosey & Lazdunski 1988). DHP receptors are found in T-tubular membranes from skeletal muscle at a density 50–100 times higher than in any other tissue (Fosset et al 1983, Glossman et al 1983). Skeletal muscle has therefore been used as the source for successful solubilization and purification of the DHP receptor. Initial reports described a single large subunit of 150–170 kDa and several smaller ones ranging from 30–50 kDa (Glossmann & Ferry 1983, Curtis & Catterall 1984, Borsotto et al 1984, 1985). More recently, several laboratories have demonstrated that two large subunits, $\alpha_1$ and $\alpha_2$, copurify with equimolar stoichiometry (Leung et al 1987, Hosey et al 1987, Takahashi et al 1987, Tanabe et al 1987, Vaghy et al 1987). Under nonreducing conditions, the two subunits have very similar apparent molecular masses of 165–175 kDa. Disulfide reducing agents do not alter the apparent molecular mass of the $\alpha_1$ polypeptide but decrease that of the $\alpha_2$ polypeptide to 140–150 kDa and lead to the appearance of small polypeptides (24–30 kDa, see e.g. Schmid et al 1986, Takahashi et al 1987), which Takahashi et al (1987) call $\delta$ subunit. Thus $\alpha_2$ and $\delta$ subunits appear to be disulfide linked. There is general consensus (for review see Hosey & Lazdunski 1988) that the $\alpha_1$ polypeptide contains the DHP as well as the phenylalkylamine receptor. Two low molecular weight subunits of $\sim 50$ kDa ($\beta$) and $\sim 30$ kDa ($\gamma$) are present under nonreducing and reducing conditions (Glossmann & Ferry 1983, Curtis & Catterall 1984, Borsotto et al 1985, Leung et al 1987, Flockerzi et al 1986a, Vaghy et al 1987). $\alpha_1$ and $\beta$ subunits can be phosphorylated by cyclic AMP dependent kinase (Curtiss & Catterall 1985, Hosey et al 1986, Flockerzi et al 1986b, Takahashi et al 1987), an intrinsic kinase present in triads (Imagawa et al 1987), a multifunctional Ca-calmodulin-dependent kinase (Hosey et al 1986), and by protein kinase C (Glossman et al 1988). Hydrophobic labeling (Takahashi et al 1987) has been found to stain $\alpha_1$ and $\gamma$ subunits strongly and the $\alpha_2$ and $\delta$ polypeptide weakly, but not the $\beta$ subunit. $\alpha_2$, $\gamma$, and $\delta$ subunits are glycosylated, whereas $\alpha_1$ and $\beta$ subunits are not. Based on these biochemical observations and on results from the amino acid sequence data of the $\alpha_1$ subunit (Tanabe et al 1987, Ellis et al 1988), Takahashi et al (1987) proposed a model for the subunit structure of the DHP-sensitive Ca channel in skeletal muscle, in which the $\alpha_1$ subunit forms the ion channel and is surrounded by the $\alpha_2$ and $\gamma$ subunits. The $\beta$ subunit, because of its phosphorylation sites and lack of hydrophobicity, is postulated to attach noncovalently to the cytoplasmic side of $\alpha_1$, whereas the $\delta$ subunit is disulfide-linked to the $\alpha_2$ subunit and extends to the external surface, because of its glycosylation sites.

Amino acid sequences have been obtained by molecular cloning techniques for the $\alpha_1$ and $\alpha_2$ subunits (Tanabe et al 1987, Ellis et al 1988). The $\alpha_1$ subunit bears striking homology to the voltage-activated Na channel. The hydropathy profile suggests that the two channels have a similar general architecture consisting of four repeated homologous domains. Tanabe et al (1987) propose that each domain has six membrane-spanning segments (S1–S6). Segment S4 in each repeat contains 5 or 6 Arg or Lys residues at every third position, a sequence that is highly conserved in the corresponding S4 segment of Na channels (Noda et al 1984) and K channels (Papazian et al 1987, Pongs et al 1988). The S4 segment must be regarded as a good candidate for the voltage-sensing region of the channel protein (Noda et al 1984, Tanabe et al 1987). Site-directed mutagenesis of the positive charges in S4 should soon corroborate or disprove this hypothesis. The sequence of the $\alpha_1$ subunit contains seven potential phosphorylation sites for cyclic AMP–dependent kinase, all of which are located at the cytoplasmic side according to the folding scheme proposed on the basis of the hydropathy profile (Tanabe et al 1987). No further inferences about the tertiary protein structure are obvious from the primary sequence. Notably there are no Ca-binding sites corresponding to the EF-hand structure on intracellular Ca-binding proteins (Kretsinger 1980).

The $\alpha_2$ subunit shows no significant homology to ion channel or receptor proteins (Ellis et al 1988). As expected from the biochemical analysis (Takahashi et al 1987), the $\alpha_2$ subunit has a higher overall hydrophilicity than the $\alpha_1$ subunit and may contain as little as three transmembrane domains (Ellis et al 1988).

The finding that the $\alpha_1$ subunit contains the receptors for DHP and phenylalkylamines and that its structure resembles that of known channel proteins, strongly suggests that the $\alpha_1$ subunit represents an essential part of the DHP-sensitive Ca channel. Little is known at this point, however, about the subunit composition of functional DHP receptor Ca channels. Purified DHP receptors, presumably containing all subunits apparent after denaturation on SDS gels, have been shown to be functional in assays of Ca and Ba uptake into reconstituted membrane vesicles (Curtis & Catterall 1986) and in single channel measurements from artificial bilayers after incorporation of purified DHP receptors (Flockerzi et al 1986b, Smith et al 1987, Talvenheimo et al 1987, Horne et al 1988, Hymel et al 1988). The reconstituted purified channels, however, showed a variety of features not expected for native DHP-sensitive Ca channels, such as persistent activation at negative membrane potentials and a variety of channel conductances not commonly found in native channels. Possible explanations for the functional differences between native and reconstituted purified channels include variable subunit stoichiometry of reconstituted channels, partial denaturation during the purification steps, and effects of detergent not completely removed prior to the functional reconstitution.

No published reports exist on functional expression of Ca channels in frog oocytes after injection of mRNA obtained from the cloned cDNA of the $\alpha_1$ or $\alpha_2$ subunits. When cDNA coding for the $\alpha_1$ subunit was injected into the nuclei of myotubes from mice with muscular dysgenesis, however, functional DHP sensitive Ca channels as well as normal excitation-contraction (EC) coupling were restored (Tanabe et al 1988). Dysgenic mice had previously been shown to lack EC-coupling and DHP-sensitive Ca currents (Pincon-Raymond et al 1985, Beam et al 1986, Rieger et al 1987), probably because of mutation of a single locus in the gene coding for the $\alpha_1$ subunit (Tanabe et al 1988). The observation that EC-coupling depends on functional DHP receptors, but not on their ability to conduct an inward Ca current (Tanabe et al 1988), strongly supports the hypothesis proposed by Rios & Brum (1987) of a dual role of the DHP receptor as a Ca channel and as the voltage sensor for EC-coupling. One of the important questions that remain to be answered is whether the same molecule actually serves as Ca channel and voltage sensor for EC-coupling or whether the DHP receptor can exist in two forms, functioning either as Ca channel or as voltage sensor. Since in dysgenic mice the $\alpha_1$ subunit is probably the only missing subunit (Knudson et al 1989), restitution of normal function by the $\alpha_1$ subunit does not allow further conclusions about the subunit composition of a functional DHP receptor channel.

## Dihydropyridine-Sensitive Ca Channels in Cells Other than Skeletal Muscle (L-Type Ca Channels)

Several lines of evidence suggest that DHP-sensitive Ca channels from non-skeletal muscle cells are distinct structural and functional entities.

1. Dysgenic mice lack DHP receptors in muscle, but have normal DHP receptor densities and DHP-sensitive Ca channels in heart and sensory neurons (Pincon-Raymond et al 1985, Beam et al 1986), a finding that strongly suggests that different genes are coding for the two channel types (Tanabe et al 1988).
2. RNA blot analysis with probes to the $\alpha_1$ subunits revealed the expected band at 6.5 kb for total RNA from skeletal muscle but showed only a very weak hybridizing signal for RNA from heart and none for RNA from ileum or brain (Ellis et al 1988), despite the well-known presence of DHP receptors in these tissues. In contrast, a probe to the $\alpha_2$ subunit revealed a hybridizing signal in the total RNA from each of these sources (Ellis et al 1988).

3. Although DHP-sensitive Ca currents from skeletal muscle and other tissues share a generally similar pharmacology and a similar mechanism by which they select for Ca over other ions (Tsien et al 1987), the skeletal muscle channels differ significantly from those in other cells in the two major functional characteristics of ion channels: The unitary conductance is smaller (Coronado & Affolter 1986, Rosenberg et al 1986) and channel gating is at least a factor of ten slower in the skeletal muscle channel than in the DHP-sensitive channel from other tissues.

Significant sequence homology between the DHP receptors in skeletal muscle and other tissue, is, however, likely, since some poly- and monoclonal antibodies raised against skeletal muscle DHP receptors recognize cardiac and brain channels (Schmid et al 1986, Cooper et al 1987, Takahashi & Catterall 1987). Furthermore, hybridization of cardiac total mRNA with cDNA encoding two short (57 and 80 nucleotides, respectively) segments of the $\alpha_1$ subunit of the skeletal muscle DHP receptor specifically abolishes the capability of the cardiac mRNA to induce expression of cardiac DHP-sensitive Ca currents in frog oocytes (Lotan et al 1989).

Although most of the biochemical characterization of DHP receptors has been carried out on receptors from skeletal muscle, most functional studies, particularly at the level of single channels in intact cells, were performed on cardiac, secretory, and neuronal cells. DHP-sensitive channels in these tissues share most of the important functional properties, and I refer to them as L-type Ca channels (Nowycky et al 1985a, Nilius et al 1985). Alternative names include "high voltage activated channels" (HVA; see Carbone & Lux 1984), "$I_{slow}$" (Bean 1985), and "fast deactivating channels" (Armstrong & Matteson 1985). This classification is based on functional similarities and should not imply that L-type Ca channels are all products of the same gene.

The most striking functional similarities among L-type Ca channels are (a) the unitary conductances with mono- and divalent charge carriers; (b) sensitivities to inorganic and organic blockers; (c) activation range at positive test potentials; (d) very slow inactivation when Ca is not the charge carrier; (e) steady-state inactivation only at positive holding potentials; (f) a stereotypical response to DHP Ca channel agonists; (g) lability of channel activity in a cell-free environment. Whether other characteristics described for L-type channels in certain tissues will hold for all types of L-type Ca channels remains to be seen. Such properties include Ca-dependent inactivation (Eckert & Chad 1984) and the increase of L-type channel activity by cyclic AMP-dependent protein kinase, known to exist in L-type channels from cardiac and $GH_3$ cells (e.g. Cachelin et al 1983, Brum et al

1984, Kameyama et al 1985, Tsien et al 1986, Armstrong & Eckert 1987). The voltage-dependent inactivation of L-type channels appears to be one of the most variable parameters between individual L-type channels. Generally, the most rapidly inactivating L-type Ca currents are found in the heart, whereas neuronal or secretory L-type Ca channels may not inactivate in a voltage-dependent way at all. Even among individual L-type Ca channels from the same source, however, great variability in the rate of inactivation has been reported (Cavalie et al 1986).

## T-Type Ca Channels

T-type Ca channels, originally described in vertebrate cells by Carbone & Lux (1984), have now been recognized in a variety of excitable and nonexcitable cells. Alternative names for these T-type channels include "low threshold activated" (Carbone & Lux 1984), "$I_{fast}$" (Bean 1985), and "slowly deactivating channels" (Armstrong & Matteson 1985). T-type channels from all tissues share a small unitary conductance with similar Ca and Ba permeabilities, rapidly transient channel kinetics, slow deactivation of tail currents, activation at negative potentials, steady inactivation even for small depolarizations, and stability of channel activity in cell-free conditions. T-type channels have mainly been described as resistant to DHP drugs, but recently DHP sensitivity was reported (Van Skiver et al 1989). Although a few drugs like tetramethrin (Hagiwara et al 1988), amiloride (Tang et al 1988), diphenylhydantoin (Yaari et al 1987), and octanol (Llinas & Yarom 1986) have been shown to act on T-type Ca channels, none of them is selective for T-type channels (C. Chen and P. Hess, unpublished results). Because of the lack of specific ligands, no information is available about the biochemical composition of T-type Ca channels. They appear to be separate molecular entities, since functional expression of T-type Ca channels can be selectively inhibited by transforming oncogenes under conditions where L-type currents remain unchanged (Chen et al 1988). Similarly, mRNA from heart retains the ability to induce the expression of a transient inward Ca current in oocytes after hybridization with segments of cDNA that eliminate L-type channel expression (Lotan et al 1989). The time course of inactivation of the transient current induced in the oocytes, however, is much slower than that expected for T-type current, and whether this current is indeed carried by T-type Ca channels remains to be seen.

## N-Type Ca Channels

N-type Ca channels, first described in chick dorsal root ganglion (DRG) cells by Nowycky et al (1985a), are a heterogenous group of Ca channels that are activated at positive potentials, insensitive to DHP drugs, and substantially, although not completely, inactivated at positive holding potentials (Fox et al 1987a,b, Kostyuk et al 1988, Plummer et al 1989, Lemos & Nowycky 1989). N-type Ca channels may be neuron specific, since they have only been described in cells of neuronal origin, and N-type channel expression in PC12 cells is enhanced by nerve growth factor (Plummer et al 1989, Porzig et al 1989, but also see Streit & Lux 1987). Initially N-type channels had been reported to inactivate rapidly and completely during a strong depolarization (Nowycky et al 1985a). Therefore, the decaying component of whole cell Ca or Ba currents was taken as N-type current and any maintained component was ascribed to current through L-type channels (Fox et al 1987a). This separation procedure assigns relatively large components (40–70%) of whole cell Ba current elicited from negative potentials to current carried by L-type channels. Recently, it has become clear, however, that in many neuronal cells, current carried by N-type Ca channels can contain a large maintained current component even at the end of test depolarizations of several hundred miliseconds (Plummer et al 1989), and that N-type Ca channels can give rise to a non-inactivating current component when elicited from depolarized holding potentials (Plummer et al 1989). Thus, separation of current components based only on the kinetics of inactivation may lead to erroneous conclusions about the proportion of whole cell current carried by N- or L-type Ca channels. For example, the initial conclusion that $\omega$-conotoxin (see Kerr & Yoshikami 1984, Olivera et al 1985) blocks both N- and L-type Ca channels persistently in chick DRG, rat sympathetic (superior cervical ganglion, SCG), and hippocampal neurons (McCleskey et al 1987) was based on the finding that in all these cells, $\omega$-conotoxin blocks a decaying as well as a maintained component of current. However, when DHP agonists were used to identify selectively a current component exclusively carried by L-type channels, then $\omega$-conotoxin was found not to block L-type Ca channels in SCG and PC12 cells (Plummer et al 1989). Similarly, Kasai et al (1987) found no evidence for persistent block of $\omega$-conotoxin on L-type channels in chick DRG neurons, and recently Aosaki & Kasai (1989) showed at the single channel level that L-type channels are indeed not blocked irreversibly by the toxin. Possible explanations for the different $\omega$-conotoxin sensitivities postulated by different groups might include the following:

1. The use of synthetic (Kasai et al 1987, Plummer et al 1989, Aosaki & Kasai 1989) vs. purified native $\omega$-conotoxin (McCleskey et al 1987), might affect the results. The purified material might contain other toxins, or, alternatively, the synthetic toxin might have lost the ability of the native toxin to recognize L-type channels.

2. Millimolar concentrations of divalent charge carriers might anta-

gonize ω-conotoxin binding to L-type channels more than N-type channels. Since ω-conotoxin is highly basic, its binding to the channel protein is likely to involve electrostatic interactions that can be expected to be weakened by divalent ions and increasing ionic strength, as shown both in electrical recordings (Oyama et al 1987) and in ω-conotoxin binding studies (Abe et al 1986, Knaus et al 1987, Barhanin et al 1988, Marqueze et al 1988). However, the experimental conditions in the studies by McCleskey et al (1987), Kasai et al (1987), Aosaki & Kasai (1989), and Plummer et al (1989) are sufficiently similar to make this argument an unlikely explanation for the claimed differences.

3. The most likely interpretation of all the published data is that N-type Ca channels predominate the whole cell currents, contributing both steady and decaying current components, and that L-type channels are not blocked by ω-conotoxin and only carry a small fraction of the whole cell Ba current. This view is also consistent with the sensitivity of whole cell Ca currents to DHP agonists. Although currents at the peak of the current-voltage relationship are quite insensitive to DHP agonists (Plummer et al 1989, Jones & Marks 1989, Aosaki & Kasai 1989), thus suggesting a small proportion of L-type channels, the presence of normally DHP sensitive L-type Ca channels is revealed with appropriate pulse protocols: At the foot of the activation curve, where N-type channels are barely opening, DHP agonists induce a large increase of the current (Holz et al 1988, Jones & Marks 1989). Similarly, the tail currents following repolarization are greatly prolonged by DHP agonists (Plummer et al 1989), as expected from the known effects of DHP agonists on single L-type channel tail currents (Hess et al 1984, Nowycky et al 1985b, Fox et al 1987b, Plummer et al 1989).

Even if ω-conotoxin does not block L-type Ca currents from any neuronal source, ω-conotoxin cannot be used as a selective blocker for N-type Ca channels, since in addition to the ω-conotoxin sensitive N-type channels, a class of high threshold activated, DHP resistant (N-type) Ca channels has been described that is only reversibly and weakly inhibited by ω-conotoxin at high concentrations (Plummer et al 1989).

The interpretation that N-type channels can carry most of the macroscopic inward Ba current in DRG or SCG neurons is also consistent with the observations of Swandulla & Armstrong (1988), who failed to find evidence for more than one high-threshold-activated current component in chick DRG, based on careful analysis of tail deactivation. All the current in their experiments probably was carried by N-type channels, since Swandulla & Armstrong (1988) reported that the DHP agonist Bay K 8644 had no effect on the currents.

The rate of inactivation appears to be the property that differs most between N-type currents from different neuronal tissues, with time constants varying from 25 to several hundred milliseconds (Fox et al 1987a,b, Wanke et al 1987, Gross & MacDonald 1987, Kostyuk et al 1988, Hirning et al 1988, Lipscombe et al 1988, Plummer et al 1989, Jones & Marks 1989, Lemos & Nowycky 1989). As pointed out above, even among equal Ca channel types in the same cell (cardiac L-type Ca channels), the rate of inactivation of individual, otherwise identical, channels has been found to vary by more than a factor of ten (Cavalie et al 1986). Thus the rate of inactivation must be expected to be a particularly poor criterion for the identification of a channel type.

Recently, Llinas et al (1989) described a channel from cerebellar Purkinje cells, which they studied in bilayers. The channel is neither DHP- nor $\omega$-conotoxin sensitive, but is blocked by a toxin fraction from a spider venom (FTX, Llinas et al 1989). The unitary properties of this channel differ from those of other neuronal Ca channels studied in their native membranes. Whether the functional differences reflect differences in experimental conditions or whether this channel really represents a novel neuronal Ca channel type will be interesting to determine. (Llinas et al 1989).

Although the existence of neuronal Ca channels with properties distinct from L- or T-type channels is well established, the validity of lumping together non-DHP-sensitive, high-threshold-activated neuronal Ca channels from various cell types as N-type Ca channels remains questionable. Only the elucidation of the biochemical composition will establish the degree of heterogeneity between individual Ca channel types. So far, little information is available about the subunit composition of neuronal L-type channels (Takahashi & Catterall 1987). Iodinated $\omega$-conotoxin labels one or more polypeptides ranging from 135–310 kDa (Abe et al 1986, Abe & Saisu 1987, Cruz et al 1987, Marqueze et al 1988, Barhanin et al 1988, Glossman et al 1988). Apparently, the agent used to cross-link the toxin to the receptor determines which subunit(s) are labeled.

## Mechanism of Ion Permeation in Ca Channels

The topic of the mechanism by which Ca channels select for Ca ions over other competing ions has recently been reviewed (Tsien et al 1987), and the reader is referred to that paper for the experimental background for the two proposed models discussed here.

The first model (Kostyuk et al 1983) postulates that an externally located high-affinity binding site for Ca controls the selectivity of the channel ("allosteric model"). At Ca concentrations higher than 1 $\mu$M, this site is mostly occupied and hence the channel conformation is changed to one in which only Ca ions can pass. Permeating Ca ions transiently bind to a site in the pore itself, but only weakly, with an apparent half saturation of several millimolars. When the external regulatory binding site is not occupied by Ca (in the presence of Ca chelators), the channel pore becomes nonselective and allows passage of monovalent ions.

The second model (Hess & Tsien 1984, Almers & McCleskey 1984) postulates two intrachannel Ca binding sites with high affinity for Ca and ion-ion interactions in the doubly occupied channel ("two-site model"). Only the divalent free channel can pass monovalent ions freely. At high concentrations of Ca, double occupancy of the pore promotes accelerated Ca flux because the two ions repel each other electrostatically and each Ca ion speeds up the departure rate of the other one. The selectivity sequence of the intra-channel binding sites directly determines the selectivity of the channel.

Both models have strengths and weaknesses (Kostyuk & Mironov 1986, Tsien et al 1987). The most attractive feature of the allosteric model is that it treats the Ca channel in a way analogous to a Ca-binding protein. Ca binds to a regulatory site, which then triggers a conformational change, in this case changing channel selectivity. The conformational change, however, also represents the greatest weakness of the model. Its nature cannot be assessed and thus the actual mechanism by which the channel selects among permeant and blocking ions in its ion-conducting pathway remains unsolved. An extension of the allosteric model (Kostyuk & Mironov 1986) in which each ion, when interacting with the regulatory external site, induces a specific conformational change favoring its own passage (Kostyuk & Mironov 1986) can explain experimental data like the anomolous mole fraction effect for mixtures of Ca and Ba (see Tsien et al 1987, Kostyuk & Mironov 1986). Yet the mechanism operating on the timescale of the passage of a single ion that would allow such an adaptation of the selectively filter remains unknown.

The two-site model is capable of explaining most experimental results obtained on DHP-sensitive Ca channels with the least number of ad hoc assumptions (Tsien et al 1987). A weakness of the model is that it assumes symmetry of the channel energy profile, an assumption that has not been tested vigorously, mainly because of the well-known difficulties of studying L-type channels in cell-free conditions, where access to the internal channel side would be possible. Clear hints of channel asymmetry are beginning to emerge, however. Although internal Ca ions can block monovalent currents, the voltage and concentration dependence differ from the case of externally applied Ca (Ma & Coronado 1988, Hess et al 1989). Furthermore, in a preliminary report, Standen et al (1988) found that internal Cd at concentrations up to 1 mM failed to block L-type Ca channels in smooth muscle. The experiments were obtained at negative potentials on inward currents carried by Ba. Whether this represents another expression of channel asymmetry or whether internal Cd cannot access its binding site at all from the inside remains to be seen.

Another general weakness of the two-site model is that it views the channel as an "inert" structure that does not interact with permeant ions other than through binding and unbinding reactions. This must be a simplistic view of ion permeation, since several lines of independent evidence suggest that permeant ions do induce conformational changes in Ca channels (see below). Kostyuk & Mironov (1986) argue that electrostatic repulsion is unlikely to be of importance, since the sequential filling of multiple binding sites on Ca-binding proteins perturbs the apparent affinity of an individual Ca binding site by less than 0.5 pK units, much less than the 3–4 pK units needed in the two-site model. This argument has some merit, but it ignores the possibility that the dielectric constant in the channel could be much lower than in the aqueous solution surrounding a Ca binding protein. Indeed, values of the dielectric constant of $\sim 20$ are needed in the calculation of the two-site model to yield sufficient repulsion with a reasonable electrical distance of the two sites (Almers & McCleskey 1984, Hess et al 1986).

Interactions between permeant ions and the kinetics of channel gating which are not mediated by alteration of the negative surface charge can be taken as direct evidence for conformational changes induced by permeant ions. Such effects have not been reported in DHP-sensitive channels, but were found in Ca channels from *Aplysia* neurons (Chesnoy-Marchais 1985) and from rat brain Ca channels (Ca channel type remains unclear) reconstituted into planar lipid bilayers (Nelson 1986). In both cases, channel conductance and mean open time were inversely related for different permeant ions, as though occupancy of the channel promoted channel closing.

Evidence for conformational changes induced by Ca ions has also been presented to explain the block of monovalent single channel currents by micromolar concentrations of external Ca in T-type channels (Lux et al 1988). In contrast to the typical pattern of open channel block, which individual Ca ions produce in L-type channels that carry monovalent ions (Lansman et al 1986), Lux et al (1988) report that Ca ions prolong the closed-time durations of Na carrying T-type channels.

Recently, conformational changes in L-type Ca channels resulting from allosteric interactions between an external proton-binding site and one (or several) intra-pore ion-binding site(s) have been described (Prod'hom et al 1987, 1989, Pietrobon et al 1988, 1989). Prod'hom et al (1987) found that in the presence of monovalent charge carriers, external protons induced very rapid current fluctuations between a high and a low conductance level of the open L-type channel. They concluded that the two conductances represent different conformations of the channel molecule (Pietrobon et al 1988, 1989, Prod'hom et al 1989). The equilibrium between the two conductance states varied with the species of permeant ion in such a way that the higher an ion's affinity to the conduction pathway, the more the low conductance state was destabilized. Although this allosteric interaction between the channel permeation path and the conformational state of the channel has little importance for the mechanism with which the L-type channel selects for Ca over monovalent ions (Prod'hom et al 1989), it demonstrates that permeant ions interact with the channel and can induce conformational changes. Ca, the physiologically preferred ion, appears to stabilize the channel in a favored open conformation, and to prevent it from assuming conformations that occur only when the channel is occupied by less tightly binding ions.

Konnerth et al (1987) and Davies et al (1988) have described a different effect of protons, which they interpret as a transformation of a Ca channel from its Ca-conducting, voltage-activated state into a Na-conducting, non-voltage-dependent state. The proton-induced, "transformed" current activates in response to a rapid step decrease of $pH_0$ and relaxes within several seconds. The authors do not explicitly say which type of Ca channel is being transformed. L-type Ca channels are unlikely candidates, since L-type Ca channels remain dihydropyridine-sensitive and voltage-gated when conducting monovalent ions, whereas the proton-activated currents are not. In addition, proton-activated currents cannot be elicited in cardiac cells (Davies et al 1988), whereas monovalent currents through L-type channels are identical in channels from cardiac and noncardiac cells (Prod'hom et al 1989). Whether the proton-induced currents are carried by another type of Ca channel remains to be seen. A more likely possibility is that they are not carried by Ca channels, but represent a proton-activated nonselective cation channel, as originally proposed by Kristhal & Pidoplichko (1980, 1981).

## Ca Channel Modulation

The best-established mechanism of Ca channel modulation is that of L-type Ca channels mediated by cAMP-dependent kinase. This pathway has been dissected in an exemplary way in cardiac cells (Sperelakis & Schneider 1976, Reuter & Scholz 1977, Cachelin et al 1983, Brum et al 1984, Kameyama et al 1985, Harzell & Fischmeister 1986, Tsien et al 1987) and probably involves phosphorylation of the channel itself as the last step of the cascade. Although a similar mechanism operates in L-type Ca channels from other cells (Armstrong & Eckert 1987), whether all L-type Ca channels are sensitive to elevations of intracellular cAMP remains unclear. Recently, three additional pathways have been postulated to modulate DHP-sensitive Ca channels.

1. Activation of protein kinase C has been reported to increase (Doesemeci et al 1988, Fish et al 1988, Lacerda et al 1988) or decrease (during prolonged application; see Lacerda et al 1988) L-type Ca current in cardiac and smooth muscle cells.

2. Internal GTP-binding proteins, specifically $G_s$ and its $\alpha$ subunit, have been shown to increase the activity of skeletal muscle DHP-sensitive channels (Yatani et al 1988) and cardiac L-type Ca channels (Yatani et al 1987), and a pertussis-toxin-sensitive GTP binding protein enhances L-type Ca currents in Y1 cells (Hescheler et al 1988).

3. Internal inositol trisphosphate ($IP_3$) increased the activity of skeletal muscle DHP-receptor channels (Vilven & Coronado 1988).

Since most signalling mechanisms use at least one, and often several, of the above pathways, specificity of signaling is puzzling. A striking example is the modulation by cAMP-dependent kinase, which is normally preceded by activation of $G_s$, and couples beta receptors to adenylate cyclase. Since the cAMP-dependent kinase alone can fully mimic the effects of externally applied beta adrenergic transmitter (Kameyama et al 1985, Flockerzi et al 1986b), the role, if any, of direct activation of the channel by $G_s$ is unclear (Yatani et al 1989). Also puzzling is why direct application of the catalytic subunit of cAMP-dependent protein kinase was ineffective in studies that reported a direct effect of GTP-binding proteins (Yatani et al 1987).

T-type Ca channels are not modulated by interventions that lead to elevated cAMP (Bean 1985, Bonvallet 1987, Tytgat et al 1988). Two directly conflicting reports have been made about the modulation of L- and T-type channels by angiotensin II in adrenal cortical cells. Cohen et al (1988), separating current components on the basis of tail deactivation, concluded that angiotensin II increased T-type currents with no effect on L-type channels, whereas Hescheler et al (1988), using peak currents to separate L- and T-type currents, found angiotensin II enhanced L-type current in a pertussis-toxin-sensitive way, with no effect on T-type channels. The differences underlying the two conclusions are not clear, but perhaps the analysis of tail currents misled Cohen et al (1988), if L-type channels also contribute a component of slowly deactivating tail current. This possibility would explain why Cohen et al (1988) also concluded that T-type currents are inhibited by 300 nM nitrendipine, a result contrary to the DHP insensitivity generally found for T-type Ca channels. An inhibitory effect of a diacylglycerol analogue on T-type channels has recently been reported (Marchetti & Brown 1988). As for all effects obtained by diacylglycerol analogues and phorbol esters, whether this is a direct effect of the drug (Hockberger et al 1989) or is mediated by activation of protein kinase C remains to be seen.

MODULATION OF NEURONAL CA CHANNELS    Neuronal L-type Ca channels may be up-regulated by cAMP-dependent phosphorylation like their counterparts in heart muscle (Gray & Johnson 1987, Lipscombe et al 1988). The most striking modulation of neuronal Ca channels, however, is inhibition. This effect was first demonstrated by Dunlap & Fischbach (1978) and was subsequently shown to be induced by a large variety of neurotransmitters and effectors, ranging from noradrenaline to acetylcholine, GABA, serotonin, and opioids (for review see Miller 1987a, b). Whether all these agents exert their inhibitory effects on the same Ca channel type remains unclear, but in most cases N-type Ca channels appear to be the target (Lipscombe & Tsien 1987, Wanke et al 1987, Gross & MacDonald 1987, Ewald et al 1988, Plummer et al 1989), particularly if one takes into account the properties of N-type channels described above. In all cases where it has been studied, the inhibitory action of neurotransmitters involve GTP-binding proteins (Holz et al 1986, Scott & Dolphin 1987, Dolphin & Scott 1987, Wanke et al 1987, Hescheler et al 1987, Ewald et al 1988, Plummer et al 1989), in particular $G_0$ (Hescheler et al 1987, Ewald et al 1988). An important question remains: How does coupling between the GTP-binding proteins and the channels occur? In chick DRG cells this coupling involves activation of protein kinase C (Rane & Dunlap 1986), whereas in other cells GTP-binding proteins exert their effects apparently independently of C kinase (Wanke et al 1987). The answer to the question is made more difficult by the suggestion (Hockberger et al 1989) that some activators of protein kinase C can directly inhibit neuronal Ca channels independently of their effects on the kinase. In the case of oleoylacetylglycerol (OAG), this direct effect is, however, only observed after storage of the OAG for 3–4 weeks in the refrigerator (Hockberger et al 1989), and might therefore be related to a breakdown product rather than to OAG itself.

A different effect of GTP-binding proteins has been described by Scott & Dolphin (1987), who reported that in the presence of internal GTP-$\gamma$-S, both DHP and phenylalkylamine Ca channel antagonists at high concentrations produced a large increase of neuronal Ca current. Scott & Dolphin attributed this effect to a conversion of drug antagonism to agonism mediated by GTP-binding proteins on L-type currents. This attractive hypothesis awaits confirmation in single channels, but so far, more recent studies on whole cell currents (Plummer et al 1989) have failed to reproduce the original effects described by Scott & Dolphin (1987).

Although GTP-binding proteins clearly couple receptors to neuronal Ca channels, a direct action of GTP-binding proteins on the channel(s) cannot yet be claimed. So far, GTP-binding proteins have only been shown to reconstitute the receptor-coupling to the channel.

The modification of channel function that underlies the decrease in whole cell current also remains unclear. Bean (1989b) has presented evidence that the modulation by Noradrenaline reduces neuronal Ca current by shifting the activation curve toward more positive potentials. Thus, at a given test potential, a modulated channel would be less likely to open in response to a depolarization. The detailed functional consequences of the modulation remain to be investigated by single-channel recordings.

ACKNOWLEDGEMENTS

I am grateful for support by the US Public Health Service, the American Cancer Society, and the Marquee Trust and thank B. Bean, C. Chen, E. Liman, D. Logothetis, D. Pietrobon, and M. Plummer for helpful discussions.

*Literature Cited*

Abe, T., Saisu, H. 1987. Identification of the receptor for omega-conotoxin in brain. *J. Biol. Chem.* 262: 9877–82

Abe, T., Koyano, K., Saisu, H., Nishiuchi, Y., Sakakibara, S. 1986. Binding of $\omega$-conotoxin to receptor sites associated with the voltage-sensitive calcium channel. *Neurosci. Lett.* 71: 203–308

Almers, W., McCleskey, E. W. 1984. Nonselective conductance in calcium channels in frog muscle: Calcium selectivity in a single-file pore. *J. Physiol.* 353: 585–608

Aosaki, T., Kasai, H. 1989. Characterization of two kinds of high-voltage-activated Ca-channel currents in chick sensory neurons: Differential sensitiviy to dihydropyridines and $\omega$-conotoxin GVIA. *Pfluegers Arch.* 414: 150–56

Armstrong, C. M., Matteson, D. R. 1985. Two distinct populations of calcium channels in a clonal line of pituitary cells. *Science* 227: 65–67

Armstrong, D., Eckert, R. 1987. Voltage activated calcium channels that must be phosphorylated to respond to membrane depolarization. *Proc. Natl. Acad. Sci. USA* 84: 2518–22

Barhanin, J., Schmid, A., Lazdunski, M. 1988. Properties of structure and interaction of the receptor for $\omega$-conotoxin, a polypeptide active on Ca channels. *Biochem. Biophys. Res. Commun.* 150: 1051–62

Beam, K. G., Knudson, C. M., Powell, J. A. 1986. A lethal mutation in mice eliminates the slow calcium current in skeletal muscle cells. *Nature* 320: 168–70

Bean, B. P. 1985. Two kinds of calcium chan-

nels in canine atrial cells. *J. Gen. Physiol.* 86: 1–30

Bean, B. P. 1989a. Classes of calcium channels in vertebrate cells. *Annu. Rev. Physiol.* 51: 367–84

Bean, B. P. 1989b. Neurotransmitter inhibition of neuronal calcium currents by changes in channel voltage-dependence. *Nature* 340: 153–56

Bonvallet, R. 1987. A low threshold calcium current recorded at physiological Ca concentration in single frog atrial cells. *Pfluegers Arch.* 408: 540–42

Borsotto, M., Barhanin, J., Fosset, M., Lazdunski, M. 1985. The 1,4-dihydropyridine receptor associated with the skeletal muscle voltage dependent Ca channel. Purification and subunit composition. *J. Biol. Chem.* 260: 14255–63

Borsotto, M., Barhanin, J., Norman, R. I., Lazdunski, M. 1984. Purification of the dihydropyridine receptor of the voltage dependent Ca channel from skeletal muscle transverse tubules using (+) [3H]PN 200-110. *Biochem. Biophys. Res. Commun.* 122: 1357–66

Brum, G., Osterrieder, W., Trautwien, W. 1984. $\beta$-adrenergic increase in the calcium conductance of cardiac myocytes studied with the patch clamp. *Pfluegers Arch.* 401: 111–18

Cachelin, A. B., dePeyer, J. E., Kokubun S., Reuter, H. 1983. Calcium channel modulation by 8-bromo-cyclic AMP in cultured heart cells. *Nature* 304: 402–4

Carbone, E., Lux, H. D. 1984. A low voltage activated, fully inactivating Ca channel in verebrate sensory neurones. *Nature* 310: 501–11

Cavalie, A., Pelzer, D., Trautwien, W. 1986. Fast and slow gating behaviour of single calcium channels in cardiac cells. *Pfluegers Arch.* 406: 241–58

Chen, C., Corbley, M. J., Roberts, T. M., Hess, P. 1988. Voltage-sensitive calcium channels in normal and transformed 3T3 fibroblasts. *Science* 239: 1024–26

Chesnoy-Marchais, D. 1985. Kinetic properties and selectivity of calcium-permeable single channels in Aplysia neurones. *J. Physiol.* 367: 457–88

Cohen, C. J., McCarthy, R. T., Barrett, P. Q., Rasmussen, H. 1988. Calcium channels in adrenal glomerulosa cells: K and angiotensin II increase T-type Ca channel current. *Proc. Natl. Acad. Sci. USA* 85: 2412–16

Cooper, C. L., Vandaele, S., Barhanin, J., Fosset, M., Lazdunski, M., Hosey, M. M. 1987. Purification and characterization of the dihydropyridine-sensitive voltage-dependent calcium channel from cardiac tissue. *J. Biol. Chem.* 262: 509–12

Coronado, R., Affolter, H. 1986. Insulation of the conduction pathway of muscle transverse tubule calcium channels from the surface charge of bilayer phospholipid. *J. Gen. Physiol.* 87: 933–53

Cruz, L. J., Johnson, S. D., Olivera, B. M. 1987. Characterization of the ω-conotoxin target: Evidence for tissue-specific heterogeneity in calcium channel types. *Biochemistry* 26: 820–24

Curtis, B. M., Catterall, W. A. 1984. Purification of the calcium antagonist receptor of the voltage-sensitive calcium channel from skeletal muscle transverse tubules. *Biochemistry* 23: 2113–17

Curtis, B. M., Catterall, W. A. 1985. Phosphorylation of the calcium antagonist receptor of the voltage-sensitive Ca channel by cAMP-dependent protein kinase. *Proc. Natl. Acad. Sci. USA* 82: 2528–32

Curtis, B. M., Catterall, W. A. 1986. Reconstitution of the voltage-sensitive calcium channel purified from skeletal muscle transverse tubules. *Biochemistry* 25: 3077–83

Davies, N. W., Lux, H. D., Morad, M. 1988. Site and mechanism of activation of proton-induced sodium current in chick dorsal root ganglion neurones. *J. Physiol.* 400: 159–87

Dolphin, A. C., Scott, R. H. 1987. Calcium channel currents and their inhibition by (−)-baclofen in rat sensory neurones: Modulation by guanine nucleotides. *J. Physiol.* 386: 1–17

Dosemeci, A., Dhallan, R. S., Cohen, N. M., Lederer, W. J., Rogers, T. B. 1988. Phorbol ester increases calcium current and simulates the effects of angiotensin II

on cultured neonatal rat heart myocytes. *Circulat. Res.* 62: 347–57

Dunlap, K., Fischblach, G. D. 1978. Neurotransmitters decrease the calcium component of sensory neurones action potentials. *Nature* 276: 837–39

Eckert, R., Chad, J. E. 1984. Inactivation of Ca channels. *Prog. Biophys. Mol. Biol.* 44: 215–67

Ellis, S. B., Williams, M. E., Ways, N. R., Brenner, R., Sharp, A. H., Leung, A. T., Campbell, K. P., McKenna, E., Koch, W. J., Hui, A., Schwartz, A., Harpold, M. M. 1988. Sequence and expression of mRNAs encoding $\alpha_1$ and $\alpha_2$ subunits of a DHP-sensitive calcium channel. *Science* 241: 1661–64

Ewald, D. A., Sternweis, P. C., Miller, R. 1988. Guanine nucleotide-binding protein $G_0$-induced coupling of neuropeptide Y receptors to $Ca^{2+}$ channels in sensory neurons. *Proc. Natl. Acad. Sci. USA* 85: 3633–37

Fish, R. D., Sperti, G., Colucci, W. S., Clapham, D. E. 1988. Phorbol ester increases the dihydropyridine-sensitive calcium conductance in a vascular smooth muscle cell line. *Circulat. Res.* 62: 1049–54

Flockerzi, V., Oeken, H. J., Hofmann, F. 1986a. Purification of a functional receptor for calcium channel blockers from rabbit skeletal muscle microsomes. *Eur. J. Biochem.* 161: 217–24

Flockerzi, V., Oeken, H. J., Hofmann, F., Pelzer, D., Cavalie, A., Trautwein, W. 1986b. The purified dihydropyridine-binding site from skeletal muscle T-tubules is a functional Ca channel. *Nature* 323: 66–68

Fosset, M., Jaimovich, E., Delpont, E., Lazdunski, M. 983. [$^3$H]Nitrendipine receptors in skeletal muscle. Properties and preferential localization in transverse tubules. *J. Biol. Chem.* 258: 6086–92

Fox, A. P., Nowycky, M. C., Tsien, R. W. 1987a. Kinetic and pharmacological properties distinguishing three types of calcium currents in chick sensory neurones. *J. Physiol.* 394: 149–72

Fox, A. P., Nowycky, M. C., Tsien, R. W. 1987b. Single-channel recordings of three types of calcium channels in chick sensory neurones. *J. Physiol.* 394: 173–200

Glossmann, H., Ferry, D. R. 1983. Solubilization and partial purification of putative calcium channels labelled with [$^3$H]-nimodipine. *Naunyn-Schmied. Arch. Pharmacol.* 323: 279–91

Glossmann, H., Ferry, D. R., Boschek, C. B. 1983. Purification of the putative calcium channel from skeletal muscle. *Naunyn-Schmied. Arch. Pharmacol.* 323: 1–11

Glossmann, H., Striessnig, J., Kymel, L.,

Zernig, G., Knaus, H. G., Schindler, H. 1988. The structure of the Ca channel: Photoaffinity labeling and tissue distribution. In *The Calcium Channel: Structure, Function and Implications*, ed. M. Morad, W. Naylor, S. Kazda, M. Schramm. pp. 168–92. Berlin: Springer

Gray, R., Johnston, D. 1987. Noradrenaline and β-adrenoreceptor agonists increase activity of voltage dependent calcium channels in hippocampal neurons. *Nature* 327: 620–22

Gross, R. A., MacDonald, R. L. 1987. Dynorphin A selectively reduces a large transient (N-type) calcium current of mouse dorsal root ganglion neurons in cell culture. *Proc. Natl. Acad. Sci. USA* 84: 5469–73

Hagiwara, N., Irisawa, H., Kameyama, M. 1988. Contribution of two types of calcium currents to the pacemaker potentials of rabbit sino-atrial node cells. *J. Physiol.* 395: 233–53

Hagiwara, S., Byerly, L. 1981. Calcium channel. *Annu. Rev. Neurosci.* 4: 69–125

Hartzell, H. C., Fischmeister, R. 1986. Opposite effects of cyclic GMP and cyclic AMP on Ca current in single heart cells. *Nature* 323: 273–75

Hescheler, J., Rosenthal, W., Hinsch, K.-D., Wulfern, M., Trautwein, W., Schultz, G. 1988. Angiotensin-II induced stimulation of voltage dependent Ca$_{2+}$ currents in an adrenal cortical cell line. *EMBO J.* 7: 619–24

Hescheler, J., Rosenthal, W., Trautwein, W., Schultz, G. 1987. The GTP-binding protein, G$_0$, regulates neuronal calcium channels. *Nature* 325: 445–47

Hess, P., Lansman, J. B., Tsien, R. W. 1984. Different modes of Ca channel gating behaviour favoured by Ca agonists and antagonists. *Nature* 311: 538–44

Hess, P., Lansman, J. B., Tsien, R. W. 1986. Calcium channel selectivity for divalent and monovalent cations. Voltage and concentration-dependence of single channel current in guinea pig ventricular heart cells. *J. Gen. Physiol.* 88: 293–319

Hess, P., Prod'hom, B., Pietrobon, D. 1989. Mechanisms of interaction of permeant ions and protons with dihydropyridine sensitive calcium channels. *NY Acad. Sci.* 560: 80–93

Hess, P., Tsien, R. W. 1984. Mechanism of ion permeation through calcium channels. *Nature* 309: 453–56

Hirning, L. D., Fox, A. P., McCleskey, E. W., Olivera, B. M., Thayer, S. A., Miller, R. J., Tsien, R. W. 1988. Dominant role of N-type Ca$^{2+}$ channels in evoked release of norepinephrine from sympathetic neurons. *Science* 239: 57–61

Hockberger, P., Toselli, M., Swandulla, D., Lux, H. D. 1989. A diacylglycerol analogue reduces neuronal calcium currents independently of protein kinase C activation. *Nature* 338: 340–42

Holz, G. G., Dunlap, K., Kream, R. M. 1988. Characterization of the electrically evoked release of substance p from dorsal ganglion neurons: Methods and dihydropyridine sensitivity. *J. Neurosci.* 8: 463–71

Holz, G. G., Rane, S. G., Dunlap, K. 1986. GTP-binding proteins mediate transmitter inhibition of voltage-dependent calcium channels. *Nature* 319: 670–72

Horne, W. A., Abdel-Ghany, M., Racker, E., Weiland, G. A., Oswald, R. E., Cerione, R. A. 1988. Functional reconstitution of skeletal muscle Ca channels: Separation of regulatory and channel components. *Proc. Natl. Acad. Sci. USA* 85: 3718–22

Hosey, M. M., Barhanin, J., Schmid, A., Vandaele, S., Ptasienski, J., O'Callahan, C., Cooper, C., Lazdunski, M. 1987. Photoaffinity labelling and phosphorylation of a 165 kilodalton peptide associated with dihydropyridine and phenylalkylamine sensitive calcium channels. *Biochem. Biophys. Res. Commun.* 147: 1137–45

Hosey, M. M., Borsotto, M., Lazdunski, M. 1986. Phosphorylation and dephosphorylation of the major component of the voltage dependent Ca channel in skeletal muscle membranes by cyclic AMP and Ca-dependent processes. *Proc. Natl. Acad. Sci. USA* 83: 3733–37

Hosey, M. M., Lazdunski, M. 1988. Calcium channels: Molecular pharmacology, structure and regulation. *J. Membr. Biol.* 104: 81–105

Hymel, L., Striessnig, J., Glossman, H., Schindler, H. 1988. Purified skeletal muscle 1,4-dihydropyridine receptor forms phosphorylation-dependent oligomeric calcium channels in planar bilayers. *Proc. Natl. Acad. Sci. USA* 85: 4290–94

Imagawa, T., Leung, A. T., Campbell, K. P. 1987. Phosphorylation of the 1,4-dihydropyridine receptor of the voltage-dependent Ca channel by an intrinsic protein kinase in isolated triads from rabbit skeletal muscle. *J. Biol. Chem.* 262: 8333–39

Janis, R. A., Silver, P. J., Triggle, D. J. 1987. Drug action and cellular calcium regulation. *Adv. Drug Res.* 16: 309–591

Jones, S. W., Marks, T. N. 1989. Calcium currents in bullfrog sympathetic neurons. I. Activation kinetics and pharmacology. *J. Gen Physiol.* 94: 151–67

Kameyama, M., Hoffmann, F., Trautwein, W. 1985. On the mechanism of β-adrenergic regulation of the Ca channel in the guinea-pig heart. *Pfluegers Arch.* 405: 285–93

Kasai, H., Aosaki, T., Fukuda, J. 1987. Presynaptic Ca-antagonist ω-conotoxin irreversibly blocks N-type Ca-channels in chick sensory neurons. *Neurosci. Res.* 4: 228–35

Kerr, L. M., Yoshikami, D. 1984. A venom peptide with a novel presynaptic blocking action. *Nature* 308: 282–84

Knaus, H. G., Striessnig, J., Koza, A., Glossmann, H. 1987. Neurotoxic aminoglycoside antibiotics are potent inhibitors of [$^{125}$I]-omega-conotoxin binding to guinea-pig cerebral cortex membranes. *Naunyn Schmiedeberg's Arch. Pharmacol.* 336: 583–86

Knudson, C. M., Chaudhari, N., Sharp, A. H., Powell, J. A., Beam, K. G., Campbell, K. P. 1989. Specific absence of the α$_1$ subunit of the dihydropyridine receptor in mice with muscular dysgenesis. *J. Biol. Chem.* 264: 1345–48

Konnerth, A., Lux, H. D., Morad, M. 1987. Proton-induced transformation of calcium channel in chick dorsal root ganglion cells. *J. Physiol.* 386: 603–33

Kostyuk, P. G. 1981. Calcium channels in the neuronal membrane. *Biochim. Biophys. Acta* 650: 128–50

Kostyuk, P. G., Mironov, S. L. 1986. Some predictions concerning the calcium channel model with different conformational states. *Gen. Physiol. Biophys.* 6: 649–59

Kostyuk, P. G., Mironov, S. L., Shuba, Y. M. 1983. Two ion-selecting filters in the calcium channel of the somatic membrane of mollusc neurones. *J. Membr. Biol.* 76: 83–93

Kostyuk, P. G., Shuba, Y. M., Savchenko, A. N. 1988. Three types of calcium channels in the membrane of mouse sensory neurons. *Pfluegers Arch.* 411: 661–69

Kretsinger, R. H. 1980. Structure and evolution of calcium modulated proteins. *CRC Crit. Rev. Biochem.* 8: 119–74

Krishtal, O. A., Pidoplichko, V. I. 1980. A receptor for protons in the nerve cell membrane. *Neuroscience* 5: 2325–27

Krishtal, O. A., Pidoplichko, V. I. 1981. A receptor for protons in the membrane of sensory neurones may participate in nociception. *Neuroscience* 6: 2599–2601

Lacerda, A. E., Rampe, D., Brown, A. M. 1988. Effects of protein kinase C activators on cardiac Ca channels. *Nature* 335: 249–51

Lansman, J. B., Hess, P., Tsien, R. W. 1986. Blockade of current through single calcium channels by Cd$^{2+}$, Mg$^{2+}$ and Ca$^{2+}$. Voltage- and concentration-dependence of Ca entry into the pore. *J. Gen. Physiol.* 88: 321–47

Lemos, J. R., Nowycky, M. C. 1989. Two types of Ca channels coexist in vertebrate nerve terminals. *Neuron* 2: 1419–26

Leung, A. T., Imagawa, T., Campbell, K. P. 1987. Structural characterization of the 1,4-dihydropyridine receptor of the voltage dependent calcium channel from rabbit skeletal muscle. Evidence for two distinct heigh molecular weight subunits. *J. Biol. Chem.* 262: 7943–46

Lipscombe, D., Bley, K., Tsien, R. W. 1988. Modulation of neuronal Ca channels by cAMP and phorbol esters. *Soc. Neurosci. Abstr.* 14: 153 (Abstr.)

Lipscombe, D., Madison, D. V., Poenie, M., Reuter, H., Tsien, R. Y., Tsien, R. W. 1988. Spatial distribution of calcium channels and cytosolic calcium transients in growth cones and cell bodies of sympathetic neurons. *Proc. Natl. Acad. Sci. USA* 85: 2398–2402

Lipscombe, D., Tsien, R. W. 1987. Noradrenaline inhibits N-type Ca channels in frog sympathetic neurones. *J. Physiol.* 377: 97P (Abstr.)

Llinas, R., Sugimori, M., Lin, J.-W., Cherksey, B. 1989. Blocking and isolation of a calcium channel from neurons in mammals and cephalopods utilizing a toxin fraction (FTX) from funnel-web spider poison. *Proc. Natl. Acad. Sci. USA* 86: 1689–93

Llinas, R., Yarom, Y. 1986. Specific blockade of the low threshold calcium channel by high molecular weight alcohols. *Soc. Neurosci. Abstr.* 12: 174 (Abstr.)

Lotan, I., Goelet, P., Gigi, A., Dascal, N. 1989. Specific block of calcium channel expression by a fragment of dihydropyridine receptor cDNA. *Science* 243: 666–69

Lux, H. D., Carbone, E., Zucker, H. 1988. Block of sodium currents through a neuronal calcium channel by external calcium and magnesium ions. See Glossman et al 1988, pp. 129–37

Ma, J., Coronado, R. 1988. Conductance-activity, current-voltage, and mole fraction relationships for calcium, barium and sodium in the T-tubule calcium channel. *Biophys. J.* 53: 556a (Abstr.)

Marchetti, C., Brown, A. M. 1988. Protein kinase activator 1-oleoyl-2-acetyl-sn-glycerol inhibits two types of calcium currents in GH$_3$ cells. *Am. J. Physiol.* 254: C206–10

Marqueze, B., Martin-Moutot, N., Leveque, C., Couraud, F. 1988. Characterization of the ω-conotoxin-binding molecule in rat brain synaptosomes and cultured neurones. *Mol. Pharmacol.* 34: 87–90

McCleskey, E. W., Fox, A. P., Feldman, D. H., Cruz, L. J., Olivera, B. M., Tsien, R.

W., Yoshikami, D. 1987. ω-conotoxin: Direct and persistent blockade of specific types of calcium channels in neurons but not muscle. *Proc. Natl. Acad. Sci. USA* 84: 4327–31

Miller, R. 1987a. Calcium channels in neurones. In *Structure and Physiology of the Slow Inward Calcium Channel*, ed. D. J. Triggle, J. C. Venter. pp. 161–246. New York: Liss

Miller, R. 1987b. Multiple calcium channels and neuronal function. *Science* 235: 46–52

Nelson, M. T. 1986. Interactions of divalent cations with single calcium channels from rat brain synaptosomes. *J. Gen. Physiol.* 87: 201–22

Nilius, B., Hess, P., Lansman, J. B., Tsien, R. W. 1985. A novel type of cardiac calcium channel in ventricular cells. *Nature* 316: 443–46

Noda, M., Shimizu, S., Tanabe, T., Takai, T., Kayano, T., et al. 1984. Primary structure of *Electrophorus Electricus* sodium channel deduced from cDNA sequence. *Nature* 312: 121–27

Nowycky, M. C., Fox, A. P., Tsien, R. W. 1985a. Three types of neuronal calcium channels with different calcium agonist sensitivity. *Nature* 316: 440–43

Nowycky, M. C., Fox, A. P., Tsien, R. W. 1985b. Long-opening mode of gating of neuronal calcium channels and its promotion by the dihydropyridine calcium agonist Bay K 8644. *Proc. Natl. Acad. Sci. USA* 82: 2178–82

Olivera, B. M., Gray, W. R., Zeikus, R., McIntosh, J. M., Victoria di Santos, J. R., Cruz, L. J. 1985. Peptide neurotoxins from fish-hunting cone snails. *Science* 230: 1338–43

Oyama, Y., Tsuda, Y., Sakakibara, S., Akaike, N. 1987. Synthetic ω-conotoxin: A potent calcium channel blocking neurotoxin. *Brain Res.* 424: 58–64

Papazian, D. M., Schwarz, T. L., Tempel, B. L., Jan, Y. N., Jan, L. Y. 1987. Cloning of genomic and complementary DNA from *Shaker*, a putative potassium channel gene from *Drosophila*. *Science* 237: 749–53

Pietrobon, D., Prod'hom, B., Hess, P. 1988. Conformational changes associated with ion permeation in L-type Ca channels. *Nature* 333: 373–76

Pietrobon, D., Prod'hom, B., Hess, P. 1989. Interactions of protons with single open L-type calcium channels. pH dependence of proton induced current fluctuations with $Cs^+$, $K^+$ and $Na^+$ as charge carriers. *J. Gen. Physiol.* 94: 1–21

Pincon-Raymond, M., Rieger, F., Fosset, M., Lazdunski, M. 1985. Abnormal transverse tubule system and abnormal amount of receptors for Ca channel inhibitors of the dihydropyridine family in skeletal muscle from mice with embryonic muscular dysgenesis. *Dev. Biol.* 112: 458–66

Plummer, M. R., Logothetis, D. E., Hess, P. 1989. Elementary properties and pharmacological sensitivities of calcium channels in mammalian peripheral neurons. *Neuron* 2: 1453–63

Pongs, O., Kecskemethy, N., Mueller, R., Krah-Jentzens, I., Baumann, A., Kiltz, H. H., Canal, I., Llamazares, S., Ferrus, A. 1988. Shaker encodes a family of putative potassium channel proteins in the nervous system of *Drosophila*. *EMBO J.* 7: 1087–96

Porzig, H., Becker, C., Reuter, H. 1989. Effects of NGF-induced differentiation on two classes of Ca channels in living PC12 cells. *Experientia* 45: A28 (Abstr.)

Prod'hom, B., Pietrobon, D., Hess, P. 1987. Direct measurement of proton transfer rates to a group controlling the dihydropyridine sensitive Ca channel. *Nature* 329: 243–46

Prod'hom, B., Pietrobon, D., Hess, P. 1989. Interactions of protons with single open L-type calcium channels. Location of protonation site and dependence of proton induced current fluctuations on concentration and species of permeant ion. *J. Gen. Physiol.* 94: 23–42

Rane, S. G., Dunlap, K. 1986. Kinase C activator 1,2-oleoylacetylglycerol attenuates voltage dependent calcium current in sensory neurones. *Proc. Natl. Acad. Sci. USA* 83: 184–88

Reuter, H. 1973. Divalent cations as charge carriers in excitable membranes. *Prog. Biophys. Mol. Biol.* 26: 1–43

Reuter, H., Scholz, H. 1977. The regulation of the Ca conductance of cardiac muscle by adrenaline. *J. Physiol.* 264: 49–62

Rieger, F., Bournaud, R., Shimahara, T., Garcia, L., Pincon-Raymond, M., Romey, G., Lazdunski, M. 1987. Restoration of dysgenic muscle contraction and calcium channel function by co-culture with normal spinal cord neurons. *Nature* 330: 563–66

Rios, E., Brum, G. 1987. Involvement of dihydropyridine receptors in excitation-contraction coupling in skeletal muscle. *Nature* 325: 717–20

Rosenberg, R. L., Hess, P., Reeves, J. P., Smilowitz, H., Tsien, R. W. 1986. Calcium channels in planar lipid bilayers: Insights into mechanisms of ion permeation and gating. *Science* 231: 1564–66

Schmid, A., Barhanin, J., Coppola, T., Borsotto, M., Lazdunski, M. 1986. Immunochemical analysis of subunit structures of 1,4-dihydropyridine receptors associated with voltage dependent Ca channels in skeletal, cardiac and smooth muscles. *Biochemistry* 25: 3492–95

Scott, R. H., Dolphin, A. C. 1987. Activation of a G protein promotes agonist responses to calcium channel ligands. *Nature* 330: 760–62

Smith, J. S., McKenna, E. J., Ma, J., Vilven, J., Vaghy, P. L., Schwartz, A., Coronado, R. 1987. Calcium channel activity in a purified dihydropyridine receptor preparation of skeletal muscle. *Biochemistry* 26: 7182–88

Sperelakis, N., Schneider, J. 1976. A metabolic control mechanism for calcium ion influx that may protect the ventricular myocardial cell. *Am. J. Cardiol.* 37: 1079–85

Standen, N. B., Worley, J. F., Nelson, M. T. 1988. Effects of internal divalent cations on single calcium channels in vascular smooth muscle. *Biophys. J.* 53: 231a

Streit, J., Lux, H. D. 1987. Voltage dependent calcium currents in PC12 growth cones and cells during NGF-induced cell growth. *Pfluegers Arch.* 408: 634–41

Swandulla, D., Armstrong, C. M. 1988. Fast-deactivating calcium channels in chick sensory neurons. *J. Gen. Physiol.* 92: 197–218

Takahashi, M., Catterall, W. A. 1987. Identification of an α subunit of dihydropyridine sensitive brain calcium channels. *Science* 236: 88–91

Takahashi, M., Seagar, M. J., Jones, J. F., Reber, B. F. X., Catterall, W. A. 1987. Subunit structure of dihydropyridine-sensitive calcium channels from skeletal muscle. *Proc. Natl. Acad. Sci. USA* 84: 5478–82

Talvenheimo, J. A., Worley, J. F. III, Nelson, M. T. 1987. Heterogeneity of calcium channels from a purified dihydropyridine receptor preparation. *Biophys. J.* 52: 891–99

Tanabe, T., Beam, K. G., Powell, J. A., Numa, S. 1988. Restoration of excitation-contraction coupling and slow calcium current in dysgenic muscle by dihydropyridine receptor complementary DNA. *Nature* 366: 134–39

Tanabe, T., Takeshima, H., Mikami, A., Flockerzi, V., Takahashi, H., Kangawa, K., Kojima, M., Matsuo, H., Hirose, T., Numa, S. 1987. Primary structure of the receptor for calcium channel blockers from skeletal muscle. *Nature* 328: 313–18

Tang, C.-M., Presser, F., Morad, M. 1988. Amiloride selectively blocks the low threshold (T) calcium channel. *Science* 240: 213–15

Tsien, R. W. 1983. Calcium channels in excitable cell membranes. *Annu. Rev. Physiol.* 45: 341–58

Tsien, R. W., Bean, B. P., Hess, P., Lansman, J. B., Nilius, B., Nowycky, M. C. 1986. Mechanisms of calcium channel modulation by beta-adrenergic agents and dihydropyridine calcium agonists. *J. Mol. Cell. Cardiol.* 18: 691–710

Tsien, R. W., Hess, P., McCleskey, E. W., Rosenberg, R. L. 1987. Calcium channels: Mechanisms of selectivity, permeation and block. *Annu. Rev. Biophys. Biophys. Chem.* 16: 265–90

Tsien, R. W., Lipscombe, D., Madison, D. V., Bley, K. R., Fox, A. P. 1988. Multiple types of neuronal Ca channels and their selective modulation. *Trends Neurosci.* 11: 431–38

Tytgat, J., Nilius, B., Vereecke, J., Carmeliet, E. 1988. The T-type Ca channel in guinea-pig ventricular myocytes is insensitive to isoproterenol. *Pfluegers Arch.* 411: 704–6

Vaghy, P. L., Striessnig, J., Miwa, K., Knaus, H. G., Itagaki, K., McKenna, E., Glossmann, H., Schwartz, A. 1987. Identification of a novel 1,4-dihydropyridine- and phenylalkylamine-binding polypeptide in calcium channel preparations. *J. Biol. Chem.* 262: 14337–42

Van Skiver, D. M., Spires, S., Cohen, C. J. 1989. High affinity and tissue specific block of T-type Ca channels by felodipine. *Biophys. J.* 55: 593a (Abstr.)

Vilven, J., Coronado, R. 1988. Opening of dihydropyridine calcium channels in skeletal muscle membranes by inositol trisphosphate. *Nature* 336: 587–89

Wanke, E., Ferroni, A., Malgaroli, A., Ambrosini, A., Pozzan, T., Meldolesi, J. 1987. Activation of a muscarinic receptor selectively inhibits a rapidly inactivated $Ca^{2+}$ current in rat sympathetic neurons. *Proc. Natl. Acad. Sci. USA* 84: 4313–17

Yaari, Y., Hamon, B., Lux, H. D. 1987. Development of two types of calcium channels in cultured mammalian hippocampal neurons. *Science* 235: 680–82

Yatani, A., Imoto, Y., Codina, J., Hamilton, S. L., Brown, A. M., Birnbaumer, L. 1988. The stimulatory G protein of adenylyl cyclase, $G_s$, also stimulates dihydropyridine sensitive Ca channels. *J. Biol. Chem.* 263: 9887–95

Yatani, A., Codina, J., Imoto, Y., Reeves, J. P., Birnbaumer, L., Brown, A. M. 1987. A G protein directly regulates mammalian cardiac calcium channels. *Science* 238: 1288–92

Yatani, A., Okabe, K., Brown, A. M. 1989. Time course of cardiac Ca current after a concentration jump of isoproterenol. *Biophys. J.* 55: 36a (Abstr.)

*Annu. Rev. Neurosci. 1990. 13:357–71*

# COCHLEAR PROSTHETICS

*Gerald E. Loeb*

Bio-Medical Engineering Unit, Queen's University, Kingston, Ontario K7L 3N6, Canada

## THE STATE OF THE ART

Cochlear prostheses are being used clinically to restore functional hearing in patients suffering from profound sensorineural deafness. The devices include one or more electrodes implanted in or near the cochlea to provide electrical stimulation of the remaining auditory nerve fibers, thereby bypassing the defective sensory hair-cells (Figure 1). There are many different designs in various stages of development, testing, and availability, with widely differing therapeutic benefit both among devices and among individual patients receiving a given device.

This review focuses on those aspects of the current development and evaluation processes that involve researchers from the fundamental neuroscience community. In the past, such a review might have focused on issues of tissue damage and stability of evoked percepts; while still important, these problems have been largely resolved by improved materials and designs, functional testing in animals (Walsh & Leake-Jones 1982, Maslan & Miller 1987), mechanical testing in cadaver temporal bones (Kennedy 1987, Webb et al 1988), and clinical experience (Yin & Segerson 1986, Waltzman et al 1986, Clark et al 1988, Terr et al 1988). Current research is geared toward integrating the clinical results from these implants with theories of auditory perception, in the hope of developing efficient approaches to improving therapeutic benefit.

In addition to the specific literature citations and the list of recent book-length treatments of the field provided here (Table 1), the opinions and evaluations expressed here are based on a survey of 120 leading researchers active in this field, 54 of whom provided detailed responses to a seven page questionnaire that was circulated in January, 1989 (De Foa and Loeb, in preparation).

0147–006X/90/0301–0357$02.00

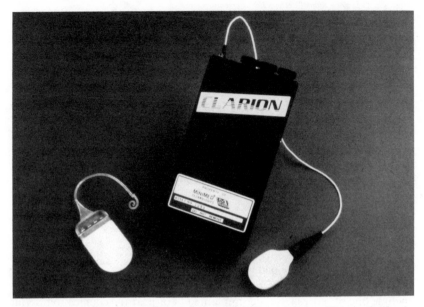

*Figure 1*   Clarion™ cochlear prosthesis produced by Minimed Corp. of Sylmar, California, showing spirally molded electrode and ceramic case (*lower left*) containing custom IC chip and related circuitry for driving the 16-contact electrode array with signals specified by the external speech processor unit (*center*) and transmitted along with power by an inductive coil (*bottom right*). (Photograph courtesy of Minimed Technologies.)

**Table 1**   Recent review volumes

---

*Cochlear Prostheses*, ed. C. W. Parkins, S. W. Anderson, *Annals of the New York Academy of Science*, Vol. 405, 1983

*Cochlear Implants*, ed. R. A. Schindler, M. M. Merzenich, New York: Raven Press, 1985

*British Journal of Audiology*, Vol. 20, number 1, February 1986

*Otolaryngology Clinics of North America*, Vol. 19, number 2, May, 1986

*Cochlear Implant: Current Situation*, ed. P. Banfai, Erkelenz, GDR: Bermann GmbH, 1988

*Cochlear Implants in Young Deaf Children*, ed. E. Owens, D. K. Kessler, Boston: College Hill Press, 1989

*Models of the Electrically Stimulated Cochlea*, ed. J. M. Miller, F. A. Spelman, New York: Springer-Verlag, 1989

---

## Socio-Economic Factors

There are at least two million functionally deaf individuals in the USA alone (defined as being unable to understand speech or most ambient sounds even with the use of an acoustic-amplification hearing aid; DiPietro 1984). Most mechanical problems with middle ear conduction of sound are now treatable surgically; most of the remaining deafness represents sensorineural pathophysiology. Although there are many proven non-aural techniques for restoring functional communication (e.g. speech-reading and sign language), a large proportion of these individuals are excluded, for various reasons, from most of the social and economic interactions that virtually define human society. Yet many of these patients have lesions confined primarily to the cochlear hair cells, with largely intact auditory nerves and central pathways. In such patients, direct electrical stimulation of the remaining auditory nerve fibers can restore this communication channel without interfering with other sensory and motor activities, in contradistinction to visual and vibrotactile displays of acoustic information, which have had very limited clinical acceptance (Rose et al 1988, Skinner et al 1988, Thornton 1988).

For the purposes of this discussion, it is useful to define three broad categories of deaf individuals:

1. POST-LINGUALLY DEAFENED ADULTS These have been the most successful and enthusiastic users of cochlear prostheses. Their communicative skills and social ties are strongly based on hearing and their acquired deafness is often a severe psychological blow. Even so, there is a wide range of awareness and attitudes toward prostheses among these patients and their therapists.

2. PRE-LINGUALLY DEAFENED ADULTS These have been the least successful or enthusiastic users because they have acclimated to life without hearing and because their mature nervous systems seem to lack the ability to learn to deal with auditory sensations.

3. PRE-LINGUALLY DEAFENED, YOUNG CHILDREN Only a limited number of implants have been performed in this group, which poses a host of ethical, technical, medical, and scientific quandries. In addition to the problems and potential of stimulating and evaluating the undeveloped auditory system (discussed below), there is much controversy on whether and how to integrate this new technology with the several different, fiercely competitive approaches to developing language skills in deaf children.

## Technology

Table 2 provides an overview of the myriad devices that have evolved (and often become extinct) during 30 years of active research and development.

**Table 2**   Cochlear prostheses

| Device Name | P.I./Institute/City | Manufacturer/City | First Desc. | Est. # Implants | Approx. Cost US $ | Availability [a] |
|---|---|---|---|---|---|---|
| - | Djourno & Eyries/Paris | - | 1957 | 1 | - | never |
| - | Doyle/Univ. Southern California/Los Angeles | - | 1964 | 1 | - | never |
| - | Simmons/Stanford/Calif. | - | 1966 | 4 | - | superceded |
| Bioear | Simmons, White/Stanford | Biostim/Brooksville, Fl | 1984 | 7 | 4,500 | defunct |
| UCSF | Michelson/Univ. California/San Francisco | - | 1971 | 5 | - | superceded |
| UCSF-Storz | Merzenich, Schindler/UCSF | Storz/St. Louis, MO | 1984 | 18 | 12,000 | withdrawn |
| Clarion | Schindler, Merzenich/UCSF + Wilson/Research Triangle/ Durham, NC | Minimed/Sylmar, CA | 1988 | 0 | 13,000 | R&D |
| 3M-House | House/House Hearing Inst./Los Angeles | 3M Corp./St.Paul, MN | 1973 | 3000 | 6,500 | withdrawn |
| Chorimac | Chouard/CHU Saint-Antoine/Paris | Bertin/Paris | 1973 | 150 | 9,000 | superseded |
| Monomac | Chouard/CHU Saint-Antoine/Paris | Bertin/Paris | 1987 | ? | ? | local |
| Minimac | Chouard/CHU Saint-Antoine/Paris | Bertin/Paris | 1988 | ? | ? | local |
| EPI | Douek, Fourcin/ Guy's Hosp./London | - | 1978 | 9 | ? | local |
| UCH-RNID | Fraser/Univ. College Hosp./London | Fine Tech./London | 1988 | 45 | 1,100 | local |
| Vienna | Hochmair/Innsbruck | - | 1978 | 70 | ? | local |
| 3M-Vienna | Burian/Hochmair/Tech. Univ./Vienna | 3M Corp./ St. Paul, MN | 1978 | 80 | 9,500 | withdrawn |
| Implex | Banfai, Hortmann/ Cologne-Duren Group | Hortmann, GmbH/ Cologne | 1978 | 100 | 6,000 | Europe superseded |
| Exco-16 | Banfai, Hortmann/ Cologne-Duren Group | Hortmann, GmbH/ Cologne | 1985 | 60 | ? | Europe |
| Ineraid | Eddington/Univ. Utah/ Salt Lake City | Symbion/ Salt Lake City, UT | 1980 | 100+ | 11,000 | IDE |
| Nucleus | Clark/Univ. Melbourne | Cochlear-Nucleus/ Sidney, Aust | 1980 | 1500 | 12,200 | PMA |
| - | Dillier, Spillman/ Universitatspital/Zurich | - | 1982 | 10 | ? | local |
| MSR-UCL | Gersdorff, Sneppe/Catholic Univ.Leuven/Brussels | Siemens | 1985 | 9 | ? | local |
| Laura | Marquet Peeters/Univ. Instelling Antwerp | Forelec/Antwerp | 1988 | 2 | ? | R&D |
| Prelco | Cazals/INSERM/Bordeaux | Racia/Bordeaux | 1985 | 23 | 2,000 | local |
| Medtronic | Fraysse/Toulouse | - | - | 17 | ? | local |
| - | Gerhardt/Humboldt Univ./Berlin | - | 1987 | 10 | ? | local |
| Omina | Bosch/Hosp.de la Cruz Roja/ Barcelona | - | - | 12 | ? | local |
| ECME | Bochanek/Acad. Med./Warsaw | - | - | 3 | ? | local |
| - | Valvoda, Tichy/Acad Science/Prague | - | 1988 | 3 | ? | local |

Compiled from ASHA (1986), Chouard et al (1988), market survey by Cochlear A.G. (1988) and references cited.

**a**   U.S. Food and Drug Administration designations: IDE=investigational implants only at approved centers; PMA=approved for marketing to licensed practitioners

**b**   Number of electrically separate, active contacts, excluding reference or ground.

**c**   M=directly in auditory nerve in modiolus; X= extracochlear

**d**   Number of channels commonly activated in parallel; Nucleus device sweeps 2 output channels among 21 bipolar sites.

| Contacts [b] | Electrode Depth mm [c] | Channels [d] | Stimulus Waveform [e] | Coupling [f] | Reference |
|---|---|---|---|---|---|
| 1 | M | 1 | pulse | perc. | Djourno & Eyries (1957) |
| 1 | ? | 1 | AM | trans. | Doyle et al (1964) |
| 8 | M | 8 | pulse | trans. | Simmons (1983) |
| 1 | X-3 | 1 | analog or pulse | trans. | Dent et al (1988) |
| 2 | 15 | 1 | analog | trans. | Michelson (1971) |
| 16 | 24 | 4 | analog | trans. | Schindler and Kessler (1987) |
| 16 | 24 | 8 | analog | trans. | Wilson et al (1988) |
| 1 | 6 | 1 | AM or pulse | trans. | Fretz and Fravel (1985) |
| 8, 12 | 0-3 [g] | 8, 12 | analog | trans. | Chouard (1978) |
| 1 | X-10 | 1 | pulse | trans. | Chouard et al (1988) |
| 15 | ? [h] | 15 | pulse | trans. | Chouard et al (1988) |
| 1 | X | 1 | pulse | ext. | Walliker et al (1985) |
| 1 | X | 1 | analog | ext. | Conway (1988) |
| 8 | 14 | 1 | analog | trans. | Burian et al (1986) |
| 1 | X | 1 | analog | trans. | Burian et al (1986) |
| 1,4,8 | X | 1,4,8 | analog | perc./trans. | Banfai et al (1979) |
| 16 | X | 16 | pulse | perc./trans. | Banfai et al (1986) |
| 6 | 22 | 4 | analog | perc. | Parkin and Stewart (1988) |
| 22 | 25 | 2/21 | pulse | trans. | Franz et al (1987) |
| 1 | X | 1 | pulse | trans. | Spillman et al (1982) |
| 1,2 | X-5 | 1,2 | analog | trans. | Gersdorff et al (1988) |
| 16 | 18 | 8, 16 | analog or pulse | trans. | Marquet (1986) |
| 1 | X | 1 | analog | trans. | Negrevergne et al (1988) |
| 1 | X-2 | 1 | pulse | trans. | - |
| 2 | X | 2 | analog | trans. | Gerhardt and Wagner (1987) |
| 1,6,8 | ? | 1,6,8 | ? | ? | - |
| 1 | X | 1 | ? | ? | - |
| 1 | X | 1 | analog | trans. | Valvoda et al (1988) |

e   Analog=band-pass filtered acoustic signal; Pulse=brief stimuli, usually at $F_0$ rate, amplitude-modulated by acoustic envelope; AM=unrectified carrier modulated by acoustic signal

f   perc.=percutaneous plug; trans.=transcutaneous inductive coupling; ext.=exteriorized, removable appliance.

g   Electrodes inserted through individual fenestrations in the lateral wall of the cochlea.

h   In Scala tymponi via round window.

The landscape has changed significantly within the past six to eight years, with commercially designed and manufactured devices supplanting the "home-built" efforts of university shops but still undergoing attrition among themselves. For the moment, the commercial field is dominated by the multichannel device manufactured by Nucleus Corp. from Australia, which is the only device that is now approved by the Food and Drug Administration and marketed in the United States (as of July 1989).

The various designs are distinguished by the set of decisions that they embody regarding the following options:

SINGLE VS. MULTICHANNEL ELECTRODE    Speech perception requires a signal with at least 3 kHz bandwidth, whereas the ability of neurons to follow each cycle of acoustic or electrical stimulation is limited to about 300–500 pps. (Curiously, the phase-locking that appears to contribute to acoustic pitch discrimination in the 500–5000 Hz band has no clear influence on electrically evoked percepts; Loeb et al 1983b). The intact auditory system solves this by spatial filtering and parallel processing along the basilar membrane. Multichannel prostheses attempt to replicate this function as well as simple transduction by having multiple, independently addressable electrode-contacts located near different subpopulations of auditory nerve fibers. In general, the percept elicited by each has a noisy timbre but a distinct pitch related to the tonotopic map of the auditory nerve fibers in the cochlea.

INTRACOCHLEAR VS. EXTRACOCHLEAR ELECTRODE    Both single and multi-channel devices have had their electrode contacts positioned outside of the cochlea (usually on the round window and bone overlying the cochlear turns; "X" in Table 2), within the cochlea (usually on a flexible, slender probe inserted along the scala tympani), and in the auditory nerve itself as it passes through the modiolar bone ("M" in Table 2). More recently, feasibility studies have been conducted on placing electrodes on and in the cochlear nucleus, thereby bypassing the auditory nerve entirely (McElveen et al 1985, 1987).

PERCUTANEOUS VS. TRANSCUTANEOUS COUPLING    The acoustic signals are picked up and processed by a wearable unit similar to a hearing aid, which formats the stimulus waveforms and transmits them to the electrode. This can be done via a percutaneous connector affixed to the skull and passing through the scalp or via transcutaneous inductive-coupling of radio-frequency signals. For multichannel systems, such RF signals often include sophisticated digital encoding schemes and power transmission to drive custom integrated circuitry in a hermetically encapsulated implant (e.g. see Figure 1).

SPEECH PROCESSING    Three general approaches have emerged for transforming the acoustic signal into stimulus waveforms. The simplest is to use the acoustic waveform itself (so-called "analog" stimulation), suitably band-filtered and compressed in its dynamic range to conform to the sensitivity of neurons to electrical stimulation (White 1986). Multichannel versions of such an approach employ a bank of filters, driving each electrode in parallel according to the output of its corresponding bandpass filter. The rationale is that the nervous system may be able to make some use of the information contained in the raw acoustic waveforms, although the biophysical events leading to spike initiation suggest that the resultant fine temporospatial patterning is highly unphysiological (Kiang et al 1979). More recently, improved channel isolation and overall speech performance have been achieved by converting the envelope of the signal from each filter into a set of narrow pulses that can be delivered in a basally to apically sequenced, non-overlapping pattern to each electrode at a repetition rate that is determined by the voiced-fundamental pitch (Wilson et al 1988a,b). These approaches, based on the frequency-channel vocoder, contrast with the formant-extraction method (e.g. Nucleus) in which an on-line microprocessor tracks the spectral location and relative amplitude of one or two vowel formants and selects one or two electrode contacts for stimulation based on a previously stored map of the pitch sensations that are elicited at each available site (Franz et al 1987).

## ISSUES IN NEUROSCIENCE

### Biophysics of Auditory Nerve Stimulation

Obviously, a larger number of functionally separate stimulation channels permits a more faithful representation of the speech signal in the evoked neural activity and, generally, provides better speech comprehension. The problem is in defining and obtaining "functional separation." In the typical scala tympani approach, the electrode contacts lie 0.5–1.0 mm from the spiral ganglion cells that are embedded in the medial wall of the scala. Stimulus current injected through one such contact tends to spread diffusely in the volume-conductive fluids and tissues that surround it. At threshold, the first auditory nerve fibers to be recruited will be those closest to the contact, but any attempt to provide a useful dynamic range of stimulus intensities produces rapid spread of activation to nearby neurons in the same and even adjacent turns of the cochlear spiral. These more distant neurons are intended to be under the separate, perhaps simultaneous control of other stimulating electrodes.

Theoretical considerations (Finley et al 1989) and empirical studies (Merzenich & White 1977) both indicate that the best spatial selectivity can be achieved by using bipolar pairs of contacts that are oriented radially (perpendicularly) to the axis of the cochlear spiral. Such a configuration produces a potential gradient that is oriented parallel to the long axis of the overlying spiral ganglion cells, whose apical dendrites (formerly innervating hair cells in the organ of Corti) and central axons are similarly radially oriented. Under ideal conditions, with the contacts oriented optimally against the medial wall of the canal, such an electrode provides a space-constant of about 0.87 mm (10 dB attenuation per millimeter distance away) for stimulus spread both apically and basally from the bipolar pair. This space constant would be compatible with the use of eight such pairs arranged at 2 mm intervals in the region 10–24 mm from the round window where the critical speech frequencies are normally transduced (Vivion et al 1981). In contrast monopolar electrodes have a space constant of 13.0 mm (0.67 dB/mm) and longitudinally oriented bipolar pairs (Black & Clark 1980) have a space constant of 2–4 mm (4.3–2.2 dB/mm); extra-cochlear contacts are even more severely compromised in their selectivity. Attempts to focus the electrical fields by introducing antiphasic waveforms on adjacent electrodes have produced relatively little improvement (Ifukube & White 1987), presumably because the large anodal currents that are required result in virtual cathodes and new sites of spike initiation at more distant nodes of Ranvier along the auditory nerve fibers (Ranck 1975).

Considerable circumstantial evidence suggests that much of the variability in the functional benefit realized by individual patients using a multichannel implant is related to the histopathology of the spiral ganglion cells, which is highly variable among patients with the same nominal etiology of their deafness and often heterogeneous in different regions of the same cochlea (Hinojosa et al 1987). Patients with poorer clinical results generally have higher current thresholds to achieve any auditory sensation, narrower dynamic ranges before reaching maximal loudness, and psychophysical test results suggestive of overlapping recruitment when adjacent channels are activated simultaneously (Shannon 1983, White 1984). Detailed modeling work using finite-element analysis is under way to determine the optimal electrode configuration(s) for different regions and conditions of the cochlea (Finley et al 1989).

Unfortunately, most of the design features that make an electrode array biophysically ideal also make it extremely tedious to fabricate and difficult to insert atraumatically. To get the electrode contacts to lie on the medial wall of the canal, either the electrode must fit snugly into the scala tympani or it must spring spontaneously into a tightly curved spiral. To get the electrode into the scala tympani without rupturing the overlying basilar membrane (which sequesters potassium-rich endolymph that is toxic to neurons), the surgeon must slide it in without pushing, while working through the long, narrow access afforded by the external auditory canal. To date, multichannel prostheses have opted either for electrodes that are known to be far from optimal (e.g. monopolar—Chorimac, Ineraid, Exco-16; longitudinal bipolar—Nucleus) or have paid the price of very slowly progressing research and development programs (e.g. near-radial bipolar—University of California at San Francisco; Loeb et al 1983a).

Now that reasonable numbers of patients have been implanted with various electrode designs, it should be possible to determine which electrode design features are most important for psychophysical function. This knowledge, it is hoped, will lead to design compromises that are based on informed trade-offs rather than historical accident.

## Prognostic Testing

The above-noted patient variability in functional results now represents the single largest hindrance to the widespread clinical application and commercial success of cochlear prosthetics. The problem is that the number of post-lingually deafened adults with profound deafness (no detectable acoustic threshold) is quite small, whereas a much larger number have some residual hearing but are unable to make effective use of a well-fitted acoustic hearing aid. Implantation of a multichannel prosthesis generally results in the loss of any residual acoustic hearing, and therefore is not justifiable unless its function in that patient can be guaranteed to be better than what will be lost. At present, the results from such a prosthesis can vary from sufficient speech recognition for conversation over the telephone to only a general awareness of the rhythm of ambient sounds (Gantz et al 1988).

Obviously, a preoperative evaluation of the condition of the spiral ganglion cells and CNS auditory pathways is highly to be desired, particularly if, as expected, it can be shown to correlate with prosthetic function. Unfortunately, the few attempts to develop and validate such tests to date have encountered technological problems and paradoxical results.

AUDITORY SENSATIONS FROM PROMONTARY STIMULATION   It is a relatively simple office procedure to pass a needle electrode through the tympanic membrane so that its tip rests against the boney promontory next to the round window. Electrical stimulation in deaf subjects usually produces auditory sensations, confirming the presence of a functional auditory nerve (Chabolle et al 1988). Paradoxically, such stimulation in subjects with normal hearing produces no auditory sensations (Eddington et al 1978, Liard et al 1988). In the absence of a plausible mechanism for this phenomenon, there seems to be little enthusiasm for developing detailed psychophysical tests based on this procedure.

ELECTRICALLY EVOKED AUDITORY BRAINSTEM RESPONSES (EABR)    The highly organized, tonotopic projections throughout the auditory CNS result in coherent, remote field potentials in response to both acoustic and electrical stimulation. Because of their temporal coherence, even the very small signals recordable noninvasively as scalp potentials can be revealed by stimulus-triggered signal averaging (Chouard et al 1979, Dobie & Kimm 1980, Starr & Brackmann 1981, Waring et al 1985, Kileny & Kemink 1987). This has been used to evaluate prosthesis function in children who are too young to provide psychophysical data (Miyamoto & Brown 1987). The rate of growth of such potentials with increasing stimulus strength has been used to evaluate the spatial selectivity of intracochlear electrodes (Gardi 1985) and has been suggested as a potentially useful way to interpret promontory stimulation in terms of auditory nerve survival (Simmons & Smith 1983). Unfortunately, the earliest waves of the EABR are the most likely to correlate with nerve survival (particularly in view of the perceptual paradox noted above), and these are easily obscured by electrical artifact from the stimulation. Special techniques are needed to alternate stimulus phases to obtain cancellation of the artefact (Gardi 1985, Banfai et al 1986) and to protect high-gain amplifiers from saturation; these have not proven easy to integrate into the normal repertoire of audiometric procedures. In theory, it should be possible to obtain fairly detailed representations of the integrity of the auditory pathways by constructing three-dimensional vectors of the EABR (based on orthogonal lead placements) and using current-source-density analysis to relate this to the tonotopic representation in various projections (Gardi 1985). If these could be shown to relate well to the function of cochlear prostheses, they would easily justify substantial developmental costs (Stypulkowski et al 1986).

SCANNING TECHNIQUES    Computed X-ray tomography is now routinely used to screen intracochlear prosthesis candidates for patency of the scala tympani. Resolution is now marginally adequate and likely to continue to improve. However, there have been and will probably continue to be reports of both unexpected obstructions to electrode insertion and successful insertions obtained by drilling through clearly ossified barriers (Balkany et al 1988, Gantz et al 1988). Atrophy of the cochlear nucleus has been noted in experimentally deafened animals (Webster & Webster 1979); it might be detectable with high-resolution magnetic resonance imaging (MRI). Magnetoencephalogram (MEG) scanning (Hari et al 1988) is also a distant but promising way to get around technical problems with the EABR.

ETIOLOGICAL FACTOR ANALYSIS    Attempts to correlate prosthetic outcome with patient-history items such as cause and duration of deafness, age and rate of onset, audiometric pattern, etc have been discouraging (Chouard et al 1987, Fritze & Eisenwort 1988, Geers & Moog 1988). Many deaf patients have no defined etiology and many probably had gradual hearing loss that went undetected for many years. Furthermore, even when these factors are well known, there seems to be little correlation with the histopathology of spiral ganglion cells (Hinojosa et al 1987).

## Development and Plasticity

The application of cochlear prostheses to prelingually deaf children is undoubtedly the greatest challenge and research opportunity for neuroscientists in this field (reviewed by Loeb 1989). Two hypotheses underlie the urgency and the promise:

CRITICAL PERIOD HYPOTHESIS    The auditory and speech centers in the CNS (like the visual system) have critical periods in their natural development, at which times a sustained absence of organized sensory input will lead to a permanent and irreversible deficit in their ability to process information.

PLASTICITY HYPOTHESIS    During these critical periods, the plasticity of the nervous system permits it to develop information processing strategies that are, in some way, optimized to extract the information that is embedded in complexly encoded and often noisy sensory channels.

If valid, these hypotheses suggest that cochlear prostheses should be implanted in deaf children at the earliest possible age and that such children will learn to make much better use of the distorted input thus provided than would an adult with or without previous hearing. Unfortunately, evidence supporting these hypotheses is still limited and difficult to obtain. Evidence from animals with experimentally induced auditory deprivation supports the notion of a post-natal critical period and the efficacy of chronic electrical stimulation in preventing disuse atrophy (Wong-Riley et al 1981, Lousteau 1987; although see Balkany et al 1986). Certainly the results of cochlear implants in adults who were deafened prelingually are quite poor (Tong et al 1988). However, in the absence of animal models for speech comprehension and production, questions about functional plasticity will probably have to be resolved through carefully designed psychophysical experiments on children who have received cochlear implants.

Long-term data from such children are just starting to be available but are very difficult to interpret. In addition to the obvious problems of designing and administering appropriate tests, the only two devices used to date in young children are the least likely to benefit from plasticity, at least in the precortical brain. The 3M-House implant provides a single-channel signal consisting of an unrectified 16 kHz carrier that is over-modulated by the acoustic envelope; this permits very simple implanted electronics but provides almost no neural signal beyond prosody. However, even this device seems to work much better in some children than it has in adults (Berliner & Eisenberg 1987). The multi-channel Nucleus system works much better than the 3M-House device in post-lingually deafened adults (Gantz et al 1988), but its formant-extraction approach is intended specifically to evoke percepts related to the speech-recognition strategies of individuals with previously normal hearing. None of the raw spectral information that might support alternative speech-recognition strategies is represented in the output stimuli. Even so, the few reports to date from young children suggest excellent results, with slower initial learning curves than in adults but continued improvement over longer periods of time, particularly regarding the high-level cognitive and speech-production skills that are particularly difficult to teach to deaf children (Luxford et al 1988, Berliner et al 1988).

## CONCLUSIONS

In the early days, the clinical pioneers ignored and were usually ignored by the basic research community. It is now widely recognized that the resultant blind empiricism is no longer an effective way to advance this complex art. Cochlear prosthetics provides neuroscientists with an almost unprecedented opportunity to apply basic knowledge to an important and rapidly evolving area of clinical practice, while at the same time offering unique, high-technology tools to conduct psychophysical experiments on human CNS functions that have no animal models. For the individual investigator, the problem is to identify a suitable niche in the complex and shifting social order of commercial manufacturers, clinical trials, and engineering support.

*Literature Cited*

American Speech-Language Hearing Association. 1986. Cochlear implants. Report of the Ad Hoc Committee on Cochlear Implants. *ASHA* 28(4): 29–52

Balkany, T., Gantz, B., Nadol, J. B. Jr. 1988. Multichannel cochlear implants in partially ossified cochleas. *Ann. Otol. Rhinol. Laryngol. Suppl.* 135: 3–7

Balkany, T., Reite, M., Rasmussen, K., Stypulkowski, P. 1986. Behavioral effects of cochlear prosthesis on deafened mon-

keys. *Otolaryngol. Clin. North Am.* 19(2): 435–46

Banfai, P., Hortmann, G., Kubik, S., Wustrow, F. 1979. Cochlear implant mit multielektroden ohne eröffnung der innenohrräume. *Laryngol. Rhinol. Otol.* 58: 526–34

Banfai, P., Karczag, A., Kubik, S., Lüers, P., Sürth, W. 1986. Extracochlear sixteen-channel electrode system. *Otolaryngol. Clin. North Am.* 19(2): 371–408

Berliner, K. I., Eisenberg, L. S. 1987. Our experience with cochlear implants: Have we erred in our expectations? *Am. J. Otol.* 8(3): 222–29

Berliner, K. I., Tonokawa, L. L., Brown, C. J., Dye, L. M. 1988. Cochlear implants in children: Benefits and concerns. *Am. J. Otol.* 9: 86–92

Black, R. C., Clark, G. M. 1980. Differential electrical stimulation of the auditory nerve. *J. Acoust. Soc. Am.* 67: 868–74

Burian, K., Hochmair-Desoyer, I. J., Eisenwort, B. 1986. *Otolaryngol. Clin. North Am.* 19(2): 313–28

Chabolle, F., Garabedian, N., Meyer, B., Chouard, C. H. 1988. Resultats et indications therapeutiques fournis par la stimulation electrique de la fenetre ronde dans 581 cas de surdite totale bilaterale. *Ann. Oto-Laryng.* 105: 237–41

Chouard, C. H. 1978. Multiple intra-cochlear electrodes for rehabilitation in total deafness. *Oto-Laryngol. Clin. North Am.* 11: 217–53

Chouard, C. H., Fugain, C., Meyer, B., Chabolle, F. 1987. Indications for multi- or single-channel cochlear implant for rehabilitation for total deafness. *PACE* 10(1 Pt. 2): 237–39

Chouard, C. H., Meyer, B., Danadieu, F. 1979. Auditory brainstem potentials in man evoked by electrical stimulation of the round window. *Acta Otolaryngol.* 87: 287–93

Chouard, C. H., Weber, J. L., Meyer, B., Chabolle, F., Fugain, C. 1988. Les implants cochléaires monocanal monomac et multicanaux minimac. *Ann. Oto-Laryng.* 105: 227–36

Clark, G. M., Shepherd, R. K., Franz, B. K., Dowell, R. C., Tong, Y. C., et al. 1988. The histopathy of the human temporal bone and auditory central nervous system following cochlear implantation in a patient. Correlation with psychophysics and speech perception results. *Acta Otolaryngol. Suppl.* 448: 1–65

Conway, J. M. 1988. UCH/RNID Cochlear Implant Programme—Description of sound processor and implant. In *Cochlear Implant: Current Situation*, ed. P. Banfai, pp. 279–82. Erkeleng, FRG: Rudolf Bermann GmbH

Dent, L. J., Simmons, F. B., Roberts, L. A., Lusted, H. S., White, R. L., Goesele, K. H. 1988. Four years' experience with a single-channel scala tympani implant system and percutaneous connector. See Conway 1988, pp. 523–45

DiPietro, L. 1984. *Deafness: A Fact Sheet.* Washington, DC: Gallaudet College

Djourno, A., Eyries, C. 1957. Prothèse auditive par excitation electrique à distance du nerf sensoriel à l'aide d'un bobinage inclus à demeure. *Presse Med.* 35: 14–17

Dobie, R. A., Kimm, J. 1980. Brainstem responses to electrical stimulation of the cochlea. *Arch. Otolaryngol.* 106: 573–77

Doyle, J. H., Doyle, J. B., Turnbull, F. M. 1964. Electrical stimulation of the eighth cranial nerve. *Arch. Otolaryngol.* 80: 388–91

Eddington, D. K., Dobell, W. H., Mladejousky, M. G., Brackmann, J. E., Parkin, J. L. 1978. Auditory prostheses research with multiple channel intracochlear stimulation in man. *Am. Otol. Rhinol. Laryngol. Suppl. 53* 87: 5–38

Finley, C. C., Wilson, B. S., White, M. W. 1989. Models of neural responsiveness to electrical stimulation. In *Models of the Electrically Stimulated Cochlea*, ed. J. M. Miller, F. A. Spelman. New York: Springer-Verlag. In press

Franz, B. K., Dowell, R. C., Clark, G. M., Seligman, P. M., Patrick, J. F. 1987. Recent developments with the nucleus 22-electrode cochlear implant: A new two formant speech coding strategy and its performance in background noise. *Am. J. Otol.* 8(6): 516–18

Fretz, R. J., Fravel, R. P. 1985. Design and function: A physical and electrical description of the 3M-House cochlear implant system. *Ear Hear Suppl. 3* 6: 205–35

Fritze, W., Eisenwort, B. 1988. Predictability of the result following cochlear implantation. *HNO* 36(8): 332–34

Gantz, B. J., McCabe, B. F., Tyler, R. S. 1988. Use of multichannel cochlear implants in obstructed and obliterated cochleas. *Otolaryngol. Head Neck Surg.* 98(1): 72–81

Gantz, B. J., Tyler, R. S., Knutson, J. F., Woodworth, G., Abbas, P., et al. 1988. Evaluation of five different cochlear implant designs: Audiologic assessment and predictors of performance. *Laryngoscope* 98(10): 1100–6

Gardi, J. N. 1985. Human brain stem and middle latency responses to electrical stimulation: Preliminary observations. In *Cochlear Implants*, ed. R. A. Schindler, M. M. Merzenich, pp. 351–63. New York: Raven

Geers, A. E., Moog, J. S. 1988. Predicting long-term benefits from single-channel cochlear implants in profoundly hearing-impaired children. *Am. J. Otol.* 9(2): 169–76

Gerhardt, H. J., Wagner, H. 1987. The Berlin cochlear implant—surgical concept. *Acta Otolaryngol.* 103(5–6): 628–31

Gersdorff, M., Sneppe, R., Wittemans, S., Vanderbemden, S., Montmirail, C. 1988. The MSR-UCL cochlear implant: New adaptations. *Ann. Otolaryngol. Chir. Cervicofac.* 105(4): 261–63

Hari, R., Pelizzone, M., Makela, P., Hallstrom, J., Huttunen, J., Knuutila, J. 1988. Neuromagnetic responses from a deaf subject to stimuli presented through a multichannel cochlear prosthesis. *Ear Hear* 9(3): 148–52

Hinojosa, R., Blough, R. R., Mhoon, E. E. 1987. Profound sensorineural deafness: A histopathological study. *Ann. Otol. Rhinol. Laryngol. Suppl. 128* 96: 43–46

Ifukube, T., White, R. L. 1987. A speech processor with lateral inhibition for an eight channel cochlear implant and its evaluation. *IEEE Trans. Biomed. Eng.* 34(11): 876–82

Kennedy, D. W. 1987. Multichannel intracochlear electrodes: Mechanism of insertion trauma. *Laryngoscope* 97(1): 42–49

Kiang, N. Y., Eddington, D., Delgutte, B. 1979. Fundamental considerations in designing auditory implants. *Acta Otolaryngol.* 87: 204–18

Kileny, P. R., Kemink, J. L. 1987. Electrically evoked middle-latency auditory potentials in cochlear implant candidates. *Arch. Otolaryngol. Head Neck Surg.* 113(10): 1072–77

Liard, P., Pelizzone, M., Rohr, A., Montandon, P. 1988. Noninvasive extratympanic electrical stimulation of the auditory nerve. *ORL J. Otorhinolaryngol. Relat. Spec.* 50(3): 156–61

Loeb, G. E. 1989. Neural prosthetic strategies for young children. In *Cochlear Implants in Young Deaf Children*, ed. E. Owens, D. K. Kessler, pp. 137–52. Boston: College-Hill

Loeb, G. E., Byers, C. L., Rebscher, S. J., Casey, D. E., Fong, M. M., Schindler, R. A., Gray, R. F., Merzenich, M. M. 1983a. The design and fabrication of an experimental cochlear prosthesis. *Med. Biol. Eng. Comput.* 21: 241–54

Loeb, G. E., White, M. W., Merzenich, M. M. 1983b. Spatial cross-correlation—A proposed mechanism for acoustic pitch perception. *Biol. Cybern.* 47: 149–63

Lousteau, R. J. 1987. Increased spiral ganglion cell survival in electrically stimulated, deafened guinea pig cochleae. *Laryngoscope* 97(7 Pt. 1): 836–42

Luxford, W. M., House, W. F., Hough, J. V., Tonokawa, L. L., Berliner, K. I., Martin, E. 1988. Experiences with the nucleus multichannel cochlear implant in three young children. *Ann. Otol. Rhinol. Laryngol. Suppl.* 135: 14–16

Marquet, J., Van Durme, M., Lammens, J., Collier, R., Peeters, S., Bosiers, W. 1986. Acoustic simulation experiments with pre-processed speech for an 8-channel cochlear implant. *Audiology* 25(6): 353–62

Maslan, M. J., Miller, J. M. 1987. Electrical stimulation of the guinea pig cochlea. *Otolaryngol. Head Neck Surg.* 96(4): 349–61

McElveen, J. T. Jr., Hitselberger, W. E., House, W. F. 1987. Surgical accessibility of the cochlear nuclear complex in man: Surgical landmarks. *Otolaryngol. Head Neck Surg.* 96(2): 135–40

McElveen, J. T. Jr., Hitselberger, W. E., House, W. F., Mobley, J. P., Terr, L. I. 1985. Electrical stimulation of cochlear nucleus in man. *Am. J. Otol.* 6: 81–88

Merzenich, M. M., White, M. W. 1977. Cochlear implant. The interface problem. In *Functional Electrical Stimulation: Applications in Neural Prostheses*, ed. F. T. Hambrecht, J. B. Reswick, pp. 321–40. New York: Marcel Dekker

Michelson, R. P. 1971. Electrical stimulation of the human cochlea. *Arch. Otolaryng.* 93: 317–23

Miyamoto, R. T., Brown, D. D. 1987. Electrically evoked brainstem responses in cochlear implant recipients. *Otolaryngol. Head Neck Surg.* 96(1): 34–38

Negrevergne, M., Dauman, R., Lagourgue, P., Bourdin, M. 1988. The Prelco monocanal extra-cochlear implant. *Rev. Laryngol. Otol. Rhinol.* 109(3): 273–76

Parkin, J. L., Stewart, B. E. 1988. Multichannel cochlear implantation: Utah design. *Laryngoscope* 98(3): 262–65

Ranck, J. B. Jr. 1975. Which elements are excited in electrical stimulation of mammalian central nervous system: A review. *Brain Res.* 98: 417–40

Rose, D. E., Haymond, J., Facer, G. W. 1988. Vibrotactile device in patients with single-channel cochlear implant: A preliminary study. *Otolaryngol. Head Neck Surg.* 99(1): 42–45

Schindler, R. A., Kessler, D. K. 1987. The UCSF/Storz cochlear implant: Patient performance. *Am. J. Otol.* 8(3): 247–55

Shannon, R. V. 1983. Multichannel electrical stimulation of the auditory nerve in man. II. Channel interaction. *Hearing Res.* 12: 1–16

Simmons, F. B. 1983. Percepts from modioiar (eighth nerve) stimulation. *Ann. NY Acad. Sci.* 405: 259–63

Simmons, F. B., Smith, L. 1983. Estimating nerve survival by electrical ABR. *Ann. NY Acad. Sci.* 405: 422–23

Skinner, M. W., Binzer, S. M., Fredrickson, J. M., Smith, P. G., Holden, T. A., Holden, L. K., Juelich, M. F., Turner, B. A. 1988. Comparison of benefit from vibrotactile aid and cochlear implant for postlinguistically deaf adults. *Laryngoscope* 98(10): 1092–99

Spillman, T., Dillier, N., Guntensperger, J. 1982. Electrical stimulation of hearing by implanted cochlear electrodes in humans. *Appl. Neurophysiol.* 45: 32–37

Starr, A., Brackmann, D. 1981. Brain stem potentials evoked by electrical stimulation of the cochlea in human subjects. *Ann. Otol. Rhinol. Laryngol.* 88: 550–56

Stypulkowski, P. H., van den Honert, C., Kvistad, S. D. 1986. Electrophysiologic evaluation of the cochlear implant patient. *Otolaryngol. Clin. North Am.* 19(2): 249–57

Terr, L. I., Linthicum, F. H. Jr., House, W. F. 1988. Histopathologic study of the cochlear nuclei after 10 years of electrical stimulation of the human cochlea. *Am. J. Otol.* 9(1): 1–7

Thornton, A. R. 1988. The acceptability of cochlear implants and vibrotactile aids. *Br. J. Audiol.* 22(2): 105–12

Tong, Y. C., Busby, P. A., Clark, G. M. 1988. Perceptual studies on cochlear implant patients with early onset of profound hearing impairment prior to normal development of auditory, speech and language skills. *J. Acoust. Soc. Am.* 84(3): 951–62

Valvoda, M., Betka, J., Hurby, J. 1988. The first experience with cochlear implants in Czechoslovakia. See Conway 1988, pp. 235–36

Vivion, M. C., Merzenich, M. M., Leake-Jones, P. A., White, M. W., Silverman, M. 1981. Electrode position and excitation patterns for a model cochlear prosthesis. *Ann. Otol. Rhinol. Laryngol. Suppl. 82* 90: 19–20

Walliker, J. R., Douek, E. E., Frampton, S., Abberton, E., Fourcin, A. J., Howard, D. M., Nevard, S., Rosen, S., Moore, B. C. 1985. Physical and surgical aspects of external single channel electrical stimulation of the totally deaf. See Gardi 1985, pp. 143–55

Walsh, S. M., Leake-Jones, P. A. 1982. Chronic electrical stimulation of auditory nerve in cat: Physiological and histological results. *Hearing Res.* 7: 181–304

Waltzman, S. B., Cohen, N. L., Shapiro, W. H. 1986. Long-term effects of multichannel cochlear implant usage. *Laryngoscope* 96(10): 1083–87

Waring, M. D., Don, M., Brimacombe, J. A. 1985. ABR assessment of stimulation in induction coil implant patients. See Gardi 1985, pp. 375–78

Webb, R. L., Clark, G. M., Shepherd, R. K., Franz, B. K., Pyman, B. C. 1988. The biologic safety of the Cochlear Corporation multiple-electrode intracochlear implant. *Am. J. Otol.* 9(1): 8–13

Webster, D., Webster, M. 1979. Effects of neonatal conductive hearing loss on brainstem auditory nuclei. *Ann. Otol. Rhinol. Laryngol.* 88: 684–88

White, M. W. 1984. The multichannel cochlear prosthesis: Channel interactions. In *IEEE Front. Eng. Comput. Health Care*, pp. 396–400

White, M. W. 1986. Compression systems for hearing aids and cochlear prostheses. *J. Rehabil. Res. Dev.* 23(1): 25–39

Wilson, B. S., Finley, C. C., Farmer, J. F. Jr., Lawson, D. T., Weber, B. A., Wolford, R. D., Kenan, P. D., White, M. W., Merzenich, M. M., Schindler, R. A. 1988a. Comparative studies of speech processing strategies for cochlear implants. *Laryngoscope* 98: 1069–77

Wilson, B. S., Schindler, R. A., Finley, C. C., Kessler, D. K., Lawson, D. T., Wolford, R. D. 1988b. Present status and future enhancements of the UCSF cochlear prosthesis. See Conway 1988, pp. 395–427

Wong-Riley, M. T. T., Leake-Jones, P., Walsh, S., Merzenich, M. M. 1981. Maintenance of neuronal activity by electrical stimulation of unilaterally deafened cats demonstable with cytochrome oxidase technique. *Ann. Otol. Rhinol. Laryngol. Suppl. 82* 90: 30–32

Yin, L., Segerson, D. A. 1986. Cochlear implants: Overview of safety and effectiveness. The FDA evaluation. *Otolaryngol. Clin. North Am.* 19(2): 423–33

*Annu. Rev. Neurosci. 1990. 13:373–85*

# A NEURAL MODEL FOR CONDITIONED TASTE AVERSIONS

*Kathleen C. Chambers*

Department of Psychology, University of Southern California,
Los Angeles, California 90089-1061

From the time the parameters defining conditioned food aversions (CFAs, learned aversions to a food or fluid when consumption of that substance is followed by illness) were determined, this learning situation has fallen outside the main conceptualizations of the traditional forms of classical and instrumental learning (Garcia & Koelling 1966, Garcia et al 1966). The two main characteristics that distinguish CFAs from the traditional learning paradigms are learning after a long delay between the food stimulus and the illness (up to several hours) and strong and persistent learning after a single pairing of the food stimulus and illness. It is history, now, that these differences altered the theoretical framework of learning and memory. After the publication and acceptance of the seminal papers of Garcia (Garcia & Koelling 1966, Garcia et al 1955, 1966), there was a flurry of research on CFAs (Riley & Baril 1976). But as is apparently true of all things novel, habituation set in and interest in this area waned. Now with the growth of the field of behavioral neuroscience and the successful application of neurobiological techniques to the study of learning and memory, interest in this maverick of learning is again increasing.

## NEURAL MODEL FOR CLASSICAL CONDITIONING: A POOR MODEL FOR CONDITIONED TASTE AVERSIONS

The most significant progress in identifying and characterizing the neuronal substrates of learning and memory has been made for classically conditioned situations, e.g. autonomic conditioning of heart rate (Cohen 1982, Kapp et al 1982) and eye blink conditioning (Thompson 1986). In

373

0147–006X/90/0301–0373$02.00

these Pavlovian paradigms, a stimulus (unconditioned stimulus, US) that elicits a response (unconditioned response, UR) is paired with another stimulus (conditioned stimulus, CS) that does not elicit the UR. The two stimuli are paired so that they are contiguous and so that the CS can provide information about the US (Rescorla 1988). Learning is inferred when presentation of the CS produces a response similar to the UR. The determination of the neural substrates for this learning situation has involved the identification of four pathways: the US, UR, CS, and CR pathways.

Although CFA learning is thought to be a form of classical conditioning, it does not fit within this four-pathway model. The food stimulus has been identified as the CS, the illness as the US, and the avoidance of the food as the CR. There is, however, no clearly identifiable UR; a CFA can occur without an overt UR (Garcia et al 1972). A three-pathway conceptualization has been implicit in most discussions of the neural basis of CFAs. Discussions have focused on how the food stimulus and illness are integrated neurally to produce a new response to the food. But a detailed analysis of the learning situation for one form of CFA learning, conditioned taste aversion (CTA), suggests that this conceptualization is not adequate.

Taste stimuli have been known to elicit behavioral responses prior to food absorption in addition to the well-known physiological responses, salivation and increased insulin release (Fischer et al 1972, Grill & Berridge 1985, Pavlov 1927, Steiner 1979). The most preferred tastes, such as sweet, evoke increased consummatory responses and the least preferred tastes, such as bitter, evoke reduced consummatory responses and food spillage (Carpenter 1956, Rozin 1967). More recently, Grill & Norgren (1978) have described more complex behavioral responses elicited by different novel taste stimuli in rats. The rats were fitted with an intraoral catheter and the tastes were delivered directly into the mouth. The animals exhibited essentially two different patterns of stereotyped mouth, tongue, head, paw, and forearm movements that reflected hedonic responsiveness to taste. Preferred substances elicited a series of rhythmic mouth movements and alternations between tongue protrusions and tongue retractions that resulted in swallowing and paw licking (ingestive responses). Nonpreferred substances elicited mouth gaping with tongue retraction followed by long duration tongue protrusion and then tongue retraction and mouth closure. This sequence of responses was repeated several times, resulted in a reduction in swallowing, and was often followed by a sequence of fixed action patterns, which included chin rubbing, head shaking, paw wiping, and forelimb flailing (aversive responses). Other tastes elicited a mixture of these two patterns.

When rats are poisoned after consuming a preferred sweet taste such as sucrose, their subsequent behavioral responses to sucrose resemble those exhibited after consumption of nonpreferred bitter tastes such as quinine. They exhibit decreases in consumption levels, spillage of food and stereotyped aversive responses (Berridge et al 1981, Garcia & Koelling 1966, Rozin 1967). Illness, then, alters the response elicited by taste.

In most classical conditioning situations, the response elicited by the CS is altered because of its association with the US. In some cases the CR resembles the UR quite closely. The CR generally does not, however, become identical to the UR (Holland 1984, Rescorla 1988). The CR can lack the intensity and some of the response repertoire observed for the UR and can include some responses that are not part of the UR. In some cases the CR produced by a given CS is opposite that of the UR, e.g. increases in activity and heart rate elicited by a shock US and decreases in activity and heart rate elicited by a tone CS (de Toledo & Black 1966, Rescorla 1988). Holland (1984) has suggested that the CR is composed of two behavioral elements: one that is similar to or at least in some way appropriate to the US and one that is similar to the response elicited by the CS prior to conditioning (Figure 1).

What distinguishes the CTA learning situation from many other forms of classical conditioning is that the CR is entirely part of the repertoire of elicited responses for the CS sensory modality, in this case, taste. Although the response is appropriate to the US in that illness often produces decreases in food consumption, the decrease in consumption in a CTA situation is not general as in the case of illness, but is specific to the CS. The CR is not similar to the response elicited by the CS prior to conditioning but is opposite that response. Consequently, the function of the US also is different in CTAs than in traditional classical learning. The US does not act as the elicitor of what will become the essential characteristics of the CR. Instead it changes the response elicited by the CS from one form (ingestive) to another (aversive).

# PROPOSED NEURAL MODEL FOR ACQUISITION OF CONDITIONED TASTE AVERSIONS

The determination of the neural substrates for CTAs, then, should involve the identification of the following pathways (Figure 1): the US pathway, the CS pathway, the pathway for the elicited response to the CS prior to conditioning ($UR_{cs}$), and the pathway for the elicited response to the CS after conditioning ($CR_{cs}$). Each taste is connected to both the ingestive and aversive responses. These connections are probably innate, since

TRADITIONAL CLASSICAL CONDITIONING

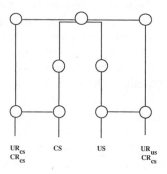

CONDITIONED TASTE AVERSIONS

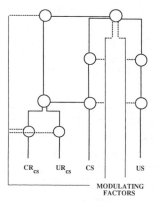

*Figure 1*   Simplified schematic of a neural model for traditional classical conditioning and conditioned taste aversion. Abbreviations: $CR_{cs}$, conditioned response to the CS; CS, conditioned stimulus; $UR_{cs}$, unconditioned response to the CS; $UR_{us}$, unconditioned response to the unconditioned stimulus; US, unconditioned stimulus.

hedonic reactions to taste have been observed in prematurely born and full-term neonatal individuals (Steiner 1973, 1979). The relative strengths of the two innate connections are dependent on the given taste. In the case of sucrose, the innate connection to the ingestive response is stronger than the innate connection to the aversive response.

If exposure to a taste is followed by illness, the connection to the ingestive response system will weaken and the connection to the aversive response system will strengthen. It is most likely that the illness-induced changes involve two processes rather than one. Grill & Berridge (1985) have suggested that palatability processing involves two mechanisms and have provided evidence that the ingestive and aversive response systems can change independently. Thus, in order for the aversive response system to be expressed solely, a weakening of the ingestive response system would have to occur. If exposure to a taste is not followed by negative consequences, a stronger connection to the ingestive response system will result. A stronger connection to the ingestive response system also will occur if a given taste is associated with positive reinforcement or if it is followed by recuperation from illness (Garcia et al 1977, Revusky 1967, 1974, Rozin 1969). So, experiential factors can alter the strengths of the innate connections to the ingestive and aversive response patterns. Thus, after a given taste is experienced, the relative strengths of the ingestive and aversive response systems are a function of the original innate connections, the number of exposures to the taste with illness, and the number of exposures to the taste without illness. This hypothesis is supported by the findings that CTAs to nonpreferred tastes are stronger than to preferred tastes (Etscorn 1973), that repeated pairings of a taste with illness strengthens an aversion, and that repeated pairings of a taste without illness reduces the strength of an aversion (Kalat & Rozin 1973).

Other factors associated with the CS and US can influence the strength of an aversion and therefore must be taken into account when developing a neural model for CTAs. The strength of an aversion has been found to be a function of the intensity of the taste as measured by concentration (Dragoin 1971) and the amount consumed on the first exposure (Bond & Di Guisto 1975), the intensity of the US (Revusky 1968), and prior experience with the US (Cannon et al 1975).

Several factors that are not essential or critical for aversion learning can modulate the development and strength of CTAs. The development and strength of an aversion are dependent on the hormonal milieu and deprivation state of the animal. The presence of testosterone (T) increases the proportion of animals that develop a CTA (Chambers et al 1981), and the presence of dexamethasone attenuates the strength of an aversion (Hennessy et al 1976). Water deprivation reduces the proportion of male rats that develop an aversion (Chambers et al 1981). It is interesting that deprivation can alter the hedonic value of tastes. Foods are reported to be highly palatable with deprivation and unpleasant with satiety (Cabanac 1971). Also, the number of ingestive responses decreases and the number of aversive responses increases as meal termination approaches (Grill & Berridge 1985). So, the relative strengths of the ingestive and aversive response systems are also a function of modulating factors. A complete understanding of the neural mechanisms controlling CTAs would have to include a determination of the neural circuitry for the modulating factors (Figure 1).

# KNOWN NEURAL CIRCUITRY FOR CONDITIONED TASTE AVERSIONS

## The US Pathway

A number of reviews have examined the US pathways (Ashe & Nachman 1980, Borison & Wang 1953, Coil & Garcia 1977, Kiefer 1985). The vagus nerve conveys information from the gastric-intestinal mucosa primarily to the caudal region of the nucleus of the solitary tract (NST; Torvik 1956). It is then conveyed to the pontine parabrachial nucleus (PBN; Norgren 1978) and the insular cortex (Cechetto & Saper 1987). The area postrema, an area of the brain on the floor of the fourth ventricle that lacks a blood-brain barrier, detects chemicals in the blood. As there are reciprocal neural connections between the area postrema and the NST (Morest 1960, 1967), information about these blood-borne chemicals is probably conveyed to the NST.

A wide variety of substances can be used as the US. The route by which information about these substances is conveyed to the brain varies with the particular chemical and the route of administration. LiCl, a widely used illness-inducing agent, appears to act primarily by way of the area postrema. Lesions of the dorsolateral region of this area attenuate or abolish the learning of taste aversions induced by LiCl (Ritter et al 1980). Vagotomized rats, however, develop essentially normal taste aversions (Martin et al 1978). The vagus nerve mediates copper sulfate–induced aversions when this substance is administered intraperitoneally or intra-gastrically, but when it is given intravenously the area postrema mediates the aversion (Coil & Norgren 1981, Coil et al 1978).

Although the vagus nerve and the area postrema are important routes for many different chemicals, they may not be the only means by which information is conveyed to the brain. The area postrema is an important structure for the induction of emesis when apomorphine is administered, but neither lesions of this area nor vagotomy has an effect on the ability of an animal to learn CTAs (Kiefer et al 1981, Van der Kooy et al 1983). It is not known what effect disruption of both systems has on CTA learning.

## The CS Pathway

The CS for CFA learning involves stimuli that are normally used by a given species for the identification of food. For many species taste is the primary stimulus for identification. But it must be noted that other stimuli such as odor can serve as weak cues (Kiefer 1985) and some species use other senses, such as vision in birds, as the primary stimulus (Gaston 1977).

The gustatory pathway has been reviewed recently by Norgren (1984) and Travers, Travers & Norgren (1987). In summary, taste receptor cells are located primarily in the tongue and hard and soft palates. Taste information is transmitted primarily to three peripheral taste nerves: the chorda tympani branch of the facial nerve, the lingual branch of the glossopharyngeal nerve, and the greater superficial petrosal branch of the facial nerve. Gustatory afferent fibers from the facial and glossopharyngeal nerves terminate in the ipsilateral NST. Ascending axons from the NST terminate in the ipsilateral PBN in rodents and lagomorphs and in the ventroposteromedial nucleus of the thalamus in primates. The PBN sends projections ipsilaterally to the parvicellular division of the ventroposteromedial nucleus (VPMpc) and also projects extensively to the ventral forebrain, in particular, the lateral hypothalamus, central nucleus of the amygdala, and bed nucleus of the stria terminalis. The VPMpc projects to the insular cortex, which projects back to the VPMpc, central amygdala, PBN, and NST.

## The $UR_{cs}$ and $CR_{cs}$ Pathways

Neural areas rostral to the PBN are not critical for hedonic reactions to taste. Hedonic responsiveness remains in rats with supracollicular decerebrate preparations that leave only the first (NST) and second (PBN) central gustatory relay nuclei. Intraoral taste stimulation of these rats elicits the same ingestive and aversive patterns of taste responsiveness at the same concentrations as it does in intact rats (Grill & Berridge 1985).

Some neurons in the PBN project to oro-motor nuclei (Travers & Norgren 1983) and respond to both the hedonic dimensions of taste and oro-lingual movement (Schwartzbaum 1983). It seems likely that the ingestive and aversive behaviors are organized entirely in the brain stem and that the control of these behaviors involves the NST and PBN and their polysynaptic connections to the motor neuron pools controlling the behaviors. As the ingestive and aversive movements to taste stimuli are stereotyped, repetitive, and rhythmic, the neural circuits for these behaviors may function as pattern generators, with higher brain systems acting only as modulators. In this sense, the control of the behaviors may resemble that of vertebrate locomotion (Grillner 1985).

Since the behavioral response elicited by bitter tastes is similar to that elicited by a taste that has been paired with illness, one would expect that at least at the level of the behavioral pattern generators, the neural code for the CS would be similar to that for quinine. Chang & Scott (1984) have found that the pattern of activity of sucrose-best NST neurons in response to a sweet taste changes after rats acquire an aversion to this taste so that the activity pattern more closely resembles that of bitter tastes.

## CS-US Integration

Possible sites of taste-illness integration have been discussed in a number of recent reviews (Ashe & Nachman 1980, Gaston 1978, Grill 1985, Kiefer 1985). The search for such sites has been plagued by inconsistent findings, and unequivocal candidates have not emerged. The neural control of this primitive form of learning is clearly complex, and there are likely a number of neural routes that can be used for laying down the trace. Despite the contradictory endeavors, some findings have come out of the search that should provide more insight into the neural organization of CTAs.

One of the more consistent findings has been that lesions of the basolateral amygdala disrupt taste aversion learning and retention of an aversion learned prior to lesioning (Simbayi et al 1986, Nachman & Ashe 1974). After finding that cutting the connections between the amygdala and the temporal cortex produced the same deficits as lesions of the basolateral amygdala, however, Fitzgerald & Burton (1983) suggested that it is the destruction of the fibers of passage that produces the deficits after lesions of the basolateral amygdala, not the destruction of the nucleus itself. Recently, Dunn & Everitt (1988) found that ibotenic induced lesions that spare the fibers of passage had no effect on aversion learning. Anna Brownson, Richard Thompson, and I have confirmed and extended this finding in a preliminary study. Electrolytic lesions attenuated both the acquisition of an aversion and the retention of an aversion induced prior to lesioning; neurotoxic lesions (NMDA-induced), which also spare the fibers of passage, had no effect on either acquisition or retention (Figure 2). Clearly the issue of axons of passage is critical to an understanding of neural mechanisms.

Lesions of the PBN disrupt acquisition of a CTA when there is a delay between the CS and US (Schulkin et al 1986, Di Lorenzo 1988), but if there is no delay, animals can learn an aversion (Di Lorenzo 1988). Similar results have been found for lesions of the gustatory cortex (Lorden 1976). These findings suggest that there are different neural mechanisms for learning when CS-US intervals are short and long. Any neural model must include both pathways.

# BEYOND ACQUISITION OF CONDITIONED TASTE AVERSIONS

Although I have focused only on the acquisition process, a complete neural model of CTA should include retention or memory storage processes and extinction processes as well. There probably are neural areas that are part of the pathways for all three processes, but the pathways are different.

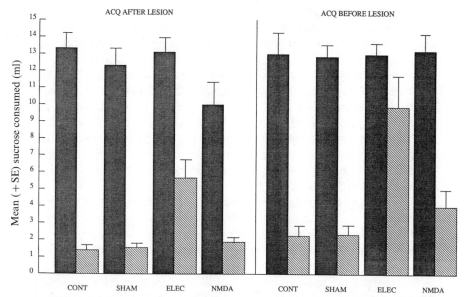

*Figure 2* Mean (+SE) sucrose consumed by control (CONT), sham, electrolytic lesioned (ELEC), and neurotoxic lesioned (NMDA) male rats on the day of acquisiton (ACQ; *dark bars*) and the first day after acquisition (*hatched bars*) when acquisition was given after and before lesions of the basolateral amygdala.

Each process has its own set of modulating factors that influence that process independently of the other.

## Retention

Little is known about the neural mechanisms for CTA, but the data so far suggest that the neural mechanisms for acquisition and retention differ. Although the gustatory neocortex is involved in acquisition, it is not essential (Braun et al 1972, Lasiter & Glanzman 1982, Lorden 1976). It is, however, critically involved in retention of aversions, as animals with these lesions do not retain a previously learned aversion (Braun et al 1981, Yamamoto et al 1981).

## Extinction

Extinction has been regarded merely as a reflection of the acquisition process. Indeed, in much of the early Pavlovian and Skinnerian literature, the strength of acquisition was indexed by the rate of extinction. If extinction was slow, acquisition was said to be strong, and if extinction was fast, acquisition was asserted to be weak. Extinction, however, at least of a

CTA, is far more complex than this. Some evidence suggests that the neural processes mediating acquisition and extinction of a CTA are different. If animals are anesthetized or under cortical spreading depression when they are given exposure to a taste, they do not acquire a CTA but they do extinguish a previously learned aversion Bureš & Burešová 1979). The vagus nerve plays a role in extinction that is independent of its involvement during acquisition (Coil et al 1978, Kiefer et al 1981). Rats that are vagotomized prior to or after acquisition of a CTA exhibit a faster extinction than nonvagotomized animals, even when a US (apomorphine) that is not vagally mediated is used. A number of other factors modulate extinction independenly of acquisition. The rate of extinction is altered when adrenocorticotropin (ACTH) levels are elevated, when T is present, and when animals are under water deprivation (Chambers 1985, Chambers & Sengstake 1979, Kendler et al 1976, Sengstake & Chambers 1979). ACTH and T decrease and water deprivation increases the rate. It is the presence of these factors during extinction that alters the rate of extinction. Their presence or absence during acquisition of the aversion has no effect.

As suggested decades ago by Clark Hull (1943), extinction is a learning process. Simply stated for CTAs, it is unlearning that the taste predicts illness and learning that it predicts safety (Chambers 1985, Kiefer et al 1981). With respect to the neural model for CTAs outlined above, extinction is a process by which connections to the aversive response system are weakened and connections to the ingestive response system are strengthened. Any information on the subsequent consequences of ingesting the CS is processed. If the consequences are neutral, that information serves to alter the relative strengths of the two response systems. Thus, after a CS has been experienced without negative consequences, the relative strengths of the ingestive and aversive response systems are a function of the relative strengths of these systems after the CTA, the number of exposures to the taste without illness, modulating factors, and probably the original innate predisposition.

## CONCLUSION

Since its discovery, students of learning have argued whether CTA fits best in a classical or instrumental learning framework. The fit seemed poor in either case. Although most have placed it within classical conditioning, the issue is still unresolved. Taste aversion learning seems to share the following with traditional kinds of Pavlovian conditioning: a neural integration of the CS and US, a change in the meaning of the CS as a result of this integration so that the CS becomes a signal that predicts the occurrence of the US, and an elicitation of a CR by the CS that is an anticipation of the occurrence of the US. There are characteristics of CTAs, however, that traditional Pavlovian conditioning does not share, i.e. the ease with which the US and CS are integrated, the ability to integrate despite the long delay between the US and CS and despite intervening CSs from the same sensory modality, the context independence of the CS, and the origin, stemming from the CS, of the CR. It is unclear how critical these differences are, but they certainly alter how one would develop a neural model of learning.

Acknowledgment

This work was supported by Grant N0014-J-1296 from the Office of Naval Research, Grant HD20970 from the National Institutes of Health, and Grant BRSG S07RR07012-21. The author wishes to thank Gerald C. Davison and Richard F. Thompson for their helpful comments.

*Literature Cited*

Ashe, J. H., Nachman, M. 1980. Neural mechanisms in taste aversion learning. *Prog. Psychobiol. Physiol. Psychol.* 9: 233–62

Berridge, K., Grill, H. J., Norgren, R. 1981. Relation of consummatory responses and preabsorptive insulin release to palatability and learned taste aversions. *J. Comp. Physiol. Psychol.* 95: 363–82

Bond, N., Di Giusto, E. 1975. Amount of solution drunk is a factor in the establishment of taste aversion. *Anim. Learn. Behav.* 3: 81–84

Borison, H. L., Wang, S. C. 1953. Physiology and pharmacology of vomiting. *Pharmacol. Rev.* 5: 193–230

Braun, J. J., Kiefer, S. W., Ouellet, J. V. 1981. Psychic ageusia in rats lacking gustatory neocortex. *Exp. Neurol.* 72: 711–16

Braun, J. J., Slick, T. B., Lorden, J. F. 1972. Involvement of gustatory neocortex in the learning of taste aversions. *Physiol. Behav.* 9: 637–41

Bureš, J., Burešová, O. 1979. Neurophysiological analysis of conditioned taste aversion. In *Brain Mechanisms in Memory and Learning: From the Single Neuron to Man*, ed. M. A. Brazier, pp. 127–38. New York: Raven

Cabanac, M. 1971. Physiological role of pleasure. *Science* 173: 1103–7

Cannon, D. S., Berman, R. F., Baker, T. B., Atkinson, C. A. 1975. Effect of preconditioning unconditioned stimulus experience on learned taste aversions. *J. Exp. Psychol.: Anim. Behav. Process.* 104: 270–84

Carpenter, J. A. 1956. Species differences in taste preferences. *J. Comp. Physiol. Psychol.* 49: 139–44

Cechetto, D. F., Saper, C. B. 1987. Evidence for a viscerotopic sensory representation in the cortex and thalamus in the rat. *J. Comp. Neurol.* 262: 27–45

Chambers, K. C. 1985. Sexual dimorphisms as an index of hormonal influences on conditioned food aversions. *Ann. NY Acad. Sci.* 443: 110–25

Chambers, K. C., Sengstake, C. B. 1979. Temporal aspects of the dependency of a dimorphic rate of extinction on testosterone. *Physiol. Behav.* 22: 53–56

Chambers, K. C., Sengstake, C. B., Yoder, R. L., Thornton, J. E. 1981. Sexually dimorphic acquisition of a conditioned taste aversion in rats: Effects of gonadectomy, testosterone replacement and water deprivation. *Physiol. Behav.* 27: 83–88

Chang, F.-C. T., Scott, T. R. 1984. Conditioned taste aversions modify neural responses in the rat nucleus tractus solitarius. *J. Neurosci.* 4: 1850–62

Cohen, D. H. 1982. Central processing time for a conditioned response in a vertebrate model system. In *Conditioning: Representation of Involved Neural Functions*, ed. C. D. Woody, pp. 517–34. New York: Plenum

Coil, J. D., Garcia, J. 1977. *Conditioned taste aversions: The illness US.* Presented at Northeast. Reg. Meet. Animal Behav. Soc., Oct. Memorial Univ. Newfoundland, St. John's, Newfoundland, Canada

Coil, J. D., Norgren, R. 1981. Taste aversions conditioned with intravenous copper sulfate: Attenuation by ablation of the area postrema. *Brain Res.* 212: 425–33

Coil, J. D., Rogers, R. C., Garcia, J., Novin, D. 1978. Conditioned taste aversions: Vagal and circulatory mediation of the toxic unconditioned stimulus. *Behav. Biol.* 24: 509–19

de Toledo, L., Black, A. H. 1966. Heart rate: Changes during conditioned suppression in rats. *Science* 152: 1404–6.

Di Lorenzo, P. M. 1988. Long-delay learning in rats with parabrachial pontine lesions. *Chem. Senses* 13: 219–29

Dragoin, W. B. 1971. Conditioning and extinction of taste aversions with variations in intensity of the CS and the UCS in two strains of rats. *Psychon. Sci.* 22: 303–4

Dunn, L. T., Everitt, B. J. 1988. Double dissociations of the effects of amygdala and insular cortex lesions on conditioned taste aversion, passive avoidance, and neophobia in the rat using the excitotoxin ibotenic acid. *Behav. Neurosci.* 102: 3–23

Estcorn, F. 1973. Effects of a preferred vs a nonpreferred CS in the establishment of a taste aversion. *Physiol. Psychol.* 1: 5–6

Fischer, U., Hommel, H., Ziegler, M., Michael, R. 1972. The mechanism of insulin secretion after oral glucose administration. I. Multiphasic course of insulin mobilization after oral administration of glucose in conscious dogs. Differences to the behaviour after intravenous administration. *Diabetologia* 8: 104–10

Fitzgerald, R. E., Burton, M. J. 1983. Neophobia and conditioned taste aversion deficits in the rat produced by undercutting temporal cortex. *Physiol. Behav.* 30: 203–6

Garcia, J., Ervin, F. R., Koelling, R. A. 1966. Learning with prolonged delay of reinforcement. *Psychon. Sci.* 5: 121–22

Garcia, J., Hankins, W. G., Coil, J. D. 1977. Koalas, men and other conditional gastronomes. In *Food Aversion Learning*, ed. N. W. Milgram, L. Krames, T. M. Alloway, pp. 195–218. New York: Plenum

Garcia, J., Kimeldorf, D. J., Koelling, R. A. 1955. Conditioned aversion to saccharin resulting from exposure to gamma radiation. *Science* 122: 157–58

Garcia, J., Koelling, R. A. 1966. Relation of cue to consequence in avoiding learning. *Psychon. Sci.* 4: 123–24

Garcia, J., McGowan, B. K., Green, K. 1972. Biological constraints on conditioning. In *Biological Boundaries of Learning*, ed. M. E. P. Seligman, J. L. Hager. New York: Meredith

Gaston, K. E. 1977. An illness-induced conditioned aversion in domestic chicks: One-trial learning with a long delay of reinforcement. *Behav. Biol.* 20: 441–53

Gaston, K. E. 1978. Brain mechanisms of conditioned taste aversion learning: A review of the literature. *Physiol. Psychol.* 6: 340–53

Grill, H. J. 1985. Introduction: Physiological mechanisms in conditioned taste aversions. *Ann. NY Acad. Sci.* 443: 67–88

Grill, H. J., Berridge, K. C. 1985. Taste reactivity as a measure of the neural control of palatability. *Prog. Psychobiol. Physiol. Psychol.* 11: 1–61

Grill, H. J., Norgren, R. 1978. The taste reactivity test. I. Mimetic responses to gustatory stimuli in neurologically normal rats. *Brain Res.* 143: 263–79

Grillner, S. 1985. Neurobiological bases of rhythmic motor acts in vertebrates. *Science* 228: 143–49

Hennessy, J. W., Smotherman, W. P., Levine, S. 1976. Conditioned taste aversion and the pituitary-adrenal system. *Behav. Biol.* 16: 413–24

Holland, P. C. 1984. Origins of behavior in Pavlovian conditioning. In *The Psychology of Learning and Motivation*, ed. G. Bower, 18: 129–74. Orlando, Fla.: Academic

Hull, C. L. 1943. *Principles of Behavior*. New York: Appleton

Kalat, J. W., Rozin, P. 1973. "Learned safety" as a mechanism in long-delay taste-aversion learning in rats. *J. Comp. Physiol. Psychol.* 83: 198–207

Kapp, B. S., Gallagher, M., Applegate, C. D., Frysinger, R. C. 1982. See Cohen 1982, pp. 581–99

Kendler, K., Hennessey, J. W., Smotherman, W. P., Levine, S. 1976. An ACTH effect on recovery from conditioned taste aversion. *Behav. Biol.* 17: 225–29

Kiefer, S. W. 1985. Neural mediation of conditioned food aversions. *Ann. NY Acad. Sci.* 443: 100–9

Kiefer, S. W., Rusiniak, K. W., Garcia, J., Coil, J. D. 1981. Vagotomy facilitates extinction of conditioned taste aversions in rats. *J. Comp. Physiol. Psychol.* 95: 114–22

Lasiter, P. S., Glanzman, D. L. 1982. Cortical substrates of taste aversion learning: Dorsal prepiriform (insular) lesions disrupt taste aversion learning. *J. Comp. Physiol. Psychol.* 96: 376–92

Lorden, J. F. 1976. Effects of lesions of the gustatory neocortex on taste aversion learning in the rat. *J. Comp. Physiol. Psychol.* 90: 665–79

Martin, J. R., Cheng, F. Y., Novin, D. 1978. Acquisition of learned taste aversion following bilateral subdiaphragmatic vagotomy in rats. *Physiol. Behav.* 21: 13–17

Morest, D. K. 1960. A study of the structure of the area postrema with Golgi methods. *Am. J. Anat.* 107: 291–303

Morest, D. K. 1967. Experimental study of the projections of the nucleus of the tractus solitarius and the area postrema in the cat. *J. Comp. Neurol.* 130: 277–300

Nachman, M., Ashe, J. H. 1974. Effects of basolateral amygdala lesions on neophobia, learned taste aversions, and sodium appetite in rats. *J. Comp. Physiol. Psychol.* 87: 622–43

Norgren, R. 1978. Projections from the nucleus of the solitary tract in the rat. *Neuroscience* 3: 207–18

Norgren, R. 1984. Central neural mechanisms of taste. In *Handbook of Physiology*, Sect. 1, *The Nervous System*, Vol. III, *Sensory Processes*, Pt. 2, ed. J. M. Brookhart, V. B. Mountcastle, I. Darian-Smith, S. R. Geiger, pp. 1087–1128. Bethesda: Am. Physiol. Soc.

Pavlov, I. P. 1927. *Conditioned Reflexes*, Trans. G. V. Anrep. London: Oxford Univ. Press

Rescorla, R. A. 1988. Pavlovian conditioning: It's not what you think it is. *Am. Psychol.* 43: 151–60

Revusky, S. H. 1967. Hunger level during food consumption: Effects on subsequent preference. *Psychon. Sci.* 7: 109–10

Revusky, S. 1968. Aversion to sucrose produced by contingent X-irradiation: Temporal and dosage parameters. *J. Comp. Physiol. Psychol.* 65: 17–22

Revusky, S. H. 1974. Retention of a learned increase in the preference for a flavored solution. *Behav. Biol.* 11: 121–25

Riley, A. L., Baril, L. L. 1976. Conditioned taste aversions: A bibliography. *Anim. Learn. Behav.* 4: 1S–13S

Ritter, S., McGlone, J. J., Kelley, K. W. 1980. Absence of lithium-induced taste aversion after area postrema lesion. *Brain Res.* 201: 501–6

Rozin, P. 1969. Adaptive food sampling patterns in vitamin deficient rats. *J. Comp. Physiol. Psychol.* 69: 126–32

Rozin, P. 1967. Specific aversions as a component of specific hungers. *J. Comp. Physiol. Psychol.* 64: 237–42

Schulkin, J., Flynn, F. W., Grill, H. J., Norgren, R. 1985. Central gustatory lesions.

II: Effects on salt appetite and taste aversion learning. *Neurosci. Abstr.* 15: 1260

Schwartzbaum, J. S. 1983. Electrophysiology of taste-mediated functions in parabrachial nuclei of behaving rabbit. *Brain Res. Bull.* 11: 61–89

Sengstake, C. B., Chambers, K. C. 1979. Differential effects of fluid deprivation on the acquisition and extinction phases of a conditioned taste aversion. *Bull. Psychon. Soc.* 14: 85–87

Simbayi, L. C., Boakes, R. A., Burton, M. J. 1986. Effects of basolateral amygdala lesions on taste aversions produced by lactose and lithium chloride in the rat. *Behav. Neurosci.* 100: 455–65

Steiner, J. E. 1973. The gustofacial response: Observation on normal and anencephalic newborn infants. In *Development in the Fetus and Infant, 4th Symposium on Oral Sensation and Perception*, ed. J. F. Bosma, 4: 254–78. Bethesda: NIH DHEW

Steiner, J. E. 1979. Human facial expressions in response to taste and smell stimulation. In *Advances in Child Development and Behavior*, ed. H. W. Reese, L. P. Lipsitt, 13: 257–95. New York: Academic

Thompson, R. F. 1986. The neurobiology of learning and memory. *Science* 233: 941–47

Torvik, A. 1956. Afferent connections to the sensory trigeminal nuclei, the nucleus of the solitary tract and adjacent structures: An experimental study of the rat. *J. Comp. Neurol.* 106: 51–141

Travers, J. B., Norgren, R. 1983. Afferent projections to the oral motor nuclei in the rat. *J. Comp. Neurol.* 220: 280–98

Travers, J. B., Travers, S. P., Norgren, R. 1987. Gustatory neural processing in the hindbrain. *Annu. Rev. Neurosci.* 10: 595–632

Van Der Kooy, D., Swerdlow, N. R., Koob, G. F. 1983. Paradoxical reinforcing properties of apomorphine: Effects of nucleus accumbens and area postrema lesions. *Brain Res.* 259: 111–18

Yamamoto, T., Azuma, S., Kawamura, Y. 1981. Significance of cortical-amygdalar-hypothalamic connections in retention of conditioned taste aversion in rats. *Exp. Neurol.* 74: 758–68

*Annu. Rev. Neurosci. 1990. 13:387–401*

# NEUROTRANSMITTERS IN THE MAMMALIAN CIRCADIAN SYSTEM

*Benjamin Rusak and K. G. Bina*

Department of Psychology, Life Sciences Center, Dalhousie University, Halifax, Nova Scotia, Canada B3H 4J1

## INTRODUCTION

### Background

Circadian (daily) rhythms in behavior and physiology are regulated in mammals by a neural mechanism centered in the suprachiasmatic nucleus (SCN) of the anterior hypothalamus (Rusak & Zucker 1977, Moore 1983). Although other oscillators exist in the circadian system, the SCN remains the only one that has been localized, demonstrated to oscillate endogenously, and shown to regulate overt rhythms (Rusak 1989). Its pacemaker properties have been confirmed by the demonstration that an SCN transplanted into an animal made arrhythmic by a previous SCN lesion restores circadian organization (Sawaki et al 1984, Lehman et al 1987); most importantly, the period of the restored rhythm is determined by the genetic makeup of the donor, not the host, animal (Ralph et al 1988). As the pacemaker for the circadian system, the SCN serves two primary functions: the internal generation of circadian rhythms and their synchronization (entrainment) to local time cues, primarily those provided by environmental lighting cycles.

It has been more than 15 years since the identification of the SCN as the mammalian pacemaker sparked an interest in the anatomy and physiology of this small population of neurons ($\sim 24\,000$ in rats; Güldner 1983). The purpose of this chapter is to review and evaluate the evidence relating to one aspect of the neurobiology of the SCN; namely, the neurotransmitters that have been proposed to play a role in rhythm generation and entrainment. This task is complicated by the extraordinary variety of neurotransmitters, receptors, and transmitter-related enzymes that have

387

been identified in this small nucleus. Among the putative neurotransmitters represented in some way in the SCN are serotonin (Aghajanian et al 1969), acetylcholine (Ichikawa & Hirata 1986, Miller et al 1984), excitatory amino acids (Moffett et al 1987, Liou et al 1986), and GABA (van den Pol & Gorcs 1986, Card & Moore 1984). A large number of peptides are also represented in the SCN, including vasopressin (Vandesande et al 1975), vasoactive intestinal polypeptide (Card et al 1981), somatostatin (Card & Moore 1984), neuropeptide Y (Card & Moore 1982, Harrington et al 1985), nerve growth factor (Sofroniew et al 1989), and cholecystokinin (Miceli et al 1987), among others (van den Pol & Tsujimoto 1985).

That circadian systems are virtually immune to chemical manipulation was once a common assertion. The substances first demonstrated to affect rhythms ($D_2O$, lithium, alcohol) have such broad effects on cellular metabolism that they provided no sharp insights into the underlying mechanisms they influenced (Rusak & Zucker 1975). More recently, many neuroactive substances, each with a different cellular impact, have been shown to influence mammalian circadian rhythms (Turek 1987). Their variety has, in fact, become an impediment to the construction of models of neurochemical pathways in the circadian system.

## Analysis of Neurochemical Effects on Rhythms

Two methods are commonly used to analyze the roles of neurotransmitters in the circadian system. One is to manipulate a neurochemical system (either systemically or locally in the SCN) in an intact organism and to observe the consequences on the overt expression of rhythms. The second is to study the effects of pharmacological manipulations on cellular activity in the SCN, on the assumption that SCN neural activity reflects some aspect of pacemaker function. Both approaches are predicated on anatomical and chemical evidence that the neurochemical system being manipulated is actually represented in the SCN by appropriate receptors, enzymes, or transmitters. Unfortunately, these anatomical, behavioral, and physiological analyses of the role of a transmitter often fail to yield convergent conclusions.

The effects of drugs on overt rhythms are usually evaluated in one of two ways: (*a*) effects of chronic changes in a neurotransmitter system on the free-running period of the rhythm are measured in constant conditions, and (*b*) effects of transient manipulations on the phase of the rhythm are measured in either entrained or free-running conditions. A typical approach is to determine the shape of the phase response curve for drug administration, or to ask how altering a neurotransmitter system (acutely or chronically) affects the shape of the phase response curve for light pulses.

The phase response curve is a plot of the amount and direction (advance or delay) of phase shift generated at different circadian phases by exposure to a standard stimulus, typically a light pulse delivered against a background of continuous darkness (DeCoursey 1964). It is a uniquely useful tool in the analysis of photic entrainment of rhythms that has been extended to the study of pharmacological agents (Turek 1987). Interpreting phase response curves related to pharmacological agents can, however, be difficult. Light, as a natural entraining agent, can be assumed to activate a functionally coherent neural mechanism. There is no certainty that any drug acts similarly to influence an identifiable functional system, and the interpretation of drug effects is therefore complex. For example, activating a particular transmitter system may cause rhythm delays without implying a role for that transmitter in mediating delaying effects of light. Similarly, neurotransmitter antagonists may alter responses to light pulses, without implying that the related neurotransmitter normally plays a role in mediating photic effects on rhythms (cf. Ralph & Menaker 1988, Shinozaki 1988). An analysis of the role of neurotransmitters in the circadian system based on phase response curves is a natural choice for circadian physiologists who have used these tools so successfully in other contexts. Their use in this analysis, however, may entail unstated and untested assumptions that need careful examination.

## NEUROCHEMICALS AND THE CIRCADIAN SYSTEM

### Acetylcholine

One of the earliest suggested transmitters in the circadian system was acetylcholine (ACh), proposed to act through a nicotinic binding site to mediate the effects of light. This suggestion came from studies showing that a putative nicotinic antagonist, $\alpha$-bungarotoxin (BTX) injected near the SCN could prevent light effects on daily rhythms of pineal $N$-acetyltransferase (NAT) activity, that a nonspecific cholinergtic agonist (carbachol) could mimic the effects of light, and that SCN cells showed similar responses to light and to cholinergic agents (Nishino & Koizumi 1977, Zatz & Brownstein 1979, 1981, Zatz & Herkenham 1981). Consistent with these reports is evidence that the ACh content of the SCN increases after a light pulse (Murakami et al 1984).

The interpretation that nicotinic sites mediate the light-like effects of carbachol depended on the effectiveness of selective nicotinic agonists and antagonists and the lack of effect of muscarinic agents in modifying the influence of carbachol (Zatz & Brownstein 1981). A role for a muscarinic mechanism should not be ruled out, however, since published data indicate that treatment with atropine, a muscarinic antagonist, mimicked the effects of light on pineal NAT (Zatz & Brownstein 1981). These data suggest that carbachol exerts opposite effects on NAT activity through muscarinic and nicotinic receptors.

The dichotomy between muscarinic and nicotinic receptors that is satisfactory for peripheral sites is inadequate in the central nervous system. A variety of central cholinergic receptors do not fall neatly into either category. The peripheral nicotinic antagonist BTX, for example, does not bind to most sites in the brain that bind ACh and nicotine. BTX binds strongly at the dorsolateral edge of the SCN and beyond its Nissl-defined borders, but neither ACh nor nicotine have been shown to bind anywhere in the suprachiasmatic region (Clarke et al 1985).

It remains possible, however, that an unidentified nicotinic receptor is found on SCN neurons. Nicotinic receptors have been proposed to be made up of a beta-2 subunit combined with one or more of a family of alpha subunits (Boulter et al 1987). A recent *in situ* hybridization study showed that cells in the rat SCN express the messenger RNA for the beta-2 subunit, but not for any of the alpha subunits tested (Wada et al 1989). The potential therefore exists for the manufacture of nicotinic receptors in SCN cells. In their functional form, these might be localized in the SCN or they might be transported to processes outside the nucleus itself.

The BTX binding sites in the SCN do not appear to be important in mediating cholinergic effects on rhythms in rats. These sites are largely eliminated by ovariectomy of female rats and can be restored by estrogen treatment (Miller et al 1982, 1984). Ovariectomized rats can still be phase-shifted, however, by injection of carbachol into the third ventricle (Mistlberger & Rusak 1986). Carbachol must act on as-yet-unidentified ACh receptors in the SCN or outside it, or it acts in some way on a different class of receptors.

The influence of carbachol on circadian rhythms in hamsters has been studied extensively. Carbachol injections into the third ventricle can cause phase shifts and alter entrainment of rhythms, but carbachol effects do not closely resemble those of light pulses (Earnest & Turek 1983, 1985, Meijer et al 1988a). Carbachol injections induce large phase advances during the subjective day, whereas light is ineffective during this phase. Mecamylamine, a nicotinic antagonist, has also been reported to prevent some phase shifts by light pulses (Keefe et al 1987, Miller et al 1987).

There is, however, some evidence against a cholinergic role. The interpretation of the effects of mecamylamine is confounded by recent evidence that it also inhibits the effects of excitatory amino acids (Christensen & O'Dell 1988, Shinozaki 1988, cf. Langdon & Freeman 1987), which are themselves candidate transmitters in the photic entrainment route (Cahill & Menaker 1987, Liou et al 1986). Depletion of cholinergic stores by hemicholinium also did not affect phase shifts to light pulses, although the completeness of depletion was not established (Pauly & Horseman 1985, see Keefe et al 1987). Finally, studies measuring evoked field potentials or SCN single-unit firing in response to optic nerve stimulation found that cholinergic antagonists could not prevent these responses (Shibata et al 1986, Cahill & Menaker 1987). These studies do not provide strong support for ACh as a mediator of photic effects on the circadian clock.

The neurophysiological effects of cholinergic agents on SCN cells are also controversial. Most studies report that ACh activates a majority and suppresses a minority of SCN cells, as does retinal illumination, and two studies report positive correlations between photic and cholinergic effects (Nishino & Koizumi 1977, Miller et al 1987). One of these studies used an electrode arrangement that would permit cholinergic effects to be mediated transsynaptically, without using a control procedure to prevent synaptic transmission (Nishino & Koizumi 1977). Whether the cholinergic effects were exerted directly on the cells whose photic responses were measured therefore remains uncertain. In the second study, cholinergic agonists and antagonists were injected systemically, and their effects on SCN cells might have been mediated indirectly (Miller et al 1987). The agent used to reduce cholinergic transmission in that study was mecamylamine, which we have already noted may affect both cholinergic and other systems.

One study that used SCN slices from female rats differs from others in that it found firing rates in most SCN cells were suppressed by ACh infusion in the bathing medium (Kow & Pfaff 1984). A low-$Ca^{2+}$ artificial cerebrospinal fluid (ACSF) was used to restrict transsynaptic effects. This study also found that ovariectomy reduced and estrogen treatment restored responsiveness to ACh. This result suggests that the effects may have been mediated by the estrogen-dependent BTX receptors found in and near the SCN. If so, this finding would imply that the BTX receptors mediate suppressive effects of ACh. Alternatively, these effects may be mediated by a muscarinic receptor, since firing rate suppression by ACh was blocked by prior treatment with atropine (a muscarinic antagonist) in a small sample of cells. This result is consistent with evidence that atropine can modulate the effects of carbachol on pineal NAT activity (Zatz & Brownstein 1981). It remains to be established, however, that muscarinic receptors exist in the SCN and that they are estrogen-dependent.

The results of these several studies are difficult to reconcile or even compare because of the large number of procedural differences among them. These differences include the sex of the animal studied, the mode of drug administration, and a virtually universal failure to report the circadian phases at which studies were done. Circadian timing of drug administration may be important, since circadian variation in SCN responsiveness to drugs appears to be common (Mason 1986, Mason et al 1987, Rusak & Mason 1988, Shibata & Moore 1988).

In summary, it is unlikely that ACh is a primary mediator of the effects of light on the SCN pacemaker. The clear effects of cholinergic agents on rhythms may not be mediated in the SCN, or they may affect an unknown receptor type there. Although there is agreement that SCN cells respond to cholinergic agents (despite a lack of ACh binding in the SCN), their neurophysiological effects remain poorly understood. The studies claiming the strongest positive correlations between photic and cholinergic effects on single cells in the SCN present serious interpretative problems, and they do not make a clear case for such a correlation.

The presence in the SCN of the enzyme choline acetyltransferase (ChAT), which catalyzes production of ACh, may indicate that a cholinergic projection reaches the SCN, but probably not from the retina (Brownstein et al 1975, Ichikawa & Hirata, 1986, Tago et al 1987). An applicable model may be that suggested for the role of ACh in the goldfish tectum (Henley et al 1986, Langdon & Freeman 1987). Here a cholinergic projection, perhaps from the nucleus isthmi, reaches the retinorecipient zone of the tectum, where it may act presynaptically on the terminals of retinal fibers to modulate transmitter release. This model is plausible in rodents because cholinergic receptors are also found presynaptically on terminals in the mammalian visual system (Prusky & Cynader 1988). Cholinergic effects on mammalian rhythms may result from presynaptic modulation by ACh of transmitter release from retinal fibers reaching the SCN. Alternatively, cholinergic effects might occur later in the cascade of neural events triggered by photic input to the SCN.

## Amino Acids, Benzodiazepines, and Peptides

The case for glutamate or a related excitatory amino acid (EAA) as a transmitter in the retinal projection to the SCN is of mixed quality. There is evidence for a glutamate-like transmitter in other primary sensory afferents and in other parts of the visual system (Redburn 1981, Wenthold 1981, Anderson et al 1986) as well as the SCN (Moffet et al 1987). Glutamate and aspartate also have excitatory effects on SCN neurons, both in vivo and in vitro (Nishino & Koizumi 1977, Shibata et al 1986). EAAs are released in the SCN after optic nerve stimulation in a slice preparation (although, as with ACh release, not necessarily from the ganglion cell terminals), and blockade of EAA transmission with kynurenate and other antagonists can suppress neural responses in the SCN to optic nerve stimulation (Liou et al 1986, Cahill & Menaker 1987). The prevention of photic effects on the SCN by mecamylamine may also reflect its disruption of EAA transmission rather than its cholinergic effects.

Although glutamate injections into the SCN cause phase shifts of circadian activity rhythms in hamsters, the phase response curve derived from such injections does not resemble that for light pulses. Instead large phase advances occur during the subjective day and there are few delay shifts (Meijer et al 1988b), a pattern more typical of the effects of dark pulses delivered against a background of continuous light (Boulos & Rusak 1982). These observations might imply that light normally suppresses firing of glutamatergic ganglion cells that project to the SCN, whereas darkness activates them and causes the release of EAAs in the SCN. This suggestion would be consistent with optic chiasm recordings from below the SCN, which showed increasing firing rates in the dark and decreasing rates in the light phase (Kubota et al 1981), but it would be inconsistent with the finding that glutamate application to the SCN and retinal illumination have similar effects on the firing rates of most SCN cells (Nishino & Koizumi 1977).

The shape of the glutamate phase response curve raises the general issue of what phase response curves to drug administration can tell us. Even if an EAA, for example, is the transmitter in the retinal projection to the SCN, it may also be effective at sites in the SCN and elsewhere that do not receive retinal innervation. Thus, injection of a bolus of glutamate into the SCN, by affecting targets in addition to those normally affected by photic cues, could easily fail to mimic the consequences of photic input, even if glutamate is the authentic retinal transmitter. In addition, a bolus injection of glutamate may swamp the uptake mechanisms that normally clear EAAs from the vicinity of receptors, causing a long-lasting depolarization. The result may be a depolarization block that prevents further firing, so that an "excitatory" transmitter may have a net, long-term suppressive effect on cells bathed in it. Exactly such an effect of excess glutamate has been reported for spinal cord neurons (Lipski et al 1988). These same arguments could be applied to the difference between light- and carbachol-evoked phase response curves discussed above, and to the effects of bath-applied neuropeptide Y (see below).

These alternative interpretations make strong conclusions very difficult to reach based on the behavioral data. Physiological studies suggest that EAA mechanisms play a role in the photic entrainment pathway, and may even mediate direct retinal effects on the SCN. The behavioral pharmacology raises some questions about this hypothesis but does not provide a convincing argument against it.

The inhibitory amino acid, GABA (gamma aminobutyric acid), has also been implicated in the regulation of circadian rhythms. Immunoreactivity for the enzyme, glutamic acid decarboxylase (GAD), which catalyzes GABA synthesis, is very widespread in the SCN, a finding that suggests that GABAergic neurons consititute a large proportion of SCN cells (Card & Moore 1984, Van den Pol & Tsujimoto 1985). Drugs that affect GABAergic transmission can cause phase shifts and can alter phase shifts to light pulses differentially, depending on circadian phase (Ralph & Menaker 1985, 1986, 1989). The evidence that benzodiazepines can both phase shift rhythms and alter phase shifts in response to light stimuli is consistent with a model that postulates a role for a $GABA_A$-benzodiazepine receptor complex in entrainment. Pharmacological evidence indicates, however, that the chloride conductance generally thought to mediate the functional consequences of activating this complex is not responsible for these effects (Ralph & Menaker 1989).

Baclofen, an agonist at the $GABA_B$ receptor, also blocks light-induced phase delays and advances (Ralph & Menaker 1989), and reduces evoked field potentials in the SCN following optic nerve stimulation (Shibata et al 1986). Ralph & Menaker (1989) have proposed that GABA receptor agonists may influence phase shifting because of cellular effects they have in common, such as reducing cyclic AMP synthesis and voltage-dependent $Ca^{2+}$ conductance. These effects might alter synaptic transmission as well as intracellular events critical to phase shifting the pacemaker.

Treating the various transmitters hypothesized to be involved in photic effects on circadian rhythms as separate actors is undoubtedly an over-simplification. Benzodiazepines and GABA, for example, may exert their effects by influencing glutamate transmission from retinal ganglion cells (Ralph & Menaker 1989). Cholinergic effects on phase shifting may turn out to depend in part on the same pathway. Nicotine has been shown, for example, to enhance neural activity in the hippocampus by reducing GABAergic trasmission (Freund et al 1988). So cholinergic neurons outside the SCN might affect the photic pathway partly by projecting to GABAergic neurons, inhibition of which would enhance the release or postsynaptic effects of glutamate, thereby mimicking the effects of light. Cholinergic neurons might also directly facilitate release of an EAA from retinal ganglion cells.

The benzodiazepine triazolam has potent phase-advancing effects when injected into hamsters during the subjective day (Turek & Losee-Olson 1986, Turek 1987). Triazolam also causes increased locomotor activity in hamsters, and intense activity at this phase has been shown to cause similar phase shifts (Reebs & Mrosovsky 1989). A test of triazolam effects in hamsters that were physically restrained so they could not increase activity indicated that they also did not phase shift in response to the drug (Van

Reeth et al 1988). Increased activity probably does not mediate all of the effects of triazolam, however, since it does not increase activity in SK mice, yet it phase shifts their rhythms, as do other drugs that act on the GABA-benzodiazepine complex (Ebihara et al 1988).

Destruction of the lateral geniculate area prevented triazolam-induced phase shifts in hamsters (Johnson et al 1988). This result might indicate that triazolam acts in this area or that the projection from the geniculate to the SCN mediates the effects of benzodiazepines. Benzodiazepines might act in the geniculate area, or presynaptically on geniculate projections to the SCN, or in the raphe nuclei, which are responsive to triazolam (Mendelson et al 1987) and project to both geniculate and SCN (Aghajanian et al 1969, Pasquier & Villar 1982).

The geniculate projection to the SCN is formed at least in part by cells in the intergeniculate leaflet that are immunoreactive for neuropeptide Y (Card & Moore 1982, Harrington et al 1987). During the subjective day, neuropeptide Y injections into the SCN (Albers & Ferris 1984), electrical stimulation in the intergeniculate leaflet (Rusak et al 1989), and chemical excitation of the intergeniculate leaflet (Johnson et al 1988) all cause phase advances, and ablation of the intergeniculate leaflet reduces phase advances to light pulses and alters responses of the circadian system to continuous illumination (Harrington & Rusak 1986, 1988, Pickard et al 1987). These results suggest that neuropeptide Y–containing afferents to the SCN might serve as a common pathway mediating effects on SCN function of a variety of photic and nonphotic inputs, all of which generate a phase response curve resembling that for dark pulses (Rusak et al 1989).

Neurophysiological effects of neuropeptide Y again are difficult to reconcile with the behavioral data. In hypothalamic slices from hamsters, pressure-ejected neuropeptide Y has excitatory effects on most SCN cells, as does light (Mason et al 1987). On the other hand, bath perfusion of neuropeptide Y onto rat hypothalamic slices causes a biphasic response in most cells, with suppression predominating (Shibata & Moore 1988). Low-$Ca^{2+}$ artificial cerebrospinal fluid was used to control for transsynaptic effects in the latter study, but depressed firing rates might be an indirect consequence of sustained exposure of the cells to an initially excitatory peptide (cf. Lipski et al 1988). These studies were conducted on different species and with different methods, so the reasons for the discrepancy remain uncertain. Nonetheless, we can state that hamsters respond to bolus injections of neuropeptide Y into the SCN with a phase response curve resembling that for dark pulses, whereas application of neuropeptide Y to single cells has generally the same neurophysiological effects as light stimuli in hamsters. Since these statements are based on average responses of a population of cells, the conflict may be apparent rather than real. To interpret these results, correlation of individual cells' photic and pharmacological responsiveness remains a critical task.

## Serotonin and Melatonin

The raphe nuclei provide a heavy serotonergic projection to the SCN (Aghajanian et al 1969), terminating largely on vasocative intestinal polypeptide–containing neurons (Bosler & Beaudet 1985). Ablation of the raphe in rats has been reported to have little effect on rhythm generation (Block & Zucker 1976), or to produce severe disruptions in the expression of rhythms in constant conditions but not in lighting cycles (Levine et al 1986). Serotonin applied to SCN cells usually suppresses firing (Nishino & Koizumi 1977, Meijer & Groos 1988), and the SCN shows a diurnal rhythm of serotonin uptake (Meyer & Quay 1976) and sensitivity (Mason 1986). Treatment of hamsters with clorgyline, which increases brain serotonin levels, lengthens the free-running period and reduces the size of advance shifts to light (Duncan et al 1988). Fluoxetine, a serotonin reuptake blocker, phase shifts rhythms in sparrows (Cassone & Menaker 1985). In an in vivo neurophysiological study, serotonin affected nonphotic cells in the SCN, but did not alter the firing rates of a small sample of photically responsive units (Meijer & Groos, 1988).

Melatonin also affects firing rates of SCN cells in rats and hamsters (Mason & Brooks 1988, Rusak & Mason 1988), perhaps via a high-affinity binding site recently identified in the SCN and a few other brain regions (Vaněček et al 1987). Consistent with these data is the observation that melatonin reduces the normally high level of metabolic activity in SCN slices during the subjective day (Cassone et al 1987). These effects of melatonin might be related to regulation of either circadian or seasonal rhythms; melatonin injections can entrain circadian rhythms in some rat strains, although the mechanism involved may be different from that responsible for entrainment to photic cues (Redman et al 1983).

## The Role of Methamphetamine

One of the most striking effects of a drug on a circadian system is the reported restoration of robust free-running circadian rhythms in rats treated with chronic methamphetamine after being rendered arrhythmic by SCN ablation (Honma et al 1987). It has yet to be established whether these effects are related to the release of catecholamines known to be caused by methamphetamine treatment (Moore 1978).

The restored rhythms were not entrainable by lighting cycles but could be entrained by temporally restricted daily access to food (Honma et al 1989). These results suggest that the rhythm represents the output of the well-studied food-entrainable oscillator, which is known to be outside the

SCN and to survive its ablation (Stephan et al 1979). This oscillator is typically expressed as an overt activity rhythm only under food deprivation, a finding that is consistent with the fact that chronic methamphetamine treatment reduces food intake and body weight. Although catecholamines have not been demonstrated to play a significant role in generation or photic entrainment of circadian rhythms, these results suggest that they might be involved in the regulation of the food-entrainable oscillators outside the SCN.

In an attempted replication of this phenomenon, we have not seen a restoration of free-running rhythms in either SCN-ablated rats or hamsters (R. Silver and B. Rusak, unpublished observations). The differences in results obtained might suggest that the rat strain used by Honma et al (1987, 1989) has a particularly robust food-entrainable oscillator that is readily expressed as a free-running rhythm under conditions of reduced food intake.

## SUMMARY

This discussion of the roles of transmitters in the circadian system has focused mostly on the entrainment mechanism because it is not clear to what extent neurotransmission is important to the other major function of circadian systems, rhythm generation. Schwartz et al (1987) have presented evidence that the circadian pacemaker in the SCN continues to run, but cannot be entrained by light, when tetrodotoxin is used to block sodium-dependent action potentials. Although other forms of intercellular communication are not ruled out, these results suggest that classical synaptic neurotransmission is important for entrainment but not for rhythm generation.

That the SCN contains a cholinergic marker like ChAT and is responsive to cholinergic agents but does not bind nicotine or ACh reflects a general problem in reconciling functional, physiological, and anatomical markers of neurotransmission. A mismatch between the anatomical distributions of transmitters and their receptor-binding sites is a common observation, the meaning of which remains enigmatic (Herkenham 1987).

Also, the neurophysiological consequences of injections of drugs into parts of the brain involved in rhythm regulation remain largely unknown. Interpretations of the effects of these treatments on rhythms are predicated on assumptions that may not be valid; e.g. that a bolus injection of an excitatory substance has its primary effect by activating neurons. Still to be established is whether the effects of drugs when they are administered in behavioral pharmacology studies reflect their effects on cellular functions and on neuronal responses to photic cues when they are delivered at near-physiological levels to single neurons.

ACKNOWLEDGMENTS

We are grateful to Martin Ralph and Rob Mason for their comments on an earlier version of this manuscript. Preparation of the manuscript was supported by grants from the Medical Research Council (MA8929) and the Natural Sciences and Engineering Research Council (A0305) of Canada. KGB was supported by a Killam Memorial Scholarship.

*Literature Cited*

Aghajanian, G. K., Bloom, F. E., Sheard, M. H. 1969. Electron microscopy of degeneration within the serotonin pathway of rat brain. *Brain Res.* 13: 266–73

Albers, H. E., Ferris, C. F. 1984. Neuropeptide Y: Role in light-dark cycle entrainment of hamster circadian rhythms. *Neurosci. Lett.* 50: 163–68

Anderson, K. J., Monaghan, D. T., Cangro, C. B., Namboodiri, M. A. A., Neale, J. H., Cotman, C. W. 1986. Localization of N-acetylaspartylglutamate-like immunoreactivity in selected areas of the rat brain. *Neurosci. Lett.* 72: 14–20

Block, M., Zucker, I. 1976. Circadian rhythms of rat locomotor activity after lesions of the midbrain raphe nuclei. *J. Comp. Physiol.* 109: 235–47

Bosler, O., Beaudet, A. 1985. VIP neurons as prime synaptic targets for serotonin afferents in rat suprachiasmatic nucleus: A combined radioautographic and immunocytochemical study. *J. Neurocytol.* 14: 749–63

Boulos, Z., Rusak, B. 1982. Circadian phase response curves for dark pulses in the hamster. *J. Comp. Physiol.* 146: 411–17

Boulter, J., Connolly, J., Deneris, E., Goldman, D., Heinemann, S., Patrick, J. 1987. Functional expression of two neuronal nicotinic acetylcholine receptors from cDNA clones identifies a gene family. *Proc. Natl. Acad. Sci. USA* 84: 7763–67

Brownstein, M., Kobayashi, R., Palkovits, M., Saavedra, J. 1975. Choline acetyltransferase levels in diencephalic nuclei of the rat. *J. Neurochem.* 24: 35–38

Cahill, G. M., Menaker, M. 1987. Kynurenic acid blocks suprachiasmatic nucleus responses to optic nerve stimulation. *Brain Res.* 410: 125–29

Card, J. P., Moore, R. Y. 1982. Ventral lateral geniculate nucleus efferents to the rat suprachiasmatic nucleus exhibit avian pancreatic polypeptide-like immunoreactivity. *J. Comp. Neurol.* 206: 390–96

Card, J. P., Moore, R. Y. 1984. The suprachiasmatic nucleus of the golden hamster: Immunohistochemical analysis of cell and fiber distribution. *Neuroscience* 13: 415–31

Cassone, V. M., Menaker, M. 1985. Circadian rhythms of house sparrows are phase shifted by pharmacological manipulations of brain serotonin. *J. Comp. Physiol.* 156: 145–52

Cassone, V. M., Roberts, M. H., Moore, R. Y. 1987. Melatonin inhibits metabolic activity in the rat suprachiasmatic nuclei. *Neurosci. Lett.* 81: 29–34

Christensen, B. N., O'Dell, T. J. 1988. Mecamylamine antagonizes the N-methyl-D-aspartate activated channel in isolated horizontal cells from the catfish retina. *Soc. Neurosci. Abstr.* 14: 95

Clarke, P. B. S., Schwartz, R. D., Paul, S. M., Pert, C. B., Pert, A. 1985. Nicotinic binding in rat brain: Autoradiographic comparison of [$^3$H]acetylcholine, [$^3$H]nicotine, and [$^{125}$I]α-bungarotoxin. *J. Neurosci.* 5: 1307–15

DeCoursey, P. J. 1964. Function of a light response rhythm in hamsters. *J. Cell. Comp. Physiol.* 63: 189–96

Duncan, W. C., Tamarkin, L., Sokolove, P. G., Wehr, T. A. 1988. Chronic clorgyline treatment of syrian hamsters: An analysis of effects on the circadian pacemaker. *J. Biol. Rhythms* 3: 305–22

Earnest, D. J., Turek, F. W. 1983. Role for acetylcholine in mediating effects of light on reproduction. *Science* 219: 77–79

Earnest, D. J., Turek, F. W. 1985. Neurochemical basis for the photic control of circadian rhythms and seasonal reproductive cycles: Role for acetylcholine. *Proc. Natl. Acad. Sci. USA* 82: 4277–81

Ebihara, S., Goto, M., Oshima, I. 1988. Different responses of the circadian system to GABA-active drugs in two strains of mice. *J. Biol. Rhythms* 3: 357–64

Freund, R. K., Jungschaffer, D. A., Collins, A. C., Wehner, J. M. 1988. Evidence for modulation of GABAergic neurotransmission by nicotine. *Brain Res.* 453: 215–20

Güldner, F.-H. 1983. Numbers of neurons and astroglial cells in the suprachiasmatic nucleus of male and female rats. *Exp. Brain Res.* 50: 373–76

Harrington, M. E., Nance, D. M., Rusak, B. 1987. Double-labeling of neuropeptide Y–immunoreactive neurons which project from the geniculate to the suprachiasmatic nuclei. *Brain Res.* 410: 275–82

Harrington, M. E., Rusak, B. 1986. Lesions of the thalamic intergeniculate leaflet alter hamster circadian rhythms. *J. Biol. Rhythms* 1: 309–25

Harrington, M. E., Rusak, B. 1988. Ablation of the geniculo-hypothalamic tract alters circadian activity rhythms of hamsters housed under constant light. *Physiol. Behav.* 42: 183–89

Henley, M., Lindstrom, J. M., Oswald, R. E. 1986. Acetylcholine receptor synthesis in retina and transport to optic tectum in goldfish. *Science* 232: 1627–28

Herkenham, M. 1987. Mismatches between neurotransmitter and receptor localizations in brain: Observations and implications. *Neuroscience* 23: 1–38

Honma, K., Honma, S., Hiroshige, T. 1987. Activity rhythms in the circadian domain appear in suprachiasmatic nuclei lesioned rats given methamphetamine. *Physiol. Behav.* 40: 767–74

Honma, S., Honma, K., Hiroshige, T. 1989. Methamphetamine induced locomotor rhythm entrains to restricted daily feeding in SCN lesioned rats. *Physiol. Behav.* 45: 1057–65

Ichikawa, T., Hirata, Y. 1986. Organization of choline acetyltransferase-containing structures in the forebrain of the rat. *J. Neurosci.* 6: 281–92

Johnson, R. F., Smale, L., Moore, R. Y., Morin, L. P. 1988. Lateral geniculate lesions block circadian phase-shift responses to a benzodiazepine. *Proc. Natl. Acad. Sci. USA* 85: 5301–4

Keefe, D. L., Earnest, D. J., Nelson, D., Takahashi, J. S., Turek, F. W. 1987. A cholinergic antagonist, mecamylamine, blocks the phase shifting effects of light on the circadian rhythm of locomotor activity in the golden hamster. *Brain Res.* 403: 308–12

Kow, L., Pfaff, D. W. 1984. Suprachiasmatic neurons in tissue slices from ovariectomized rats: Electrophysiological and neuropharmacological characterization and the effects of estrogen treatment. *Brain Res.* 297: 275–86

Kubota, A., Inouye, S.-I. T., Kawamura, H. 1981. Reversal of multi-unit activity within and outside the suprachiasmatic nucleus in the rat. *Neurosci. Lett.* 27: 303–8

Langdon, R. B., Freeman, J. A. 1987. Pharmacology of retinotectal transmission in the goldfish: Effects of nicotinic ligands, strychnine, and kynurenic acid. *J. Neurosci.* 7: 760–73

Lehman, M. N., Silver, R., Gladstone, W. R., Kahn, R. M., Gibson, M., Bittman, E. L. 1987. Circadian rhythmicity restored by neural transplant: Immunocytochemical characterization of the graft and its integration with the host brain. *J. Neurosci.* 7: 1626–38

Levine, J. D., Rosenwasser, A. M., Yanovski, J. A., Adler, N. T. 1986. Circadian activity rhythms in rats with midbrain raphe lesions. *Brain Res.* 384: 240–49

Liou, S. Y., Shibata, S., Iwasaki, K., Ueki, S. 1986. Optic nerve stimulation induced increase of release of $^3$H-glutamate and $^3$H-aspartate but not $^3$H-GABA from the suprachiasmatic nucleus in slices of rat hypothalamus. *Brain Res. Bull.* 16: 527–31

Lipski, J., Bellingham, M. C., West, M. J., Pilowsky, D. 1988. Limitations of the technique of pressure microinjection of excitatory amino acids for evoking responses from localized regions of the CNS. *J. Neurosci. Meth.* 26: 169–79

Mason, R. 1986. Circadian variation in sensitivity of suprachiasmatic and lateral geniculate neurons to 5-hydroxytryptamine in the rat. *J. Physiol.* 377: 1–13

Mason, R., Brooks, A. 1988. The electrophysiological effects of melatonin and a putative melatonin antagonist (*N*-acetyltryptamine) on rat suprachiasmatic neurones in vitro. *Neurosci. Lett.* 95: 296–301

Mason, R., Harrington, M. E., Rusak, B. 1987. Elecrophysiological responses of hamster suprachiasmatic neurones to neuropeptide Y in the hypothalamic slice preparation. *Neurosci. Lett.* 80: 173–79

Meijer, J. H., Groos, G. A. 1988. Responsiveness of suprachiasmatic and ventral lateral geniculate neurons to serotonin and imipramine: A micro-iontophoretic study in normal and imipramine-treated rats. *Brain Res. Bull.* 20: 89–96

Meijer, J. H., van der Zee, E., Dietz, M. 1988a. The effects of intraventricular carbachol injections on the free-running activity rhythm of the hamster. *J. Biol. Rhythms* 4: 333–48

Meijer, J. H., van der Zee, E., Dietz, M. 1988b. Glutamate phase shifts circadian activity rhythms in hamsters. *Neurosci. Lett.* 86: 177–83

Mendelson, W. B., Martin, J. V., Perlis, M., Wagner, R. 1987. Arousal induced by injection of triazolam into the dorsal raphe nucleus of rats. *Neuropsychopharmacology* 1: 85–88

Meyer, D. C., Quay, W. B. 1976. Hypothalamic and suprachiasmatic uptake of serotonin *in vitro*: Twenty-four hour changes in male and próestrous female rats. *Endocrinology* 98: 1160–65

Miceli, M. O., van der Kooy, D., Post, C. A., Della-Fera, M. A., Baile, C. A. 1987. Differential distribution of cholecystokinin in hamster and rat forebrain. *Brain Res.* 402: 318–30

Miller, J. D., Murakami, D. M., Fuller, C. A. 1987. The response of suprachiasmatic neurons of the rat hypothalamus to photic and nicotinic stimuli. *J. Neurosci.* 7: 978–86

Miller, M. M., Silver, J., Billiar, R. B. 1982. Effects of ovariectomy on the binding of [$^{125}$I]-alpha-bungarotoxin (2.2 and 3.3) to the suprachiasmatic nucleus of the hypothalamus: an in vivo autoradiographic analysis. *Brain Res.* 247: 355–64

Miller, M. M., Silver, J., Billiar, R. B. 1984. Effects of gonadal steroids on the in vivo binding of [$^{125}$I]-alpha-bungarotoxin to the suprachiasmatic nucleus. *Brain Res.* 290: 67–75

Mistlberger, R. E., Rusak, B. 1986. Carbachol phase shifts circadian activity rhythms in ovariectomized rats. *Neurosci. Lett.* 72: 357–62

Moffett, J. R., Namboodiri, M. A. A., Neale, J. H. 1987. Identification of *N*-acetylaspartylglutamate-like immunoreactivity in the anterior hypothalamus of the rat. *Soc. Neurosci. Abstr.* 13: 860

Moore, K. E. 1978. Amphetamines: Biochemical and behavioral actions in animals. In *Handbook of Psychopharmacology*, Vol. 11, *Stimulants*, ed. L. L. Iversen, S. D. Iversen, S. H. Snyder, pp. 41–98. New York: Plenum

Moore, R. Y. 1983. Organization and function of a central nervous system circadian oscillator: The suprachiasmatic hypothalamic nucleus. *Fed. Proc.* 42: 2783–89

Murakami, N., Takahashi, K., Kawashiwa, K. 1984. Effect of light on the acetylcholine concentrations of the suprachiasmatic nucleus in the rat. *Brain Res.* 311: 358–60

Nishino, H., Koizumi, K. 1977. Responses of neurons in the suprachiasmatic nuclei of the hypothalamus to putative neurotransmitters. *Brain Res.* 120: 167–72

Pasquier, D. A., Villar, M. J. 1982. Specific serotonergic projections to the lateral geniculate body from the lateral cell groups of the dorsal raphe nucleus. *Brain Res.* 249: 142–46

Pauly, J. R., Horseman, N. D. 1985. Anti-

cholinergic agents do not block light-induced circadian phase shifts. *Brain Res.* 348: 163–67

Pickard, G. E., Ralph, M. R., Menaker, M. 1987. The intergeniculate leaflet partially mediates effects of light on circadian rhythms. *J. Biol. Rhythms* 2: 35–56

Prusky, G. T., Cynader, M. S. 1988. [$^3$]nicotine binding sites are associated with mammalian optic nerve terminals. *Visual Neurosci.* 1: 245–48

Ralph, M. R., Davis, F. C., Menaker, M. 1988. Suprachiasmatic transplantation restores donor specific circadian rhythms to arrhythmic hosts. *Soc. Neurosci. Abstr.* 14: 462

Ralph, M. R., Menaker, M. 1985. Biculculine blocks circadian phase delays but not advances. *Brain Res.* 325: 362–65

Ralph, M. R., Menaker, M. 1986. Effects of diazepam on circadian phase advances and delays. *Brain Res.* 372: 405–8

Ralph, M. R., Menaker, M. 1989. GABA regulation of circadian responses to light I. Involvement of GABA[A] benzodiazepine and GABA[B] receptors. *J. Neurosci.* 9: 2858–65

Redburn, D. A. 1981. GABA and glutamate as retina neurotransmitters in rabbit retina. In *Glutamate as a Neurotransmitter*, ed. G. Di Chiara, G. L. Gessa, pp. 79–89. New York: Raven

Redman, J., Armstrong, S., Ng, K. T. 1983. Free-running activity rhythms in the rat: Entrainment by melatonin. *Science* 219: 305–28

Reebs, S. G., Mrosovsky, N. 1989. Effects of induced wheel-running on the circadian activity rhythms of Syrian hamsters: Entrainment and phase response curve. *J. Biol. Rhythms* 4: 39–48

Rusak, B. 1989. The mammalian circadian system: models and physiology. *J. Biol. Rhythms* 4: 121–34

Rusak, B., Mason, R. 1988. Short daylengths alter electrophysiological responses of hamster suprachiasmatic nucleus (SCN) neurons to melatonin. *Soc. Neurosci. Abstr.* 14: 1298

Rusak, B., Meijer, J. H., Harrington, M. E. 1989. Hamster circadian rhythms are phase-shifted by electrical stimulation of the geniculo-hypothalamic tract. *Brain Res.* 493: 283–91

Rusak, B., Zucker, I. 1975. Biological rhythms and animal behavior. *Annu. Rev. Psychol.* 26: 137–71

Rusak, B., Zucker, I. 1979. Neural regulation of circadian rhythms. *Physiol. Rev.* 59: 449–526

Sawaki, Y., Nihonmatsu, I., Kawamura, H. 1984. Transplantation of the neonatal suprachiasmatic nuclei into rats with complete bilateral suprachiasmatic lesions. *Neurosci. Res.* 1: 67–72

Schwartz, W. J., Gross, R. J., Morton, M. T. 1987. The suprachiasmatic nuclei contain a tetrodotoxin-resistant pacemaker. *Proc. Natl. Acad. Sci. USA* 84: 1694–98

Shibata, S., Liou, S. Y., Ueki, S. 1986. Influence of excitatory amino acid receptor antagonists and of baclofen on synaptic transmission in the optic nerve to the suprachiasmatic nucleus in slices of rat hypothalamus. *Neuropharmacology* 28: 403–9

Shibata, S., Moore, R. Y. 1988. Neuropeptide Y and vasopressin effects on rat suprachiasmatic nucleus neurons in vitro. *J. Biol. Rhythms* 3: 265–76

Shinozaki, H. 1988. Pharmacology of the glutamate receptor. *Prog. Neurobiol.* 30: 399–435

Stephan, F. K., Swann, J. M., Sisk, C. L. 1979. Entrainment of circadian rhythms by feeding schedules in rats with suprachiasmatic lesions. *Behav. Neural. Biol.* 25: 545–54

Tago, H., McGeer, P. L., Bruce, G. Hersh, L. B. 1987. Distribution of choline acetyltransferase-containing neurons of the hypothalamus. *Brain Res.* 415: 49–62

Turek, F. W. 1987. Pharmacological probes of the mammalian circadian clock: Use of the phase response curve approach. *Trends Pharmacol. Sci.* 8: 212–17

Turek, F. W., Losee-Olson, S. 1986. A benzodiazepine used in the treatment of insomnia phase-shifts the mammalian circadian clock. *Nature* 321: 167–68

van den Pol, A. N., Tsujimoto, K. L. 1985. Neurotransmitters of the hypothalamic suprachiasmatic nucleus: Immunocytochemical analysis of 25 neuronal antigens. *Neuroscience* 15: 1049–86

van den Pol, A. N., Gorcs, T. 1986. Synaptic relationships between neurons containing vasopression, gastrin-releasing peptide, vasoactive intestinal polypeptide and

glutamate decarboxylase immunoreactivity in the suprachiasmatic nucleus: Dual ultrastructural immunocytochemistry with gold-substituted silver peroxidase. *J. Comp. Neurol.* 252: 507–21

Vandesande, F., Dierickx, K., De Mey, J. 1975. Identification of the vasopressin-neurophysin producing neurons of the rat suprachiasmatic nuclei. *Cell Tiss. Res.* 156: 377–80

Vaněček, J., Pavlik, A., Illnerová, H. 1987. Hypothalamic melatonin receptor sites revealed by autoradiography. *Brain Res.* 435: 359–62

Van Reeth, O., Vanderhaeghen, J. J., Turek, F. W. 1988. Phase advancing effects of benzodiazepine on the hamster circadian clock can be blocked by preventing the associated hyperactivity. *Soc. Res. Biol. Rhythms Abstr.* 1: 29

Wada, E., Wada, K., Boulter, J., Deneris, E., Heinemann, S., Patrick, J., Swanson, L. W. 1989. Distribution of Alpha2, Alpha3, Alpha4, and Beta2 neuronal receptor subunit mRNAs in the central nervous system: A hybridization histochemical study in the rat. *J. Comp. Physiol.* 284: 314–35

Wenthold, R. J. 1981. Glutamate and aspartate as neurotransmitters for the auditory nerve. In *Glutamate as a Neurotransmitter*, ed. G. Di Chiara, G. L. Gessa, pp. 69–78. New York: Raven

Zatz, M., Brownstein, M. J. 1979. Intraventricular carbachol mimics the effects of light on the circadian rhythm in the rat pineal gland. *Science* 203: 358–60

Zatz, M., Brownstein, M. J. 1981. Injection of alpha-bungarotoxin near the suprachiasmatic nucleus blocks the effects of light on nocturnal pineal enzyme activity. *Brain Res.* 213: 438–42

Zatz, M., Herkenham, M. A. 1981. Intraventricular carbachol mimics the phase shifting effects of light on the circadian rhythm of wheel-running activity. *Brain Res.* 212: 234–38

*Annu. Rev. Neurosci. 1990. 13:403–14*

# DO INSECTS HAVE COGNITIVE MAPS?

*Rüdiger Wehner*

Department of Zoology, University of Zürich, 8057 Zürich, Switzerland

*Randolf Menzel*

Department of Animal Physiology and Zoology, University of Berlin, 1000 Berlin 41, Germany

As cognitive maps have come into fashion for behavioral neurobiologists and experimental psychologists (for mammals see O'Keefe & Nadel 1978, Olton 1978, Boesch & Boesch 1983, Sherry 1985, Cheng 1986, Gallistel 1989; for birds see Balda & Tureck 1984, Sherry 1985, Stevens & Krebs 1986, Krebs et al 1989) it is not astounding that recently even insects have been claimed to possess cognitive maps (Gould 1986). However unwarranted this claim might be—see e.g. the critical remarks on Gould's paper (1986) in Cartwright & Collett (1987) and Dyer & Seeley (1989)— we have taken it as a starting point for the present investigation. Of course, any such claim depends crucially on how a "cognitive map" is defined in operational terms, and finally on information about how the map is assembled and how it is used. Even though the term *cognitive map* is often defined rather vaguely and applied to various kinds of animal orientation, we hope to be in line with most workers in the field—and especially with Tolman (1948), who coined the term—when we use it in the way a human navigator does. Seen in this light, a cognitive map is the mental analogue of a topographic map, i.e. an internal representation of the geometric relations among noticeable points in the animal's environment. In operational terms, this means that an animal using such a map must be able to compute the shortest distance between two charted points without ever having traveled along that route. More generally, it must be able to determine its position, say, relative to home, or to any other charted point,

403

0147–006X/90/0301–0403$02.00

even when it has been displaced unexpectedly to an arbitrary place within its environment. Do insects have cognitive maps?

Our aim in this chapter is to tackle, and we hope, to answer, this question in those insects that exhibit the most impressive navigational capabilities, namely the social insects, particularly bees and ants. These insects are central place foragers (Orians & Pearson 1976, Houston & McNamara 1985). After each particular foraging trip, which might lead them up to 10,000 meters (Visscher & Seeley 1982) away from their "central place", they invariably return to the center, i.e. the nesting site of their colony. Do they actually use cognitive maps in meeting the navigational requirements necessarily involved in that task, or can their potentially map-based behavior be explained by simpler computational abilities that ants and bees are already known to possess?

Two navigational systems have been found to be of prime importance in foraging bees and ants: path integration (or dead reckoning, to borrow the human navigator's term), and goal localization by using some kind of stored snapshot of the landmark panorama around the goal. First and foremost, while searching for food, bees and ants always know their position relative to their starting point (home) through path integration. During their circuitous outward journeys, they continuously monitor the angles steered and the distances traveled and integrate these data, so that they are always informed about the vector pointing from their current position toward home. Consequently, the foraging insect possesses a continually updated representation of its spatial position relative to its starting point. In the process of path integration, bees and ants use skylight compasses for measuring directions (Wehner 1982, 1989). The final problem of how they continuously compute the mean home vector has recently been tackled and solved in desert ants, *Caiaglyphis fortis* (Wehner & Wehner 1986, Müller & Wehner 1988). As all computations involved in path integration are done within a self-center system of coordinates, the mechanism of path integration is inherently subject to cumulative errors. Here, the second navigational system (goal localization) comes to the fore. Ants (Wehner & Raeber 1979, Wehner 1983) and bees (Wehner 1981) finally pinpoint the position of their goal, the nest entrance, by using nearby landmarks surrounding the goal. Before setting out for a foraging trip, they store something akin to a two-dimensional snapshot of the landmarks taken from that point. Upon return, they continuously compare this remembered snapshot with their current retinal images and move so as to reduce the discrepancy between the two (Cartwright & Collett 1983). If even this system fails in localizing the nesting site, the insect resorts to a third mode of navigation by engaging in systematic searching behavior (Wehner & Srinivasan 1981).

With these navigational capabilities in mind, let us now inquire about the insect's putative mental map. Here we focus especially on experiments with honey bees, *Apis mellifera*, because it was in this hymenopteran species that Gould (1986) invoked his concept of the insect's map.

There are three possible ways of testing for map-based behavior (Figure 1). In all three cases, bees that have been trained to a food source *F* are

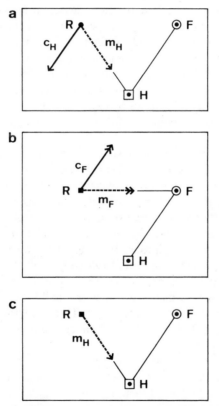

*Figure 1*  Three ways of testing for map-based behavior in central place foragers, e.g. honey bees. *H*, hive (central place); *F*, feeding site to which the bees have been trained prior to displacement; *R*, release site to which the bees were displaced from either *F* (Figure 1a) or *H* (Figure 1b: displacement prior to departure from *H*, Figure 1c: displacement after arrival at *H*); *c* (*solid arrows*), compass courses the bees should take if they relied exclusively on their path integration system; $c_F$ and $c_H$, direction of feeding site and hive, respectively, as indicated by the bee's self-center system of coordinates; *m* (*hatched arrows*), map courses; $m_F$ and $m_H$, directions of feeding site and hive, respectively, as indicated by a geocentric system of coordinates.

displaced to an arbitray release site $R$ within the colony's foraging range, 100 to 300 meters away from the hive. Note that this distance between release site and hive is less than 10% of the flight range the bees exhibit in the area in which the experiments described below are performed. The individually marked bees are displaced to site $R$ either from the food source $F$ (experiment A, Figure 1a) or from the hive $H$—and in the latter case either before departure (experiment B, Figure 1b) or after arrival from a successful foraging trip (experiment C, Figure 1c).

In experiments A and B there are two different predictions about how the displaced bees could behave. If, after release, the bees relied exclusively on their path integration system, they should select the compass courses $c_H$ and $c_F$ leading to the hive and the feeding station, respectively. On the other hand, if they realized the geometric relations among sites $R$, $H$, and $F$, i.e. used map information, they should fly along the map courses $m_H$ and $m_F$ (in experiments A and B, respectively). In all versions of experiments A and B performed at two localities by two research groups, the bees highly significantly selected the compass rather than the map courses (Figures 2, 3).

In this context one additional result is extremely interesting. At release site $R_1$ of Figure 3, the behavior of bees released under sunny and overcast conditions is compared. Under the overcast sky the bees could see neither the sun nor parts of the pattern of polarized light (as indicated by control experiments not described here). In spite of this lack of celestial information the bees selected the compass course much in the same way as they did under sunny conditions. It is already known from the elegant experiments of Dyer (Dyer & Gould 1981, Dyer 1987) that bees can infer the azimuthal position of the sun on completely overcast days from familiar landmark panoramas, which they have originally used as frames of reference for learning the sun's daily course. Thus it follows that in those of our experiments which were performed under a fully overcast sky, the bees did use landmark information, but they used this information not to compute the true position of the release site within some kind of mental map; they used the landmarks merely to determine the compass course $c_H$ that normally would have led them back to the hive.

One further observation is worth mentioning. The bees displaced from either the feeding station or the hive certainly realized that they had been displaced and that they had been released at an unusual site, because after release they circled around for a short while, spiraling higher and higher up in the air, before they flew off along a straight course. Such circling was never observed when the trained bees, prior to displacement, left either the hive or the feeding station. Notwithstanding this often intensive circling behavior, the displaced bees finally selected the proper compass course

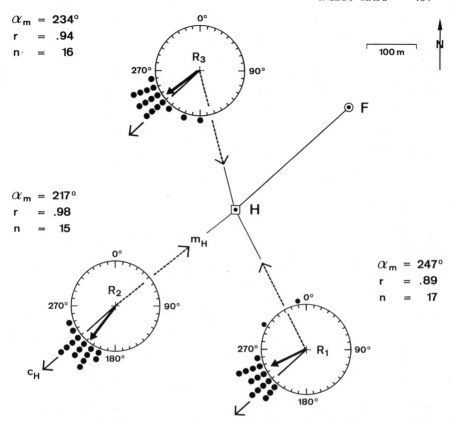

$\alpha_m = 234°$
r = .94
n = 16

$\alpha_m = 217°$
r = .98
n = 15

$\alpha_m = 247°$
r = .89
n = 17

100 m    N

*Figure 2*  Displacement experiments in honey bees, *Apis mellifera mellifera*. Experimental site: Zürich. The bees were displaced from the feeding site (*F*) to the release sites $R_1$–$R_3$. For experimental paradigm and conventions see Figure 1a. $\alpha_m$ and r, direction and length of mean orientation vector (*heavy black arrow*), respectively; n, number of individually tested bees.

and never showed any sign of map-based behavior. In conclusion, all the results of our displacement experiments—of which Figures 2 and 3 provide only some examples—are at variance with both Gould's data and his conclusion (Gould 1986, Gould & Towne 1987).

Even if the bees, as reported by Gould (1986), had finally selected what looked like a map course rather than a compass course, this would not conclusively demonstrate that the bees had used landmark-based mental maps to reach their intended goal. Dyer & Seeley (1989), one of whom (F. C. D.) is well familiar with Gould's study area, clinch a strong argument

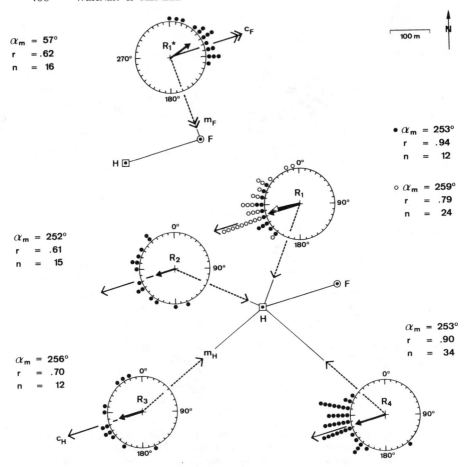

$\alpha_m = 57°$
r = .62
n = 16

$\alpha_m = 253°$
r = .94
n = 12

$\circ \alpha_m = 259°$
r = .79
n = 24

$\alpha_m = 252°$
r = .61
n = 15

$\alpha_m = 253°$
r = .90
n = 34

$\alpha_m = 256°$
r = .70
n = 12

*Figure 3* Displacement experiments in honey bees. *Apis mellifera ligustica*. Experimental site: Woods Hole, Massachusetts. The bees were displaced from the feeding site (*F*) to the release sites $R_1$–$R_4$. The data presented by the *open symbols* in $R_1$ were obtained under fully overcast conditions. For experimental paradigm and conventions see Figures 1a and 2; for release $R_1$* see Figure 1b.

against Gould's interpretation by stressing that "it would be well within the visual abilities of bees to see a forested upland on being released in a pasture, or vice versa. Bees and many other insects can find a specific familiar site relative to a small array of landmarks by moving to get a better match between their current visual image, and the image learned with previous experience at the site (Wehner 1981, Collett & Cartwright

1983). This matching process can be performed by bees no matter how they enter the array (for example via novel routes)". Cartwright & Collett (1987) add another point by arguing that novel routes might be generated by manipulating vectors rather than by referring to a landmark-based topographic map.

In a second paper Gould goes even one step further in assuming that mental maps of the landscape form the frame of reference for the honey-bee's dance communication system, and that the communication dances performed by successful foragers within the hive have evolved "to provide a direct readout" of landmark-based maps (Gould & Towne 1987). In their rebuttal of Gould's argument, Dyer & Seeley (1989) convincingly show that a communication system based on landmark maps rather than on vector information obtained via the bee's path integration system would be very inefficient indeed. As the arrangement of landmarks around any particular hive is unpredictable to bees starting their foraging lives, a map-based communication system would first require numerous flights to assemble the landmark-map, and would then require that the maps of dancers and recruits overlap. In the context of the present account, we note that a map-based communication system in the way proposed by Gould & Towne (1987) is not supported by experimental evidence.

Let us finally turn to displacement experiment C (Figure 1c). This experiment was designed in such a way that the bees had returned to the hive and therefore had reset their path integration system to zero before being displaced. Left with no compass information, they must now resort exclusively to landmark cues. Many classical experiments dating back even to the last century (Fabre 1879, Romanes 1885) have been performed in this way, but general conclusions are difficult to draw from the results. The homing success rates calculated on the basis of the times taken by the bees to return, or on the basis of the number of successful returns, varied widely even among individuals of the same colony. These variations are mainly due to the fact that the bees' foraging experiences prior to the displacement experiments were neither controlled nor known. In only one study has at least the age dependence of the homing success rate of honey bees been investigated to some extent (Lewtschenko 1959; for studies of other hymenopterans see Rau 1929, Ugolini 1986). Nevertheless, taken together, the results of all type-C experiments cited in the literature (for summary see Wehner 1981) are consistent with the hypothesis that after release the bees first perform extensive circling flights until they reach what has been called (Cartwright & Collett 1987) the catchment area of the snapshot they have taken at the hive. Thus searching behavior and mech-anisms of matching-to-memory as outlined in the beginning of this chapter are entirely adequate to explain the results of the type-C experiments.

Recent investigations, in which the foraging range of the colony and the foraging experience of individually marked bees were known (R. Wehner, unpublished), also clearly showed that the bees pick a straight path to the hive, and return quickly, only when the release site $R$ is positioned along the line $HF$. Angular deviations from that line result in a dramatic increase of the flight times needed for return. It seems as though the bees had become familiar only with a pattern of landmarks they had experienced while traversing specific routes. This conclusion is in full accord with similar experiments performed with ants, in which, unlike bees, the search trajectories can be recorded in detail (Wehner & Flatt 1972, Wehner et al 1983).

When bees and ants do not assemble and use mental topographic maps, what kind of internal representation of their foraging terrain should we attribute to them? One of us (R.W.) has published a survey of various ways whereby central place foragers might assemble spatial representations of their foraging sites (Wehner et al 1983). Let us briefly review these ways in the light of more recent data.

*Vector information*   The predominant piece of spatial information bees and ants acquire and use is vector information. By referring to the sky as a directional reference (Wehner 1989), measuring flight and walking distances, and integrating directions and distances, the insect obtains information about the radial coordinates of a foraging site as well as how to return from that foraging site to the starting point (home). For example, desert ants (*Cataglyphis*) can leave the nest and proceed along circuitous routes for more than 200 meters. After having located a food item they do not retrace their outward path, but walk directly back toward the nest. They locate the nest entrance with fair precision by using only their path integration system (Wehner 1982, Wehner & Wehner 1986, Müller & Wehner 1988). Path integration and the use of vector information have been described for bees (von Frisch 1967) and other arthropods as well (Goerner & Zeppenfeld 1980, Seyfarth et al 1982, Hoffman 1983, Goerner & Claas 1985, Mittelstaedt 1985, Ugolini 1987).

*Route information*   Depending on the degree of patchiness of the food supply within their nest environs, many central place foragers frequently revisit the same foraging site (Rosengren 1971, Fresneau 1985, Seeley 1985, Wehner 1987, Harkness & Isham 1988). In these cases the vector course leading to that site can become increasingly "decorated" with snapshots of landmarks, i.e. with landmark-based route information added to the original vector information (Rosengren & Pamilo 1978, David & Wood 1980, Kaul 1983, Fourcassie 1986, Klotz 1987). In their final approaches toward the nest, walking *Ammophila* wasps (Baerends 1941) and *Cata-*

*glyphis* ants (Wehner 1987) often select idiosyncratic and stereotyped routes, which differ slightly from the straight vector courses, and which can be changed experimentally by manipulating the arrangement of landmarks around the routes. Further evidence for the use of landmark-based route information comes from the marked trap-lining behavior of many solitary bees (Barrows 1976), bumble bees (Haas 1967, Heinrich 1976), and orchid bees (Janzen 1971). The latter have been shown to visit the same set of plants day after day in the same order along a feeding route that may extend up to 23 km from the nest.

With this additional piece of evidence, we can now extend the snapshot model introduced above. Let us propose that the homing insect uses a stack of snapshots activated sequentially as the insect approaches its nesting site. As recent experiments in *Cataglyphis* show (R. Wehner, unpublished), these shapshots are not completely unprocessed retinal images. The snapshots pertaining to points along the route that are far away from the nest contain only distant landmarks. As the ant moves closer to its final destination, it progressively refers to nearby landmarks. For simple geometrical reasons, the functional implications of this behavior are easy to understand. While distant landmarks guide the ant to the broad area of the goal, close landmarks allow for finally pinpointing the exact position of the goal. Model bees using just two snapshots (the one used first with the close landmarks filtered out) behave in exactly this way (Cartwright & Collett 1987), and real bees have been shown to discriminate between nearby and remote landmarks of equal angular (retinal) size. In their final search for the goal, they weigh near landmarks more heavily than distant ones (Cheng et al 1987). Although insects seem to be unable to measure the absolute sizes of objects, and do not resolve the optical size-distance ambiguity (Cartwright & Collett 1979, Wehner & Raeber 1979), they can obtain information about relative distances by exploiting motion parallax cues (Collett 1978, Eriksson 1980, Goulet et al 1981, Lehrer et al 1988).

*Vector maps*    When a bee forages at more than one feeding site (von Frisch 1967) and thus possesses vector information about, say, sites *A* and *B*, it could in principle devise the novel route *AB* through vector addition. Furthermore, if the two routes were linked with snapshots taken at *A* and *B* (Cartwright & Collett 1987; for birds Wallraff 1974, Fueller et al 1983), the bee could define the direct trajectory *AB* even after displacement from the hive to *A* or *B* (Figure 1c). No data are available that support this hypothesis in any insect. Furthermore, it might be worth mentioning in this context that even in path integration ants do not perform what could be called vector addition (Müller & Wehner 1988). The way they integrate their paths would be completely inadequate for assembling a vector map.

*Topographic maps*    Similarly to constructing a vector map from individual vectors, the animal could combine landmark-based routes within a geocentric system of coordinates and thus form the analogue of a topographic map as defined in the beginning. However, the results and discussions presented in this chapter make it quite clear that the question posed in the title cannot be answered in the affirmative. At present there is no convincing evidence for map-based behavior in insects. Certainly, bees cannot "make use of novel and efficient routes on the basis of map-like cognitive representations" (p. 863 in Gould 1986). The insect's navigational system is able to cope with many different ecological situations, e.g. obscured sun (Wehner & Rossel 1985, Rossel & Wehner 1986), overcast sky (Dyer 1987), apparent movement of the sun (New & New 1962, Wehner & Lanfranconi 1981, Dyer 1987), and drift by wind (Heran & Lindauer 1963), but unexpected displacements in dark boxes carried by human experimenters have obviously not been an evolutionary force that has shaped the insect's navigational system.

*Literature Cited*

Baerends, G. P. 1941. Fortpflanzungsverhalten und Orientierung der Grabwespe *Ammophila campestris. Tijdschr. Entomol. Deel* 84: 68–275

Balda, R. P., Tureck, R. J. 1984. The cache recovery system as an example of memory capabilities in Clark's Nutcrackers. In *Animal Cognition*, ed. H. L. Roitblat, T. G. Bever, H. S. Terrace, pp. 513–32. New York: Erlbaum

Barrows, E. M. 1976. Mating behavior in halictine bees (*Hymenoptera, Halictidae*): I. Patrolling and age-specific behavior in males. *J. Kansas Entomol. Soc.* 49: 105–19

Boesch, C., Boesch, H. 1983. Optimisation of nut-cracking with natural hammers by wild chimpanzees. *Behaviour* 3: 265–86

Cartwright, B. A., Collett, T. S. 1979. How honeybees know their distance from a near-by visual landmark. *J. Exp. Biol.* 82: 367–72

Cartwright, B. A., Collett, T. S. 1983. Landmark learning in bees. Experiments and models. *J. Comp. Physiol.* 151: 521–43

Cartwright, B. A., Collett, T. S. 1987. Landmark maps for honeybees. *Biol. Cybern.* 57: 85–93

Cheng, K. 1986. A purely geometric module in the rat's spatial representation. *Cognition* 23: 149–78

Cheng, K., Collett, T. S., Pickhard, A., Wehner, R. 1987. The use of visual landmarks by honeybees: Bees weight landmarks according to their distance from the goal. *J. Comp. Physiol. A* 161: 469–75

Collett, T. S. 1978. Peering—a locust behavior pattern for obtaining motion parallax information. *J. Exp. Biol.* 76: 237–41

Collett, T. S., Cartwright, B. A. 1983. Eidetic images in insects: Their role in navigation. *Trends Neurosci.* 6: 101–5

David, C. T., Wood, D. L. 1980. Orientation to trails by a carpenter ant, *Camponotus modoc* (*Hymenoptera: Formicidae*), in a giant sequoia forest. *Can. Entomol.* 112: 993–1000

Dyer, F. C. 1987. Memory and sun compensation in honey bees. *J. Comp. Physiol. A* 160: 621–33

Dyer, F. C., Gould, J. L. 1981. Honey bee orientation: A backup system for cloudy days. *Science* 214: 1041–42

Dyer, F. C., Seeley, T. D. 1989. On the evolution of the dance language. *Am. Nat.* In press

Eriksson, E. S. 1980. Movement parallax and distance perception in the grasshopper (*Phaulacridium vittatum*). *J. Exp. Biol.* 86: 337–40

Fabre, J. H. 1879. *Souvenirs Entomologiques.* 1. Sér. Paris: Delagrave

Fourcassie, V. 1986. Terrestrial cues in a polycalic colony of the red wood ants (*Formica rufa* group) according to the localization of the nest-hills. In *Orientation in*

*Space*, ed. G. Beugnon, pp. 89–96. Toulouse: Privat I.E.C.

Fresneau, D. 1985. Individual foraging and path fidelity in a ponerine ant. *Insectes Soc.* 32: 109–16

Fuller, E., Kowalski, U., Wiltschko, R. 1983. Orientation of homing pigeons: Compass orientation vs piloting by familiar landmarks. *J. Comp. Physiol.* 153: 55–58

Gallistel, C. R. 1989. Animal cognition: The representation of space, time, and number. *Annu. Rev. Psychol.* 40: 155–89

Goerner, P., Claas, B. 1985. Homing behavior and orientation in the funnelweb spider, *Agelena labyrinthica*. In *Neurobiology of Arachnids*, ed. F. G. Barth, pp. 275–97. Berlin, Heidelberg, New York: Springer

Goerner, P., Zeppenfeld, C. 1980. The runs of *Pardosa amentata* (Araneae, Lycosidae) after removing its cocoon. *Proc. Int. Congr. Arachnol.* 8: 243–48

Gould, J. L. 1986. The locale map of honey bees: Do insects have cognitive maps? *Science* 232: 861–63

Gould, J. L., Towne, W. F. 1987. Evolution of the dance language. *Am. Nat.* 130: 317–38

Goulet, M., Campan, R., Lambin, M. 1981. The visual perception of relative distances in the wood-cricket *Nemobius sylvestris*. *Physiol. Entomol.* 6: 357–67

Haas, A. 1967. Vergleichende Verhaltensstudien zum Paarungsschwarm der Hummeln (*Bombus*) und Schmarotzer-hummeln (*Psithyrus*). I. Teil. *Z. Tierpsychol.* 24: 257–77

Harkness, R. D., Isham, V. 1988. Relations between nests of *Messor wasmanni* in Greece. *Insectes Soc.* 35: 1–18

Heinrich, B. 1976. Bumblebee foraging and the economics of sociality. *Am. Sci.* 64: 384–95

Heran, H., Lindauer, M. 1963. Windkompensation und Seitenwindkorrektur der Biene beim Flug über Wasser. *Z. Vergl. Physiol.* 47: 39–55

Hoffmann, G. 1983. The search behavior of the desert isopod *Hemilepistus reaumuri* as compared with a systematic search. *Behav. Ecol. Sociobiol.* 13: 93–106

Houston, A. I., McNamara, J. M. 1985. A general theory of central place foraging for single-prey loaders. *Theor. Popul. Biol.* 28: 233–62

Janzen, D. H. 1971. Euglossine bees as long-distance pollinators of tropical plants. *Science* 171: 203–5

Kaul, R. M. 1983. Orientation of *Formica pratensis* (Hymenoptera: Formicidae). *Zool. Zh.* 62: 240–44

Klotz, J. H. 1987. Topographic orientation in two species of ants (Hymenoptera: Formicidae). *Insectes Soc.* 34: 236–51

Krebs, J. R., Hilton, S. C., Healy, S. D. 1989. Memory in food-storing birds: Adaptive specialization in brain behavior? In *Signal and Sense: Local and Global Order in Perceptual Maps*, ed. G. M. Edelmann, W. E. Gall, W. M. Cowan. New York: Neuroscience Inst. In press

Lehrer, M., Srinivasan, M. V., Zhang, S. W., Horridge, G. A. 1988. Motion cues provide the bee's visual world with a third dimension. *Nature* 332: 356–57

Lewtschenko, I. A. 1959. The return of bees to the hive. *Pchelovodstvo* 36: 38–40 (in Russian)

Mittelstaedt, H. 1985. Analytical cybernetics of spider navigation. See Goerner & Claas 1985, pp. 298–316

Müller, M., Wehner, R. 1988. Path integration in desert ants, *Cataglyphis fortis*. *Proc. Natl. Acad. Sci. USA* 85: 5287–90

New, D. A. T., New, J. K. 1962. The dances of honeybees at small zenith distances of the sun. *J. Exp. Biol.* 39: 271–91

O'Keefe, J., Nadel, L. 1978. *The Hippocampus as a Cognitive Map*. Oxford: Clarendon

Olton, D. S. 1978. Characteristics of spatial memory. See Menzel 1978, pp. 341–73

Orians, G. H., Pearson, N. E. 1979. On the theory of central place foraging. In *Analysis of Ecological Systems*, ed. D. J. Horn, G. R. Stairs, R. D. Mitchell, pp. 155–77. Columbus: Ohio State Univ. Press

Rau, P. 1929. Experimental studies in the homing of carpenter and mining bees. *J. Comp. Psychol.* 9: 35–70

Romanes, G. J. 1885. Homing faculty of Hymenoptera. *Nature* 32: 630

Rosengren, R. 1971. Route fidelity, visual memory and recruitment behaviour in foraging wood ants of the genus *Formica* (Hymenoptera, Formicidae). *Acta Zool. Fennica* 133: 1–106

Rosengren, R., Pamilo, P. 1978. Effect of winter timber felling on behaviour of foraging wood ants (*Formica rufa* group) in early spring. *Mem. Zool.* 29: 143–55

Rossel, S., Wehner, R. 1986. Polarization vision in bees. *Nature* 323: 128–31

Seeley, T. D. 1985. *Honeybee Ecology. A Study of Adaptation in Social Life*. Princeton: Princeton Univ. Press

Seyfarth, E. A., Hergenroeder, R., Ebbes, H., Barth, F. G. 1982. Idiothetic orientation of a wandering spider: Compensation of detours and estimates of goal distance. *Behav. Ecol. Sociobiol.* 11: 139–48

Sherry, D. F. 1985. Food storage by birds and mammals. *Adv. Study Behav.* 15: 153–88

Stevens, T. A., Krebs, J. R. 1986. Retrieval of stored seeds by marsh tits *Parus palustris* in the field. *Ibis* 128: 513–25

Tolman, E. C. 1948. Cognitive maps in rats and men. *Psychol. Rev.* 55: 189–208

Ugolini, A. 1986. Homing ability in *Polistes gallicus* (Hymenoptera, Vespidae). *Monitore Zool. Ital. (NS)* 20: 1–15

Ugolini, A. 1987. Visual information acquired during displacement and initial orientation in *Polistes gallicus* (Hymenoptera, Vespidae). *Anim. Behav.* 35: 590–95

Visscher, P. K., Seeley, T. D. 1982. Foraging strategy of honeybee colonies in a temperate deciduous forest. *Ecology* 63: 1790–1801

von Frisch, K. 1967. *The Dance Language and Orientation of Bees*. Cambridge, Mass: Harvard Univ. Press

Wallraff, H. G. 1974. *Das Navigationssystem der Vögel. Ein theoretischer Beitrag zur Analyse ungeklärter Orientierungsleistungen*. München/Wien: Oldenbourg

Wehner, R. 1981. Spatial vision in arthropods. In *Handbook of Sensory Physiology*, ed. H. Autrum, 7/6C: 287–616. Berlin/Heidelberg/New York: Springer

Wehner, R. 1982. Himmelsnavigation bei Insekten. Neurophysiologie und Verhalten. *Neujahrsbl. Naturforsch. Ges. Zürich* 184: 1–132

Wehner, R. 1983. Celestial and terrestrial navigation: Human strategies—insect strategies. In *Neuroethology and Behavioural Physiology*, ed. F. Huber, H. Markl, pp. 366–81. Berlin/Heidelberg/New York: Springer

Wehner, R. 1987. Spatial organization of foraging behaviour in individually searching desert ants, *Cataglyphis* (Sahara desert) and *Ocymyrmex* (Namib desert). In *From Individual to Collective Behaviour in Social Insects*, ed. J. M. Pasteels, J.-L. Deneubourg, pp. 15–42. Basel/Boston: Birkhäuser

Wehner, R. 1989. Neurobiology of polarization vision. *Trends Neurosci.* 12: 353–59

Wehner, R., Flatt, I. 1972. The visual orientation of desert ants, *Cataglyphis bicolor*, by means of terrestrial cues. In *Information Processing in the Visual Systems of Arthropods*, ed. R. Wehner, pp. 295–302. Berlin/Heidelberg/New York: Springer

Wehner, R., Harkness, R. D., Schmid-Hempel, P. 1983. Foraging strategies in individually searching ants, *Cataglyphis bicolor* (Hymenoptera: Formicidae). *Akad. Wiss. Lit. Mainz, Math.-Naturwiss. Kl.*

Wehner, R., Lanfranconi, B. 1981. What do the ants know about the rotation of the sky? *Nature* 293: 731–33

Wehner, R., Raeber, F. 1979. Visual spatial memory in desert ants. *Cataglyphis bicolor* (Hymenoptera, Formicidae). *Experientia* 35: 1569–71

Wehner, R., Rossel, S. 1985. The bee's celestial compass—A case study in behavioural neurobiology. *Fortschr. Zool.* 31: 11–53

Wehner, R., Srinivasan, M. V. 1981. Searching behaviour of desert ants, genus *Cataglyphis* (Formicidae, Hymenoptera). *J. Comp. Physiol.* 142: 315–38

Wehner, R., Wehner, S. 1986. Path integration in desert ants. Approaching a long-standing puzzle in insect navigation. *Monitore Zool. Ital. (NS)* 20: 309–31

*Annu. Rev. Neurosci. 1990. 13:415–40*

# DOPAMINE CELL REPLACEMENT: Parkinson's Disease

*David M. Yurek and John R. Sladek, Jr.*

Department of Neurobiology and Anatomy, University of Rochester School of Medicine and Dentistry, Rochester, New York 14642

## INTRODUCTION

Over the past several years considerable attention has focused on the numerous clinical attempts at neural transplantation in patients afflicted with Parkinson's disease. These clinical experiments essentially evolved from basic scientific research using various animal models of parkinsonism as recipients of either fetal embryonic nerve cell or paraneuronal tissue grafts to brain-damaged areas; animal models provided encouraging evidence that neural grafts could survive and reverse functional disorders. Although initial research in neural transplantation dates back to the turn of the last century (cf Gash 1984, Björklund & Stenevi 1985), major advances in this field have occurred only within the last two decades; accordingly, numerous questions remain to be examined and answered. These include the potential long-term effectiveness of neural grafts to restore and maintain normalized function in animal models of a variety of disorders. The question of circuit reconstruction is of paramount importance with respect to the concept of integration into a neurological control system. Alternatively, if grafted cells act more in a manner of biological pumps of missing transmitters, then a fuller understanding of diffusion mechanisms within graft targets is needed. Ultimately, it may transpire that graft-mediated improvements are the result of a trophic influence leading to the preservation or stimulation of residual or compensatory neurons.

415

0147–006X/90/0301–0415$02.00

# ANATOMY OF THE MESENCEPHALIC DOPAMINERGIC SYSTEM AND FETAL GRAFTS

One distinguishing characteristic of single dopamine (DA) neurons located in the substantia nigra is their vast ramification of terminal processes within their target site, the striatum. Each DA neuron gives rise to an estimated 250,000 terminal varicosities (Andén et al 1966, Björklund & Lindvall 1984), which suggests that a single DA neuron can have extensive postsynaptic influence on striatal neurons. Terminal arborizations of DA neurons appear to have widespread contacts with striatal neurons, since incomplete lesions of nigral DA neurons produced nondiscrete, homogenous reductions of DA and terminal degeneration throughout large portions of the striatum in rats (Shimizu & Ohnishi 1973), cats (Moore et al 1971), and primates (Carpenter & McMasters 1964). The impact that individual DA neurons have on striatal activity can be appreciated when we examine the compensatory properties that a relatively small number of DA neurons may have on overall DA activity in the striatum following pathophysiological or experimentally induced losses of DA neurons. On the average, only seven thousand DA cell bodies are located within the intact substantia nigra and project to the ipsilateral striatum in the rat. Experimental studies have determined that a loss of three quarters of these nigral DA neurons, because of a lesion of the substantia nigra, can produce considerable reductions in striatal DA. Surviving DA neurons, however, are able to compensate for this loss, and lesioned animals do not typically manifest symptoms associated with hypo-dopaminergic tone until a threshold of 80–90%, reduction of striatal DA is exceeded (Zigmond & Stricker 1972, Creese & Iversen 1975, Hefti et al 1980). Clinical studies have provided analogous findings in patients with Parkinson's disease; major signs and symptoms of parkinsonism are not usually detected in patients until DA levels drop to less than 20% of their normal level (Hornykiewicz 1988). These data indicate that surviving DA neurons are highly adaptive and are capable of exerting sufficient control of striatal activity while existing in relatively few numbers.

## Mesostriatal Dopaminergic System

Mesencephalic cell bodies appear to be the most profoundly affected locus of cells in idiopathic Parkinson's disease. Early studies of parkinsonism noted relationships between the pathophysiology of certain DA-related brain regions and movement disorders in patients diagnosed as parkinsonian (Denny-Brown 1962). Subsequent studies revealed severe deficits of DA in postmortem brains of Parkinson patients (cf Hornykiewicz 1966). The loss of brain DA in Parkinson patients was traced to the neuronal degeneration of DA cells located in the mesencephalon. The mesencephalic region of the brain consists of two groups of DA cell bodies; alphanumeric characters represent regions of dopamine cell bodies as designated by Dahlström & Fuxe (1964). Dopamine cell bodies located in the pars compacta and pars lateralis of the substantia nigra comprise region A9 and contribute axons to a rostrally coursing projection that forms dense terminal fields in the caudate nucleus and putamen (collectively known as the striatum), with sparse innervation of the globus pallidus and entopeduncular nucleus. Dopamine cell bodies located in the ventral tegemental area of the mesencephalon, an area more medial and directly adjacent to the substantia nigra, comprise region A10 and also project axons rostrally to form a major terminal field in the nucleus accumbens, with additional projections to the olfactory tubercle and other telencephalic regions. Dopaminergic projections from the mesencephalic region to the caudate nucleus, putamen, and nucleus accumbens collectively have been referred to as the mesostriatal system (Lindvall & Björklund 1983).

Dopaminergic neurons projecting into the striatum interact with a variety of targets, including interneurons, striatal efferents, and corticostriatal terminals. Striatal interneurons are characterized as short-axoned aspiny neurons, which account for less than 4% of the total striatal neurons. The neurotransmitters associated with striatal interneurons are acetylcholine (aspiny II) and $\gamma$-aminobutyric acid (GABA) (aspiny I). The striatum is organized into compartments of acetylcholinesterase-poor "patches" and acetylcholinesterase-rich "matrix," in which patches receive input primarily from prelimbic cortex while the matrix receives inputs largely from the sensorimotor cortex (Graybiel et al 1981, Gerfen 1984). The principal targets of DA neurons are striatal efferent neurons, which are morphologically characterized into two types of long-axoned spiny neurons that comprise the primary striatal output pathways. Spiny I neurons are associated with the neurotransmitter GABA and spiny II neurons are associated with substance P. The classical striatal output pathway consists of both GABAergic and substance P fibers that project to the globus pallidus. A second output pathway projects to both subdivisions of the substantia nigra. A third target for DA neurons are the terminals of corticostriatal neurons. Figure 1 summarizes the major pathways of the rodent basal ganglia.

Dopaminergic activity is regulated by two mechanisms: (a) the striatonigral feedback loop and (b) self-regulation via autoreceptor stimulation (Roth 1979). The striatonigral pathway is primarily a projection of substance P and GABAergic fibers that impinge upon DA dendritic fields in the pars reticulata of the substantia nigra and form a negative feedback

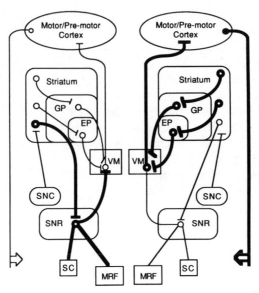

*Figure 1*   Neuronal circuitry of the rodent basal ganglia system. *Heavy lines* indicate the two major output pathways from the striatum associated with motor behavior. Abbreviations: *GP* = globus pallidus, *EP* = entopenduncular nucleus, *VM* = ventromedial nucleus of the thalamus, *SNC* = substantia nigra pars compacta, *SNR* = substantia nigra pars reticulata, *SC* = superior colliculus, *MRF* = mesencephalic reticular formation.

system; this system essentially modulates activity of DA neurons and maintains DA tone within the striatum. Substance P neurons exert an excitatory action on target neurons while GABAergic neurons have an inhibitory effect on target neurons. Groves (1983) postulates that degenerative changes in the DA system, which produce decreased striatal DA activity, favor decreased excitability of inhibitory output and increased excitability of excitatory output. Therefore, during diseased or lesioned states of the nigrostriatal DA system, the regulatory function of the striatonigral pathway stimulates increased nigral DA activity and thereby increases impulse flow in DA neurons, which in turn activates the release of DA from surviving DA neurons. Moreover, DA released from surviving DA neurons after partial destruction of the nigrostriatal pathway should have a longer half-life because of (*a*) an increase in DA release from surviving neurons, (*b*) a decrease in DA uptake sites following degeneration of presynaptic uptake sites, and (*c*) a decrease in metabolic breakdown with the elimination of presynaptic metabolic enzymes.

## Dopamine Nerve Cell Grafts

The nigrostriatal pathway and fetal DA grafts have several properties in common. First, both normal and grafted DA neurons form polysynaptic connections with host striatal neurons. As stated above, each DA neuron can have synaptic contact with many striatal neurons, which suggest that the maintenance of striatal activity is not a function of strict one-on-one connections between DA and striatal neurons; this aspect of interactions between normal DA and striatal neurons is significant when we consider that grafted DA neurons most likely establish comparatively less ordered connections with host striatal neurons. Second, as developing DA neurons in the intact nigrostriatal system show a preference for innervating their host striatal target cells, fetal DA neurons grafted into cortex overlaying the striatum also show a preference for striatal innervation and form graft-to-host synapses (Freund et al 1985, Mahalik et al 1985). Third, terminal fields originating from DA grafts appear to develop normal patterns and histofluoresce with the same intensity as they do in the intact nigrostriatal DA system (Björklund et al 1980b, Schmidt et al 1981).

There are conflicting reports to whether or not host fibers are capable of innervating DA grafts and consequently regulating DA function. Freund et al (1985) injected the retrograde label horseradish peroxidase-wheat-germ agglutinin into DA grafts and did not observe labeling in either the striatum, globus pallidus, or mesencephalic raphe of the host. The lack of labeling in these sites argues against reciprocal innervation and suggests that grafted DA neurons function without regulatory input from the host. On the other hand, Bolam et al (1987) reported the presence of afferent synaptic boutons immunoreactive for substance P and glutamate decarboxylase (GAD) within DA grafts this study, however, could not determine whether these terminals originated in cell bodies located intrinsically or extrinsically to the graft. Similarly, Doucet et al (1989) demonstrated apposition of DA-positive and DARPP-32-positive neurons at the graft-host interface, and also between DA-positive neurons and cortical afferents after injection of *Phaseolus vulagris* leucoagglutinin into anteromedial cortex. These data suggest that host afferents may provide regulatory input into mesencephalic grafts. If neurons extrinsic to the graft are capable of innervating and regulating DA activity, then pharmacological manipulation of the extrinsic neurons should alter this activity, as nigral afferents are capable of modulating activity in the normal system. It has been demonstrated that DA-grafted animals treated with DA receptor antagonists have an increased turnover of DA within the graft (Meloni et al 1988), which suggests that a regulatory feedback mechanism may exist between the host striatum and the grafts. However, the observed increase in DA turnover may not have been related to pharmacologic events occurring extrinsically to the graft and quite conceivably could have been a consequence of either (*a*) blockade of the self-regulatory autoreceptors located on the grafted DA neurons or (*b*) local circuits (i.e. within the graft) that regulate DA activity and are sensitive to DA agonists or antagonists. The data in support of host innervation and regulation of graft activity are inconclusive at this time. Lack of a verifiable regulatory input into grafts suggests that grafted DA neurons may not be receiving an adequate signal for transmitter release into the DA-denervated striatum. It is conceivable that DA grafts, co-grafted with "excitatory" neurons, may provide a source of continual stimulation and a more abundant release of DA from these co-grafts; enhanced release of DA from co-cultured or co-grafted embryonic nigral and striatal cells may be attributed to the release of excitatory agents within the co-culture or co-graft (see below).

## MODELS OF PARKINSONISM

### Unilateral Nigrostriatal Lesions: Hemiparkinsonian Model

Unilateral lesions of the nigrostriatal pathway produce persistent changes in neurochemistry and behavior associated with the function of this system. This type of lesion can be produced by a variety of methods, including electrolytic lesions, knife cut, and neurotoxic agents. Because of the relatively large nonspecific damage associated with electrolytic and knife-induced lesions, neurotoxic agents selective for DA neurons are widely used to lesion the nigrostriatal system; both 6-hydroxydopamine (6-OHDA) and 1-methyl-4-phenyl-1,2,3,6-tetrahydropyridine (MPTP) are known to produce selective lesions of mesencephalic DA neurons. The compound 6-hydroxydopamine must be stereotaxically injected in brain sites containing DA neurons while MPTP can be administered systemically. In rodents, unilateral 6-OHDA lesions produce (a) sensory neglect to stimuli applied to the side of the body contralateral to the lesion and (b) spontaneous motor and postural asymmetry (Jungberg & Ungerstedt 1976). A similar syndrome is observed in monkeys rendered hemiparkinsonian by unilaterally administering MPTP via the internal carotid artery (Bankiewicz et al 1986). Both rodents and primates that are unilaterally lesioned show distinct behavioral syndromes when administered apomorphine or amphetamine. It is generally accepted that, as a consequence of unilateral nigral lesions, striatal DA receptors up-regulate as a compensatory response to the reduced levels of striatal DA on the side ipsilateral to the nigral lesion. Therefore, following a systemic administration of apomorphine, a direct postsynaptic DA receptor agonist, these animals rotate in a direction away from the lesioned side (contralateral rotation) because apomorphine-induced stimulation of DA receptors produces an imbalanced, hyper-dopaminergic response on the side of the lesion relative to the intact side. Amphetamine stimulates the release of DA from intact neurons; therefore its action reflects the stimulation of DA receptors in the intact striatum; consequently, rotational behavior is directed towards the side of the lesion (ipsilateral rotation). Unilateral lesions of ascending A9 neurons with either 6-OHDA or MPTP produce spontaneous and drug-induced circling in rodents (Ungerstedt 1971a) and primates (Crossman & Sambrook 1978, Bankiewicz et al 1986). Duvoisin & Marsden (1975) reported a positive correlation between postural deviations (spinal curvature) and laterality of parkinsonian signs and symptoms in patients diagnosed with parkinsonism. The unilateral postural deviations in these patients may be analogous to the postural asymmetries observed in animals with unilateral lesions of the nigrostriatal pathway. Furthermore, clinical studies of Parkinson patients that show unilateral deficits indicate that these patients demonstrate spontaneous tendencies for unidirectional motor behavior and asymmetrical postural deviations (Bracha et al 1987).

## Bilateral Lesions

Bilateral lesions of the nigrostriatal pathway produce a syndrome in experimental animals that is quite similar to the observed motor dysfunctions observed in Parkinson patients: resting tremor, rigidity, akinesia, and postural abnormalities. Bilateral 6-OHDA lesions of the nigrostriatal pathway produce profound akinesia, adipsia, aphagia, and sensory neglect in rodents (Ungerstedt 1971b, Marshall et al 1971). Until recently, bilateral lesions of the substantia nigra were produced primarily by electrolytic lesions in primates (Schultz 1982, DeLong & Georgopoulos 1981); lesions that spare surrounding structures produce hypokinesia, tremor, and rigidity in monkeys. The introduction of MPTP as a selective DA neurotoxic agent offered an alternative means for selective lesioning of nigral DA neurons.

The neurotoxicity of MPTP varies considerably among species; primates appear to be most susceptible (Davis et al 1979, Burns et al 1983, Langston et al 1983). Many rodents and carnivores, by comparison, are fairly resistant to MPTP neurotoxicity (Boyce et al 1984), as described below. Monkeys treated with MPTP display persistent bradykinesia, hunched posture, limb rigidity, difficulty initiating movement, and resting and postural tremor (Burns et al 1983). Furthermore, MPTP neurotoxicity in humans also induces a behavioral syndrome strikingly similar to that observed in patients diagnosed with Parkinson's disease (Davis et al 1979, Langston et al 1984). In both the MPTP model and idiopathic Parkinson's disease, there is severe DA depletion in the central regions of the substantia nigra while DA cells in the medial regions are relatively spared (Deutch et al 1986, Hornykiewicz 1988). In contrast to the similarities between MPTP-treated animals and parkinsonism, there are several distinctions between the MPTP animal model and human parkinsonism. First, dopamine depletion is fairly homogeneous throughout the striatum and the rostral-caudal subdivisions of substantia nigra in MPTP-treated monkeys, whereas there are distinct subregional differences in patients suffering from idiopathic parkinsonism (Hornykiewicz 1988). Second, in nonhuman primates, MPTP induces severe depletions of DA ($\sim 95\%$) in the striatum, with the caudate nucleus slightly more affected than putamen; in humans, the putamen is more severely affected than the caudate nucleus in idiopathic parkinsonism.

Although most rodents are resistant to MPTP neurotoxicity, it has been demonstrated that MPTP induces enduring DA depletions in the C57 BL strain of mouse (Heikkila et al 1984, Hallman et al 1985). Moreover, the neurodegenerative effects of MPTP are greater in older than in younger mice as revealed by the Fink-Heimer method (Ricaurte et al 1987). In most cases, the behavioral effects of MPTP treatment in rodents are transient. Initially, striking changes can be observed: mice become tremulous, akinetic, and remain in a hunched posture. These acute symptoms disappear within a few weeks after administration of MPTP, after which normal behavior returns. In mice, the DA-depleting effects of MPTP are enhanced by pretreating mice with diethyldithiocarbamate (Corsini et al 1985). The combined treatment of diethyldithiocarbamate and MPTP yields the highest striatal DA depletion levels in young mice. After two weeks, these mice have persistently low levels of striatal DA and severely reduced levels of tyrosine hydroxylase (TH)-immunoreactivity in the striatum and ascending fibers of the nigrostriatal pathway, while maintaining only a slight reduction of TH-immunoreactivity in the substantia nigra (Yurek et al 1989b).

## Dopamine Receptor Changes in Authentic Parkinsonism and Models of Parkinsonism

Losses of striatal DA are associated with an alteration in the number of target receptors located on striatal cells. In parkinsonism, changes in the status of DA receptors may be dependent on the stage of progression of the disease. In early stages of parkinsonism, there appears to be a compensatory increase in DA receptors to accommodate the initial loss of DA neurons (Rinne et al 1983, Hägglund et al 1987). As the disease progresses into later stages, DA receptor numbers decrease (Rinne et al 1983, Hägglund et al 1987); this reduction may be related to the concomitant degeneration of DA target sites (dendritic spines) on striatal neurons observed in advanced parkinsonism (McNeill et al 1988). A similar discrepancy in DA receptors has been observed in MPTP-treated monkeys. Increases in DA receptors have been observed on the lesioned side of hemiparkinsonian, asymptomatic monkeys (Joyce et al 1986), while bilateral decreases in DA receptors have been observed in MPTP-treated monkeys that exhibited severe parkinsonian symptoms (Hantraye et al 1986, Steece et al 1987). In rodents, lesions of A9 neurons produce a compensatory and enduring increase in the number of striatal DA receptors (Creese et al 1977). Grafts of embryonic substantia nigra placed into the DA-denervated striatum of rats appear to normalize striatal DA receptors in the vicinity of the graft (Freed et al 1983).

## Neurochemical Changes in Parkinsonism

The hallmark of parkinsonism is severe reduction of dopamine in all components of the basal ganglia (Hornykiewicz 1988). Dopamine and its metabolites are depleted in the caudate nucleus, putamen, globus pallidus, and pars compacta of the substantia nigra. Moderate losses of DA are found in the nucleus accumbens, lateral hypothalamus, medial olfactory region, and amygdaloid nucleus. Changes in non-DAergic neuronal systems included decreases in tissue concentrations of norepinephrine, serotonin, substance P, neurotensin, and several neuropeptides in most basal gangliar structure, cerebellar cortex, and spinal cord. Increases in GABA-ergic and cholinergic activities are found in the striatum.

# DOPAMINE NERVE CELL GRAFTS

## Neural vs Paraneural Sources

Over the past decade, research in the field of DA nerve cell grafts traditionally has examined two sources of DA-producing cells: fetal nerve cells and chromaffin cells of the adrenal medulla. Initially, animal experimentation with fetal DA nerve cell grafts provided encouraging evidence that these grafts could reverse DA deficits and restore motor function in animals with experimental lesions of the nigrostriatal DA system. The projected ethical, legal, and safety issues attendant to the use of fetal tissue in clinical research prompted a search for alternative sources of DA-containing tissue for neural transplantation. Subsequently, investigators tested the feasibility of using the DA-containing adrenal chromaffin cells as a paraneuronal source of DA for transplantation (Backlund et al 1985, Madrazo et al 1987). Adrenal chromaffin cells autologously derived from rodent adrenal gland and transplanted into the DA-denervated rodent brain appeared to have a beneficial effect on the recovery of DA function in initial animal experiments. Based primarily on the observed salutary effects that adrenal medullary autografts had in the rodent model of parkinsonism, clinical trials of adrenal medullary autografts were performed in patients suffering from intractable parkinsonism. Concurrently, animal experimentation in rodents and nonhuman primates revealed that several parameters associated with the viability of adrenal medullary grafts indicated low survival rates and immunological rejection. For example, contrary to the favorable survival rates for intraventricular autografts of adrenal chromaffin cells in rodents (Freed et al 1985), survival of intraparenchymal grafts of adrenal chromaffin cells after 1–6 months appeared to be extremely low (Freed 1986). This finding corresponds with the immediate loss of catecholamine fluorescence observed in intrastriatal adrenal chromaffin autografts (Herrera-Marschitz et al 1984, Strömberg et al 1984). Moreover, adrenal medullary autografts transplanted into the brains of monkeys with and without experimentally induced parkinsonism displayed overt signs of graft degeneration, low survivability, and were infiltrated with phagocytic macrophages (Morihisa et al 1984, Hansen et al 1988). Although initial clinical evaluations of Parkinson patients receiving adrenal medullary autografts are mixed, there is evidence that the observed functional recoveries in these patients are only transient (Lindvall et al 1987), that antiparkinsonian medication cannot be decreased in patients following transplantation (Goetz et al 1989), and that graft survivability is poor (Peterson et al 1988). Animal experimentation data, taken together with the initial clinical assessments of Parkinson patients receiving transplanted adrenal medullary autografts, indicate that further basic scientific research on neural transplantation is needed (Sladek & Shoulson 1988).

There is renewed interest in the use of fetal nerve cells in neural transplant studies primarily because of enduring experimental evidence that fetal grafts have favorable survival, integrate with the host, and provide functional relief from lesion-induced syndromes in rodent and primate models of parkinsonism. At present, several clinical evaluations of Parkinson patients receiving transplants of human fetal mesencephalic tissue have been initiated (Freed et al 1989, Hitchcock et al 1989, Lindvall et al 1989). Fetal nerve cells are typically transplanted in one of two forms of grafts: (a) as solid mesencephalic tissue or (b) as dissociated cell suspensions of mesencephalic tissue (Björklund et al 1983a).

## Optimal Parameters: Rodent

Solid grafts of embryonic mesencephalic tissue obtained from rodent fetuses have been successfully transplanted as homografts or heterografts without immunosuppressive treatment. The mammalian brain appears to be an immunologically privileged site; therefore, immunological rejection of homografts or heterografts is typically not observed in neural transplantation experiments. Murine embryonic mesencephalic tissue transplanted into the adult striatum of rats shows good survival and integration with the host (Björklund et al 1982).

The gestational age of the donor is an important determinant for graft survival. Mesencephalic cells obtained from rat fetuses aged 14–16 days appear to be more viable than cells obtained from later gestating donors; this is approximately the same gestational age at which DAergic cells undergo their final cell division (Lauder & Bloom 1974). Grafts of embryonic mesencephalic tissue obtained from fetuses aged 14–16 days and transplanted into the striatum of unilaterally lesioned rats contain higher numbers of surviving DA neurons and are more effective in reversing motor asymmetry than grafts obtained from rat fetuses older than 16 days (Brundin et al 1987).

Solid grafts of embryonic mesencephalic tissue can also be dissociated into cell suspensions for transplantation. Studies on the survivability of dissociated cell suspensions of embryonic mesencephalic neurons have determined that approximately 10% of grafted DA cells survive (Brundin et al 1987). Dissected embryonic mesencephalic tissue is not a pure source of DA cells; a typical yield of surviving DA cells has been estimated to be 0.1 to 1.0% of the total viable cells in a suspended graft (Brundin et al 1985). It was previously determined that the mesencephalic region (A9 & A10) of the rat contains approximately 15,000 to 20,000 DA neurons on one side of the brain (Björklund & Lindvall 1984); rodents with severe unilateral nigrostriatal lesions display functional recovery in rotational behavior after receiving mesencephalic grafts in which there were approximately 1% or more surviving DA neurons in the graft (Björklund et al 1983b). Given that surviving DA neurons range between 0.1 to 1.0% of the total cell suspension and that approximately 150 surviving DA neurons are capable of restoring functional recovery, then a graft of dissociated mesencephalic cells should contain a minimum of 100,000 to 150,000 viable cells in order to effectively compensate for DA loss in rodents with unilateral lesions of the nigrostriatal DA system. Refining cell suspension techniques in order to reduce the number of nonspecific cells in a cell suspension graft, i.e. selectively sorting DA cells from other cell types, would provide a means for reducing the volume of cells needed to be implanted into one brain site.

## Optimal Parameters: Primate

To evaluate the possibility of using human fetal mesencephalic tissue for DA cell replacement in patients with Parkinson's disease, several studies have examined the viability of human fetal mesencephalic tissue in xenografts to rodent and nonhuman primate brain. Fetal mesencephalic tissue obtained from aborted human fetuses during the first trimester and transplanted into immunosuppressed rats has been shown to survive, preferentially reinnervate striatum, and attenuate apomorphine-induced rotational behavior when grafted into immunosuppressed rats with unilateral nigrostriatal lesions (Strömberg et al 1986). Xenografts of dissociated human mesencephalic cells obtained from 7–8 week old aborted fetuses placed into the rodent DA-denervated striatum have better survival rates and patterns of striatal reinnervation than fetal tissue obtained from 12 week or older fetuses (Brundin et al 1988). Moreover, human fetal mesencephalic tissue obtained after 9–12 weeks gestation from elective abortions can be cryopreserved and subsequently transplanted into monkey brain with good survival in the host striatum (Redmond et al 1988).

Beside using fresh embryonic mesencephalic tissue for solid or cell suspension grafts, other techniques have been devised for preserving DA cells. Of paramount importance in future clinical applications of neural transplants will be the ability to store and preserve nerve cells for use in transplant surgery. Houle & Das (1980) first demonstrated that neural embryonic tissue could be cryopreserved and successfully transplanted into the rat brain. Subsequently, cryopreserved embryonic mesencephalic tissue was successfully stored for periods up to 70 days and transplanted as homografts in rodent (Collier et al 1988) and primate (Collier et al 1987). These finding have been extended to the preservation of human fetal tissue; human fetal mesencephalic tissue has been cryopreserved for up to two months and transplanted as a xenograft into monkey brain with remarkable survival and excellent morphological development (Redmond et al 1988). Collier et al (1988) also demonstrated that embryonic mesencephalic cells could be successfully cultured after cryopreservation. Alternatively, mesencephalic tissue can be stored short-term (2–5 days) in preservation medium at 4°C and subsequently transplanted with surviving graft volumes similar to those found in grafts of fresh tissue (Sauer et al 1989). The ability to culture cryopreserved embryonic cells will allow testing sample aliquots of preserved cells for cell viability, immunocompatibility, and other safety factors associated with the use of these cells for implantation.

## Alternative Cell Sources and Techniques

The future of neural transplantation as a therapeutic tool may depend on the development of cultured or genetically engineered cells that are capable of synthesizing trophic factors or on the development of a line of surrogate cells for degenerating neurons in CNS neurodegenerative disorders. While neural explants can be maintained for periods of several weeks to several months, cultured tumorogenic cell lines theoretically can be sustained for indefinite periods. Cell lines with neural characteristics have been used for neural transplantation studies (Bottenstein 1981, Gupta et al 1985). Gash et al (1986) transplanted amitotic, differentiated cells from a human neuroblastoma cell line (IMR-32) into hippocampal-temporal lobe of fimbria-fornix lesioned monkeys and found surviving [$^3$H]thymidine-labeled grafted cells without evidence of neoplastic growth; previously it has shown that these cells synthesize several neurotransmitters, including acetylcholine and that acetylcholinesterase positive cells could be found within the IMR-32 graft. Transplanting developed cell lines with DA phenotypes may be the next step in cell replacement therapy for Parkinson's disease. Several human neuroblastoma cell lines have dopaminergic properties, including the SK-N-SH, CHP126, BN69, and LAN5 cell lines (Bottenstein 1981). Biochemical and immunological characteristics of the LAN5 human neuroblastoma cell line indicate that these cells may be the best candidates for transplantation in the parkinsonian brain (Mena et al 1987).

Another approach toward restoring neural function in trauma-associated or neurodegenerative disorders is to implant cells that are capable of releasing trophic factors in brain-damaged areas. In a pioneering study grafted fibroblasts, which had been genetically modified to synthesize and secrete nerve growth factor (NGF), prevented further degeneration of cholinergic neurons in animals with surgical lesions of the fimbria-fornix (Rosenberg et al 1988). The insertion, into implantable cells, of retroviral vectors that genetically code for the synthesis of trophic factors or neurotransmitters (known as transgene products) may have many applications in future neural grafting studies (Gage et al 1987). Another alternative is to co-graft embryonic cells with other cells that may exert a trophic influence on the development of the immature cells. For instance, both growth and development of grafted embryonic mesencephalic neurons are enhanced when co-cultured with embryonic striatal cells (Prochiantz et al 1979, DiPorzio et al 1980, Hemmendinger et al 1981, Hoffman et al 1983), striatal membranes (Prochiantz et al 1981), or soluble striatal extracts (Tomozawa & Appel 1986). The first attempts to co-graft embryonic mesencephalic and striatal tissue in the mammalian central nervous system were performed by Olson et al (1979) in oculo; an intriguing result was that DA neurons survived and innervated their embryonic striatal target cells within this CNS environment. Subsequently, Brundin et al (1986) demonstrated that intrastriatal grafts of mixed embryonic mesencephalic and striatal cell suspensions stimulated a greater density of host reinnervation than did mesencephalic grafts alone. Similarly, in vivo studies in which animals received co-grafts of embryonic mesencephalic and striatal cells placed into the lateral ventricles of rats (DeBeaurepaire & Freed 1987, Yurek et al 1988) or placed into separate sites within the parenchyma of rat striatum (Yurek et al 1989a) have demonstrated that embryonic striatal cells stimulate robust development of mesencephalic cells and that the mesencephalic cells innervate both host striatum and embryonic striatal grafts, with a preference towards innervating their embryonic target cells (Figure 2). Although most co-grafting studies have examined the trophic influence that embryonic striatal cells have on the development of DA nigral cells, several co-culture studies have determined that striatal tissue obtained from 2–3 week old neonatal rats may provide more optimal stimulation for in vitro development of embryonic DA cells than is provided by embryonic striatal cells (Prochiantz et al 1981, Tomozawa & Appel 1986). This age of neonatal striatal tissue coincides with the same time period in the development of the nigrostriatal DA system in which striatal DA levels, DA uptake, and histofluorescence of DA terminals in striatum increase from approximately 10 to 75% of their adult levels (Coyle & Campochiaro 1976). This rapid appearance of striatal DA par-

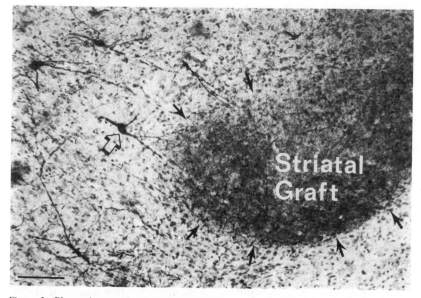

*Figure 2* Photomicrograph of an intrastriatal co-graft of embryonic mesencephalic and striatal cells six weeks after implantation. Cell suspensions of embryonic mesencephalic and striatal cells were implanted into the DA-denervated striatum of a rat with an approximate 1-mm separation between co-grafts. Brain slice was immunocytochemically stained for tyrosine hydroxylase and counterstained with cresyl violet. *Solid arrows* indicate implantation site of embryonic striatal cells; note dense TH-immunoreactivity within the embryonic striatal graft and lack of TH-immunoreactivity within the parenchyma of the host DA-denervated striatum. *Open arrow* points to a TH-positive cell body located on the periphery of the embryonic mesencephalic graft. Fibers extending from TH-positive cell bodies appear to project preferentially into the site of the embryonic striatal co-graft. Calibration bar is 100 μm.

ameters may be attributed to trophic influences associated with the developmental state of striatal neurons. In future studies, co-grafts of fetal mesencephalic and striatal tissue may be used to (*a*) stimulate the development of grafted DA neurons and synthesis of DA, and/or (*b*) promote fiber outgrowth from DA grafts.

## TARGET SITES FOR DOPAMINE NERVE CELL GRAFTS

Both intraventricular and intraparenchymal grafts of fetal nigral neurons survive and proliferate in the host brain. The ventricular environment is rich in cerebral spinal fluid (CSF) nutrients and is advantageous for the survival of grafted tissue. Nearly 99% of animals receiving ventricular homografts, including several different sources of DA-containing grafts, were found to have surviving tissue 1–18 months after implantation (Freed et al 1984). Plugs of fetal nigral tissue implanted into the lateral ventricles of monkeys showed extensive growth and arborization (Sladek et al 1986). The CSF also provides minimal resistance for the expansion and growth of grafted tissue. In addition, the anatomical location of the lateral ventricles provides an excellent base from which DA grafts can readily interact with areas directly adjacent to ventricles, such as the striatum and nucleus accumbens. Furthermore, the ventricular system may act as a conduit for substances secreted from grafts and thereby extend the influence of grafts to distal periventricular regions. Although the ventricular environment favors graft survival, the ependymal wall provides a barrier for fiber outgrowth and reinnervation of the DA-denervated striatum. Thus, when intraventricular grafts lack a direct reinnervation of the striatum, the beneficial effects associated with the release of DA probably are limited to the passive diffusion of DA into the adjacent striatum. Moreover, it appears that the ventricular environment may not fall into the category of an "immunologically privileged site," thus intraventricular grafts are susceptible to immunological surveillance by the host immune system (Freed et al 1988).

Grafting neural tissue directly into the parenchyma of the brain has also been demonstrated. For example, intraparenchymal grafts of fetal substantia nigra show excellent survival and growth after placement directly into the striatum. Björklund et al (1980a) demonstrated that dispersed embryonic nigral cells implanted within the rodent striatum survive and are functional. Intraparenchymal cell suspensions of embryonic mesencephalic tissue obtained from monkey fetus survive and provide functional improvement in MPTP-treated monkeys (Bakay et al 1987). Similar success has been reported for intraparenchymal grafts of solid mesencephalic tissue implanted in the monkey brain (Sladek et al 1986). Relative to the ventricular environment, intraparenchymal grafts appear to have the advantage of existing in a more "immunological privileged" environment. For instance, heterografts of fetal monkey tissue placed into the parenchyma of the striatum of monkeys can survive without immunosuppressive treatment (Bakay & King 1986), and there is evidence that xenografts of human tissue placed into monkey striatum may not be rejected after withdrawal of immunosuppressive therapy (Redmond et al 1988).

## Dopamine Graft Locations in Striatum and Their Functional Effects

Several studies have demonstrated that components of the DA denervation syndrome are compensated for, to a limited extent, by several DA grafting techniques. In the unilaterally DA-denervated animal, cortically placed DA grafts in proximity to the dorsomedial striatum as well as intraventricular DA grafts have been shown to reduce motor asymmetry; these grafts, however, have little or no effect on sensory neglect (Björklund et al 1980a, Dunnett et al 1981). In contrast, sensory neglect is effectively reversed when nigral grafts are placed in proximity to lateral striatum in bilaterally DA-denervated animals (Dunnett et al 1983). Furthermore, the reversal of akinesia in bilaterally lesioned animals is observed only when the nucleus accumbens is reinnervated by the DA graft (Dunnett et al 1983, Nadaud et al 1984). These studies indicate that graft placement, and the extent to which the graft reinnervates striatum, are critical factors for the functional recovery of the mesostriatal DA system.

Although sites for DA grafts in proximity to the ventricular region have been chosen because of the favorable CSF environment these regions provide for graft survival, in most cases of idiopathic Parkinson's disease the degree of DA neuronal degeneration within the striatum is more pronounced in the putamen than in the caudate nucleus. To date, most clinical attempts to graft DA-containing tissue into late stage Parkinson patients have placed the graft in periventricular regions within the head of the caudate. Although the caudate nucleus is severely affected in Parkinson's disease, DA levels in the putamen have been found to be 10–15% lower than in the caudate nucleus (Bernheimer et al 1973, Nyberg et al 1983). In addition, caudal putamen is more severely DA-depleted than rostral putamen (Kish et al 1986). Of the two striatal components (caudate nucleus and putamen), the putamen receives the majority of motor input via a cortical-thalamic-putamen pathway (DeLong & Georgopoulos 1983), thus, the putamen may be a more favorable site for DA grafts targeting motor disorders associated with Parkinson's disease. Attempts have been made to implant chunks of adrenal medullary tissue in the putamen of two patients with intractable parkinsonism (Lindvall et al 1987); however, postoperative assessment of these patients showed only transient improvements in motor function, and the survival of adrenal medullary autografts in the brains of humans remains a question. Implantation of graft tissue in deep brain sites, such as caudal putamen, may have a beneficial effect on the recovery of striatal DAergic function; surgical implantation of grafts into these sites is technically more demanding, however, and will require improvements in current techniques (see Sladek & Gash 1988, Sladek & Shoulson 1988).

## Loss of Dendritically Released Dopamine in the Substantia Nigra: Is Replacement Necessary?

Transplantation studies of DA nerve cell grafts in animal models of Parkinson's disease have focused primarily on restoring DA function in the striatum in order to reverse motor dysfunction; they have not addressed the issue of how losses of DA in the substantia nigra also might affect motor function. Although a substantial quantity of nigral DA is located within cell bodies and ultimately is destined for axonal transport to terminal fields in the striatum, a portion of nigral DA is released dendritically within the substantia nigra and modulates neuronal activity in the pars reticulata. Morphological evidence of DA-containing substrates in the dendrites of nigral neurons was concurrently reported by Björklund & Lindvall (1975) and Sladek & Parnavelas (1975) in the rat and monkey, respectively, and dendritic release of DA in the substantia nigra subsequently was established (Geffen et al 1976, Korf et al 1976, Hefti & Lichtensteiger 1978). Physiological studies of DA receptors revealed target receptors for DA in the substantia nigra and showed that these receptors have properties of the D1 receptor subtype (Gale et al 1977, Matthews & German 1986). Dendritic release from DA neurons may serve two purposes: (*a*) as a local feedback mechanism for nigral DA neurons and (*b*) to modulate activity of non-DA neurons in the pars reticulata of the substantia nigra. Two non-DA pathways associated with striatal output and nigrothalamic fibers appear to be influenced by DA release in pars reticulata. First, striatonigral output neurons may have presynaptic D1 receptors located on terminals in the substantia nigra. This hypothesis is based on evidence that kainic acid lesions of the striatum produce substantial reductions of D1 receptor binding sites in the nigra (Porceddu et al 1986). Second, neurons of the nigrothalamic pathway can be stimulated by iontophoretically applying DA to neurons in pars reticulata (Ruffieux & Schultz 1980), which indicate that dendritic release of DA in the nigra has a stimulatory effect on nigrothalamic neurons. Nigrothalamic neurons projecting to the thalamus have primary targets in the ventralis lateralis (VL) and ventralis anterior (VA) nuclei; these thalamic nuclei, in turn, have major projections to frontal, motor, and premotor cortex (DeLong & Georgopoulos 1981). The significance of DA release in the substantia nigra and its effect on motor function can be clearly discerned by its effect on nigrothalamic activity, which ultimately influences the activity of neurons in the motor regions of the cortex. Dunnett et al (1983b) placed multiple grafts of mesencephalic tissue into the 6-OHDA lesioned substantia nigra and did not observe significant behavioral recovery in rodents. Preliminary data from rodent studies in our laboratory indicate that simultaneous placement of mesencephalic grafts into the 6-OHDA lesioned substantia nigra and ipsilateral striatum may provide a slight enhancement in behavioral recovery as compared to lesioned animals receiving intrastriatal mesencephalic grafts only. The contribution of nigral DA to overall motor function is not fully understood at this time; however, it does appear to play an integral part in DAergic modulation of motor activity. Although replacement of DAergic neurons cells within the substantia nigra may restore nigral DA tone, the effect this might have on motor activity in parkinsonism is not fully understood.

## INJURY-INDUCED NEUROTROPHISM

When assessing the recovery of the DA system following intracerebral grafts, one must carefully interpret to what extent recovery can be attributed to the graft itself or, perhaps more importantly, to the plasticity of the host system. Catecholaminergic neurons are capable of compensating morphologically for the loss of neighboring neurons by sprouting additional terminal processes (Björklund & Stenevi 1979a). The property of sprouting in catecholaminergic neurons has been suggested as one possible explanation for enhanced recovery of DAergic fibers in several studies that have grafted paraneuronal sources of DA into the DA-denervated brain. For instance, transplantation of adrenal medullary tissue into the brains of DA-denervated mice has resulted in increased catecholamine fluorescence in regions surrounding the graft, despite morphological evidence of poor survival of the adrenal medullary grafts (Bohn et al 1987). Similarly, Fiandaca et al (1988) noted increased tyrosine hydroxylase immunoreactivity in striatal regions surrounding nonsurviving autografts of adrenal medullary tissue in MPTP-treated monkeys. This growth may have been stimulated by the surgical procedures attendant to the grafting technique. Catecholaminergic neurons axotomized with blade cuts have been shown to induce sprouting (Björklund & Stenevi 1979a), and surgical implantation of grafts resulted in some damage to regions immediately surrounding the implantation site. Another possibility for sprouting of residual host neurons may be the release of trophic substances from injured neurons. Whether or not diffusible substances released from the injured neurons, as observed in motor neurons (Brown et al 1981), or from the graft mediate the enhanced recovery of host neurons remains unclear. Fetal DA grafts placed in brain regions in which prior cavitations had been made provide a favorable environment for graft survival in rat brain (Björklund & Stenevi 1979b); a similar finding was observed in a delayed stereotactic transplantation technique tested in monkey brain (Bakay et al 1988). These data suggest that trophic factors released from host tissue may contribute to the survival of both graft and surviving host neurons.

It is also important to mention that surgical implantation of intra-striatally placed grafts will result in some destruction to the host striatum, which in itself may provide relief from symptoms of parkinsonism. Symptoms of parkinsonism reportedly were ameliorated by lesioning portions of the caudate nucleus (Meyers 1951, Spiegel et al 1952). Van Manen & Speelman (1988) speculate that the positive results observed in Parkinson patients receiving autografts of adrenal medullary tissue (Madrazo et al 1987) may have been related to a surgically induced caudate lesion. Unpublished autopsy reports from several patients initially diagnosed as showing partial recovery provided no evidence that the autograft survived (Peterson et al 1988).

## SUMMARY

Significant progress in neural transplantation has been observed over the last several decades. As a neuroanatomical tool, neural transplantation studies are able to examine the mechanisms involved in the development and integration of neurons into the complex neural circuitries of the brain. Today, embryonic neural tissue can be successfully transplanted as solid tissue chunks or as dissociated cell suspensions. Within the parenchyma of the brain, transplanted embryonic neurons develop mature morphology and do not appear to invoke an immunological response by the lost immune system. Not only do these neurons exhibit robust development but there is also evidence that transplanted neurons restore some degree of function to neurologically damaged circuitry; however, the extent of reintegration into the host neural circuitry still remains unclear. Moreover, the long-term survival and functioning of transplanted nerve cells also remains an unanswered question. Advances in the emerging field of genetic engineering may eventually lead to genetically modified neurons that are capable of synthesizing neurotrophic factors or missing neurotransmitters and restoring function in brain-damaged areas.

The use of neural transplantation to replace damaged nerve cells in neurodegenerative disorders, such as Alzheimer's or Parkinson's disease, is promising based on our current knowledge. However, our basic scientific knowledge of neural transplants is incomplete and warrants a prudent approach toward application of neural transplantation techniques in clinical research.

ACKNOWLEDGMENTS

This review was supported by grants from the Pew Foundation and the United Parkinson Foundation. We are grateful to our many colleagues at Rochester and Yale, including Charles Bradberry, Steve Bunney, Tim Collier, Ariel Deutch, John Elsworth, David Felten, Don Gash, Suzanne Haber, John Hansen, Eugene Redmond, Robert Roth, Ira Shoulson, Kathleen Steece, and Jane Taylor, for their thoughtful insights on the question of graft-mediated recovery of dopamine deficiency.

*Literature Cited*

Andén, N.-E., Fuxe, K., Hamberger, B., Hökfelt, T. 1966. A quantitative study on the nigro-neostriatal dopamine neuron system in the rat. *Acta Physiol. Scand.* 67: 306–12

Backlund, E.-O., Granberg, P.-O., Hamberger, B., Knutsson, E., Mårtensson, A., Sedvall, G., Seiger, A., Olson, L. 1985. Transplantation of adrenal medullary tissue to striatum in parkinsonism. *J. Neurosurg.* 62: 169–73

Bakay, R. A. E., Barrow, D. L., Fiandaca, M. S., Iuvone, P. M., Schiff, A., Collins, D. C. 1987. Biochemical and behavioral correction of MPTP parkinson-like syndrome by fetal cell transplantation. *Ann. NY Acad. Sci.* 495: 623–40

Bakay, R. A. E., Fiandaca, M. S., Sweeney, K. M., Colbassani, H. J., Collins, D. C. 1988. Delayed stereotactic transplantation technique in non-human primates. See Gash & Sladek 1988, pp. 463–71

Bakay, R. A. E., King, F. A. 1986. Transplanted fetal monkey neurons. *Lancet* 2: 163

Bankiewicz, K. S., Oldfield, E. H., Chiueh, C. C., Doppman, J. L., Jacobowitz, D. M., Kopin, I. J. 1986. Hemiparkinsonism in monkeys after unilateral internal carotid artery infusion of 1-methyl-4-phenyl-1,2,3,6-tetrahydropyridine (MPTP). *Life Sci.* 39: 7–16

Bernheimer, H., Birkmayer, W., Hornykiewicz, O., Jellinger, K., Seitelberger, F. 1973. Brain dopamine and the syndromes

of Parkinson and Huntington. *J. Neurol. Sci.* 20: 415–55

Björklund, A., Dunnett, S. B., Stenevi, U., Lewis, M. E., Iversen, S. D. 1980a. Reinnervation of the denervated striatum by substantia nigra transplants: functional consequences as revealed by pharmacological and sensorimotor testing. *Brain Res.* 199: 307–33

Björklund, A., Lindvall, O. 1975. Dopamine in dendrites of the substantia nigra neurons: suggestions for a role in dendritic terminals. *Brain Res.* 83: 531–37

Björklund, A., Lindvall, O. 1984. Dopamine-containing systems in the CNS. In *Classical Transmitters in the CNS*, ed. A. Björklund, T. Hökfelt, pp. 55–122. New York: Elsevier

Björklund, A., Schmidt, R. H., Stenevi, U. 1980b. Functional reinnervation of the neostriatum in the adult rat by use of intraparenchymal grafting of dissociated cells suspensions from the substantia nigra. *Cell Tissue Res.* 212: 39–45

Björklund, A., Stenevi, U. 1979a. Regeneration of monoaminergic and cholinergic neurons in the mammalian central nervous system. *Physiol. Rev.* 59: 62–100

Björklund, A., Stenevi, U. 1979b. Reconstruction of the nigrostriatal dopamine pathway by intracerebral nigral transplants. *Brain Res.* 177: 555–60

Björklund, A., Stenevi, U. 1985. Intracerebral neural grafting: A historical perspective. In *Neural Grafting in the Mam-*

*malian CNS*, ed. A. Björklund, U. Stenevi, pp. 3–14. Amsterdam: Elsevier

Björklund, A., Stenevi, U., Dunnett, S. B., Gage, F. H. 1982. Cross-species neural grafting in a rat model of Parkinson's disease. *Nature* 298: 652–54

Björklund, A., Stenevi, U., Schmidt, R. H., Dunnett, S. B., Gage, F. H. 1983a. Intracerebral grafting of neuronal cell suspensions. I. Introduction and general methods of preparation. *Acta Physiol. Scand. Suppl.* 522: 1–7

Björklund, A., Stenevi, U., Schmidt, R. H., Dunnett, S. B., Gage, F. H. 1983b. Intracerebral grafting of neuronal cell suspensions. II. Survival and growth of nigral cell suspensions implanted in different brain sites. *Acta Physiol. Scand. Suppl.* 522: 9–16

Bohn, M. C., Cupit, L. C., Marciano, F., Gash, D. M. 1987. Adrenal medulla grafts enhance recovery of striatal dopaminergic fibers. *Science* 237: 913–16

Bolam, J. P., Freund, T. F., Björklund, A., Dunnett, S. B., Smith, A. D. 1987. Synaptic input and local output of dopaminergic neurons in grafts that functionally reinnervate the host neostriatum. *Exp. Brain Res.* 68: 131–46

Bottenstein, J. E. 1981. Differentiated properties of neuronal cell lines. In *Functionally Differentiated Cell Lines*, ed. G. H. Sato, pp. 155–84. New York: Liss

Boyce, S., Kelly, E., Reavill, C., Jenner, P., Marsden, C. D. 1984. Repeated administration of N-methyl-4-phenyl-1,2,3,6-tetrahydropyridine to rats is not toxic to striatal neurons. *Biochem. Pharmacol.* 1: 1747–52

Bracha, H. S., Shults, C., Glick, S. D., Kleinman, J. E. 1987. Spontaneous asymmetric circling behavior in hemi-Parkinsonism: a human equivalent of the lesioned-circling rodent behavior. *Life Sci.* 40: 1127–30

Brown, M. C., Holland, R. L., Hopkins, W. G. 1981. Motor nerve sprouting. *Annu. Rev. Neurosci.* 4: 17–42

Brundin, P., Isacson, O., Björklund, A. 1985. Monitoring of cell viability in suspensions of embryonic CNS tissue and its use as a criterion for intracerebral graft survival. *Brain Res.* 331: 251–59

Brundin, P., Isacson, O., Gage, F. H., Björklund, A. 1986. Intrastriatal grafting of dopamine-containing neuronal cell suspensions: effects of mixing with target or nontarget cells. *Dev. Brain Res.* 24: 77–84

Brundin, P., Strecker, R. E., Lindvall, O., Isacson, O., Nilsson, O. G., Bardin, G., Prochiantz, A., Forni, C., Nieoullon, A., Widner, H., Gage, F. H., Björklund, A. 1987. Intracerebral grafting of dopamine neurons. *Ann. NY Acad. Sci.* 495: 473–96

Brundin, P., Strecker, R. E., Widner, H., Clarke, D. J., Nilsson, O. G., Åstedt, B., Lindvall, O., Björklund, A. 1988. Human fetal dopamine neurons grafted in a rat model of Parkinson's disease: immunological aspects, spontaneous and drug-induced behavior, and dopamine release. *Exp. Brain Res.* 70: 192–208

Burns, R. S., Chiueh, C. C., Markey, S. P., Ebert, M. H., Jacobwitz, D. M., Kopin, I. J. 1983. A primate model of Parkinsonism: Selective destruction of DA neurons in the pars compacta of the substantia nigra by n-methyl-4-phenyl-1,2,3,6-tetrahydropyridine. *Proc. Natl. Acad. Sci. USA* 80: 4546–50

Carpenter, M. B., McMasters, R. E. 1964. Lesions of the substantia nigra in the rhesus monkey. *Am. J. Anat.* 114: 293–319

Collier, T. J., Redmond, D. E., Sladek, C. D., Gallagher, M. J., Roth, R. H., Sladek, J. R. Jr. 1987. Intracerebral grafting and culture of cryopreserved primate dopamine neurons. *Brain Res.* 436: 363–66

Collier, T. J., Sladek, C. D., Gallagher, M. J., Blanchard, B. C., Daley, B. F., Foster, P. N., Redmond, D. E. Jr., Roth, R. H., Sladek, J. R. Jr. 1988. Cryopreservation of fetal rat and non-human primate mesencephalic neurons: viability in culture and neural transplant. See Gash & Sladek 1988, pp. 631–36

Corsini, G. U., Pintus, S., Chiueh, C. C., Weiss, J. F., Kopin, I. J. 1985. 1-Methyl-4-phenyl-1, 2, 3, 6-tetrahydropyridine (MPTP) neurotoxicity in mice is enhanced by pretreatment with diethyldithiocarbamate. *Eur. J. Pharmacol.* 119: 127–28

Coyle, J. T., Campochiaro, P. 1976. Ontogenesis of dopaminergic-cholinergic interactions in the rat striatum: a neurochemical study. *J. Neurochem.* 27: 673–78

Creese, I., Burt, D. R., Snyder, S. H. 1977. Dopamine receptor binding enhancement accompanies lesion-induced behavioral supersensitivity. *Science* 197: 596–98

Creese, I., Iversen, S. D. 1975. The pharmacological and anatomical substrates of the amphetamine response in the rat. *Brain Res.* 83: 419–36

Crossman, A. R., Sambrook, M. A. 1978. Experimental torticollis in the monkey produced by unilateral 6-hydroxydopamine brain lesions. *Brain Res.* 149: 498–502

Dahlström, A., Fuxe, K. 1964. Evidence for the existence of monoamine-containing neurons in the central nervous system. *Acta Physiol. Scand.* 62(Suppl. 232): 5–55

Davis, G. C., Williams, A. C., Markey, S. P., Ebert, M. H., Caine, E. D., Beichert, C. M., Kopin, I. J. 1979. Chronic parkinsonism secondary to intravenous injections of meperidine analogues. *Psychiatry Res.* 1: 249–54

DeBeaurepaire, R., Freed, W. J. 1987. Embryonic substantia nigra grafts innervate embryonic striatal co-grafts in preference to mature host striatum. *Exp. Neurol.* 95: 448–54

DeLong, M. R., Georgopoulos, A. P. 1981. Motor functions of the basal ganglia. In *Handbook of Physiology, Section I: The Nervous System*, Vol. 2, ed. J. M. Brookhart, V. B. Mountcastle, S. R. Geiger, pp. 1017–61. Bethesda, Md: Am. Physiol. Soc.

Denny-Brown, D. 1962. In *The Basal Ganglia and Their Relation to Disorders of Movement*. London: Oxford Univ. Press

Deutch, A. Y., Elsworth, J. D., Goldstein, M., Fuxe, K., Redmond, D. E., Sladek, J. R. Jr., Roth, R. H. 1986. Preferential vulnerability of A8 dopamine neurons in the primate to the neurotoxin 1-methyl-4-phenyl-1, 2, 3, 6-tetrahydropyridine. *Neurosci. Lett.* 68: 51–56

DiPorzio, U., Daguet, M.-C., Glowinski, J., Prochiantz, A. 1980. Effect of striatal cells on in vitro maturation of mesencephalic dopaminergic neurones grown in serum-free conditions. *Nature* 288: 370–73

Doucet, G., Murata, Y., Brundin, P., Bosler, O., Mons, N., Geffard, M., Ouimet, C., Björklund, A. 1989. Host afferents into intrastriatal grafts of ventral mesencephalon. *Restor. Neurol. Neurosci.* (Suppl: *3rd Int. Symp. Neural Transplantation*): 51

Dunnett, S. B., Björklund, A., Stenevi, U., Iversen, S. D. 1981. Behavioral recovery following transplantation of substantia nigra in rats subjected to 6-OHDA lesions of the nigrostriatal pathway. I. Unilateral lesions. *Brain Res.* 215: 147–61

Dunnett, S. B., Björklund, A., Schmidt, R. H., Stenevi, U., Iversen, S. D. 1983a. Intracerebral grafting of neuronal cell suspensions. V. Behavioral recovery in rats with bilateral 6-OHDA lesions following implantation of nigral cell suspensions. *Acta Physiol. Scand. Suppl.* 522: 39–47

Dunnett, S. B., Björklund, A., Schmidt, R. H., Stenevi, U., Iversen, S. D. 1983b. Intracerebral grafting of neuronal cell suspensions. IV. Behavioral recovery in rats with unilateral 6-OHDA lesions following implantation of nigral cell suspensions in different forebrain sites. *Acta Physiol. Scand. Suppl.* 522: 29–37

Duvoisin, R. C., Marsden, C. D. 1975. Note on the scoliosis of Parkinsonism. *J. Neurol. Neurosurg. Psychiatry* 38: 787–93

Fiandaca, M. S., Kordower, J. H., Hansen, J. T., Jiao, S.-S., Gash, D. M. 1988. Adrenal medullary autografts into the basal ganglia of cebus monkeys: injury-induced regeneration. *Exp. Neurol.* 102: 76–91

Freed, C. R., Breeze, R. E., Rosenberg, J. N., Barrett, J. N., Rottenberg, D. A. 1989. Therapeutic effects of human fetal dopamine cells transplanted in a patient with Parkinson's disease. *Restor. Neurol. Neurosci.* (Suppl.: *3rd Int. Symp. Neural Transplantation*): 64

Freed, W. J., Cannon-Spoor, H., Krauthamer, E. 1985. Factors influencing the efficacy of adrenal medulla and embryonic substantia nigra grafts. See Björklund & Stenevi, pp. 491–504

Freed, W. J., Cannon-Spoor, H., Krauthamer, E. 1986. Intrastriatal adrenal medulla grafts in rats. *J. Neurosurg.* 65: 664–70

Freed, W. J., Dymecki, J., Poltorak, M., Rodgers, C. 1988. Intraventricular brain allografts and xenografts: studies of survival and rejection with and without systemic sensitization. See Gash & Sladek, pp. 233–41

Freed, W. J., Hoffer, B. J., Olson, L., Wyatt, R. J. 1984. Transplantation of catecholamine-containing tissue to restore the functional capacity of the damaged nigrostriatal system. See Sladek & Gash 1984, pp. 373–406

Freed, W. J., Ko, G. N., Niehoff, D. L., Kuhar, M. J., Hoffer, B. J., Olson, L., Cannon-Spoor, H. E., Morishisa, J. M., Wyatt, R. J. 1983. Normalization of spiroperidol binding in the denervated rat striatum by homologous grafts of substantia nigra. *Science* 222: 937–39

Freund, T. F., Bolam, J. P., Björklund, A., Stenevi, U., Dunnett, S. B., Powell, J. F., Smith, A. D. 1985. Efferent synaptic connections of grafted dopaminergic neurons reinnervating the host neostriatum: a tyrosine hydroxylase immunocytochemical study. *J. Neurosci.* 5: 603–16

Gage, F. H., Wolff, J. A., Rosenberg, M. B., Xu, L., Yee, J.-K., Shults, C., Friedmann, T. 1987. Grafting genetically modified cells to the brain: possibilities for the future. *Neuroscience* 23: 795–807

Gale, K., Guidotti, A., Costa, E. 1977. Dopamine-sensitive adenylate cyclase: location in substantia nigra. *Science* 195: 503–5

Gash, D. M. 1984. Neural transplants: A historical overview. See Sladek & Gash, pp. 1–12

Gash, D. M., Notter, M. F. D., Okawara, S. H., Kraus, A. L., Joynt, R. J. 1986. Amitotic neuroblastoma cells used for neural implants in monkeys. *Science* 233: 1420–22

Gash, D. M., Sladek, J. R. Jr., eds. 1988.

*Progress in Brain Research*, Vol. 78. New York: Elsevier

Geffen, L. B., Jessell, T. M., Cuello, A. C., Inversen, L. L. 1976. Release of dopamine from dendrites in rat substantia nigra. *Nature* 260: 258–60

Gerfen, C. R. 1984. The neostriatal mosaic: compartmentalization of corticostriatal input and striatonigral output systems. *Nature* 311: 461–63

Goetz, C. G., Olanow, C., Koller, W. C., Penn, R. D., Cahill, D., Morantz, R., Stebbins, G., Tanner, C. M., Klawans, H. L., Shannon, K. M., Comella, C. L., Witt, T., Cox, C., Waxman, M., Gauger, L. 1989. Multicenter study of autologous adrenal medullary transplantation to the corpus striatum in patients with advanced Parkinson's disease. *N. Engl. J. Med.* 320: 337–41

Graybiel, A. M., Pickel, V. M., Joh, T. J., Reis, D. J., Ragsdale, C. W. 1981. Direct demonstration of a correspondence between dopamine islands and acetylcholinesterase patches in the developing striatum. *Proc. Natl. Acad. Sci. USA* 77: 1214–19

Groves, P. M. 1983. A theory of the functional organization of the neostratum and the neostriatal control of voluntary movement. *Brain Res. Rev.* 5: 109–32

Gupta, M., Notter, M. F. D., Felten, S., Gash, D. M. 1985. Differentiation characteristics of human neuroblastoma cells in the presence of growth modulators and antimitotic drugs. *Dev. Brain Res.* 19: 21–29

Hägglund, J., Aquilonius, S.-M., Eckernäs, S.-Å., Hartvig, P., Lundquist, H., Gullberg, P., Långstrom, B. 1987. Dopamine receptor properties in Parkinson's disease and Huntington's chorea evaluated by positron emission tomography using [11]C-N-methyl-spiperone. *Acta Neurol. Scand.* 75: 87–94

Hallman, H., Lange, J., Olson, L., Strömberg, I., Jonsson, G. 1985. Neurochemical and histochemical characterization of neurotoxic effects of 1-methyl-4-phenyl-1,2,3,6-tetrahydropyridine on brain catecholamine neurones in the mouse. *J. Neurochem.* 44: 117–27

Hansen, J. T., Kordower, J. H., Fiandaca, M. S., Jiao, S.-S., Notter, M. F. D., Gash, D. M. 1988. Adrenal medullary autografts into the basal ganglia of cebus monkeys: Graft viability and fine structure. *Exp. Neurol.* 102: 65–75

Hantraye, P., Loc'h, C., Tacke, U., Ricke, D., Stulzaft, O., Doudet, D., Guibert, B., Naquet, R., Maziere, B., Maziere, M. 1986. "In vivo" visualization by positron emission tomography of the progressive

striatal dopamine receptor damage occurring in MPTP-intoxicated non-human primates. *Life Sci.* 39: 1375–82

Heikkila, R. E., Hess, A., Duvoisin, R. C. 1984. Dopamine neurotoxicity of 1-methyl-4-phenyl-1, 2, 3, 6-tetrahydropyridine in mice. *Science* 224: 1451–53

Hefti, F., Lichtensteiger, W. 1978. Dendritic dopamine: Studies on the release of endogenous dopamine from subcellular particles derived from dendrites of rat nigrostriatal neurons. *Neurosci. Lett.* 10: 65–70

Hefti, F., Melamed, E., Wurtman, R. J. 1980. Partial lesions of the dopaminergic nigrostriatal system in rat brain: biochemical characterization. *Brain Res.* 195: 123–37

Hemmendinger, L. M., Garber, B. B., Hoffman, P. C., Heller, A. 1981. Target neuron-specific process formation by embryonic mesencephalic dopamine neurons *in vitro*. *Proc. Natl. Acad. Sci. USA* 78: 1264–68

Herrera-Marschitz, M., Strömberg, I., Olsson, D., Ungerstedt, U., Olson, L. 1984. Adrenal medullary implants in dopamine-denervated rat striatum. II. Acute behavior as a function of graft amount and location and its modulation by neuroleptics. *Brain Res.* 296: 53–61

Hitchcock, E. R., Kenny, B. G., Clough, C. G., Hughes, R. C., Henderson, B. T. H., Detta, A. 1989. Stereotactic implantation of foetal mesencephalon (STIM). *Restor. Neurol. Neurosci.* (Suppl.: *3rd Int. Symp. Neural Transplantation*): 64

Hoffman, P. C., Hemmendinger, L. M., Kotake, C., Heller, A. 1983. Enhanced dopamine cell survival in reaggregates containing telencephalic target cells. *Brain Res.* 274: 275–81

Hornykiewicz, O. 1966. Dopamine (3-hydroxytryamine) and brain function. *Pharmacol. Rev.* 18: 925–64

Hornykiewicz, O. 1988. Neurochemical pathology and the etiology of Parkinson's disease: Basic facts and hypothetical possibilities. *Mt. Sinai J. Med.* 55: 11–20

Houle, J. D., Das, G. D. 1980. Freezing of embryonic neural tissue and its transplantation in the rat brain. *Brain Res.* 192: 570–74

Joyce, J. N., Marshall, J. F., Bankiewicz, K. S., Kopin, I. J., Jacobowitz, D. M. 1986. Hemiparkinsonism in a monkey after unilateral internal carotid artery infusion of 1-methyl-4-phenyl-1, 2, 3, 6-tetrahydropyridine (MPTP) is associated with regional ipsilateral changes in striatal dopamine D-2 receptor density. *Brain Res.* 382: 360–64

Jungberg, T. I., Ungerstedt, U. 1976. Sensory inattention produced by 6-hydroxy-dopamine-induced degeneration of ascending dopamine neurons. *Exp. Neurol.* 53: 585–600

Kish, S. J., Rajput, A., Gilbert, J., Rozdilsky, B., Chang, L.-J., Shannak, K., Hornykiewicz, O. 1986. Elevated γ-aminobutyric acid level in striatal but not extrastriatal brain regions in Parkinson's disease: correlation with striatal dopamine loss. *Ann. Neurol.* 20: 26–31

Korf, J., Zielman, M., Westerink, B. H. C. 1976. Dopamine release in substantia nigra? *Nature* 260: 257–58

Langston, J. W., Ballard, P., Tetrud, J. W., Irwin, I. 1983. Chronic parkinsonism in humans due to a product of meperidineanalog synthesis. *Science* 219: 979–80

Langston, J. W., Forno, L. S., Rebert, C. S., Irwin, I. 1984. Selective nigral toxicity after systemic administration of 1-methyl-4-phenyl-1, 2, 3, 6-tetrahydropyridine (MPTP) in the squirrel monkey. *Brain Res.* 292: 390–94

Lauder, J. M., Bloom, F. E. 1974. Ontogeny of monoamine neurons in the locus coeruleus, raphe nuclei and substantia nigra of the rat. *J. Comp. Neurol.* 155: 469–82

Lindvall, O., Backlund, E. O., Farde, L., Sedvall, G., Freeman, R., Hoffer, B., Nobin, A., Seiger, Å., Olson, L. 1987. Transplantation in Parkinson's disease: two cases of adrenal medullary grafts to the putamen. *Ann. Neurol.* 22: 457–68

Lindvall, O., Björklund, A. 1983. Dopamine- and norepinephrine-containing neurons systems: Their anatomy in the rat brain. In *Chemical Neuroanatomy*, ed. P. C. Emson. New York: Raven

Lindvall, O., Rehncrona, S., Brundin, P., Gustavii, B., Åstedt, B., Widner, H., Lindholm, T., Björklund, A., Leenders, K. L., Rothwell, J. C., Frackowiak, R., Marsden, C. D., Johnels, B., Steg, G., Freedman, R., Hoffer, B. J., Seiger, Å., Bygdeman, M., Strömberg, I., Olson, L. 1989. Human fetal dopamine neurons grafted into the striatum in two patients with severe Parkinson's disease. *Arch. Neurol.* 46: 615–31

Madrazo, I., Drucker-Colín, R., Díaz, V., Martínez-Mata, J., Torres, C., Becerril, J. J. 1987. Open microsurgical autografts of adrenal medulla to the right caudate nucleus in two patients with intractable Parkinson's disease. *N. Engl. J. Med.* 316: 831–34

Mahalik, T. J., Finger, T. E., Strömberg, I., Olson, L. 1985. Substantia nigra transplants into denervated striatum of the rat: ultrastructure of graft and host interconnections. *J. Comp. Neurol.* 240: 60–70

Marshall, J. F., Turner, B. H., Teitelbaum, P. 1971. Sensory neglect produced by lateral hypothelamic damage. *Science* 174: 523–25

Matthews, R. T., German, D. C. 1986. Evidence for a functional role of dopamine type-1 (D-1) receptors in the substantia nigra of rats. *Eur. J. Pharmacol.* 120: 87–93

McNeill, T. H., Brown, S. A., Rafols, J. A., Shoulson, I. 1988. Atrophy of medium spiny-I striatal dendrites in advanced Parkinson's disease. *Brain Res.* 455: 148–52

Meloni, R., Childs, J., Gerogan, F., Yurkofsky, S., Gale, K. 1988. Effect of haloperidol on transplants of fetal substantia nigra: evidence for feedback regulation of dopamine turnover in the graft and its projections. See Gash & Sladek 1988, pp. 457–61

Mena, M. A., de Yebenes, J. G., Latov, N., Fahn, S. 1987. Biochemical and immunological properties of human neuroblastoma cell lines. *Schmitt Neurol. Sci. Symp. Abstr.*

Meyers, R. 1951. Surgical experiments in the therapy of certain "extrapyramidal" diseases: A current evaluation. *Acta Psychiatry Neurol. Scand. Suppl.* 67: 7–41

Moore, R. Y., Bhatnagar, R. K., Heller, A. 1971. Anatomical and chemical studies of nigro-neostriatal projection in the cat. *Brain Res.* 30: 119–35

Morihisa, J. M., Nakamura, R. K., Freed, W. J., Mishkin, M., Wyatt, R. J. 1984. Adrenal medulla grafts survive and exhibit catecholamine-specific fluorescence in the primate brain. *Exp. Neurol.* 84: 643–53

Nadaud, D., Herman, J. P., Simon, H., LeMoal, M. 1984. Functional recovery following transplantation of ventral mesencephalic cells in animals subjected to 6-OHDA lesions of the mesolimbic DA neurons. *Brain Res.* 304: 137–41

Nyberg, P., Nordberg, A., Webster, P., Winblad, B. 1983. Dopaminergic deficiency is more pronounced in the putamen than in nucleus caudatus in Parkinson's disease. *Neurochem. Pathol.* 1: 193–202

Olson, L., Seiger, Å., Hoffer, B., Taylor, D. 1979. Isolated catecholaminergic projections from substantia nigra and locus coeruleus to caudate, hippocampus and cerebral cortex formed by intraocular sequential double brain grafts. *Exp. Brain Res.* 35: 47–67

Peterson, D. I., Price, M. L., Small, C. S. 1988. Autopsy findings in a patient that had an adrenal-to-brain transplant for Parkinson's disease. *Neurology* 39(Suppl. 1): 144

Porceddu, M. L., Giorgi, O., Ongini, E., Mele, S., Biggio, G. 1986. ³H-SCH-23390 binding-sites in the rat substantia nigra: Evidence for a presynaptic localization and innervation by dopamine. *Life Sci.* 39: 321–28

Price, M. T. C., Fibiger, H. C. 1974. Apomorphine and amphetamine stereotypy after 6-hydroxydopamine lesions in the substantia nigra. *Eur. J. Pharmacol.* 29: 249–52

Prochiantz, A., Daguet, M.-C., Herbet, A., Glowinski, J. 1981. Specific stimulation of *in vitro* maturation of mesencephalic dopaminergic neurones by striatal membranes. *Nature* 293: 570–72

Prochiantz, A., DiPorzio, U., Kato, A., Berger, B., Glowinski, J. 1979. In vitro maturation of mesencephalic dopaminergic neurons from mouse embryos is enhanced in presence of their striatal target cells. *Proc. Natl. Acad. Sci. USA* 76: 5387–91

Redmond, D. E. Jr., Naftolin, F., Collier, T. J., Leranth, C., Robbins, R. J., Sladek, C. D., Roth, R. H., Sladek, J. R. Jr. 1988. Cryopreservation, culture, and transplantation of human fetal mesencephalic tissue into monkeys. *Science* 242: 768–71

Ricaurte, G. A., Irwin, I., Forno, L. S., DeLanney, L. E., Langston, E., Langston, J. W. 1987. Aging and MPTP-induced degeneration of dopamine neurons in the substantia nigra. *Brain Res.* 403: 43–51

Rinne, U. K., Rinne, J. O., Rinne, J. K., Laasko, K., Laihinen, A., Lönnberg, P. 1983. Brain receptor changes in Parkinson's disease in relation to the disease process and treatment. *J. Neural Trans.* 1983(Suppl. 18): 279–86

Rosenberg, M. B., Friedman, T., Robertson, R. C., Tuszynski, M., Wolff, J. A., Breakefield, X. O., Gage, F. H. 1988. Grafting genetically modified cells to the damaged brain: restorative effects of NGF expression. *Science* 242: 1575–78

Roth, R. H. 1979. Dopamine autoreceptors: pharmacology, function and comparison with post-synaptic receptors. *Commun. Psychopharmacol.* 3: 429–45

Ruffieux, A., Schultz, W. 1980. Dopaminergic activation of reticulata neurones in the substantia nigra. *Nature* 285: 240–41

Sauer, H., Brundin, P., Odin, P., Widner, H., Björklund, A. 1989. Effects of cool storage on effectiveness and survival of intrastriatal ventral mesencephalic grafts. *Restor. Neurol. Neurosci.* (Suppl.: *3rd Int. Symp. Neural Transplantation*): 56

Schmidt, R. H., Björklund, A., Stenevi, U. 1981. Intracerebral grafting of dissociated CNS tissue suspensions: a new approach for neuronal transplantation to deep brain sites. *Brain Res.* 218: 347–56

Schultz, W. 1982. Depletion of dopamine in the striatum as an experimental model of parkinsonism: direct effects and adaptive mechanisms. *Prog. Neurobiol.* 18: 121–66

Shimizu, N., Ohnishi, S. 1973. Demonstration of nigro-neostriatal tract by degeneration silver method. *Exp. Brain Res.* 17: 133–38

Sladek, J. R. Jr., Collier, T. J., Haber, S. N., Roth, R. H., Redmond, D. E. Jr. 1986. Survival and growth of fetal catecholamine neurons transplanted into the primate brain. *Brain Res. Bull.* 17: 809–18

Sladek, J. R. Jr., Gash, D. M., eds. 1984. *Neural Transplants: Development and Function.* New York: Plenum

Sladek, J. R. Jr., Gash, D. M. 1988. Nerve-cell grafting in Parkinson's disease. *J. Neurosurg.* 68: 337–51

Sladek, J. R. Jr., Parnavelas, J. C. 1975. Catecholamine-containing dendrites in primate brain. *Brain Res.* 100: 657–62

Sladek, J. R. Jr., Shoulson, I. 1988. Neural transplantation: a call for patience rather than patients. *Science* 240: 1386–88

Spiegel, E. A., Wycis, H. T., Freed, H. 1952. Stereoencephalotomy, thalamotomy, and related procedures. *J. Am. Med. Assoc.* 148: 446–51

Steece, K. A., Tayrien, M. W., Collier, T. J., Loy, R., Roth, R. H., Redmond, D. E. Jr., Sladek, J. R. Jr. 1987. Dopamine receptor binding in MPTP treated monkeys: correlation with neurochemistry and behavior. *Soc. Neurosci. Abstr.* 13: 71

Strömberg, I., Bygdenman, M., Goldstein, M., Seiger, Å., Olson, L. 1986. Human fetal substantia nigra grafted to the dopamine-denervated striatum of immunosuppressed rats: Evidence for functional reinnervation. *Neurosci. Lett.* 71: 271–76

Strömberg, I., Herrera-Marschitz, M., Hultgren, L., Ungerstedt, U., Olson, L. 1984. Adrenal medullary implants in the dopamine-denervated rat striatum. I. Acute catecholamine levels in grafts and host caudate as determined by HPLC-electrochemistry and fluorescence histochemical image analysis. *Brain Res.* 297: 41–51

Tomozawa, Y., Appel, S. H. 1986. Soluble striatal extracts enhance development of mesencephalic dopaminergic neurons in vitro. *Brain Res.* 399: 111–24

Ungerstedt, U. 1971a. Striatal dopamine release after amphetamine or nerve degeneration revealed by rotational behavior. *Acta Physiol. Scand.* 1971(Suppl. 367): 49–93

Ungerstedt, U. 1971b. Adipsia and aphagia after 6-hydroxydopamine induced degeneration of the nigro-striatal dopamine system. *Acta Physiol. Scand.* 1971(Suppl. 367): 95–121

Van Manen, J., Speelman, J. D. 1988. Caudate lesions as surgical treatment in Parkinson's disease. *Lancet* 1(8578): 175

Yurek, D. M., Collier, T. J., Daley, B. F., Sladek, J. R. Jr. 1988. Co-transplants: effect of fetal striatal tissue on fetal dopaminergic nigral neurons. *Soc. Neurosci. Abstr.* 14: 888

Yurek, D. M., Collier, T. J., Sladek, J. R. Jr. 1989a. Intrastriatal co-grafts of embryonic mesencephalic and striatal nerve cells: enhanced morphological development of grafted dopamine neurons, target-specific interactions, and functional recovery. *Exp. Neurol.* In press

Yurek, D. M., Deutch, A. Y., Roth, R. H., Sladek, J. R. Jr. 1989b. Morphological, neurochemical, and behavioral characterizations associated with the combined treatment of diethyldithiocarbamate and 1-methyl-4-phenyl-1, 2, 3, 6-tetrahydropyridine in mice. *Brain Res.* 497: 250–59

Zigmond, M. J., Stricker, E. M. 1972. Deficits in feeding behavior after intraventricular injection of 6-hydroxydopamine in rats. *Science* 177: 1211–14

*Annu. Rev. Neurosci. 1990. 13:441–74*

# ION CHANNELS IN VERTEBRATE GLIA

*Barbara A. Barres,*[1,2,4] *Linda L. Y. Chun,*[1,2,3] *and David P. Corey*[1,2,4]

[1] Program in Neuroscience, Harvard Medical School, Boston, Massachusetts 02115, [2] Departments of Neurology and [3] Neurosurgery, Massachusetts General Hospital, Boston, and [4] Howard Hughes Medical Institute, Boston, Massachusetts 02114

## INTRODUCTION

More than 20 years ago, Kuffler (1967) began a review of glial electrophysiology with a summary of hypotheses of glial functions; these functions were much the same as those proposed by Nageotte, Golgi, Lugaro, and Ramón y Cajal 70 years before him (reviewed by Somjen 1988) and are not substantially changed today. Yet the timelessness of these views may result more from our continued ignorance than from the prescience of these pioneers. Ramón y Cajal (1909) suggested that this poor understanding of function originated because physiologists did not have the tools to study glial cells directly. Only recently has this changed.

The last 30 years of glial electrophysiology can be divided into three periods of study. Classical studies (prior to 1970, beginning most prominently with the work of Kuffler & Potter 1964, and reviewed by Kuffler 1967 and Somjen 1975) primarily consisted of microelectrode voltage recording from the large glial cells of invertebrates and lower vertebrates in situ. During the second period (beginning with Dennis & Gerschenfeld 1969; reviewed by Kuffler et al 1984), the microelectrode technique was used to study mammalian glia, either in situ or in vitro. Since 1982 (Kettenmann et al 1982), the patch-clamp technique has made small cells accessible to voltage-clamp recording.

Among the many findings of the first two periods, three of the general principles to emerge, thought to hold for all glial cells in all species, were: (*a*) glial membranes are mainly permeable to potassium, (*b*) glial

441

membranes are "passive" in that their current-voltage relationships are linear, and (c) glia appear to lack sodium channels and are not excitable. This review focuses on the findings of the most recent period of study, which have challenged each of these three principles.

## NOMENCLATURE

A disturbing feature of the glial literature is that no two authors use the same nomenclature. In grey matter and in white matter, two main types of glial cells are found: astrocytes and oligodendrocytes, initially distinguished by morphological criteria. However, nomenclature based on morphology can be confusing: Both cell types differ in morphology in grey matter and in white matter. Astrocytes in white matter contain many filaments, have stellate processes, and have been called "fibrous" or "fibrillary," whereas astrocytes in grey matter contain few glial filaments, have sheet-like processes, and have been called "protoplasmic" (Peters et al 1976). Similarly, many oligodendrocytes in grey matter do not myelinate and have been called "perineuronal," whereas in white matter, many oligodendrocytes myelinate and have been called "interfascicular" (Penfield 1932). Whether these differences in appearance between grey matter and white matter astrocytes (or oligodendrocytes) result from the same cell type being in a different environment, or arise because they are actually distinct cell types, is as yet unclear. Nor is it clear whether such environmental differences affect glial electrophysiological properties. Thus, in this review a distinction is maintained, when possible, between grey and white matter glia.

In the optic nerve, which is part of central white matter, three glial types have been clearly identified and extensively characterized by the work of Raff and associates: oligodendrocytes and type-1 and type-2 astrocytes. These three glial cell types are antigenically, structurally, and developmentally distinct (reviewed by Miller et al 1989a and Raff 1989b). Equivalent cell types are found in white matter throughout the mammalian CNS (Raff et al 1983a, 1984a, Liuzzi & Miller 1987); and Raff and associates' nomenclature for white matter is used in this review. Grey matter glial cell types have not yet been so clearly classified, and may vary regionally. Because of the many similarities between the type-1 astrocyte in white matter and the predominant type of astrocyte in grey matter (Raff et al 1983a), astrocytes in cultures derived from grey matter have been increasingly referred to in the literature as "type-1 astrocytes." The cerebral hemisphere tissue used to prepare these cultures, however, invariably contains both grey and white matter.

In this review, the terms type-1 and type-2 astrocyte are used exclusively to refer to white matter astrocytes. Astrocytes in cortical cultures are referred to as cortical "type-1-like" astrocytes, as suggested by Raff (1989a).

# SURVEY OF ION CHANNELS IN GLIAL CELL TYPES

## Peripheral Glia

SCHWANN CELLS    The first direct electrophysiological evidence that glial cells could have nonlinear I(V) relationships was reported by Chiu et al (1984); their unexpected findings stimulated many studies that have followed. They observed, using whole-cell patch-clamp recording, that Schwann cells in culture express both voltage-dependent sodium and potassium currents. These currents occur at a somewhat lower density than in many neurons, but are qualitatively similar to neuronal currents, a similarity that extends to single-channel properties (Shrager et al 1985).

The voltage-dependent outward current is composed of at least three components: a chloride current and two types of potassium current (Howe & Ritchie 1988). Two other types of channels appear not to be active during normal whole-cell recording, but become active in excised inside-out patches studied with single-channel recording: a calcium-dependent cation-selective channel of about 32 pS (Bevan et al 1984), and an anion-selective channel of 450 pS (Gray et al 1984). Thus mammalian Schwann cells in culture express at least six different kinds of voltage-dependent ion channels, and others may not have been detected yet. The expression of these voltage-dependent ion channels is generally not an artifact of tissue culture: Schwann cells acutely isolated from rabbit sciatic nerve still exhibit sodium currents and potassium currents (Chiu 1987).

Neurotransmitter-activated channels have not been reported in vertebrate Schwann cells yet. They will probably soon be detected as they are present on Schwann cells in invertebrates (Ballanyi & Schlue 1988, Villegas 1975, Abbott et al 1988, Lieberman et al 1989).

MYELINATING VS NONMYELINATING SCHWANN CELLS    Schwann cells are either myelinating or nonmyelinating; which phenotype they display appears to be governed by signals from the axon (reviewed in Bray et al 1981 and Mirsky & Jessen 1988). Do such axonal signals also influence which ion channels are expressed by Schwann cells? Chiu (1987, 1988) recorded ionic currents in acutely isolated Schwann cells from sciatic nerve still bound to their axons. The bound axon allowed determination of the myelination state: Myelinated axons were of 15–20 $\mu$ in diameter whereas unmyelinated ones were only about 1 $\mu$. Sodium currents were observed in 100% of Schwann cells associated with nonmyelinated axons, but were never observed in myelinating Schwann cells. In addition, a component of inwardly rectifying potassium current was found mainly in myelinating Schwann cells (Chiu 1987, Wilson & Chiu 1989; more recently this current has also been found in nonmyelinating Schwann cells: G. F. Wilson, personal communication). Although nonmyelinating Schwann cells have large outward potassium currents, these are either much smaller or absent in myelinating Schwann cells.

Transection results in the appearance of sodium channels in myelinating Schwann cells and in the loss of sodium channels from nonmyelinating Schwann cells. These results provide further evidence for an effect of neuronal signals on Schwann cell sodium channel expression (Chiu 1988).

## Grey Matter Glia

MULLER CELLS    Muller cells, the major glial cell type found in the retina, have been especially convenient for electrophysiological studies of CNS glia; unlike cortical astrocytes, they are easy to identify by morphology alone in cell suspensions. An extensive characterization of conductances in salamander Muller cells performed with whole-cell patch recording revealed voltage-dependent calcium current and at least three components of potassium current, including $K_{Ca}$, $K_a$, $K_{ir}$ (Newman 1985b). Most (95%) of the potassium conductance of the cell is found in the endfoot, thus suggesting that the endfoot conductance participates in a special type of spatial buffering mechanism, termed "siphoning," that allows shunting of potassium through the endfoot and into the vitreous (Newman et al 1984, Newman 1984, 1985a, 1986a, 1987, Karwoski et al 1989). In these cells, a small component of voltage-dependent sodium current is also present (E. A. Schwartz, personal communication).

Only a single type of voltage-dependent ion channel was observed in cell-attached patches of salamander Muller cells: an inwardly-rectifying potassium current, mainly localized to the endfoot by Brew et al (1986). These authors suggested that this potassium channel mediated the proposed spatial buffering process, although their conditions may have precluded observation of the other ion channel types observed in whole-cell recordings by Newman. By using two-electrode voltage-clamping, the endfoot conductance in salamander Muller cells has been confirmed to be inwardly rectifying (Newman 1989).

Voltage-dependent ion channels are just beginning to be characterized in mammalian Muller cells (Nilius & Reichenbach 1988). In these rabbit Muller cells, two types of inwardly rectifying potassium channels occur along the soma, and a nonrectifying conductance (360 pS in 140 mm symmetrical potassium solutions) is in the endfoot (although the total number of patches studied was not large).

An electrogenic glutamate uptake mechanism is also spatially localized, mainly outside of the endfoot region (Brew & Attwell 1987). Although electrogenic carrier mechanisms are not the subject of this review, it is interesting that this mechanism can carry significant currents across the membrane; its conductance is close to that of the potassium conductance in the same region of the cell in which the carrier is located. This conductance is still only a small proportion, however, of the whole-cell conductance (Schwartz & Tachibana 1990).

Neurotransmitters can activate ion channels in glia. GABA activates a chloride channel in skate Muller cells that appears to be identical to the GABA$_A$ receptor-channel complex in neurons (Malchow et al 1989). These currents can be as large as several nanoamps. Because these currents were observed in acutely isolated Muller cells, their results provide the first evidence that neurotransmitter-activated ion channels are found in vivo in glia.

CORTICAL TYPE-1-LIKE ASTROCYTES    Much glial electrophysiology has been directed at cortical type-1-like astrocytes, probably because of the ease of obtaining highly purified cultures (McCarthy & deVellis 1980) and because of the lack of a specific surface marker that would allow the identification of acutely dissociated cells. As the entire cerebral hemisphere is typically used for the preparation of these cultures, the cultures probably contain both cortical astrocytes and type-1 astrocytes from white matter— although the possibility that all of the astrocytes in these cultures derive from white matter has not been ruled out (type-2 astrocytes are lost with the removal of the top layer of cells).

Perhaps because of this intensive scrutiny, these cells have been found to express more channel types than any other glial cell. At least 14 types of voltage-dependent ion channels have been observed! These include a sodium channel, two types of calcium channels, up to four types of potassium channels, at least three types of chloride channels, and four types of stretch-sensitive channels (references in Table 1).

Many of these channels are not functional in the resting cell. Thus all of the chloride channels appear subject to an inhibitory modulation and are not normally active (Gray & Ritchie 1986, Sonnhof 1987), and a sustained calcium current is only seen after incubation in substances that increase intracellular cAMP (MacVicar & Tse 1988, Barres et al 1989a). In contrast to most examples of modulatory effects, the calcium channel in this case is entirely absent prior to exposure to these substances (Barres et al 1989a). Induction may involve recruitment of channels to the membrane from intracellular stores (as occurs for hormone induction of certain ion transporters outside of the nervous system), rather than a post-trans-

**Table 1**  Summary of ion channel types in vertebrate glia

| Location | Species | Prep[a] | Channel[b] | Va[c] | References |
|---|---|---|---|---|---|
| Schwann cells | | | | | |
| Sciatic nerve | Rabbit | TC | Na 20 pS | −40 | Chiu et al 1984, Shrager et al 1985 |
| | | | $K_d$ 19 pS | −25 | |
| Sciatic nerve | Rat | TC | Cation$_{Ca}$ 32 pS | | Bevan et al 1984 |
| Sciatic nerve | Rat | TC | Anion 450 pS | | Gray et al 1984 |
| Sciatic nerve | Rabbit | TC | Cl | 0 | Howe & Ritchie 1988 |
| | | | $K_a$ | −50 | |
| | | | $K_d$ | 0 | |
| Sciatic nerve | Rabbit | AD | Na | −40 | Chiu 1987, Wilson & Chiu 1989 |
| | | | $K_d$ | −20 | |
| | | | $K_{ir}$ | | |
| Muller cells | | | | | |
| Retina | Salamander | AD, SL | Ca | −20 | Newman 1985b |
| | | | $K_{Ca}$ | −40 | |
| | | | $K_a$ | −50 | |
| | | | $K_{ir}$ | | |
| Retina | Axolotl | AD | $K_{ir}$ 40 pS | | Brew et al 1986 |
| Retina | Rabbit | AD | $K_{ir}$ 60 pS | | Nilius & Reichenbach 1988 |
| | | | $K_{ir}$ 105 pS | | |
| | | | K? 360 pS | | |
| Retina | Skate | AD | GABA$_A$ Cl | | Malchow et al 1989 |
| Ependymal cells | | | | | |
| Optic nerve | Rat | AD | Na | −40 | Barres et al 1985, 1989b |
| | | | K | −30 | |

## Cortical astrocytes

| Region | Species | Treatment | Channel | Reversal (mV) | Reference |
|---|---|---|---|---|---|
| Cerebrum | Rat | TC | Na | −40 | Bevan et al 1985, 1987 |
| | | | $K_a$ | −40 | |
| | | | $K_d$ | −40 | |
| | | | Cl | −40 | |
| Cerebrum | Rat | TC | Cl | −40 | Gray & Ritchie 1986 |
| Cerebrum | Mouse | TC, dbcAMP | Na 14 pS | −50 | Nowak et al 1987 |
| | | | $K_a$ | −40 | |
| | | | $K_{Ca}$ 230 pS | −30 | |
| | | | $K_d$ 7 pS | −30 | |
| | | | $K_d$ 20 pS | −30 | |
| | | | Cl 385 pS | | |
| | | | $Cl_{ir}$ 5 pS | < −40 | |
| Cerebrum | Rat | TC, dbcAMP | $K_{Ca}$ 25 pS | −60 | Quandt & MacVicar 1986 |
| Cerebrum | Rat | TC | Cl 400 pS | | Sonnhof 1987 |
| Cerebrum | Rat | TC | $Ca_L$/cAMP | −20 | MacVicar 1984, MacVicar & Tse 1988 |
| Cerebrum | Rat | TC | $Ca_L$/cAMP | −30 | Barres et al 1989a |
| | | | $Ca_T$ | −60 | |
| Cerebrum | Rat | TC | $K_{SAC}$ 140 pS | | Ding et al 1988, 1989 |
| | | | $K_{SAC}$ 80 pS | | |
| | | | $K_{SAC}$ <80 pS | | |
| | | | $Cation_{SIC}$ 50 pS | | |
| Cerebrum | Rat | TC | $GABA_A$ Cl 12, 21, 29, 43 pS | | Bormann & Kettenmann 1988 |

## Type 1 astrocytes

| Region | Species | Treatment | Channel | Reversal (mV) | Reference |
|---|---|---|---|---|---|
| Optic nerve | Rat | TC | $K_d$ | −20 | B. A. Barres et al, in preparation |
| | | | Na (10%) | −40 | |
| | | | $K_{ir}$ | | |
| Optic nerve | Rat | AD | $K_{Ca}$ | −30 | B. A. Barres et al, in preparation |
| | | | Na (20%) | −40 | |
| | | | $K_a$ (20%) | −30 | |
| | | | $K_{ir}$ | | |

**Table 1**  Summary of ion channel types in vertebrate glia

| Location | Species | Prep[a] | Channel[b] | $Va^c$ | References |
|---|---|---|---|---|---|
| **Type 2 astrocytes** | | | | | |
| Optic nerve | Rat | TC | Na | −40 | Bevan et al 1987, Bevan & Raff 1985 |
|  |  | TC | $K_a$ | −40 |  |
|  |  |  | $K_d$ | −40 |  |
| Optic nerve | Rat | TC | $Na_n$ | −30 | Barres et al 1988, 1989b, 1990 |
|  |  |  | $Na_g$ | −40 |  |
|  |  |  | $Ca_T$ | −60 |  |
|  |  |  | $Ca_L$ | −30 |  |
|  |  |  | $K_a$ | −30 |  |
|  |  |  | $K_d$ | 0 |  |
|  |  |  | $K_{ir}$ |  |  |
|  |  |  | $Cl_{or}$ 25 pS |  |  |
|  |  |  | Cl 260 pS |  |  |
| Cerebellar | Rat | TC | Glu (and Quis): 47, 39, 15, 10, 6 pS | | Usowicz et al 1989 |
|  |  |  | Kainate: 15, 10, 6, 2 pS | | |
| **Oligodendrocytes** | | | | | |
| "Brain" | Lamb | TC | $K_a$ | −40 | Soliven et al 1988a |
|  |  |  | $K_d$ | 0 |  |
|  |  |  | $K_{ir}$ |  |  |
| Spinal cord | Mouse | TC | $K_{lk}$ 6–125 pS |  | Kettenmann 1982, 1984a |
| Cerebrum | Mouse | TC | $K_{ir}$ |  | Sontheimer & Kettenmann 1988 |
|  |  |  | $K_a$ |  |  |
| Spinal cord | Mouse | TC | $K_d$ |  | Sontheimer & Kettenmann 1988 |
|  |  |  | $K_{ir}$ |  |  |

| | | | Channel | $V_a$ | Reference |
|---|---|---|---|---|---|
| Optic nerve | Rat | TC | $K_{ir}$ 30 pS | $-130$ | Barres et al 1988, 1989, 1990 |
| | | | $K_{ir}$ 120 pS | $-130$ | |
| | | | $Cl_{or}$ 25 pS | | |
| | | | $K_d$ | $-30$ | |
| | | | $K_a$ | $-30$ | |
| Optic nerve | Rat | AD | $K_{ir}$ | $-130$ | Barres et al 1987, in preparation |
| | | | $K_a$ | $-30$ | |
| **O2A progenitors** | | | | | |
| Optic nerve | Rat | TC | Na | $-30$ | Bevan et al 1987 |
| | | | $K_a$ | $-10$ | |
| | | | $K_d$ | $-10$ | |
| Optic nerve | Rat | AD and TC | $Na_n$ | $-30$ | Barres et al 1990 |
| | | | $K_{d/Ca}$ | $-30$ | |
| | | | $K_d$ | $-30$ | |
| | | | $K_a$ | $-30$ | |
| | | | $K_{ir}$ | | |
| | | | Glu, Kainate, Quisqualate | | |
| Cerebellum | Rat | TC | Glu, Kainate, and Quisqualate | | S. G. Cull-Candy et al 1989 |
| "Brain" | Mouse | TC | Na | $-40$ | Sontheimer et al 1989 |
| | | | $K_{a/Ca}$ | $-60$ | |
| | | | $K_d$ | $-40$ | |

[a] TC: tissue culture; AD: acutely dissociated; SL: slice; dbcAMP: dibutyryl cAMP; cAMP: cyclic AMP.

[b] $K_{Ca}$: calcium-dependent potassium current; $K_a$: decaying potassium current; $K_d$: "delayed rectifier" loosely used here to mean a sustained potassium current; $K_{ir}$: inwardly-rectifying potassium current; $K_{a/Ca}$: a calcium-dependent decaying potassium current; $K_{lk}$: nonrectifying potassium current; $Cl_f$: inwardly rectifying chloride current; $Cl_{or}$: outwardly rectifying chloride current; $Ca_t$: transient calcium current; $Ca_l$: long-lasting calcium current; $Na_n$: neuronal form of the sodium current; $Na_g$: glial form of the sodium current; SAC: stretch activated channel; SIC: stretch inactivated channel.

[c] $V_a$: activation voltage (most negative voltage where current is observed); for $K_{ir}$ current is generally observed negative to the potassium equilibrium potential.

lational modification of existing channels, such as phosphorylation. Since several studies of astrocyte channels involved cells first "rounded up" by dibutyrl cAMP (Quandt & MacVicar 1986, Nowak et al 1987), some of the channels observed in these cells may not be normally functional.

An unresolved issue is whether heterogeneity of channel phenotypes among cortical astrocytes occurs, since the proportion of astrocytes expressing a given current type is often not apparent (or not reported).

Neurotransmitter-activated channels have been observed in cortical type-1-like astrocytes in culture. Their presence has long been suspected because of the large number of neurotransmitters that induce either depolarization or hyperpolarization. These changes in membrane potential, however, could have been caused by at least three different mechanisms: activation of an electrogenic transmitter transport mechanism, modulation of voltage-dependent ionic currents contributing to the resting conductance, or direct activation of a ligand-gated ion channel. Examples of all three types of neurotransmitter effects in astrocytes are now known.

First, inward currents are caused by electrogenic glutamate uptake in cerebellar astrocytes in culture (Cull-Candy et al 1988). Second, $\beta$-adrenergic agonists and vasoactive intestinal peptide induce a voltage-dependent calcium current in cortical astrocytes in culture, presumably by raising cAMP (MacVicar & Tse 1988, Barres et al 1989a). Third, GABA activates a chloride conductance in 100% of cortical type-1-like astrocytes in culture (Bormann & Kettenmann 1988). As in neurons, these receptors have $GABA_A$ pharmacology, have multiple conductance substates, are blocked by bicuculline and picrotoxin, and are potentiated by barbiturates and benzodiazepines (Backus et al 1988).

Glutamate activation of cation-selective channels in cultured cortical astrocytes, acting on non-NMDA receptors, has been reported, although no evidence of single channels or of blockade by glutamate antagonists has been found (Sontheimer et al 1988). Because NMDA responses were not detected, the mechanism of the aspartate-induced depolarizations that occur in 100% of astrocytes in culture (Kettenmann & Schachner 1985) remains unknown, but is probably accounted for by the ubiquitous presence of an electrogenic amino acid uptake process (e.g. Schwartz & Tachibana 1990).

EPENDYMAL CELLS    Despite the ease of identifying ependymal cells, these cells have received little electrophysiological attention (see Connors & Ransom 1987 for review). Voltage-dependent sodium channels have recently been characterized in acutely-isolated rat ependymal cells, and voltage-dependent potassium current is also present in all ependymal cells studied (Barres et al 1985, 1989c).

## White Matter Glia

TYPE-1 ASTROCYTES    Cultures of astrocytes from pure white matter may be prepared from optic nerve. These cultures are still heterogeneous because they contain both type-1 and type-2 astrocytes, but these may be distinguished by morphology and antigenic phenotype (Raff et al 1983a, Miller & Raff 1984).

So far there has been little electrophysiological study of type-1 astrocytes. All type-1 astrocytes express both a delayed rectifying and an inwardly rectifying potassium current in culture (Barres et al 1987 and in preparation). In contrast, acutely isolated type-1 astrocytes, recognized by surface labeling with antibodies to the RAN-2 surface antigen, all express a charybdotoxin-sensitive, calcium-dependent, sustained potassium current in addition to $K_{ir}$ (Barres et al 1987 and in preparation). Sodium currents are present in 10–20% of cells in culture, 20% of acutely isolated cells lacking processes (Barres et al 1989b), but are found in 100% of cells acutely isolated by the "tissue-print" technique that preserves many of their processes (B. A. Barres et al, in preparation, and see Table 1). Thus cells in vivo may have channels localized to their processes. Co-culture experiments have demonstrated that neurons up-regulate the expression of sodium channels in type-1 astrocytes in culture (B. A. Barres et al, in preparation).

TYPE-2 ASTROCYTES    The type-2 astrocyte has only recently been recognized as a distinct astrocytic component of white matter, based on morphology, developmental appearance, and surface antigenic phenotype (reviewed in Raff 1989b, Miller et al 1989a). Most recently type-2 astrocytes have been shown to be structurally distinct: Whereas type-1 astrocytes have radial processes that terminate on blood vessels and the pia limitans, type-2 astrocyte processes are mainly longitudinal and may terminate on nodes of Ranvier (ffrench-Constant & Raff 1986, ffrench-Constant et al 1986, Miller et al 1989b; in addition some type-1 astrocytes may contribute perinodal processes: Suárez & Raff 1989).

Thus it is not surprising that type-2 astrocytes are electrophysiologically distinct from type-1 astrocytes. In culture, initial studies demonstrated the presence of a sodium current and two components of outward potassium current (Bevan & Raff 1985, Bevan et al 1987). Subsequently, five components of inward current have been observed, by using specific ion isolation solutions; these include two forms of the sodium current, two forms of calcium current, and an inwardly rectifying potassium current (Barres et al 1988, 1990). Two types of chloride channels were also detected with single-channel recording that, like many other chloride channels, only become active with excision of the patch (Barres et al 1988, Bevan et al

1987). A subset of acutely isolated astrocytes from rat optic nerve appear to have a related ion channel phenotype, thus suggesting that type-2 astrocytes in vivo are electrophysiologically similar to those in culture (Barres et al 1989f).

There are at least two fundamental differences between ion channel expression in type-2 astrocytes and type-1 astrocytes. First, although each of these cell types has sodium currents, the sodium current in type-2 astrocytes appears indistinguishable from that found in neurons, whereas the type-1 astrocytes express a "glial" form (Barres et al 1989b), which opens more slowly and has a shifted voltage sensitivity. Second, the type-2 astrocyte constitutively expresses two types of calcium current, whereas calcium current is only found in type-1 astrocytes that have elevated intracellular cAMP (Barres et al 1988, 1989a).

Glutamate-activated channels in type-2 astrocytes in culture have been demonstrated and characterized with whole-cell and single-channel recording (Usowicz et al 1989). They are cation-selective and are found in both cerebellar and optic nerve type-2 astrocytes. In neurons, these channels are usually activated by three classes of agonists—NMDA, kainate, and quisqualate—but in the type-2 astrocyte, only non-NMDA activated channels were observed. Because type-2 astrocytes are found mainly or exclusively in white matter, where neurotransmission is not thought to occur, this finding suggests the possibility of neuronal-glial signaling in white matter (Barres 1989; see below).

OLIGODENDROCYTES    Oligodendrocytes were the first glial cell type to be studied with the patch-clamp technique (Kettenmann et al 1982, 1984a). "Leakage" potassium channels of varying conductances ranging from 6 to 125 pS were observed with single-channel recording in oligodendrocytes from mouse spinal cord. These channels were initially reported to be voltage-independent, although further observation revealed that their probability of opening increased with depolarization. Yet the resting conductance of the cell is entirely composed of a potassium permeability that decreases with depolarization (Kettenmann et al 1984b).

These channels have recently been characterized further, with both whole-cell and single-channel recording (Barres et al 1988). The resting conductance of optic nerve oligodendrocytes in culture results from two types of inwardly rectifying potassium channels of 30 and 120 pS. These channels are strongly voltage-dependent, opening more frequently with increasing degree of depolarization. Their open channel I(V) relation is inwardly rectifying, so even when the channel is open, little outward potassium current occurs. This explains the increase in membrane resistance with depolarization.

The resting oligodendrocyte membrane is impermeable to chloride (Kettenmann et al 1983). An outwardly-rectifying chloride channel, however, is observed in most excised patches, but not in cell-attached patches (Barres et al 1988). This channel appears to be identical to a voltage-dependent chloride channel found in many epithelial tissues, which can be activated by $\beta$-adrenergic agonists (Welsh & Lidtke 1986, Frizzel et al 1986), thus suggesting that the chloride channel in oligodendrocytes may be activated by neurotransmitters.

Reports about the nature and density of outward potassium currents in oligodendrocytes have varied greatly. For instance, rat optic nerve oligodendrocytes were initially reported to lack outward potassium current (Barres et al 1988), but lamb brain oligodendrocytes in culture express two different components of outward potassium current (see Table 1, Soliven et al 1988b). Although heterogeneity of oligodendrocytes between species or parts of the brain is possible, there are two more likely explanations. First, it is possible that serum in the culture medium alters the ion channel phenotype or density of currents expressed. For instance, Sontheimer & Kettenmann (1988) found that mouse brain oligodendrocytes cultured in serum-free medium lack outward potassium current but that spinal cord oligodendrocytes cultured in 10% calf serum express two outward potassium channels. Optic nerve oligodendrocytes cultured in completely serum-free medium express both inwardly-rectifying potassium currents and outward potassium currents, and the density of the outward currents increases with increasing age in culture (Barres et al 1990). Second, the smaller outward currents are very difficult to isolate from the large inwardly rectifying current: When currents before and after cesium blockade were subtracted, small outward components not previously observed became apparent (Barres et al 1988, 1989e). Whatever the explanation for these differences in culture, both inwardly and outwardly rectifying potassium currents are likely to occur in vivo, as both are present in acutely dissociated optic nerve oligodendrocytes (Barres et al 1987 and in preparation).

Neurotransmitters have not yet been shown to activate ion channels in oligodendrocytes. Glutamate-activated ion channels are not present in acutely dissociated oligodendrocytes from postnatal optic nerve (Barres et al 1990) or in oligodendrocytes in culture (Cull-Candy et al 1989). Neurotransmitters have been reported, however, to modulate potassium currents in lamb oligodendrocytes. Activators of adenylate cyclase and protein kinase C significantly decrease the outward potassium current; moreover, isoproterenol at concentrations as low as 0.1 $\mu$M decreases the current (Soliven et al 1988b). More recently, phorbol esters but not forskolin have been observed to decrease the inwardly rectifying potassium current (D. Nelson, personal communication). In oligodendrocyte cultures, the attachment process activates protein kinase C, and so it has been suggested that attachment influences the behavior of oligodendrocyte ion channels (Vartanian et al 1986).

GLIAL PROGENITOR CELLS    The type-2 astrocyte and oligodendrocyte are derived from a common progenitor cell, the O2A (Raff et al 1983b, 1984b, Temple & Raff 1985), yet these two descendants have quite different channel phenotypes (Table 1). How do the electrophysiological properties of the O2A compare to those of the type-2 astrocyte and oligodendrocyte? Bevan et al (1987) demonstrated that O2A progenitors in serum-free culture also expressed voltage-dependent ion channels and these were a subset of channel types observed in type-2 astrocytes in serum-containing cultures, but differed from those found in oligodendrocytes (Bevan & Raff 1988, Barres et al 1988). O2A progenitors expressed a sodium current and both sustained and inactivating potassium currents.

The ion channel phenotype of O2A progenitors in culture has been compared with that found in acutely isolated O2As (Barres et al 1990). In addition to the current components previously observed in O2A progenitors in culture (see above), a small component of inwardly-rectifying current and a component of charybdotoxin-sensitive outward potassium current are also found. This identical phenotype occurs in all acutely isolated O2A progenitors, a finding that indicates that channel expression by O2As is not an artifact of culture (Barres et al 1990).

The acutely isolated progenitors appear "neuronal" in some respects: Sodium current density (50 pA/pF) is near that found in many neurons (e.g. retinal ganglion cells), and these cells fire single regenerative potentials with moderate amounts of depolarizing current (Barres et al 1990, also see Bevan et al 1987). O2As express a form of the sodium channel that is distinct from that occurring in type-1 astrocytes but that is indistinguishable from that observed in retinal ganglion cells when cells are studied under identical experimental conditions (Barres et al 1989b).

In order to compare the properties of cells in the O2A lineage, it is necessary to study all three cell types in the same conditions. Serum-free culture conditions in which O2A progenitors replicate many aspects of their in vivo behavior have been developed recently (Lillien et al 1988, 1990). In these cultures, O2As, plated from P0 tissue, divide and differentiate on schedule: Oligodendrocytes appear beginning in the first days of culture, and type-2 astrocytes appear after about two weeks of culture.

With these cultures, the developmental program of channel expression in the O2A lineage has been examined (Barres et al 1990). Ion channel types expressed by the O2A progenitors are not a simple subset of either descendant cell. The oligodendrocyte expresses several fewer channel types (Table 1). The type-2 astrocyte differs mainly by the additional expression of two types of calcium current and the loss of expression (or alteration of) a charybdotoxin-sensitive component of outward potassium current. Thus all three cells in the O2A lineage have distinct ion channel phenotypes.

The electrophysiological properties of cells along the developmental pathway from the O2A progenitor to the oligodendrocyte have been much more finely dissected by Sontheimer et al (1989), who studied, in addition, two transitional cell stages. The earliest transitional stage is still bipotential and has a channel phenotype identical to the O2A, whereas the later transitional stage is committed to become an oligodendrocyte and has the oligodendrocyte channel phenotype.

O2As also express neurotransmitter receptors. Glutamate receptors of the non-NMDA type were initially detected on O2A progenitors in culture with binding studies (Gallo et al 1989), and binding of glutamate to these receptors activates cation-selective ion channels (Cull-Candy et al 1989). Non-NMDA glutamate agonists have also been found to activate ion channels in acutely dissociated O2As (Barres et al 1990).

## SPECIFIC ISSUES

### Glial Cellular Phenotypes

DIVERSITY   Perhaps the most striking feature of the data summarized in Table 1 is that vertebrate glial cells express a great variety of ion channels. No major type of ion channel is found in neurons that is not observed in at least some glial cell types. Moreover, receptors for the main excitatory and inhibitory transmitters used by CNS neurons, glutamate and GABA, are also found on CNS glial cells. As in neurons, neurotransmitters may also modulate voltage-dependent ion channels, and specific ion channel types may be highly localized to specific regions of the cell membrane.

There is also a new appreciation of a diversity of ion channel phenotypes among glial cell types. Muller cells, type-1 astrocytes, type-2 astrocytes, and oligodendrocytes all have their own distinct phenotypes. No conclusive evidence has been found, however, that the ion channel phenotype of oligodendrocytes in different brain regions differ, or that cortical type-1-like astrocytes differ from type-1 astrocytes. The study of the properties of glial cells is clearly still in its infancy, and a complete picture of even the types of ion channels present in each of these cell types is still not available. For instance the apparent lack of inwardly rectifying potassium currents in cortical type-1 astrocytes or of calcium currents in Schwann cells (Table 1) may simply indicate that these currents have not yet been specifically sought.

Glial ion channels are not simply an artifact of culture: They have been observed in all acutely isolated glial cells recorded from so far. Channel expression in vivo is not just the result of "accidental" channel expression, for instance as the result of a leaky promoter, because channels often occur at the same high densities as in neurons. The existence of specific ion channel phenotypes that are homogeneously expressed in specific glial cell types also argues against "leaky" expression.

Homogeneity of ion channel phenotype among glial cells of a particular type, with diversity between glial types, argues that ion channel phenotype is linked to cellular identity, and further suggests that these different ion channel phenotypes have functional significance. Understanding that significance may take much more work.

NEURONAL VS GLIAL PHENOTYPES    Neurons can be reliably differentiated from glia by antigenic phenotype: Neurons are recognized by the presence of neurofilaments, astrocytes by the presence of glial filaments, and oligodendrocytes by the presence of myelin-specific proteins or glycolipids. Can electrophysiological phenotype also be reliably used to distinguish between neurons and glia? Clearly the presence or absence of a specific ion channel is not sufficient. Despite the presence of voltage-dependent sodium and calcium channels in many glial cells, glial cells seem less capable of generating robust action potentials. Thus it is still true that when vigorous excitability is present, the cell may be identified as a neuron. Inexcitability could be found either in interneurons or in glial cells.

Nevertheless, electrophysiological studies of glia have at least blurred the distinction between neurons and glia. All vertebrate glia contain voltage-dependent ion channels and, where they have been specifically sought, neurotransmitter-gated ion channels, and thus they are capable of dynamically sensing and responding to their environment. For instance, the type-2 astrocyte has glutamate-activated ion channels, a sodium current, and calcium currents of the types implicated in neurotransmitter release. Several retinal cell types were initially classified as glia because they lacked axons or excitability, but have since been reclassified as neurons, by defining a neuron as a cell that is an integral component of a neural circuit. Is it possible that some glial cell types will yet succumb to a neuronal reclassification?

PERIPHERAL VS CENTRAL GLIA    Are electrophysiological properties of glia in peripheral nerves similar to those of glia in central white matter? In rat, nonmyelinating Schwann cells express the RAN-2, A5E3, and GFAP antigens, whereas myelinating Schwann cells express myelin-specific proteins (Jessen & Mirsky 1984, Mirsky & Jessen 1984, 1986). These antigenic phenotypes are also found in CNS astrocytes and oligodendrocytes, respectively (Mirsky & Jessen 1987). Moreover, ion channel phenotypes seem analogous: For instance, astrocytes and nonmyelinating Schwann cells express sodium current and large outward potassium currents, whereas oligodendrocytes and myelinating Schwann cells do not. It will be interesting to know how far, developmentally and functionally, such an analogy will extend.

Along these lines, Jessen et al (1989) have recently identified a bipotential Schwann cell precursor in rat, whose differentiation is sensitive to the presence or absence of serum in the medium, generating nonmyelinating and myelinating Schwann cells, respectively. Although this suggests that type-2 astrocytes may in some way be analogous to nonmyelinating Schwann cells, this is clearly not the case for their electrophysiological properties. Instead, the available data suggest that nonmyelinating Schwann cells may be closely similar to type-1 astrocytes: Unlike type-2 astrocytes they each express the glial form of the sodium current, appear to require the presence of neurons for expression of this sodium current, and lack calcium currents.

## Specialization of Ion Channels in Glia

A general theme to emerge from studies of glial ion channels is that the channels are generally similar to their neuronal counterparts. In fact, for most channel types found in glia, there is no evidence of any difference. Three glial channels differ from those in neurons, however, a finding that suggests that they are functionally specialized for a glial role.

GABA-GATED CHANNELS    GABA$_A$-activated chloride channels in cortical astrocytes interact with inverse agonists differently than in neurons (Bormann & Kettenmann 1988). Although $\beta$-carboline decreased GABA-activated currents in chromaffin cells to 50% of control, it increased the astrocyte currents by 30% under identical conditions. This difference may be accounted for by a different receptor structure between neurons and glia; specifically the two receptor types may share a common GABA$_A$ beta subunit but have differing alpha subunits (Backus et al 1988, Casalotti et al 1987).

GLUTAMATE-GATED CHANNELS    Non-NMDA glutamate-activated channels probably differ between type-2 astrocytes and neurons (Usowicz et al 1989). In neurons, the largest conductance substate is preferentially activated by NMDA (Jahr & Stevens 1987, Cull-Candy & Usowicz 1987, Ascher & Nowak 1988). On neurons lacking NMDA receptors, glutamate does not activate the large substate (Llano et al 1988). In the type-2 astrocyte, however, which also lacks NMDA receptors, glutamate activates the largest substate as well as smaller ones. Moreover, this large substrate is blocked by magnesium in neurons but not in type-2 astrocytes (Vsowicz et al 1989).

SODIUM CHANNELS   Sodium channels in glia have been suggested to be different from those in neurons ever since the earliest studies of glial sodium channels (Shrager et al 1985, Bevan et al 1985). These investigators compared the I(V) relation of channels from Schwann cells and astrocytes with data from nodes of Ranvier in the PNS. The voltage dependence of activation of the glial channels appeared to be shifted in a depolarizing direction by about 30 mV (see Figure 4 in Shrager et al 1985). The two I(V) relations were obtained by different techniques, however; patch clamp for the glia, sucrose gap for nodes. Because only the relative degree of depolarization for the sucrose gap studies was known, the two I(V) relations were aligned for comparison by assuming the steady-state inactivation (h-infinity) curves had identical midpoints.

To examine this question further, Barres et al (1989b) directly compared the properties of sodium channels in retinal ganglion cells and in optic nerve type-1 astrocytes. They also observed a difference in voltage dependence of the two channels; however, the voltage dependence of both activation and inactivation was shifted in a hyperpolarizing direction in glia. The h-infinity curve for glia has a midpoint at $-80$ mV whereas that of the neurons has a midpoint of $-55$; also, the voltage-dependence of activation was shifted negatively by 10 mV. (Thus all of these studies are in close agreement if the node sodium channels match the CNS neurons in steady-state inactivation.) Kinetic differences between glial and neuronal channels were also found: the glial form activated more than four times more slowly than the neuronal form, inactivated twice as slowly, and reopened more frequently (Barres et al 1989b).

Thus, both glial and neuronal forms of the sodium channel appear specialized to be partially inactivated at the resting potential of the cell, since the midpoints of the h-infinity curves are near the resting potentials for both glia and neurons (see Barres et al 1989b for further discussion). Chiu et al (1984) and Bevan et al (1985) have argued that glial sodium channels in Schwann cells and cortical astrocytes would be inactivated at their normal resting potentials; however, this argument was based on their whole-cell patch resting potential measurements of $-40$ mV, in conflict with a huge body of data that astrocyte resting potentials are between $-90$ and $-70$ mV at normal extracellular K concentrations (see e.g. Bevan & Raff 1985).

TTX SENSITIVITY OF GLIAL SODIUM CHANNELS   Most reports suggest that sodium channels in astrocytes in culture are poorly TTX-sensitive, requiring micromolar concentrations to block (Bevan et al 1985, Nowak et al

1987). In contrast, when neuronal and glial TTX sensitivity were compared using acutely dissociated retinal ganglion cells, type-1 astrocytes, and O2A progenitor cells, all of these cells were found to be highly sensitive to TTX, with half of the current blocked at concentrations of 2 to 3 nM (Barres et al 1989b). Opposite results have been reported for Schwann cells. Sodium current in acutely isolated Schwann cells is poorly sensitive to TTX (Chiu 1987), whereas in culture it is very TTX sensitive, being completely blocked with concentrations of 60 nM (Shrager et al 1985).

Binding studies have revealed both low- and high-affinity binding sites (Bevan et al 1985, Yarowsky & Krueger 1989). In young cultures the binding sites have low affinity, whereas the appearance of high-affinity binding sites correlates strongly with the appearance of stellate astrocytes in cultures over time; these could be induced either with age or with a serum-free culture medium (Yarowsky & Krueger 1989). Both of these sites appear to be associated with sodium channels, since sodium influx associated with both low- and high-affinity sites was stimulated in the presence of batrachotoxin and sea anemone polypeptide toxin.

Several possibilities need to be considered: heterogeneity of astrocyte types in cortical cultures, possible access problems of TTX to Schwann cell sodium channels (Chiu 1987), and the possible effects of the culture environment on TTX binding.

## Neurotransmitter-gated Channels in White Matter Glia: Where Is the Neurotransmitter Coming From?

The structure and function of white matter have received remarkably little attention. There is some evidence that impulse-mediated release of neurotransmitters, particularly glutamate, may occur into central white matter tracts and into peripheral nerves both in vertebrates and invertebrates (Wheeler et al 1966, Weinreich & Hammerschlag 1975, Abbott et al 1988, Lieberman et al 1989). This nonsynaptic axonal release is impulse dependent and is not triggered by potassium depolarization alone. Although the presence of vesicles has been observed at some nodes (Metuzals 1965), vesicle fusion along axons or nodes has not been detected. On the other hand, impulse-mediated release is calcium-independent (Weinreich & Hammerschlag 1975), thus suggesting a nonvesicular release mechanism. A glutamate transport mechanism in squid axons has been demonstrated (Baker & Potashner 1971, 1973). Such electrogenic neurotransmitter uptake mechanisms may run in reverse to release neurotransmitters, at least under certain experimental conditions (horizontal cells: Schwartz 1987; type-2 astrocytes: Gallo et al 1986, Levi et al 1986; amacrine cells: O'Malley & Masland 1989; see Schwartz & Tachibana 1990 for further discussion of this possibility).

Where could this transmitter arise from? Some investigators appear to assume that it comes from axons, since glutamate is present at high concentrations. Yet the glial cells have not been eliminated as a source: Electrogenic glutamate transport mechanisms are found in Muller cells and astrocytes (Brew & Attwell 1987, Cull-Candy et al 1988, Schwartz & Tachibana 1990). This seems an unlikely source, however, as glial cells also contain cytoplasmic enzymes that rapidly metabolize glutamate.

What purpose would nonsynaptic neurotransmission in white matter and peripheral axon tracts serve? It has been suggested that it may facilitate potassium regulatory mechanisms (Pichon et al 1987, Barres et al 1988). Type-2 astrocytes' processes contact nodes of Ranvier (ffrench-Constant & Raff 1986); thus Usowicz et al (1989) have suggested that activation of glutamate receptors on these processes would produce an influx of sodium and an efflux of potassium around the node, and thus possibly influence electrical excitability at the node. Alternatively, glutamate-induced depolarization of type-2 astrocytes could activate their voltage-dependent calcium channels. A possibility is that type-2 astrocytes may synthesize GABA (Barres et al 1990); thus glutamate depolarization of their processes may induce the release of GABA onto nodes or paranodes.

## Differences Between Classical and Modern Studies

There is now overwhelming evidence that all vertebrate glial cells express voltage-dependent ion channels (references in Table 1). These observations are in conspicuous conflict with a large body of previous work that had found glial cells to have passive membrane properties. For instance, Kuffler et al (1966) reported that all glial cells in *Necturus* optic nerve had passive membrane properties, whereas we have reported that mammalian optic nerve glia are not passive, but that each cell type expresses its own distinct phenotype (Barres et al 1988, 1989a,b, 1990).

What is the basis for this difference? Early experimenters used invertebrates and lower vertebrates, because these animals have large glial cells that allow penetration by one or more microelectrodes for voltage-recording studies. Recent studies have tended to use mammalian cells instead, as patch-clamp recording has permitted stable recording from smaller cells. Thus the difference might depend on the recording method, or perhaps glial cells in invertebrates and lower vertebrates have different electrophysiological properties.

To resolve this issue, we recorded from isolated glial cells from *Necturus* optic nerve, using the same techniques and solutions that we have used to study glia in the rat optic nerve (Barres et al 1989c). Figure 1a demonstrates the highly process-bearing nature of the acutely isolated *Necturus* astrocytes. With patch-clamp recording, all cells recorded from had nonlinear

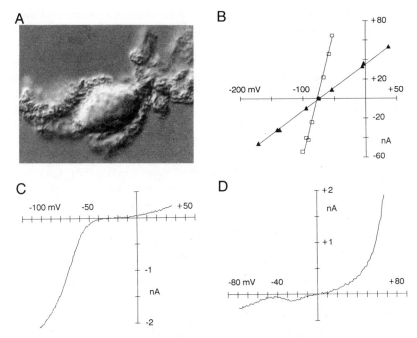

*Figure 1* Electrophysiological properties of *Necturus* optic nerve glia. (*a*) Nomarski micrograph of an astrocyte enzymatically isolated from *Necturus* optic nerve with the tissue print technique (see text). (*b*) I-V relation replotted from the data of Kuffler et al (1966). These data were obtained from microelectrode recordings of *Necturus* optic nerve glia in situ and were replotted for ease of comparison with the patch clamp data. (*c,d*) Whole-cell patch clamp records from two different acutely isolated *Necturus* optic nerve astrocytes in tissue prints, both morphologically similar to that shown in (*a*). Although all cells had a radial process-bearing morphology, all cells fell into one of two categories: those with inwardly rectifying potassium currents (*c*), and those with mainly sustained outwardly rectifying potassium currents and small transient inward currents (*d*). These currents were both voltage- and time-dependent (data not shown). The bath solution contained (in mM): KCl 20, NaCl 100, CaCl2 4, Hepes 5, Dextrose 3, pH 7.2. The pipette solution contained (in mM): KCl 120, MgCl2 1, Ca buffered to $10^{-7}$ M, EGTA 2, Hepes 5, pH 7.2. Linear leakage currents have been subtracted; however, input resistances were at least 100 times greater than those recorded with microelectrodes in situ (*b*, and Kuffler et al 1966). Note the difference in the current axes.

I(V) relationships (Figure 1*c,d*). These currents were both voltage- and time-dependent (see Barres et al 1989c for further description). Although *Necturus* optic nerve is unmyelinated, containing only astrocytes and not oligodendrocytes, we observed that half of the cells had properties similar to those of rat oligodendrocytes (Figure 1*c*), while the other half were similar to rat astrocytes in optic nerve (Figure 1*d*).

Thus we conclude that the differing results between earlier and more recent glial electrophysiological studies cannot be attributed to species differences. Instead it is likely that extensive glial coupling in situ interfered with electrophysiological measurements (see Barres et al 1989c for consideration of other, less likely, possibilities). A glial syncytium has a low input resistance, and limits the depolarization of the glial membrane. Typical glial input resistances in situ are about 10 megohms or less, whereas glial cells in culture or after acute isolation often have input resistances several orders of magnitude greater. Few investigators of glial cells in situ have achieved large depolarizations or hyperpolarizations (e.g. Trachtenberg & Pollen 1970); the largest have been reported by Kuffler & Potter (1964), who observed both inward and outward rectification with deviations from rest of greater than 50 mV. (A limited degree of depolarization may also explain the linear I-V relations measured with microelectrodes in astrocytes in culture: e.g. Ransom et al 1977, Moonen et al 1980.) No investigators studied the $I(V)$ relation of glial cells in situ in the presence of elevated extracellular potassium; it would be surprising if these conditions do not reveal a prominent inward rectification in many glial cells.

Thus simple explanations based on limitations of microelectrode recordings of glia in situ can plausibly account for the previous inability to detect these voltage-dependent ion channels. It should be interesting to return to in situ preparations, equipped with new understanding of glial membrane properties derived from patch recordings.

## Role of Glial Ion Channels in Potassium Regulation

What, if any, are the functions of ion channels in glia? A role in potassium homeostatic mechanisms has most frequently been suggested, because glial cells probably regulate extracellular potassium levels. How good is this evidence? There are only a few compelling studies, but these are exceptionally elegant. The experiments that most strongly suggest a role for glia in potassium regulation are those of Coles of Tsacapoulos (1979; also see Coles & Orkand 1983 and Coles et al 1989) on invertebrate retinal glia in situ and those of Ballanyi et al (1987) on mammalian glia in olfactory cortical slices. Ion concentrations were recorded during neural activity with ion-selective microelectrodes positioned intracellularly in neurons and glia and in the extracellular space. Both groups demonstrated that elevation of intracellular potassium in glia (and depression in neurons) occurs during neuronal firing. Their experiments also indicate that these potassium homeostatic mechanisms involve passive fluxes of ions (see below), most likely through channels, instead of active transport of ions by pumps or extracellular diffusion (Walz & Hertz 1983, Sykova 1983, Gardner-Medwin 1980, 1983a,b, Gardner-Medwin & Nicholson 1983).

SPATIAL BUFFERING    Two opposing regulatory mechanisms involving passive flux of potassium through glial channels have been suggested: "spatial buffering" and potassium accumulation. According to a spatial buffering mechanism, potassium would enter glia wherever the local potassium reversal potential was more positive than the resting potential. Potassium would be rapidly shunted by current flow from a proximal region of excess to a more distal region (Orkand et al 1966) driven primarily by a voltage gradient within the cell. Such a mechanism is widely thought to operate in the context of a strongly coupled glial syncytium (Orkand et al 1966, Gardner-Medwin 1983a,b). In hypothesizing a role for glial potassium channels in a spatial buffering mechanism, it should be recalled that except for inwardly rectifying potassium channels, most of the glial potassium channels are not activated by the degree of depolarization that would occur by the slight elevations of potassium that occurs with neuronal activity. Another difficulty with the spatial buffer mechanism in the context of a glial syncytium is that glial cell processes may be significantly longer than their own space constant (discussed by Gardner-Medwin 1983a,b, 1986).

A special kind of spatial buffering mechanism, termed potassium siphoning, has been proposed by Newman et al (1984). This hypothesis suggests that where the elevated extracellular potassium is shunted is determined by the distribution of the potassium conductance along the glial cell. Thus Newman et al (1984) have proposed that in the retina the potassium will enter Muller cells in the plexiform layers and exit from the endfeet, where the potassium conductance is greatest, into the vitreous. Because the conductances in both of these regions are inwardly rectifying potassium channels that are active at and near the resting potential, the possible involvement of voltage-dependent potassium channels in this mechanism appears plausible and is supported by new evidence. Light-evoked increases of extracellular potassium occurred in the vitreous, at the site of Muller endfeet, as predicted by the siphoning hypothesis, and these increases were entirely abolished by barium (Karwoski et al 1989).

POTASSIUM ACCUMULATION    In a spatial buffering mechanism, only a potassium permeability is required, and potassium entry is exactly balanced by potassium exit. In contrast, potassium can accumulate in glia if anion flux occurs to balance the charge of entering potassium; water would also enter to maintain osmolarity. Because of the presence of chloride channels in glia, potassium accumulation mechanisms have been proposed in which passive entry of potassium, chloride, and water could occur (as described for muscle fibers by Boyle & Conway 1941). Potassium accumulation mechanisms have the virtue of allowing local storage of the accumulated potassium for later return to the neuron. They also would remain operative in the face of spatially widespread neuronal activity, whereas such activity would tend to make spatial buffering ineffective (see Gardner-Medwin 1980 for discussion).

Bevan et al (1985) proposed that K and Cl influx into astrocytes could occur through their voltage-dependent chloride and potassium channels. The astrocyte channels they studied, however, are activated by depolarizations above $-40$ mV, a degree of depolarization that would not occur by physiological elevations in extracellular potassium (although, as they suggest, this could occur in pathophysiological conditions). More likely is potassium accumulation by oligodendrocytes through their inwardly rectifying potassium channels with compensatory chloride influx through "modulated" chloride channels. Although the inwardly-rectifying potassium channel is active at the resting potential and underlies the resting conductance of the oligodendrocyte, the chloride channels normally exist in an inhibited state. This "modulated Boyle and Conway" hypothesis proposes that neuronal activity triggers a signal, emanating from the axon, that activates the oligodendrocyte chloride channels (Barres et al 1988). An interesting aspect of the hypothesis is that the strong inwardly rectifying nature of the potassium channel would resist the rapid, potentially catastrophic release of the accumulated potassium chloride back into the extracellular space after termination of neuronal activity (as demonstrated in muscle by Hodgkin & Horowicz 1959). The return of chloride channel inhibition with the termination of neural activity would have the same effect. Although this hypothesis was originally proposed for oligodendrocytes at their paranodal regions, it is now clear that it could also apply to type-2 and type-1 astrocytes processes at the node, since they have similar inwardly rectifying potassium and chloride conductances (Barres et al 1990).

The occurrence of glial potassium accumulation in vivo is strongly supported by experimental evidence. The elevation of potassium in invertebrate and mammalian glia is accompanied by a chloride elevation (Coles et al 1989, Ballanyi et al 1987). Chloride-free solutions severely impair extracellular potassium regulation and decrease potassium uptake by glia in drone retina (Coles et al 1989). In addition, the elevation of potassium in mammalian glia in olfactory cortical slices is largely blocked by barium (Ballanyi et al 1987, Grafe & Ballanyi 1987), thus suggesting the involvement of an inwardly rectifying potassium channel active at the glial membrane potentials of their experiments, $-60$ to $-80$ mV.

Thus it is likely that glial cells regulate extracellular potassium, and that it is largely mediated by passive flux of potassium, and additionally chloride in some cases, through ion channels. Currently the evidence most strongly supports potassium accumulation mechanisms, except in the retina, where a spatial buffer process is likely to be operative (a concurrent accumulation mechanism has not been ruled out there). Possibly, glial cells in different parts of the nervous system, such as retina, parenchymal grey matter, and white matter, use different potassium regulatory mechanisms.

POSSIBLE NEUROTRANSMITTER INVOLVEMENT IN POTASSIUM REGULATION The modulated Boyle and Conway potassium accumulation mechanism suggests a modulation of a specific voltage-dependent chloride channel observed in oligodendrocytes as one possible way that neurotransmitters could be involved. A ligand-gated chloride channel could also work as well: GABA activated chloride channels are found in cultured astrocytes (Borman & Kettenmann 1988). A similar role for GABA-activated Cl channels in Muller cells has been suggested (Malchow et al 1989). On the other hand, chloride entry could tend to inhibit a spatial buffer mechanism by limiting the degree of depolarization induced by influx of potassium.

An important issue for these is the equilibrium potential for chloride in glial cells in situ. Because glial cells in culture appear to accumulate chloride actively and thus have chloride equilibrium potentials depolarized to the resting potential, it has been assumed that glial cells in situ do, too. Recent measurements of intracellular chloride in mammalian astrocytes in guinea pig olfactory slices indicate that chloride is probably passively distributed, so that reversal is near the resting potential (Ballanyi et al 1987).

## Other Possible Functions of Glial Ion Channels

EXCITABILITY    Glial cells do not generate robust action potentials. Under certain experimental conditions, regenerative behavior can be observed in glial cells (calcium regenerative potentials: Newman 1985, MacVicar 1984; sodium regenerative potentials: Barres et al 1988, 1990, Bevan et al 1987). In general, these require the injection of relatively large amounts of depolarizing current and is limited to a single action potential overshooting 0 mV but with incomplete repolarization. Glial excitability in vivo is not yet completely excluded, particularly since there has been little electrophysiological study of adult glial cells.

GLIA TO NEURON TRANSFER OF CHANNELS    It has been hypothesized that glial cells act as local channel synthetic factories for neurons (Chiu et al 1984, Bevan et al 1985). Sodium channels could be transferred to axons from Schwann cell "fingers" for astrocyte processes contacting nodes of Ranvier. Sodium channels in Schwann cells have a short lifetime (3 days); if the same lifetime holds for axonal channels, this would create a large metabolic load for neurons that might be assumed by Schwann cells (Ritchie 1988). It was also argued that the astrocyte and Schwann cell resting potential were so depolarized ($-40$ mV) that sodium channels in glia would be nonfunctional because they would be inactivated (Chiu et al 1984, Bevan et al 1985). As discussed above, however, these values are probably artifactually depolarized.

Immuno-electron microscopic localization of sodium channels in adult rat optic nerve has recently been accomplished by using a new antisera to rat-brain sodium channels. In addition to a high density of labeling at nodes, some but not all perinodal astrocyte processes exhibited an intense immunoreactivity with the antisera (Black et al 1989a). Moreover, staining of a subset of astrocytes with predominantly longitudinal processes was detected; this staining was mainly cytoplasmic, although membrane staining was also present (Black et al 1989a,b, staining of nodal axoplasm also occurred). These findings were interpreted as consistent with the transfer hypothesis (either from astrocyte to node or in the other direction).

Sodium channels are not only found at the node but in the internode, and the total number of internodal channels far exceeds the total number of nodal channels (Ritchie & Rogart 1977, Chiu & Schwarz 1987, Shrager 1987, 1989). As new nodes form in demyelinated sciatic nerves, Schwann cells bridge the immature nodal gaps; sodium currents were not found in these Schwann cells, as might have been expected if they were an important source of new nodal channels (Shrager 1989). Finally, the properties of glial and neuronal channels appear significantly different (Chiu et al 1984, Barres et al 1989b). There is little convincing support for the transfer hypothesis.

MAINTENANCE OF RESTING POTENTIAL    "Leakage" channels, permeable only to potassium but not voltage-dependent, are classically thought to underlie the resting potential. So far, no channels in glia have been found to have these properties. Recently, potassium-selective stretch-activated channels have been observed in astrocytes in culture (Ding et al 1988, 1989); such channels have been suggested to contribute to resting potential in other cell types (Medina & Bregestovski 1988).

Inwardly rectifying potassium channels form nearly the entire resting conductance of oligodendrocytes and Muller cells (Barres et al 1988, 1989e; E. A. Newman, personal communication). Recent experiments have demonstrated that these channels are also found in type-2 astrocytes, type-1 astrocytes, O2A progenitors (Barres et al 1990 and in preparation), and Schwann cells (Wilson & Chiu 1989). It now seems very likely that the inwardly rectifying potassium channel underlies the resting potential in all glial cell types.

DEVELOPMENT    Because glial progenitor cells express ion channel types differing from those in their descendants, and are excitable, it has been suggested that these may play a role in progenitor cell function (Bevan et al 1987, Barres et al 1989d, Sontheimer et al 1989). This possibility is further suggested by the recent finding that these cells synthesize GABA (Barres et al 1990). It is also possible that glutamate-activated channels on these cells may be involved in signal detection, cell migration, or process outgrowth during development, as glutamate and other transmitters influence process outgrowth in developing neurons (Mattson et al 1988).

SECRETION    Although glial cells can release preloaded, radiolabeled neurotransmitters (e.g. Shain et al 1986, Minchen & Iverson 1974, Gallo et al 1986), there is yet little evidence to support the role of ion channels, such as calcium channels in secretion. Nor is there evidence that such neurotransmitter release would occur without experimental preloading. However, glutamate can induce the release of preloaded $^3$H-GABA from cerebellar O2A progenitor cells and type-2 astrocytes in culture (Gallo et al 1986), and O2A progenitors (and possibly type-2 astrocytes) synthesize GABA (Barres et al 1990). Thus it is possible that GABA release may occur without preloading.

Glial cells contain electrogenic cotransport mechanisms for many neurotransmitters; these carriers mediate voltage-dependent uptake of transmitters but may also be capable of mediating efflux. Because these transporters can support a large flux of transmitter, they can effectively control the amino acid concentration in a restricted extracellular space (Schwartz & Tachibana 1990). Moreover, calculations have shown that the extracellular concentration of a transmitter carried by an electrogenic carrier mechanism is a steep function of membrane potential, regardless of whether the transporter can mediate efflux (Schwartz & Tachibana 1990). Thus changes of glial membrane potential, mediated by ion channels, may influence extracellular neurotransmitter concentrations.

## Are Glia in vitro Good Models of Glia in vivo?

Some properties of glial cells in vitro are different in vivo; a well-known example is that the polygonal shape of type-1-like astrocytes in neuron-free cultures is never observed in vivo, where they are instead process-bearing. Glial ion channel expression can also be altered in culture.

There are several examples of serum effects on glial ion channel expression in culture. Type-2 astrocytes cultured in serum-free medium express a vastly different density of calcium current compared to that found in serum-containing cultures (Barres et al 1990). Moreover, both a charybdotoxin-sensitive component of outward current and the glial form of the sodium current found in type-2 astrocytes in serum-containing cultures are not present in serum-free cultures. A more dramatic effect is the permissive effect of different lots of fetal-calf serum on the ability of for-skolin to induce calcium channels in cortical astrocytes (Barres et al 1989a).

Such influences of serum on glial ion channel expression probably represent examples of "modulated differentiation," since the change of serum does not alter cell morphology or its basic identifying antigenic phenotype. This sensitivity of ion channel expression to culture conditions is not unique to glia: Neurons are similarly sensitive (Bossu et al 1988).

Because of the possibility that culture conditions may affect glial behavior, many investigators have considered alternative preparations. Slices of CNS tissue or in situ recordings have been used for electro-physiological studies of glia (e.g. Walz & MacVicar 1988), but the problem of cell-type identifications is severe. Many investigators have preferred the use of acutely dissociated cells. Channels present in acutely dissociated cells are probably also present in vivo; Newman (1985b) has observed similar voltage-dependent ion channel types in acutely isolated Muller cells and in Muller cells in retinal slices. On the other hand, channels not observed in acutely isolated cells may still be present in vivo. Although there is little evidence that extracellular exposure to enzymes alters channel properties (and much evidence against it), acutely dissociated glial cells generally lack processes. Thus if glial channel types are selectively localized to processes, as has been demonstrated for salamander Muller cells (Newman 1984, 1986b), channel phenotype of acutely dissociated cells will differ from the in vivo phenotype.

The comparison of in vitro and in vivo phenotype has been directly made for glial cells in the optic nerve (Barres et al 1987, 1989d, 1990, and in preparation). The ion channel phenotypes in each of the three optic nerve glial cell types—type-1 astrocytes, type-2 astrocytes, and oligo-dendrocytes—have been studied in four preparations progressively approximating the condition in vivo. These preparations are (a) culture in serum-containing medium, (b) culture in serum-free medium, (c) acutely dissociated cells, and (d) acutely isolated cells in tissue prints. In the last technique, optic nerve (or other neural tissue) is briefly treated with papain and then gently pressed to a sticky, poly-l-lysine-coated, glass surface (Figure 1a). A thin layer of tissue remains adherent; these cells still have processes and can be labeled with antibodies and electrophysiologically studied. Because papain does not destroy any basic ion channel types, and these cells still have processes and have not been cultured, their properties may closely reflect those in vivo.

Ion-channel expression (either channel types or their densities) in each cell type was often different in each of the four conditions. These differences were not major in the case of type-2 astrocytes and oligodendrocytes, but were most marked in type-1 astrocytes. Type-1 astrocytes expressed the same basic types of ion channels in serum-free and serum-containing cultures; however, this phenotype was very different from that of acutely isolated type-1 astrocytes, even those bearing processes. For instance, sodium channels were found in 100% of acutely isolated optic nerve type-1 astrocytes with processes, but only in a fraction of type-1 astrocytes lacking their processes (B. A. Barres, in preparation). Recent experiments indicate that in co-cultures of neurons with type-1 astrocytes, sodium channels are expressed in most type-1 astrocytes (B. A. Barres et al, in preparation). If the other ion channel differences can also be accounted for by the absence of neurons, such co-culture conditions may facilitate in vitro studies of glial function.

## CONCLUSION

Vertebrate glia comprise several cell types that are structurally, developmentally, and biochemically distinct. Experiments summarized here have demonstrated they are also electrophysiologically distinct. Although earlier studies have suggested that glial cells are "passive," more recent studies demonstrate a more dynamic view: They can express a large diversity of voltage-dependent ion channels and neurotransmitter-gated channels, these channels may be modulated by neuronal signals, and these properties occur in vivo as well as in vitro. A broad caveat from such studies is that glial cells in vitro are not necessarily good models of glial cells in vivo, at least as concerns electrophysiological properties.

Despite the diverse functionality suggested by these channels, we are not much closer to understanding the role of glia in the nervous system. The best-documented function continues to be that of potassium homeostasis; here channel studies have at least suggested more specific models for potassium regulation by glia. There is suggestive but as yet no good evidence that glia participate in information processing or signaling in the nervous system. Nonetheless, electrophysiological studies of glia have blurred the distinction between neurons and glia and encourage us to consider new hypotheses of glial function.

*Literature Cited*

Abbott, N. J., Hassan, S., Lieberman, E. M. 1988. Evidence for glutamate as the mediator of axon-Schwann cell interactions in the isolated giant axon of the squid. *J. Physiol.* 398: 63P

Ascher, P., Nowak, L. 1988. The role of divalent cations in the NMDA responses of mouse central neurones in culture. *J. Physiol.* 399: 247–66

Backus, K. H., Kettenmann, H., Schachner, M. 1988. Effect of benzodiazepines and pentobarbital on the GABA-induced depolarization in cultured astrocytes. *Glia* 1: 132–40

Baker, P. F., Potashner, S. J. 1971. The dependence of glutamate uptake by crab nerve on external Na and K. *Biochim. Biophys. Acta* 249: 616–22

Baker, P. F., Potashner, S. J. 1973. Glutamate transport in invertebrate nerve: The relative importance of ions and metabolic energy. *J. Physiol.* 232: 26P–27P

Ballanyi, K., Grafe, P., Ten Bruggencate, G. 1987. Ion activities and potassium uptake mechanisms of glial cells in guinea-pig olfactory cortex slices. *J. Physiol.* 382: 159–74

Ballanyi, K., Schlue, W. R. 1988. Direct effects of carbachol on membrane potential and ion activities in leech glial cells. *Glia* 1: 165–67

Barres, B. A. 1989. A new form of neurotransmission? *Nature* 339: 343–44

Barres, B. A., Chun, L. L. Y., Corey, D. P. 1985. Voltage-dependent ion channels in glial cells. *Soc. Neurosci. Abstr.* 11: 147

Barres, B. A., Silverstein, B. E., Chun, L. L. Y., Corey, D. P. 1987. Ion channel phenotype of three glial cell types: A comparison of in vivo and in vitro channel expression. *Neuroscience* 22(Suppl.): 2068P

Barres, B. A., Chun, L. L. Y., Corey, D. P. 1988a. Ion channel expression by white matter glia: I. Type 2 astrocytes and oligodendrocytes. *Glia* 1: 10–30

Barres, B. A., Chun, L. L. Y., Corey, D. P. 1989a. A calcium current in cortical astrocytes: Induction by cAMP and neurotransmitters, and permissive effect of serum factors. *J. Neurosci.* 9: 3169–75

Barres, B. A., Chun, L. L. Y., Corey, D. P. 1989b. Glial and neuronal forms of the voltage-dependent sodium channel: Characteristics and cell-type distribution. *Neuron* 2: 1375–88

Barres, B. A., Chun, L. L. Y., Corey, D. P. 1989c. Reassessment of membrane properties in a classic glial preparation. *Proc. 12th Int. Soc. Neurochem.: Differentiation and Functions of Glial Cells.* New York: Liss. In press

Barres, B. A., Chun, L. L. Y., Corey, D. P. 1989d. Ion channel expression by white matter glia: A new approach to determination of in vivo phenotype. See Barres et al 1989c

Barres, B. A., Koroshetz, W. J., Chun, L. L. Y., Schwartz, K. J., Corey, D. P. 1990. Ion channel expression of white matter glia. II. The O2A glial progenitor cell. *Neuron.* In press

Barres, B. A., Chun, L. L. Y., Corey, D. P. 1989f. Further evidence for the existence

of type-2 astrocytes in vivo. *Soc. Neurosci. Abstr.* 15: 14

Bevan, S., Chiu, S. Y., Gray, P. T. A., Ritchie, J. M. 1985. The presence of voltage-gated sodium, potassium and chloride channels in rat cultured astrocytes. *Proc. R. Soc. London Ser. B* 225: 299–313

Bevan, S., Gray, P. T., Ritchie, J. M. 1984. A calcium-activated cation-selective channel in rat cultured Schwann cells. *Proc. R. Soc. London Ser. B* 222: 349–55

Bevan, S., Lindsay, R. M., Perkins, M. N., Raff, M. C. 1987. Voltage-gated ionic channels in rat cultured astrocytes, reactive astrocytes and an astrocyte-oligodendrocyte progenitor cell. *J. Physiol.* 82: 327–35

Bevan, S., Raff, M. 1985. Voltage-dependent potassium currents in cultured astrocytes. *Nature* 315: 229–32

Black, J. A., Friedman, B., Waxman, S. G., Elmer, L. W., Angelides, K. J. 1989a. Immuno-ultrastructural localization of sodium channels at nodes of Ranvier and perinodal astrocytes in rat optic nerve. *Proc. R. Soc. London Ser. B.* In press

Black, J. A., Waxman, S. G., Friedman, B., Elmer, L. W., Angelides, K. J. 1989b. Sodium channels in astrocytes of rat optic nerve in situ: Immuno-electron microscopic studies. *Glia* 2: 353–69

Bormann, J., Kettenmann, H. 1988. Patch-clamp study of GABA receptor Cl channels in cultured astrocytes. *Proc. Natl. Acad. Sci. USA* 85: 9336–40

Bossu, J. L., Dupont, J. L., Feltz, A. 1988. Potassium currents in rat cerebellar Purkinje neurones maintained in culture in L15 (Leibovitz) medium. *Neurosci. Lett.* 89: 55–62

Boyle, P. J., Conway, E. J. 1941. Potassium accumulation in muscle and associated changes. *J. Physiol.* 100: 1–63

Bray, G. M., Rasminsky, M., Agauyo, A. J. 1981. Interactions between axons and their sheath cells. *Annu. Rev. Neurosci.* 4: 127–62

Brew, H., Attwell, D. 1987. Electrogenic glutamate uptake is a major current carrier in the membrane of axolotl retinal glial cells. *Nature* 327: 707–9

Brew, H., Gray, P. T. A., Mobbs, P., Attwell, D. 1986. Endfeet of retinal glial cells have higher densities of ion channels that mediate K buffering. *Nature* 324: 466–68

Casalotti, S. O., Stephenson, F. A., Barnard, E. A. 1987. Separate subunits for agonist and benzodiazepine binding in the GABA-A receptor oligomer. *J. Biol. Chem.* 261: 15013–16

Chiu, S. Y. 1987. Sodium currents in axon-associated Schwann cells from adult rabbits. *J. Physiol.* 386: 181–203

Chiu, S. Y. 1988. Changes in excitable membrane properties in Schwann cells of adult rabbit sciatic nerves following nerve transection. *J. Physiol.* 396: 173–88

Chiu, S. Y., Schrager, P., Ritchie, J. M. 1984. Neuronal-type sodium and potassium channels in rabbit cultured Schwann cells. *Nature* 311: 156–57

Chiu, S. Y., Schwarz, W. 1987. Sodium and potassium currents in acutely demyelinated internodes of rabbit sciatic nerves. *J. Physiol.* 391: 631–49

Coles, J. A., Orkand, R. K. 1983. Modification of potassium movement through the retina of the drone (*apis mellifera*) by glial uptake. *J. Physiol.* 340: 157–74

Coles, J. A., Orkand, R. K., Yamate, C. L. 1989. Chloride enters glial cells and photoreceptors in response to light stimulation in the retina of the honey bee drone. *Glia.* 2: 287–97

Coles, J. A., Tsacopoulos, M. 1979. Potassium activity in photoreceptors, glial cells and extracellular space in the drone retina: Changes during photostimulation. *J. Physiol.* 290: 525–49

Connors, B. W., Ransom, B. R. 1987. Electrophysiological properties of ependymal cells (radial glia) in dorsal cortex of the turtle, *Pseudemys Scripta. J. Physiol.* 385: 287–306

Cull-Candy, S. G., Howe, J. R., Ogden, D. C. 1988. Noise and single channels activated by excitatory amino acids in rat cerebellar granule neurones. *J. Physiol.* 400: 189–222

Cull-Candy, S. G., Mathie, A., Symonds, C. J., Wyllie, F. 1989. Distribution of quisqualate and kainate receptors in rat type-2 astrocytes and their progenitors in culture. *J. Physiol.* 419: 204P

Cull-Candy, S. G., Usowicz, M. M. 1987. Multiple-conductance channels activated by excitatory amino acids in cerebellar neurons. *Nature* 325: 525–28

Dennis, M. J., Gerschenfeld, H. M. 1969. Some physiological properties of identified mammalian neuroglial cells. *J. Physiol.* 203: 211–22

Ding, J. P., Bowman, C. L., Sokabe, M., Sachs, F. 1989. Mechanical transduction in glial cells: Sacs and sics. *Biophys. J.* 55: 244a

Ding, J. P., Yang, X. C., Bowman, C. L., Sachs, F. 1988. A stretch-activated ion channel in rat astrocytes in primary cell culture. *Soc. Neurosci. Abstr.* 14: 425.5

ffrench-Constant, C., Miller, R. H., Kruse, J., Schachner, M., Raff, M. C. 1986. Molecular specialization of astrocyte processes at nodes of Ranvier in rat optic nerve. *J. Cell Biol.* 102: 844–52

ffrench-Constant, C., Raff, M. C. 1986. The oligodendrocyte-type 2 astrocyte cell lineage is specialized for myelination. *Nature* 323: 335–38

Frizzel, R. A., Rechkemmer, G., Shoemaker, R. L. 1986. Altered regulation of airway epithelial cell chloride channels in cystic fibrosis. *Science* 233: 558–60

Gallo, V., Giovannini, C., Suergiu, R., Levi, G. 1989. Expression of excitatory amino acid receptors by cerebellar cells of the type-2 astrocyte cell lineage. *J. Neurochem.* 52: 1–9

Gallo, V., Suergiu, R., Levi, G. 1986. Kainic acid stimulates GABA release from a subpopulation of cerebellar astrocytes. *Eur. J. Pharmacol.* 132: 319–22

Gardner-Medwin, A. R. 1980. Membrane transport and solute migration affecting the brain cell microenvironment. *Neurosci. Res. Prog. Bull.* 18: 208–26

Gardner-Medwin, A. R. 1983a. A study of the mechanisms by which potassium moves through brain tissue in rat. *J. Physiol.* 335: 353–74

Gardner-Medwin, A. R. 1983b. Analysis of potassium dynamics in mammalian brain tissue. *J. Physiol.* 335: 393–426

Gardner-Medwin, A. R. 1986. A new framework for assessment of potassium-buffering mechanisms. *Ann. NY Acad. Sci.* 481: 287–302

Gardner-Medwin, A. R., Nicholson, C. 1983. Changes of extracellular potassium activity induced by electric current through brain tissue in the rat. *J. Physiol.* 335: 375–92

Grafe, P., Ballanyi, K. 1987. Cellular mechanism of potassium homeostasis in the mammalian nervous system. *Can. J. Physiol. Pharmacol.* 65: 1038–42

Gray, P. T., Bevan, S., Ritchie, J. M. 1984. High conductance anion-selective channels in rat cultured Schwann cells. *Proc. R. Soc. London Ser. B* 221: 395–409

Gray, P. T. A., Ritchie, J. M. 1986. A voltage-gated chloride conductance in rat cultured astrocytes. *Proc. R. Soc. London Ser. B* 228: 267–88

Hodgkin, A. L., Horowicz, P. 1959. The influence of potassium and chloride ions on the membrane potential of single muscle fibers. *J. Physiol.* 148: 127–60

Howe, J. R., Ritchie, J. M. 1988. Two types of potassium current in rabbit cultured Schwann cells. *Proc. R. Soc. London Ser. B* 235: 19–27

Jahr, C. E., Stevens, C. F. 1987. Glutamate activates multiple single channel conductances in hippocampal neurons. *Nature* 325: 522–25

Jessen, K. R., Mirsky, R. 1984. Nonmyelin-forming Schwann cells coexpress surface proteins and intermediate filaments not found in myelin-forming cells: A study of Ran-2, A5E3 antigen and glial fibrillary acidic protein. *J. Neurocytol.* 13: 923–34

Jessen, K. R., Morgan, L., Mirksy, R. 1989. Schwann cell precursors and their development. *J. Neurochem.* In press (Abstr.)

Karwoski, C. J., Lu, H. K., Newman, E. A. 1989. Spatial buffering of light-evoked potassium increases by retinal Muller (glial) cells. *Science* 244: 578–80

Kettenmann, H., Orkand, R. K., Lux, H. D. 1984a. Some properties of single potassium channels in cultured oligodendrocytes. *Pflugers Arch.* 400: 215–21

Kettenmann, H., Orkand, R. K., Lux, H. D., Schachner, M. 1982. Single potassium channel currents in cultured mouse oligodendrocytes. *Neurosci. Lett.* 32: 41–46

Kettenmann, H., Schachner, M. 1985. Pharmacological properties of GABA-, glutamate-, and aspartate-induced depolarizations in cultured astrocytes. *J. Neurosci.* 5: 3295–3301

Kettenmann, H., Sonhof, U., Schachner, M. 1983. Exclusive potassium dependence of the membrane potential in cultured mouse oligodendrocytes. *J. Neurosci.* 3: 500–5

Kettenmann, H., Sonhof, U., Camerer, H., Kuhlmann, S., Orkand, R. K., Schachner, M. 1984b. Electrical properties of oligodendrocytes in culture. *Pflugers Arch.* 400: R43

Kuffler, S. W. 1967. The Ferrier Lecture. Neuroglial cells: Physiological properties and a potassium mediated effect of neuronal activity on the glial membrane potential. *Proc. R. Soc. London Ser. B* 168: 1–21

Kuffler, S. W., Nicholls, J. G., Martin, A. R. 1984. *Fron Neuron to Brain: A Cellular Approach to the Function of the Nervous System.* Sunderland, Mass: Sinauer Assoc.

Kuffler, S. W., Nicholls, J. G., Orkand, R. K. 1966. Physiologic properties of glial cells in the central nervous system of amphibia. *J. Neurophys.* 29: 768–87

Kuffler, S. W., Potter, D. D. 1964. Glia in the leech central nervous system: Physiologic properties and neuron-glia relationship. *J. Neurophys.* 27: 290–320

Levi, G., Gallo, V., Ciotti, M. T. 1986. Bipotential precursors of putative fibrous astrocytes and oligodendrocytes in rat cerebellar cultures express distinct surface features and "neuron-like" GABA transport. *Proc. Natl. Acad. Sci. USA* 83: 1504–8

Lieberman, E. M., Abbott, N. J., Hassan, S. 1989. Evidence that glutamate mediates axon to Schwann cell signalling in the squid. *Glia* 2: 94–102

Lillien, L., Raff, M. C. 1990. Analysis of the cell-cell interactions that control type-2 astrocyte development in vitro. *Neuron.* Submitted

Lillien, L. E., Sendtner, M., Rohrer, H., Hughes, S. M., Raff, M. C. 1988. Type-2 astrocyte development in rat brain cultures is initiated by a CNTF-like protein produced by type 1 astrocytes. *Neuron* 1: 485–94

Liuzzi, F. J., Miller, R. H. 1987. Radially oriented astrocytes in the normal adult rat spinal cord. *Brain Res.* 403: 385–88

Llano, I., Marty, A., Johnson, J. W., Ascher, P., Gahwiler, B. H. 1988. Patch-clamp recording of amino acid-activated responses in "organotypic" slice cultures. *Proc. Natl. Acad. Sci. USA* 85: 3221–25

MacVicar, B. A. 1984. Voltage-dependent calcium channels in glial cells. *Science* 226: 1345–47

MacVicar, B. A., Tse, F. W. 1988. Norepinephrine and cAMP enhance a nifedipine-sensitive calcium current in cultured rat astrocytes. *Glia* 1: 359–65

Malchow, R. P., Qian, H., Ripps, H. 1989. GABA-induced currents of skate Muller (glial) cells are mediated by neuronal-like GABA-A receptors. *Proc. Natl. Acad. Sci. USA* 86: 4326–30

Mattson, M. P., Lee, R. E., Adams, M. E., Guthrie, P. B., Kater, S. B. 1988. Interactions between entorhinal axons and target hippocampal neurons: A role for glutamate in the development of hippocampal circuitry. *Neuron* 1: 865–76

McCarthy, K. D., deVellis, J. 1980. Preparation of separate astroglial and oligodendroglial cell cultures from rat cerebral tissue. *J. Cell Biol.* 85: 890–902

Medina, I. R., Bregestovski, P. D. 1988. Stretch-activated ion channels modulate the resting membrane potential during early embryogenesis. *Proc. R. Soc. London Ser. B* 235: 95–102

Metuzals, J. 1965. Ultrastructure of the nodes of Ranvier and their surrounding structures in the central nervous system. *Z. Zellforsch.* 65: 719–59

Miller, R. H., ffrench-Constant, C., Raff, M. C. 1989. The macroglial cells of the rat optic nerve. *Annu. Rev. Neurosci.* 12: 517–34

Miller, R. H., Fulton, B. P., Raff, M. C. 1989b. A novel type of glial cell associated with nodes of Ranvier in rat optic nerve. *Eur. J. Neurosci.* 1: 172–80

Miller, R. H., Raff, M. C. 1984. Fibrous and protoplasmic astrocytes are biochemically and developmentally distinct. *J. Neurosci.* 4: 585–92

Minchen, M. C., Iverson, L. L. 1974. Release of 3H-GABA from glial cells in rat dorsal root ganglia. *J. Neurochem.* 23: 533–40

Mirsky, R., Jessen, K. R. 1984. A cell surface protein of astrocytes, Ran-2, distinguishes non-myelin-forming Schwann cells from myelin-forming Schwann cells. *Dev. Neurosci.* 6: 304–16

Mirsky, R., Jessen, K. R. 1986. The biology of non-myelin-forming Schwann cells. *Ann. NY Acad. Sci.* 486: 132–46

Mirsky, R., Jessen, K. R. 1987. Molecular properties of peripheral glia. In *Glial-Neuronal Communication in Development and Regeneration*, NATO ASI Ser. H, H. H. Althaus, W. Seifert, pp. 55–62. Berlin: Springer-Verlag

Mirsky, R., Jessen, K. R. 1988. Axonal control of Schwann cell differentiation. In *The Current Status of Peripheral Nerve Regeneration*, pp. 91–97. New York: Liss

Moonen, G., Franck, G., Schoffeniels, E. 1980. Glial control of neuronal excitability in mammals: I. Electrophysiological and isotopic evidence in culture. *Neurochem. Int.* 2: 299–310

Newman, E. A. 1984. Regional specialization of retinal glial cell membrane. *Nature* 309: 155–57

Newman, E. A. 1985a. Membrane physiology of retinal glial (Muller) cells. *J. Neurosci.* 5: 2225–39

Newman, E. A. 1985b. Voltage-dependent calcium and potassium channels in retinal glial cells. *Nature* 317: 809–11

Newman, E. A. 1986a. Regional specialization of the membrane of retinal glial cells and its importance to K spatial buffering. *Ann. NY Acad. Sci.* 481: 273–86

Newman, E. A. 1986b. High potassium conductance in astrocyte endfeet. *Science* 233: 453–54

Newman, E. A. 1987. Regulation of potassium levels by Muller cells in the vertebrate retina. *Can. J. Physiol. Pharmacol.* 65: 1028–32

Newman, E. A. 1989. Inward rectifying potassium channels in retinal glial (Muller) cells. *Soc. Neurosci. Abstr.* 15: 353

Newman, E. A., Frambach, D. A., Odette, L. L. 1984. Control of extracellular potassium levels by retinal glial cell K siphoning. *Science* 225: 1174–75

Nilius, B., Reichenbach, A. 1988. Efficient K buffering by mammalian retinal glial cells is due to cooperation of specialized ion channels. *Pflugers Arch.* 411: 654–60

Nowak, L., Ascher, P., Berwald-Netter, Y. 1987. Ionic channels in mouse astrocytes in culture. *J. Neurosci.* 7: 101–9

O'Malley, D. M., Masland, R. H. 1989. Corelease of acetylcholine and GABA by a retinal neuron. *Proc. Natl. Acad. Sci. USA* 86: 3414–18

Orkand, R. K., Nicholls, J. G., Kuffler, S.

W. 1966. Effect of nerve impulses on the membrane potential of glial cells in the central nervous system of amphibia. *J. Neurophys.* 29: 788–806

Penfield, W. 1932. Neuroglia: Normal and pathological. In *Cytology and Cellular Pathology of the Nervous System*, ed. W. Penfield, pp. 423–79. New York: Hoeber

Peters, A., Palay, S. L., Webster, H. F. 1976. *The Fine Structure of the Nervous System: The Neurons and Supporting Cells*, pp. 233–44. Philadelphia: Saunders

Pichon, Y., Abbott, N. J., Lieberman, E. M., Larmet, Y. 1987. Potassium homeostasis in the nervous system of cephalopods and crustacea. *J. Physiol.* 82: 346–56

Quandt, F. N., MacVicar, B. A. 1986. Calcium activated potassium channels in cultured astrocytes. *Neuroscience* 19: 29–41

Raff, M. C. 1989a. Subclasses of astrocytes in culture: What should we call them? See Barres et al 1989c

Raff, M. C. 1989b. Glial cell diversification in the rat optic nerve. *Science* 243: 1450–55

Raff, M. C., Abney, E. R., Cohen, J., Lindsay, R., Noble, M. 1983a. Two types of astrocytes in cultures of developing rat white matter: Differences in morphology, surface gangliosides, and growth characteristics. *J. Neurosci.* 3: 1289–1300

Raff, M. C., Abney, E. R., Miller, R. H. 1984a. Two glial cell lineages diverge prenatally in rat optic nerve. *Dev. Biol.* 106: 53–60

Raff, M. C., Miller, R. H., Noble, M. 1983b. A glial progenitor cell that develops in vitro into an astrocyte or an oligodendrocyte depending on culture medium. *Nature* 303: 390–96

Raff, M. C., Williams, B. P., Miller, R. H. 1984b. The in vitro differentiation of a bipotential glial progenitor cell. *EMBO J.* 3: 1857–64

Ramon y Cajal, S. 1909. *Histologie du Systeme Nerveux de l'homme et des Vertebres.* Madrid: Inst. Ramon y Cajal

Ransom, B. R., Neale, E., Henkart, M., Bullock, P. N., Nelson, P. G. 1977. Mouse spinal cord in cell culture. I. Morphology and intrinsic neuronal electrophysiological properties. *J. Neurophys.* 40: 1132–50

Ritchie, J. M. 1988. Sodium-channel turnover in rabbit cultured Schwann cells. *Proc. R. Soc. London Ser. B* 233: 423–30

Ritchie, J. M., Rogart, R. B. 1977. Density of sodium channels in mammalian myelinated nerve fibers and nature of the axonal membrane under the myelin sheath. *Proc. Natl. Acad. Sci. USA* 74: 211–15

Schwartz, E. A. 1987. Depolarization without calcium can release GABA from a retinal neuron. *Science* 238: 350–55

## 474    BARRES, CHUN & COREY

Schwartz, E. A., Tachibana, M. 1990. Electrophysiology of glutamate and sodium cotransport in a glial cell of the salamander retina. *J. Physiol.* Submitted

Shain, W., Madelian, V., Martin, D. L., Kimelberg, H. K., Perrone, M., Lepore, R. 1986. Activation of beta-adrenergic receptors stimulates release of an inhibitory transmitter from astrocytes. *J. Neurochem.* 46: 1298–1303

Shrager, P. 1987. The distribution of sodium and potassium channels in single demyelinated axons of the frog. *J. Physiol.* 392: 587–602

Shrager, P. 1989. Sodium channels in single demyelinated mammalian axons. *Brain Res.* 483: 149–54

Shrager, P., Chiu, S. Y., Ritchie, J. M. 1985. Voltage-dependent sodium and potassium channels in mammalian cultured Schwann cells. *Proc. Natl. Acad. Sci. USA* 82: 948–52

Soliven, B., Szuchet, S., Arnason, B. G., Nelson, D. J. 1988a. Forskolin and phorbol esters decrease the same potassium conductance in cultured oligodendrocytes. *J. Membr. Biol.* 105: 177–86

Soliven, B., Szuchet, S., Arnason, B. G., Nelson, D. J. 1988b. Voltage-gated potassium currents in cultured ovine oligodendrocytes. *J. Neurosci.* 8: 2131–41

Somjen, G. G. 1988. Nervenkitt: Notes on the history of the concept of neuroglia. *Glia* 1: 2–9

Somjen, G. G. 1975. Electrophysiology of neuroglia. *Annu. Rev. Physiol.* 37: 163–90

Sonnhof, U. 1987. Single voltage-dependent K and Cl channels in cultured rat astrocytes. *Can. J. Physiol. Pharmacol.* 65: 1043–50

Sontheimer, H., Kettenmann, H. 1988. Heterogeneity of potassium currents in cultured oligodendrocytes. *Glia* 1: 415–20

Sontheimer, H., Kettenmann, H., Backus, K. H., Schachner, M. 1988. Glutamate opens Na/K channels in cultured astrocytes. *Glia* 1: 328–36

Sontheimer, H., Trotter, J., Schachner, M., Kettenmann, H. 1989. Channel expression correlates with differentiation stage during the development of oligodendrocytes from their precursor cells in culture. *Neuron* 2: 1135–45

Suárez, I., Raff, M. C. 1989. Subpial and perivascular astrocytes associated with

nodes of Ranvier in the rat optic nerve. *J. Neurocytol.* In press

Sykova, A. 1983. Extracellular potassium accumulation in the central nervous system. *Prog. Biophys. Molec. Biol.* 42: 135–89

Temple, S., Raff, M. C. 1985. Differentiation of a bipotential glial progenitor cell in single cell microculture. *Nature* 313: 223–25

Trachtenberg, M. C., Pollen, D. A. 1970. Neuroglia: Biophysical properties and physiologic function. *Science* 167: 1248–53

Usowicz, M. M., Gallo, V., Cull-Candy, S. G. 1989. Multiple conductance channels in type-2 cerebellar astrocytes activated by excitatory amino acids. *Nature* 339: 380–83

Vartanian, T., Szuchet, S., Dawson, G., Campagnoni, A. T. 1986. Oligodendrocyte adhesion activates protein kinase C-mediated phosphorylation of myelin basic protein. *Science* 234: 1395–98

Villegas, J. 1975. Characterization of acetylcholine receptors in the Schwann cell membrane of the squid nerve fiber. *J. Physiol.* 249: 679

Walz, W., Hertz, L. 1983. Functional interactions between neurons and astrocytes. II. Potassium homeostasis at the cellular level. *Prog. Neurobiol.* 20: 133–83

Walz, W., MacVicar, B. A. 1988. Electrophysiological properties of glial cells: Comparison of brainslices with primary cultures. *Brain Res.* 443: 321–24

Weinreich, D., Hammerschlag, R. 1975. Nerve impulse-enhanced release of amino acids from non-synaptic regions of peripheral and central nerve trunks of bullfrog. *Brain Res.* 84: 137–42

Welsh, M. J., Lidtke, C. M. 1986. Chloride and potassium channels in cystic fibrosis airway epithelia. *Nature* 322: 467–69

Wheeler, D. D., Boyarsky, L. L., Brooks, W. H. 1966. The release of amino acids from nerve during stimulation. *J. Cell. Physiol.* 67: 141–48

Wilson, G. F., Chiu, S. Y. 1989. Regulation of Schwann cell K channels during initial stages of myelinogenesis. *Soc. Neurosci. Abstr.* 15: 993

Yarowsky, P. J., Krueger, B. K. 1989. Development of saxitoxin-sensitive and insensitive sodium channels in cultured neonatal rat astrocytes. *J. Neurosci.* 9: 1055–61

*Annu. Rev. Neurosci. 1990. 13:475–511*

# HEBBIAN SYNAPSES:
# Biophysical Mechanisms and Algorithms

*Thomas H. Brown, Edward W. Kairiss, and Claude L. Keenan*

Department of Psychology, Yale University, New Haven, Connecticut 06520

## INTRODUCTION

The hypothesis that memory is associated with use-dependent synaptic modifications (cf. Tanzi 1893, Ramón y Cajal 1911, Freud 1894, Ziegler 1900, Woodworth 1921, Wood-Jones & Porteus 1928, Konorski 1948, Hebb 1949, Hayek 1952, Eccles 1983) has been carefully documented in higher invertebrates (Carew & Sahley 1986, Byrne 1987, Kandel et al 1987). Here we review the type of synaptic learning mechanism originally suggested by Hebb (1949) to operate in higher vertebrates. Theoretical studies suggest that useful and potentially powerful forms of learning and self-organization can emerge in networks of elements that are interconnected by various formal representations of a Hebbian modification (Palm 1982a, Kohonen 1984, 1987, Anderson 1985, Denker 1986, Hopfield & Tank 1986, Linsker 1986, 1988a,b, Tesauro 1986, Bear et al 1987, Pearson et al 1987, Klopf 1988, 1989, Schmajuk & Moore 1988, Sejnowski & Tesauro 1989).

Interest in the computational power of networks of elements that are interconnected by Hebbian modification algorithms has been enhanced by the recent neurophysiological discovery of a synaptic mechanism in the hippocampus that produces modifications that resemble some contemporary interpretations of Hebb's original postulate for learning (Kelso et al 1986, Malinow & Miller 1986, Sastry et al 1986, Wigström et al 1986). This mechanism is responsible for a type of long-term synaptic potentiation

475

(LTP) that can be experimentally induced in parts of the hippocampus and probably in many other brain regions (Brown et al 1988a,b).

In this review we first summarize the evolution of the current concept of a Hebbian synaptic learning mechanism and propose a contemporary definition. This naturally leads to the question of whether Hebbian synapses have now been demonstrated to exist. We approach this question by reviewing recent facts and hypotheses about LTP that are pertinent to our definition of a Hebbian synaptic modification. At least one type of LTP that has been demonstrated in vitro in the hippocampus appears to qualify as a specific instance of a Hebbian modification. After reviewing some additional examples and variations of a Hebbian synaptic modification, we turn to theoretical analyses of Hebbian learning rules. Several of the modification algorithms that have been considered in studies of adaptive neural networks also satisfy our definition of a Hebbian modification. Additional experimental and theoretical research will be required to understand the conditions under which these or other algorithms can be viewed as useful abstractions of the neurophysiology of LTP.

## THE IDEA OF A HEBBIAN SYNAPSE

The idea that associative learning emerges from a Hebb-like conjunctive or correlational mechanism at the synapse has a history that can be traced back a century. Here we present highlights of the evolution of the current concept.

### The Law of Neural Habit

An antecedent to the idea of a Hebbian synapse is found in James (1890; reprinted in Anderson & Rosenfeld 1988). James argued that the laws of association reflect the laws of cerebral physiology (p. 225):

> How does a man come, after having the thought of A, to have the thought of B the next moment? or how does he come to think of A and B always together? These were the phenomena which Hartley undertook to explain by cerebral physiology. I believe he was, in essential respects, on the right track, and I propose simply to revise his conclusions by the aid of distinctions which he did not make.

In particular, James maintained that: "there is no other elementary causal law of association than the law of neural habit: When two elementary brain processes have been active together or in immediate succession, one of them, on reoccurring, tends to propagate its excitement into the other" (p. 226). James' law of neural habit thus indicates the basic conditions for change in the form of a qualitative statement about the activity-modification relationship. Anderson & Rosenfeld (1988) note that when the term "neurons" is substituted for "elementary brain processes," the habit law immediately suggests a set of learning rules that include the idea of a Hebbian synapse.

## The Synaptic Hypothesis for Learning

James did not discuss the subcellular locus and nature of the neuronal mechanism that causes coactivated "brain processes" to change. These issues were addressed by Tanzi (1893), who identified the synapse as the locus of the modification. The proposed nature of the modification involved changes in the strengths of previously existing synaptic connections, an idea that remains the best-established hypothesis today (Abrams & Kandel 1988). A relatively clear statement of the synaptic hypothesis for learning appears in Wood-Jones & Porteus (1928). In discussing synaptic "linkages" onto "dendrons," these authors concluded (p. 48) that "the power which creates these linkages must be plastic," and they suggested a biophysical mechanism for the expression of this plastic change (pp. 355–56):

> A first step toward an explanation is the recognition of the importance of the synapse, the break in the nervous chain, at the point of contact between the fibres of the affecter and the effector system. We must assume that there is a certain resistance at this break to the passage of the nervous impulse, a resistance that is capable under certain conditions of being lowered so as to either permit the passage of the impulse unimpeded or to step it down, so to speak, so that it will not call forth any, or only a very weak response in the muscles. The lowering of the synaptic resistance we have called facilitation.

This explanation of learning in terms of synaptic facilitation has a more modern sound if rephrased in terms of an increase in the synaptic conductance instead of a lowering of the synaptic resistance.

## The Conditions for Synaptic Modification

Wood-Jones & Porteus (1928) did not elaborate on the conditions that trigger or induce the enhanced synaptic efficacy. Hebb (1949) took up this matter two decades later (see also Konorski 1948) in his famous neuropsychological treatise, which contains a statement now known (Stent 1973) as "Hebb's postulate of learning" (p. 62):

> When an axon of cell A is near enough to excite cell B or repeatedly or consistently takes part in firing it, some growth process or metabolic change takes place in one or both cells such that A's efficiency, as one of the cells firing B, is increased.

Hebb proposed this change as the basis of a memory formation and storage process that would cause enduring modifications in the elicited activity patterns of spatially distributed "nerve cell assemblies." This postulate combines the law of neural habit with the synaptic hypothesis for learning. It specifies the location of the modification and it provides a qualitative statement of the conditions for change. It does not furnish much guidance in regard to the quantitative details of the activity-modification relationship (the modification rule or algorithm). In regard to the nature of the biophysical mechanism that increases synaptic efficiency, Hebb (1949) suggested a decrease in the synaptic resistance.

## The Concept of a Hebbian Synaptic Modification

Since 1949 the concept of a Hebbian synapse has evolved to include several key features, which are emphasized to a greater or lesser extent by different investigators (Stent 1973, Sejnowski 1981, Palm 1982a, Kohonen 1984, 1987, Anderson 1985, Levy & Desmond 1985, Denker 1986, Kelso et al 1986, Tesauro 1986, Bear et al 1987, Klopf 1988, 1989, Levy & Burger 1987, Brown et al 1988a,b, 1989, Reiter & Stryker 1988, Sejnowski & Tesauro 1989, Schmajuk & Moore 1988). These key features form the basis of the contemporary understanding of a Hebbian synaptic mechanism. We define a Hebbian synapse as one that uses a *time-dependent, highly local, and strongly interactive mechanism to increase synaptic efficacy* as a function of the conjunction or correlation between pre- and postsynaptic activity. The key terms in this definition are elaborated below.

TIME-DEPENDENT MECHANISM    Modifications in a Hebbian synapse depend on the exact time of occurrence of pre- and postsynaptic activity. In particular the modifications are driven by what Klopf (1988) calls a "real-time mechanism" (cf. Sutton & Barto 1981, Donegan & Wagner 1987), which he defines as follows (pp. 17–18):

> Real-time learning mechanisms emphasize the temporal association of signals: each critical event in the sequence leading to learning has a time of occurrence associated with it and this time plays a fundamental role in the computations that yield changes in the efficacy of synapses. It should be noted that "real-time," in this context, does not mean continuous time as contrasted with discrete time nor does it refer to a learning system's ability to accomplish its computations at a sufficient speed to keep pace with the environment within which it is embedded. Rather, a real-time learning mechanism, as defined here, is one for which the time of occurrence of each critical event in the sequence leading to learning is of fundamental importance with respect to the computations the learning mechanism is performing. Real-time learning mechanisms may be contrasted with nonreal-time learning mechanisms such as the perceptron . . . , adaline . . . , or back propagation . . . learning mechanisms for which error signals follow system responses and only the order of the inputs, outputs, and error signals is important, not the exact time of occurrence of each signal, relative to the others.

Because the expression *real-time mechanism* may have unintended connotations, we call this a *time-dependent mechanism*, but the meaning is the same.

LOCAL MECHANISM    The synapse is the transmission site where the signals or information representing ongoing activity in the pre- and postsynaptic elements are in spatiotemporal contiguity. A Hebbian synapse uses this locally available information to cause a local, input-specific synaptic modification. The necessary information or signal for change can be nothing more than the natural consequence of ordinary intercellular communication at the synapse. Hebbian synapses are thus said to enable an "unsupervised" form of learning—in the sense that a specific external "teacher" signal is not required to instruct change on an individual, synapse-by-synapse, basis. The idea of a local mechanism does not exclude some form of neuromodulatory control over the modification process. A "reinforcement signal" is not excluded.

INTERACTIVE MECHANISM    Whether a change occurs at a Hebbian synapse depends on activity levels on both sides of the synaptic cleft. Palm (1982a) notes that a Hebbian mechanism is based on a "true interaction" between pre- and postsynaptic activity. "Noninteractive" mechanisms or rules can be purely presynaptic, purely postsynaptic, or a superposition of these, but there is no true interaction. Palm's (1982a) distinction between interactive and noninteractive rules is similar to Finkel & Edelman's (1987) distinction between "dependent" and "independent" rules (p. 712):

> In an independent synaptic rule, presynaptic and postsynaptic modifications can occur independently, a reasonable property given the individuality of the two neurons involved. Nevertheless, since the work of Hebb (1949) and von [sic] Hayek (1952), most theoretical models have opted for some form of dependent synaptic rule. In a dependent rule, the locus of the change in synaptic efficacy can be presynaptic and/or postsynaptic (and usually is not specified), but the change is contingent upon events at both loci.

We would add only that this dependence or interaction can be statistical and that the magnitude of the statistical interaction can vary across types of synapses or physiological conditions. A noninteractive mechanism is thus one in which either there is no interaction or the statistical interaction is inconsequential or undetectable.

CONJUNCTIVE OR CORRELATIONAL MECHANISM    We have not yet addressed the logical or statistical form of the interactive mechanism (cf. Palm 1982a). One interpretation of Hebb's postulate is that the condition for change is simply the conjunction of pre- and postsynaptic activity. Thus the co-occurrence of (some level of) pre- and postsynaptic activity (within some short time interval) is sufficient to cause the synaptic enhancement. For this reason, a Hebbian synapse is sometimes called a *conjunctional synapse*. Did Hebb (1949) mean for the coincidence of pre- and postsynaptic activity to be sufficient to cause a modification? Or was there also an implied requirement for a positive correlation between pre- and postsynaptic activity? Hebb was not clear on this point. In Sejnowski's (1977a,b) formalization of Hebb's postulate, correlation over time between pre- and postsynaptic activity is responsible for changes in synaptic efficacy. This condition would be satisfied if cell A "consistently takes part in firing" (Hebb 1949) cell B. Thus a Hebbian synapse also is sometimes called a *correlational synapse* (Anderson 1985). Sejnowski (1977a,b) expresses the condition for change mathematically in terms of the covariance between pre- and postsynaptic activity. A similar idea can also be expressed in terms of conditional probabilities.

## Generalizing the Concept of a Hebbian Modification

SYNAPTIC ENHANCEMENT AND DEPRESSION    Hebb (1949) did not discuss the consequence of uncorrelated or negatively correlated pre- and postsynaptic activity. Are we to assume that uncorrelated activity reduces synaptic efficacy? Or that it has no effect on efficacy? Note that the definition of a Hebbian synapse given above does not exclude additional processes that can decrease synaptic strength. The idea that positively correlated activity causes synaptic strengthening and that either uncorrelated or negatively correlated activity causes synaptic weakening was in fact an essential part of Stent's (1973) extension of the Hebbian idea to the problem of understanding experience-dependent aspects of the development of striate cortex (cf. Bear et al 1987, Reiter & Stryker 1988, Singer 1988).

Later we consider two categories of modification rules: those that can only strengthen the synapse and those that can strengthen or weaken the synapse, depending on the relationship between pre- and postsynaptic activity levels. The latter may be considered extensions or *generalizations of the original concept* of a Hebbian modification as defined above. Palm (1982a,b) has suggested a formalism for classifying what he calls "Hebbian," "anti-Hebbian," and "non-Hebbian" synaptic modifications. According to this scheme, a Hebbian synapse increases its strength with correlated pre- and postsynaptic activity and decreases its strength with negatively correlated activity. Conversely, an anti-Hebbian synapse rewards negatively correlated activity and punishes correlated activity. In both cases, modification of synaptic efficacy involves a real-time, local, and interactive mechanism. A non-Hebbian synapse is one that does not rely on a real-time, local, and interactive mechanism. At such "non-interactive" synapses, the modifications can be expressed as simple superpositions of purely presynaptic or purely postsynaptic rules. Palm (1982a,b) has discussed vector representations of these synaptic modification rules.

We should point out in passing that the term "anti-Hebbian" is not used consistently in the literature. As just noted, for Palm (1982a), an anti-Hebb rule is one that decreases synaptic efficacy as a result of positively *correlated* pre- and postsynaptic activity. Callaway et al (1987) use the term anti-Hebbian in this sense to describe changes at neuromuscular junction during development (see also Callaway et al 1989). Kohonen & Oja (1976) use this form of rule to construct an orthogonalizing filter, and Hopfield et al (1983) use an anti-Hebb rule to remove spurious memory states. In contrast, Levy & Desmond (1985) use the phrase Hebb/anti-Hebb to refer to a rule (discussed below) that increases synaptic efficacy when pre- and postsynaptic activity are coincident, and decreases synaptic efficacy if postsynaptic activity is unaccompanied by presynaptic activity (see also Ranck 1964, Rosenblatt 1967, Stent 1973, Kohonen et al 1974, Cooper et al 1979).

As noted above, the idea of a Hebbian modification can be generalized to include the combination of an interactive synaptic enhancement and some type of activity-dependent synaptic depression. We refer to this combination as a *generalized Hebbian synaptic mechanism*. The depression can be of the noninteractive type or any of several different types of interactions between pre- and postsynaptic activity (cf. Stent 1973, Sejnowski 1977a,b, Palm 1982a, Kohonen 1984, Linsker 1986, Bear et al 1987, Levy & Burger 1987). The interactive condition that causes depression can simply be non-coincident pre- or postsynaptic activity. By this we mean activity that occurs on one side of the synaptic cleft but not the other. In this type of synapse, the occurrence of (some level of) presynaptic activity in the absence of (some level of) postsynaptic activity (Figure 1*D*) or the exact opposite activity relationship (Figure 1*C*) causes synaptic depression. Alternatively, the condition that causes synaptic weakening can be the occurrence of pre- and postsynaptic activity that is statistically uncorrelated or that is negatively correlated in time.

TEACHER SIGNALS AND REINFORCEMENT LEARNING    Although a Hebbian mechanism uses only local information, the modification process may be subject to global control signals. Such signals may enable the induction or consolidation of changes at synapses that have met the above criteria for a Hebbian modification. A global "reinforcement signal" can thus control Hebbian plasticity in a large population of activated synapses. Substances that have been demonstrated to act as neuromodulators (such as catecholamines and acetylcholine) are plausible candidates for this role. Such global modulation of a local process is different from an external "teacher" signal that explicit "instructs" selective modification on a synapse-by-synapse basis independent of local activity.

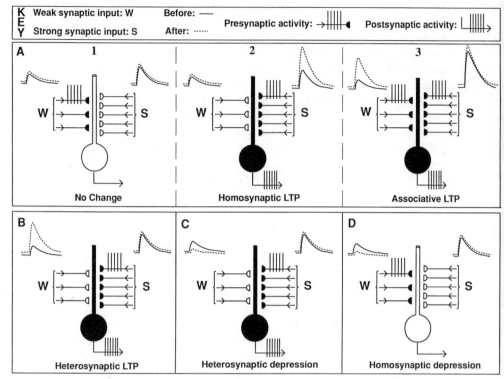

*Figure 1* Illustration of several varieties of use-dependent synaptic changes. Each neuron is shown receiving two sets of nonoverlapping synaptic inputs, one weak (*W*) and the other strong (*S*). The difference between these two sets reflects the number of afferent fibers (for a quantitative interpretation, see Brown et al 1988). The waveforms above each input illustrate schematically the excitatory postsynaptic potential produced by a single stimulation of that input before (*solid curve*) and after (*broken curve*) tetanic (high-frequency) stimulation of one or both inputs. *Filled elements* indicate activity during the conditioning (tetanic) stimulation. *A.* Associative LTP. *A1.* Tetanic stimulation of *W* alone does not cause LTP in either input. *A2.* Subsequent tetanic stimulation of *S* may cause homosynaptic LTP (synaptic enhancement in the stimulated input) but not heterosynaptic LTP (enhancement in the unstimulated input). *A3.* Concurrent tetanic stimulation of *W* and *S* causes associative LTP (enhancement in *W*). Further, LTP in the *S* input may also occur under these conditions (but is not illustrated). *B.* Heterosynaptic LTP. Tetanic stimulation of the *S* input alone causes LTP in the unstimulated *W* input. Homosynaptic LTP in the stimulated *S* input may also occur under these circumstances (but is not shown). *C.* Heterosynaptic depression. Stimulation of the *S* input alone causes depression in the unstimulated *W* input. Homosynaptic LTP in the *S* input may also occur under these conditions (but is not shown). *D.* Homosynaptic depression. High frequency stimulation of *W* alone causes depression in the *W* input. In principle, homosynaptic depression could also occur in the *S* input if tetanic stimulation of this input did not cause homosynaptic LTP.

# HEBBIAN SYNAPTIC MECHANISMS

Hebbian synapses have now been shown to exist in the sense that one can demonstrate experimentally a use-dependent form of synaptic enhancement that is governed by a time-dependent, highly local, and strongly interactive mechanism. This mechanism appears to be responsible for one form of hippocampal LTP. Whether this potential for a Hebbian synaptic modification is actually used in the development and organization of behavior through the formation of Hebb's (1949) nerve "cell assemblies" remains to be seen.

## Varieties of Hippocampal Long-Term Potentiation

LTP is a use-dependent and persistent increase in synaptic strength that can be induced by brief periods of synaptic stimulation (reviewed in Teyler & DiScenna 1987, Bliss & Lynch 1988, Brown et al 1988a,b, 1989). LTP was first described in the rabbit hippocampal formation (Bliss & Lømo 1973, Bliss & Gardner-Medwin 1973) and was subsequently demonstrated to occur in numerous excitatory synapses of the central and peripheral nervous systems of vertebrates and invertebrates (Brown & McAfee 1982, Racine et al 1983, Baxter et al 1985, Briggs et al 1985, Miller et al 1987, Walters & Byrne 1985, Teyler & DiScenna 1987, Artola & Singer 1987, Bindman et al 1988).

There are at least two and possibly many more different LTP mechanisms or forms of LTP (Brown et al 1989). This conclusion is based on the kinetics of the passive decay of the enhancement, the activity-modification relationships that govern the induction and expression of the enhancement, and the particular membrane receptors that control the modification process (cf. Barrionuevo & Brown 1983, Baxter et al 1985, Briggs et al 1985, Racine & Kairiss 1987, Brown et al 1988b, 1989, Racine & deJonge 1988, Johnston et al 1989, Cotman et al 1988). Most of our knowledge of the cellular neurophysiology of LTP in the hippocampus comes from two synaptic systems: the Schaeffer collateral/commissural (Sch/comm) inputs to the CA1 pyramidal neurons and the mossy-fiber synaptic inputs to the CA3 pyramidal neurons [reviewed in Brown et al (1989)].

The Sch/comm synapses display an associative form of LTP (Barrionuevo & Brown 1983, Kelso & Brown 1986) (see Figure 1A, panels 1–3) that is governed by an interactive mechanism (Figure 2). The enhancement depends on both presynaptic activity and the voltage across the post-synaptic membrane (Kelso et al 1986, Malinow & Miller 1986, Wigström et al 1986, Gustafsson et al 1987, Brown et al 1989). The actual statistical strength of the interaction has not been measured at these or any other

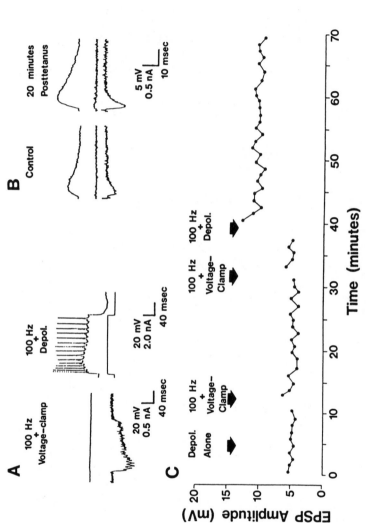

*Figure 2* Direct demonstration of the interactive mechanism. Recordings are of Schaffer collateral synaptic responses in hippocampal neurons of region CA1 (see Figure 3*A*). *A.* (*Left*) Voltage-clamp record of inward synaptic currents (*lower trace*) and membrane potential (*upper trace*) during the synaptic stimulation train. (*Right*) Current-clamp recording of postsynaptic action potentials (*upper trace*) produced by an outward current step (*lower trace*) that is paired with the synaptic stimulation train. *B.* Current-clamp (*top traces*) and voltage-clamp (*bottom traces*) records before and 20 min after pairing synaptic stimulation with the outward current step. *Middle trace* is the membrane potential during voltage clamp. *C.* EPSP amplitudes as a function of the time of occurrence (*arrows*) of three manipulations: an outward current step (depol. alone) or synaptic stimulation trains delivered while applying either a voltage clamp (100 Hz+voltage clamp) or an outward current step (100 Hz+depol.). Each point is the average of five consecutive EPSP amplitudes. Modified from Kelso et al (1986).

plastic synapses. No published data indicate whether LTP in the mossy-fiber synapses involves a strictly noninteractive mechanism, a weakly inter-active mechanism, or a strongly interactive mechanism. The spatio-temporal features of the activity-enhancement relationships have not been studied in the mossy-fiber system. Most of the rest of the discussion is therefore restricted to the associative type of LTP displayed by the Sch/comm synapses (see Figure3$A$ for the experimental preparation).

## Some Features of Associative LTP in Hippocampus

INTENSITY THRESHOLD    Bliss & Gardner-Medwin (1973) noted that the intensity of the electrical shocks delivered in a conditioning train (also called a *tetanic stimulation*) is an important determinant of LTP induction in the perforant pathway input to the dentate gyrus. Low-intensity tetanic stimulations were generally ineffective for inducing LTP. Increasing the stimulus intensity increased the effectiveness of the tetanic stimulation. This stimulus intensity effect has been called *cooperativity* (McNaughton et al 1978). As we use the term below, *cooperativity* refers to a condition in which the probability of inducing LTP in a particular set of synapses by some presynaptic stimulation pattern, or the magnitude of the enhance-ment resulting in this set of synapses from this stimulation pattern, is known or suspected to be a monotonic increasing function of the total number of synaptic inputs that are stimulated at about the same time.

This shock-intensity effect has been seen at two of the three most com-monly studied synaptic systems of the hippocampal formation: the per-forant pathway inputs to the dentate gyrus (Bliss & Gardner-Medwin 1973, McNaughton et al 1978) and the Sch/comm inputs to the CA1 region (Barrionuevo & Brown 1983, Lee 1983). In these systems, stimulation of a small number of afferent inputs [which produce a weak ($W$) postsynaptic response] commonly fails to induce LTP in that set of synapses (Figure 1$A$, panel 1), whereas the identical stimulation pattern applied to a larger total number of afferents [which collectively produce a strong ($S$) post-synaptic response] does induce LTP in that larger set of inputs (Figure 1$A$, panel 2). This intensity effect has not yet been reported to be a feature of the mossy-fiber synaptic inputs to the CA3 region.

SPATIOTEMPORAL SPECIFICITY    *Associative LTP* (Levy & Steward 1979, Barrionuevo & Brown 1983, Kelso & Brown 1986) refers to a particular type of interaction between separately and independently stimulated $W$ and $S$ synaptic inputs to a neuron or small region (see Figure 3$A$ for experimental setup). The nature of the interaction that defines associative LTP is taken from the perspective of the $W$ input. In particular, associative LTP is an enhancement that can be induced in the $W$ input if it and the $S$

input are stimulated together at about the same time, but not if it and the
S input are stimulated separately at very different times or if only one of
them is stimulated. Figure 1A illustrates associative LTP graphically
(panels 1–3). Note that the definition of associative LTP specifically
excludes the possibility of heterosynaptic LTP (Figure 1B) produced in
the W input by tetanic stimulation of the S input. A synaptic system that
displays associative LTP will exhibit cooperativity, but the reverse need
not be true.

The activity-enhancement relationships that govern associative LTP
(Figure 1A) in the hippocampus are relevant to its possible role as a
synaptic substrate for aspects of learning. Some key spatiotemporal features of the activity-enhancement relationships include the following
(Kelso & Brown 1986): (a) the induction of the functional modulation is
rapid, (b) the expression of the enhanced synaptic strength is persistent,
(c) the modification of one synaptic input can be conditionally controlled
by temporal contiguity with activity in another synaptic input to the same
region, and (d) the associative enhancement appears to be specific to just
those synapses whose activity conforms to the temporal requirement.

These features are shared by the synaptic mechanisms that have been
shown in *Aplysia* to underlie simple forms of associative memory (Carew
& Sahley 1986, Byrne 1987, Kandel et al 1987), although there are well-
known differences in the synaptic mechanisms in the two systems (cf.
Carew et al 1984, Kelso et al 1986). Sejnowski & Tesauro (1989) describe
different implementations of a Hebbian algorithm, one of which is functionally equivalent to parts of the gill-withdrawal circuitry in *Aplysia*, while
another resembles the LTP mechanism described below.

## Biophysical Models of LTP Induction in Hippocampus

The spatiotemporal features of associative LTP (cf. Barrionuevo & Brown
1983, Kelso & Brown 1986, Levy & Desmond 1985, Levy & Burger 1987,
Brown et al 1989) can be accounted for easily by a Hebb-like mechanism.
Four research groups furnished evidence for an interactive mechanism
underlying the induction of LTP in the Sch/comm synapses (Kelso et al
1986, Malinow & Miller 1986, Sastry et al 1986, Wigström et al 1986).
These studies showed that some consequence of postsynaptic depolarization is necessary to enable LTP induction at just those synapses that
are eligible to change by virtue of being active at about the same time.
This interactive mechanism can operate (in vitro) in the absence of sodium
spikes and seems to have the required spatiotemporal specificity to account
for aspects of associative LTP (Kelso et al 1986, Gustafsson et al 1987,
Brown et al 1989).

There is general agreement that the induction of associative LTP is triggered by $Ca^{2+}$ influx into the postsynaptic cell (Eccles 1983, Bliss & Lynch 1988). Several investigators have suggested that the $Ca^{2+}$ influx occurs through ionic channels gated by the $N$-methyl-D-aspartate (NMDA) subtype of glutamate receptor (Collingridge & Bliss 1987, Bliss & Lynch 1988, Cotman et al 1988, Brown et al 1988a,b, 1989). The key observation (Collingridge et al 1983; see also Muller et al 1988, Kauer et al 1988) was that antagonists of the NMDA receptor (such as AP5) block the induction of LTP but not its expression (Figure 3$B$). Much of the current debate has focused on the subcellular locus of the NMDA receptor-gated channels that are responsible for the postsynaptic $Ca^{2+}$ influx, the dynamics and compartmentalization of the resulting increase in $[Ca^{2+}]_i$, and the molecular reactions that are controlled by this transient increase in $[Ca^{2+}]_i$. The development of a caged $Ca^{2+}$ buffer that releases $Ca^{2+}$ when stimulated by a UV light pulse (Tsien & Zucker 1986) provided another means of testing the idea that an increase in $[Ca^{2+}]_i$ is sufficient to induce LTP. Recent experiments have shown that cells containing this caged compound exhibit an LTP-like synaptic enhancement when exposed to a pulse of UV light (Malenka et al 1988).

ALTERNATIVE MODELS    Several models have been proposed for the induction of LTP. Here we consider just three, beginning with one that we have been developing for the Sch/comm synapses (Brown et al 1988b, 1989, Zador et al 1990). In this model, which we denote *Model 1*, the trigger for synaptic enhancement is $Ca^{2+}$ influx through NMDA receptor-gated channels that are located on the dendritic spine head. Model 1 maintains that ($a$) the peak transient increase in $[Ca^{2+}]_i$ is localized within the dendritic spine, ($b$) the spine amplifies the local change in $[Ca^{2+}]_i$, and ($c$) the relationship between the peak transient increase in $[Ca^{2+}]_i$ and the amount of LTP is nonlinear. In regard to this last point, it may be appropriate to think of some threshold level of $[Ca^{2+}]_i$ in the spine head as having a high probability of triggering molecular "switches" (Miller & Kennedy 1986) that initiate the modification process.

Model 1 maintains that for the NMDA receptor-gated channels on the spine head to allow $Ca^{2+}$ influx, the receptor-iontophore complex must receive two local signals (Brown et al 1988b, 1989). One is a chemical signal (glutamate binding to the NMDA receptor) resulting from activity in the presynaptic terminal. The other is an electrical signal (strong depolarization of the dendritic spine head) that results from ongoing activity in the postsynaptic cell. The depolarization is needed to relieve a $Mg^{2+}$ block of the channel (Nowak et al 1984, Mayer et al 1984, Jahr & Stevens 1987). The voltage across the spine head is controlled not only by activity in synapses located on this spine, but also by ($a$) the activity in other

**A**

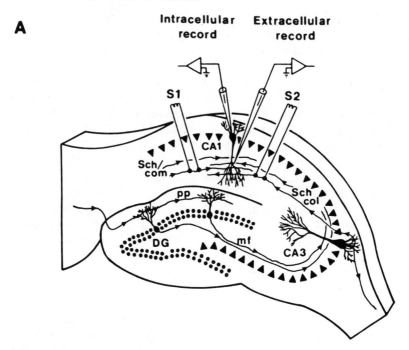

**B**

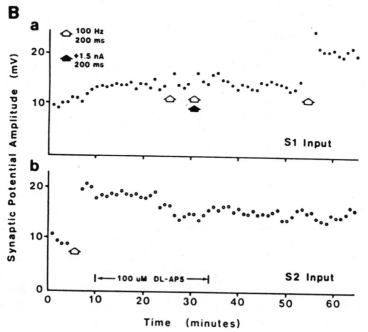

synapses located on other parts of the postsynaptic neuron, (*b*) the electro-tonic structure of the postsynaptic neuron, and the relationship of this structure to the set of active synapses, and (*c*) the spatial distribution of various types of membrane conductances throughout this structure (see (Johnston & Brown 1984, Dingledine 1986, Brown et al 1988b).

Some predictions of Model 1 have already been confirmed. One pre-diction (Brown et al 1988b) was that LTP induction should be prevented when the postsynaptic cell is depolarized to $+10$ mV during the tetanic stimulation. The reason is that $Ca^{2+}$ entry through NMDA channels is not expected at this membrane potential (MacDermott et al 1986, Mayer et al 1987). By contrast, $Ca^{2+}$ entry through voltage-gated calcium channels should not be blocked at the potential. This prediction was confirmed for the Sch/comm synapses (Malenka et al 1988).

*Model 2* differs in one major respect. In this model $Ca^{2+}$ influx is through voltage-dependent calcium channels rather than NMDA receptor-gated channels (Gamble & Koch 1987). An extension of this model, developed for the mossy-fiber synapses of the CA3 region (Hopkins & Johnston 1988, Johnston et al 1988), maintains that for the voltage-dependent calcium channels to be effective, norepinephrine must act on the postsynaptic cell. This extension is consistent with the observation that LTP induction in the mossy-fiber system is blocked by $\beta$-receptor antagonists (Hopkins & Johnston 1988, Johnston et al 1988).

*Model 3* also differs in one major respect. In this model $Ca^{2+}$ influx is through NMDA receptor-gated channels located on the dendritic shaft rather than the spine head (Bliss & Lynch 1988, Wickens 1988). This model was suggested for the type of LTP that occurs in the perforant pathway input to the dentate gyrus and possibly also the Sch/comm input to the CA1 region. The rationale underlying this hypothesis and its implications are considered at length elsewhere (Bliss & Lynch 1988, Brown et al 1989, Wickens 1988).

←⎯⎯⎯⎯⎯⎯⎯⎯⎯⎯⎯⎯⎯⎯⎯⎯⎯⎯⎯⎯⎯⎯⎯⎯⎯⎯⎯⎯⎯⎯⎯⎯⎯⎯⎯

*Figure 3*   Role of NMDA receptors in the induction but not the expression of LTP. *A.* Schematic of the hippocampal brain slice in relationship to stimulating and recording electrode (for additional detail, see Brown & Zador 1990). A cell in the CA1 region was impaled with an intracellular electrode, while field potentials were monitored with an extra-cellular electrode. Two sets of synaptic inputs were stimulated independently with two separate electrodes (S1 and S2). *B. Upper panel:* Bath application of 100 $\mu$M DL-AP5 did not attenuate the S1-produced EPSPs, but did prevent the *induction* of LTP following either tetanic stimulation (three trains 12 seconds apart, 100 Hz for 200 msec each) alone (*open arrow*) or when paired (*open and solid arrows*) with simultaneous outward current steps (1.5 nA, 200 msec). *Lower panel:* Addition of DL-AP5 did not block the expression of LTP in the S2 pathway that was *induced* in medium prior to the addition of DL-AP5. Modified from Brown & Zador (1990).

Simulations of the synaptic activity-dependent $[Ca^{2+}]_i$ dynamics have only been performed for Model 1 (Zador et al 1990) and Model 2 (Gamble & Koch 1987), but the essential features of all three models can be appreciated intuitively. Cooperativity emerges naturally in all three models from the relationship between the strength of the stimulated synaptic inputs and the resulting $Ca^{2+}$ influx. Cooperativity has not yet been demonstrated to be a feature of LTP induction at the mossy-fiber synapses. If the predicted cooperativity cannot be demonstrated at the mossy-fiber synapses, Model 2 will have to be rejected or substantially revised.

A related prediction of Model 2 is that the induction of LTP in the mossy-fiber synapses should be controllable solely by manipulations of the postsynaptic membrane potential. Specifically, LTP induction should be prevented if the postsynaptic membrane potential is maintained under voltage-clamp conditions at a relatively negative potential during tetanic stimulation of the mossy-fiber synaptic inputs. One version (Gamble & Koch 1987) of this model implies that it should be possible to induce LTP experimentally simply by strongly depolarizing the postsynaptic neuron in the absence of presynaptic stimulation. This prediction results because $Ca^{2+}$ influx into the spines is not directly dependent on transmitter release but only on the potential at the spine head, which should be controllable from the soma (Brown et al 1988b). Neither prediction has been adequately assessed in the mossy-fiber synapses.

The three models also differ in regard to the occurrence of associative (Figure 1A) versus heterosynaptic (Figure 1B) LTP. Model 1 contains strong constraints against heterosynaptic LTP (Brown et al 1988b, 1989, Zador et al 1990). Model 2 predicts that heterosynaptic should occur among mossy-fiber synapses. Because the mossy-fiber synapses are electrotonically near each other (Brown et al 1988b), tetanic stimulation of an S input should cause sufficient depolarization in spines associated with unstimulated synaptic inputs to induce LTP in them. In the absence of further constraints, Model 3 might also predict heterosynaptic LTP among closely adjacent synapses (see Brown et al 1989). Additional details and other differential and testable predictions of the models are reviewed at length elsewhere (Brown et al 1988b, 1989).

SPATIOTEMPORAL SPECIFICITY    The spatiotemporal specificity of associative LTP is easy to appreciate in Model 1. The spatial (or input) specificity results from the fact that $Ca^{2+}$ influx into the spine head requires glutamate binding to NMDA receptors on the subsynaptic membrane. The physical geometry of the spine, plus $Ca^{2+}$ buffering systems, restrict the large, synaptically produced, transient changes in $[Ca^{2+}]_i$ to the spine region

(Zador et al 1990). Therefore only active synapses can undergo enhancement. This input specificity is not a requirement of Model 2. As indicated above, the latter model is consistent with the occurrence of heterosynaptic LTP (see Figure 1*B*). Without further constraints, input specificity is also not achieved in Model 3.

In Model 1, the temporal specificity arises from the fact that the NMDA receptor-iontophore complex must receive two signals simultaneously for the iontophore to become highly permeable to $Ca^{2+}$. Glutamate must be bound to the NMDA receptor, and the membrane containing the iontophore must be sufficiently depolarized to relieve the $Mg^{2+}$ block. Thus tetanic stimulation of a *W* input fails to cause synaptic enhancement (Figure 1*A*, panel 1) unless this stimulation is accompanied by sufficiently strong postsynaptic depolarization (Figure 1*A*, panel 3). Note that the model does not require that presynaptic stimulation and postsynaptic depolarization occur simultaneously (Zador et al 1990). The maximum amount of glutamate binding occurs with a short delay after each presynaptic stimulation. Glutamate unbinding from the NMDA receptor takes additional time. After unbinding, glutamate either binds again, diffuses out of the synaptic cleft, or becomes resequestered intracellularly. Because complete removal of glutamate from the synaptic cleft after each presynaptic stimulation takes time, LTP could be induced in principle even if there were a small temporal gap—a brief "trace period"—between the end of a *W* presynaptic stimulation and the subsequent onset of a strong postsynaptic depolarization (Brown et al 1989). Simulations of Model 1 predict a trace period of up to tens of milliseconds (Zador et al 1990).

Two groups of investigators (Levy & Steward 1983, Gustafsson et al 1987, Levy & Burger 1987) have in fact reported the occurrence of LTP even after a trace period of 10 msec or longer (reviewed in Brown et al 1989). There is some basis for postulating that the removal of glutamate from the synaptic cleft is relatively slow. One reason is that the component of postsynaptic current associated with NMDA receptor activation, sometimes called the NMDA current, appears to be relatively persistent in hippocampal neurons (for review see Mayer & Westbrook 1987; see also Forsythe & Westbrook 1988). The existence of what we have called a "forward-pairing" trace period (Brown et al 1989) could be relevant to some of the computations that might be carried out in a network of such synapses (Tesauro 1986).

## Mechanism of LTP Expression and Maintenance

The mechanisms responsible for the expression and maintenance of LTP are less well understood than those described above for the induction process. The important questions concern the locus of change and its nature. Results of recent pharmacological studies have suggested that, although LTP *induction* depends on NMDA receptor activation (Collingridge et al 1983, Kauer et al 1988, Muller et al 1988), the *expression* of LTP depends on non-NMDA receptors (Kauer et al 1988, Muller & Lynch 1988, Muller et al 1988) (see Figure 3*B*). Some of these results have been interpreted as demonstrating a postsynaptic locus for the biochemical changes that underlie LTP expression and maintenance. The published results are indeed suggestive, but they are not yet sufficient to rule out all possible presynaptic mechanisms.

## Neuromodulation of LTP Induction and Maintenance

There may be multiple controls—both local and extrinsic—over the extent and nature of most if not all use-dependent synaptic modifications. All of the preceding LTP induction models can easily accommodate an external signal that controls the overall plasticity of a brain region. Such signals might set the occasion for learning. Any condition that reduces $\gamma$-aminobutyric (GABA)-mediated inhibition (and therefore increases postsynaptic depolarization) would be expected to increase the magnitude of LTP or the probability of its occurrence. Pharmacological disinhibition is known (in vitro) to have this effect in the hippocampus (Wigström & Gustafsson 1983, Pacelli et al 1987).

Disinhibition is not the only route through which a neuromodulator might affect use-dependent plasticity. Johnston and co-workers have shown that pharmacological agonists and antagonists of the noradrenergic $\beta$-receptor can control the magnitude, probability of occurrence, or persistence of LTP in the mossy-fiber synapses (Hopkins & Johnston 1988, Johnston et al 1988). Although endogenously secreted norepinephrine would appear to lack the spatial specificity to instruct change selectively at any particular synapse, in principle it could serve as an extrinsic "global teacher" signal (Brown et al 1988b) that enables or preserves changes in groups of active synapses. In this way, a neuromodulatory signal can interact with a Hebbian mechanism to produce a foundation for "reinforcement learning." To understand selective information storage in the nervous system, we need to learn more about neuromodulatory control of the induction and persistence of use-dependent synaptic modifications. After LTP has been induced, can its persistence be controlled by a subsequent neuromodulatory action? Is norepinephrine or some other neuromodulator a signal for the "consolidation" (Livingston 1967b, Kety 1972, Gold & McGaugh 1975, McGaugh 1988, 1989) of use-dependent synaptic modifications?

# GENERALIZED HEBBIAN SYNAPTIC MECHANISMS

Theoretical studies and common sense point to the utility of being able both to increase and to decrease synaptic strength. In principle, such synapses can be constructed by combining an interactive form of synaptic enhancement with either an interactive or a noninteractive form of synaptic depression (Palm 1984, Linsker 1986, 1988a,b, Kairiss et al 1988). The question is whether these different forms of use-dependent synaptic plasticity in fact co-occur at the same synapses. We know from research on *Aplysia* that use-dependent forms of enhancement and depression can combine at the same synapses in interesting ways that allow learning-related adaptive modifications to occur (Hawkins & Kandel 1984, Carew & Sahley 1986). However, the synaptic enhancement that has been demonstrated thus far in *Aplysia* is not Hebbian (Carew et al 1984). Below we first consider the possibility that an interactive mechanism for enhancement may co-occur at synapses that exhibit an interactive or noninteractive mechanism for depression.

## Associative LTP and Heterosynaptic Depression in Dentate Gyrus

Levy & Burger (1987) have suggested that the synapses of the perforant pathway inputs to the dentate gyrus undergo modifications that conform to what we have termed a generalized Hebb rule. Both associative LTP (Figure 1*A*) and heterosynaptic depression (Figure 1*C*) have been reported in these synapses (Levy & Steward 1979, 1983, Levy 1985, Levy & Desmond 1985). In abstracting their experimental results, Levy & Burger (1987) assumed an interactive form of heterosynaptic depression, which was embedded in a single modification rule of the general type given by Eq. 6 (Figure 4, discussed below). In this equation the level of presynaptic activity determines the direction of synaptic change. Synaptic depression occurs if the presynaptic activity level is low when the postsynaptic activity level is high. Synaptic enhancement occurs if presynaptic activity is high when postsynaptic activity is high. The amount of synaptic change is linearly proportional to the level of postsynaptic activity. Associative LTP induction in this system is known to involve the NMDA receptor (Collingridge & Bliss 1987). Two interesting and unresolved questions are whether the induction of heterosynaptic depression in the perforant pathway depends on NMDA receptor activation and whether it involves an interactive mechanism.

## Associative LTP and Homosynaptic Depression in Hippocampus

Stanton & Sejnowski (1989) looked for a covariance-type of rule (Sejnowski 1977a,b) in the Sch/comm synaptic input to the CA1 pyramidal neurons. In agreement with previous results, they found that the synapses display associative LTP when pre- and postsynaptic activity are positively correlated or active "in phase." The new finding was that when pre- and postsynaptic activities were negatively correlated or active "out of phase," a long-term depression was produced. A similar depression was reported when the inputs were activated while the postsynaptic membrane was hyperpolarized. The effect seemed to be long-lasting and restricted to the activated synapses. It did not depend on activation of NMDA receptors. This homosynaptic depression appears to be different from the hetero-synaptic depression reported in the dentate gyrus. These results are suggestive of the type of covariance mechanism that is discussed further below.

## Synaptic Plasticity in Developing Visual Cortex

Numerous investigators (Stent 1973, Shatz & Stryker 1978, Bienenstock et al 1982, Linsker 1986, Bear & Cooper 1989, Pearson et al 1987, Reiter & Stryker 1988, Constantine-Paton et al 1990) have proposed various extensions or generalizations of the basic Hebb rule to explain ocular dominance plasticity and other experience-dependent aspects of visual cortical development. Ocular dominance plasticity refers to an experience-dependent shift in the relative responsiveness of visual cortical neurons to input from the two eyes. The classical observation was that when vision in one eye is occluded during a critical period of development, an ocular dominance shift occurs in favor of the nonoccluded eye (Wiesel & Hubel 1963, 1965). Stent (1973) interpreted these results in terms of a synaptic modification rule that combined a simple Hebb rule with a noninteractive form of synaptic depression. Because of the interactive enhancement, postsynaptic activity driven by the nonoccluded eye strengthened all co-active synapses, which would be those from the nonoccluded eye. Because of the noninteractive depression, this same postsynaptic activity weakened all synapses that were not co-active, which would be those from the occluded eye.

The two components of Stent's generalized Hebbian modification are similar to certain of the synaptic modifications considered above. An interactive enhancement of the sort that Stent required could be implemented by the type of NMDA receptor-dependent LTP mechanism that we previously discussed in regard to the Sch/comm synapses. If the distribution of NMDA receptors is a guide to the occurrence of Hebbian synapses, we would expect to find such synapses in several regions of the neocortex, including the visual cortex (Monaghan & Cotman 1985, Cotman et al 1988). Although much less is known about the biophysics of synaptic transmission in neocortex than in the hippocampus, evidence for a Hebbian form of LTP is growing. Baranyi & Szente (1987) reported (in vivo) the involvement of postsynaptic activity in the induction of LTP in some cells of cat motor cortex (see also Baranyi & Fehér 1981). Additionally, Bindman et al (1988) discovered (in vitro) a Hebb-like mechanism in rat sensorimotor cortex—one that shows a spatial specificity similar to that found in hippocampus. Furthermore, an NMDA receptor-dependent form of LTP has recently been reported in visual cortex (Artola & Singer 1987, Connors & Bear 1988, Kimura et al 1988). Finally, pharmacological blockade of NMDA receptors has been shown to prevent certain forms of experience-dependent plasticity in visual cortex (Singer et al 1986, Bear et al 1987, Kleinschmidt et al 1987, Rauschecker & Hahn 1987; see also Cline et al 1987). The noninteractive synaptic depression could be implemented by some form of heterosynaptic depression (Figure 1C), which is unfortunately still poorly understood.

Reiter & Stryker (1988) recently discovered a very different effect of monocular occlusion when the visual cortex is pharmacologically inhibited by local application of muscimol (an agonist for the GABA-A receptor). They reported that under these conditions, ocular dominance shifted in favor of the less active synapses from the occluded eye. This finding led them to conclude that the level of postsynaptic activity controls the direction of cortical plasticity. Bear et al (1987) previously reached a similar conclusion. Reiter & Stryker (1988) explained this interaction between muscimol and monocular occlusion on ocular dominance plasticity as follows:

> This suggests the following type of "learning rule" for ocular dominance plasticity: the connections between the more-active open eye inputs and strongly inhibited postsynaptic cells are weakened relative to the inputs from the less-active closed eye, while the reverse happens when the postsynaptic cells are not inhibited completely. This learning rule suggests the presence of a threshold membrane potential that determines the direction of an ocular dominance shift.

To account for all of the above ocular dominance shifts in terms of known synaptic mechanisms, one could postulate the co-occurrence of three forms of plasticity in the visual cortex (cf. Kairiss et al 1988): an interactive synaptic enhancement, a noninteractive heterosynaptic depression, and a noninteractive homosynaptic depression (see Figure 1 and Eq. 12). Alternatively, the combination of an interactive form of synaptic enhancement and an interactive form of synaptic depression can also be made to fit the results.

## Interactive Mechanisms for Synaptic Depression

It is natural to wonder whether there are interactive mechanisms for synaptic depression. There are, of course, numerous examples of homosynaptic depression (Figure 1*D*) (Bruner & Kennedy 1970, Zucker 1972, Castelluci & Kandel 1974, Kandel 1976, Krasne 1976, 1978, Zucker & Bruner 1977, Rinaldi et al 1987) and heterosynaptic depression (Figure 1*C*) (Levy & Steward 1983, Levy & Burger 1987, Montorolo et al 1988), but these have not been explicitly shown to involve an interactive mechanism. Very recent work raises the possibility that some synapses do have an interactive mechanism for depression.

One line of evidence comes from studies of visual cortical neurons. Frégnac and co-workers have shown that an iontophoretically produced decrease in postsynaptic activity, when coupled with presynaptic activation, causes a long-lasting decrease in the response to activation of these same synapses (Frégnac et al 1988, Frégnac & Shulz 1989). Conversely, these neurons increased their response to stimuli that were paired with an iontophoretically induced increase in postsynaptic activity. Another candidate possibility is raised by the work of Stanton & Sejnowski (1989) mentioned above. The experimental tools now exist for determining whether the depression that they reported, which occurs when presynaptic activity is paired with postsynaptic hyperpolarization, involves an interactive mechanism. A third synaptic system that may be controlled by an interactive form of depression is the parallel fiber input to the Purkinje cells of the cerebellum (Ekerot & Kano 1985, Crepel & Krupa 1988, Ito 1989).

Some interesting speculations have been made about the biochemical mechanisms underlying interactive forms of synaptic depression. One proposal (Bear & Cooper 1989, in press) is that decreases in synaptic strength are mediated via second messenger systems, such as inositol triphosphate ($IP_3$) or diacyl glycerol (DG), that are activated by non-NMDA receptors, whereas increases in synaptic strength result from calcium influx through NMDA receptor-activated channels.

## ABSTRACT REPRESENTATIONS OF HEBBIAN MODIFICATIONS

Levy & Burger (1987) estimate that perhaps as many as 50–100 different algebraic equations have been used to describe theoretical synaptic activity-modification relationships or learning rules. Here we select just a few of these rules or algorithms that satisfy our definition of a Hebbian modification. The selection was further limited to simple algorithms that have or could be argued to capture abstractly some important feature of known or suspected neurophysiology. The shortcomings of these abstractions can be even more informative than any claimed successes. In addition, they provide a convenient, higher-level language for communicating and exploring theoretical relationships among changes at the synaptic, network, systems, and behavioral levels (Donegan et al 1989).

Obviously these Hebbian algorithms are not intended as substitutes for detailed and realistic biophysical models of synaptic function (cf. Zador et al 1990). But those interested in developmental and cognitive aspects of computational neuroscience can reasonably argue that some extreme simplifications will inevitably be required to appreciate self-organization and learning in the mammalian brain (see Kohonen 1984, 1987, Linsker 1986, Arbib 1987, Churchland et al 1990, Sejnowski & Churchland 1989, Sejnowski et al 1988). The challenge at every level of organization is to determine which facts are theoretically important and which are not (see Crick & Asanuma 1986, Sejnowski & Churchland 1989).

## Hebbian Algorithms

Although Hebb did not attempt to quantify the relationship between post- and presynaptic firing and the ensuing synaptic enhancement, his idea is commonly translated into discrete-time equations of the form:

$$w_{ij}(t+1) = w_{ij}(t) + \Delta w_{ij}(t),$$ 1.

where

$$\Delta w_{ij}(t) = F[a_i(t), a_j(t)],$$ 2.

$t$ is discrete time, $w_{ij}$ is the "weight" of the connection from presynaptic unit $u_j$ to postsynaptic unit $u_i$, $a_j(t)$ and $a_i(t)$ are some measures of pre- and postsynaptic activity, and the change in synaptic efficacy $\Delta w_{ij}(t)$ is some function[1] $F$ of both pre- and postsynaptic activity.

The synaptic weight $w_{ij}$ is assumed to be a suitable abstract representation of "synaptic strength." The relationship between $w_{ij}$ and the variables in contemporary biophysical models of synaptic transmission is usually unspecified. Commonly $w_{ij}$ is treated as a dimensionless variable. For concreteness we assume here that $w_{ij}$ is proportional to the expected value of a multiparameter quantal probability model (cf. Brown et al 1989) of quantal synaptic transmission. The activity terms $a_j(t)$ and $a_i(t)$ are also often treated as dimensionless variables. Here they are assumed to be proportional to some measure of a biophysical variable such as the spike frequency during a short time interval, the potential across a designated patch of membrane, the conductance to $Ca^{2+}$ or another ion in some membrane region, the current carried by $Ca^{2+}$ or another ion through a

---

[1] $F$ is most appropriately termed a functional because the activity terms themselves are functions of time.

given area of membrane, or the intracellular activity of $Ca^{2+}$ or some other second messenger in a particular subcellular compartment.

SPIKE PRODUCT RULE    Much of what follows concerns the form of the functional $F$ in Eq. 2. Perhaps the simplest functional is a product of the pre- and postsynaptic spike frequencies:

$$w_{ij}(t+1) = w_{ij}(t) + c\bar{x}_i(t)\bar{x}_j(t), \qquad\qquad 3.$$

where $\bar{x}_i$ and $\bar{x}_j$ are the mean firing frequencies of the post- and presynaptic units, respectively (averaged over the previous fraction of a second), and $c$ is a proportionality constant that determines the learning rate (cf. Levy & Desmond 1985, Klopf 1988, 1989, Sejnowski & Tesauro 1989). The three key assumptions in Eq. 3 are that the contributions of pre- and postsynaptic activity can be separated into two corresponding activity terms, that the proper measurement of activity is mean spike frequency, and that the nature of the interaction between pre- and postsynaptic spiking frequency is a simple product.

Our current understanding of associative LTP is not well-represented by these assumptions. The causally relevant postsynaptic variable may be depolarization of the spine head rather than spike frequency (Kelso et al 1986, Brown et al 1988b, 1989, Zador et al 1990). The choice of presynaptic spike frequency for the presynaptic term may also be inappropriate. Although we commonly use high-frequency presynaptic stimulation to induce LTP (Brown et al 1989), this use is not necessary. LTP can also be induced by pairing single presynaptic stimulations with properly timed postsynaptic depolarizations (Wigström et al 1986). These pairings do not have to occur at a high repetition rate to be effective. The relative efficacy of different repetition rates is still unclear. Several groups have shown that activity patterns at approximately the theta frequency (5–7 Hz) are particularly effective (Larson & Lynch 1986, Larson et al 1986, Rose & Dunwiddie 1986, Staubli & Lynch 1987, Greenstein et al 1988), probably because of disinhibition at this frequency (Pacelli et al 1987).

ACTIVITY PRODUCT RULE    We can generalize Eq. 3 by not specifying the activity measurement:

$$w_{ij}(t+1) = w_{ij}(t) + g[a_i(t)]h[a_j(t)]c, \qquad\qquad 4.$$

where $g[a_i(t)]$ reflects some measure of postsynaptic activity $a_i(t)$, and $h[a_j(t)]$ indicates some measure of presynaptic activity $a_j(t)$, both taken at the same time $t$. Equation 4 is an improvement over Eq. 3 because it leaves open the proper specification of the pre- and postsynaptic activity functions. Nevertheless, this general form of synaptic learning rule (Figure

4, *left curve*) still leaves much to be desired, not only as an abstraction of known synaptic physiology but also from a psychological and computational perspective.

The first and most obvious problem is that Eq. 4 requires successive pairings of pre- and postsynaptic activity to cause equal increments in synaptic strength. Unlike any known synaptic modification, $w_{ij}$ can therefore increase without bound. The LTP mechanism saturates after one or just a few stimulations (deJonge & Racine 1985), in sharp contrast to the behavior of Eq. 4. Clearly, networks of synapses governed by Eq. 4 might not lend themselves easily to the negatively accelerated acquisition curves of the sort captured by the learning equation of Rescorla & Wagner (1972):

Hebb rule:  $\Delta w_{ij} = g(a_i)[h(a_j)]c$

Generalized Hebb rule:  $\Delta w_{ij} = g(a_i)\{[h(a_j)]c - w_{ij}\}$

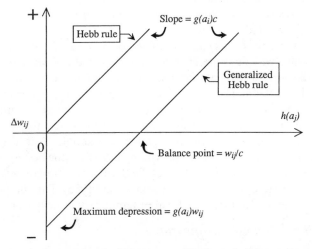

*Figure 4*  Abstract representation of Hebbian modifications. Ongoing presynaptic activity in the axonal bouton is represented by $h(a_j)$; ongoing postsynaptic activity is represented by $g(a_i)$; and the connection weight is represented by $w_{ij}$. In the Hebbian activity product rule (Eq. 4, plot on *left*) the change in weight $\Delta w_{ij}$ is the product of the presynaptic activity function $h(a_j)$, the postsynaptic activity function $g(a_i)$, and a proportionality constant $c$ that determines the rate of change. The Hebb rule only allows positive values of $\Delta w_{ij}$, the magnitude of which is proportional to the slope $g(a_i)c$. In one type of generalized Hebb activity product rule (Eq. 6, plot on *right*), the presynaptic activity function is replaced with $\{[h(a_j)]c - w_{ij}\}$. This rule allows positive or negative values of $\Delta w_{ij}$ depending on whether $h(a_j)$ exceeds or is less than the balance point $w_{ij}/c$. The magnitude of $\Delta w_{ij}$ is again proportional to $g(a_i)c$.

$$\Delta V_n = K(\lambda - V_{n-1}), \qquad \qquad \qquad \text{5.}$$

where $\Delta V_n$ is the change in associative strength on trial $n$, $K$ is a learning rate parameter reflecting stimulus salience, and $\lambda$ is the asymptotic associative strength.

The second problem is that unless ongoing pre- and postsynaptic activities are perfectly negatively correlated, $w_{ij}$ will always tend to increase due to chance coincidences (Grossberg 1976, Sejnowski 1981, Kohonen 1984). Thus $w_{ij}$ will automatically tend to grow without bound and its value can never decrease. This in turn leads to "runaway instability" (Sejnowski & Tesauro 1989). Once the $w_{ij}$ associated with some presynaptic unit $u_j$ is large enough, subsequent activity in $u_j$ will always increase $w_{ij}$ regardless of coactivity in other presynaptic units. One obvious solution is simply to clip Eq. 4 at some upper limit on $w_{ij}$ (Tesauro 1986). In the absence of passive decay, however, clipping would introduce the problem of saturation. Spurious coincidences plus clipping will cause a uniform set of $w_{ij}$ to evolve to some asymptotic level automatically. As the set of $w_{ij}$ approaches its maximum value, all information represented in the synaptic weights is of course lost. New memories could not be stored and old ones would be obliterated.

Some of these problems can be avoided by adding a passive decay term to Eq. 4. Thus we might prefer to assume that in the absence of synaptic activity, the current $w_{ij}$ is not necessarily permanent. Instead $w_{ij}$ can relax back to other states as a function of the immediate and long-term history of the synapse. Such a learning rule could be useful for temporary information storage, but this solution reintroduces the problem of how to achieve more persistent representations. In this context we should note the lack of experimental evidence that LTP is permanent. In fact, the best information suggests that experimentally induced LTP decays over the course of hours, days, or weeks (Racine et al 1983, Reymann et al 1985). Thus hippocampal LTP might be involved in some form of temporary information storage.

## Generalized Hebbian Algorithms

In the previous section we considered some problems with simple product rules that permit only increases in $w_{ij}$. Stable network function may sometimes require modification algorithms that cause both increases and decreases in $w_{ij}$. Here we consider various extensions of Hebb algorithms that allow both to occur. We refer to these as generalized Hebbian algorithms.

RENORMALIZATION    Von der Malsburg (1973) proposed that a "renormalization" process occurs each time a $w_{ij}$ increases. Renormalization forces the combined synaptic weights associated with each postsynaptic unit ($\Sigma w_{ij}$ for each $u_i$) to remain constant. This zero-sum constraint on $\Delta w_{ij}$ results in both increases and decreases in the connection strengths. At a descriptive level, one can think of renormalization in terms of presynaptic "competition" for connection strength (see also Volper & Hampson 1987), an idea that is sometimes invoked to explain aspects of neural development (Artola & Singer 1987, Constantine-Paton et al 1990) and damage-induced plasticity (Merzenich et al 1984).

ACTIVITY PRODUCT RULE    Product rules have been modified from Eqs. 3 and 4 in various ways to allow for reversible modifications in synaptic strength. One approach is to substitute for the presynaptic term in Eq. 3 an expression such as $(x_j - x_b)$, where $x_b$ is the average background activity in the afferent input (Kohonen 1984). Another approach (cf. Levy & Desmond 1985, Levy & Burger 1987, Kohonen 1988) is to replace the presynaptic term in Eq. 4 with an expression such as $\{h[a_j(t)]c - w_{ij}(t)\}$, where $c$ is a learning rate constant. We can write the generalized Hebb algorithm as follows:

$$w_{ij}(t+1) = w_{ij}(t) + g[a_i(t)] \{h[a_j(t)]c - w_{ij}(t)\}, \qquad \text{6.}$$

where the postsynaptic term $g[a_i(t)]$ can represent the activity in a particular spine on neuron $i$ at time $t$. Equation 6 implies that for afferent inputs in which $h[a_j(t)]c < w_{ij}(t)$, the synaptic strength will decrease by an amount proportional to the postsynaptic activity $g[a_i(t)]$ (Figure 4, *right curve*). The synaptic strength increases in proportion to $g[a_i(t)]$ when $h[a_j(t)]c > w_{ij}(t)$. Thus the activity "balance point" (see Figure 3) for modifying $w_{ij}$ is a variable $(w_{ij}/c)$ that is proportional to the value of $w_{ij}$ at the time of presynaptic activation. This approach eliminates the problem of runaway instability and results in a negatively accelerated synaptic modification curve.

We suggested above that the type of modification expressed by Eq. 6 could be relevant to Stent's (1973) account of the classical experiments of Wiesel & Hubel (1963, 1965). Other results (Reiter & Stryker 1988), however, could not be explained by a modification rule with a presynaptic balance point. In their studies of visual cortical development, Cooper and co-workers considered an algorithm in which the balance point for synaptic modification was associated with the postsynaptic term (Bienenstock et al 1982, Bear et al 1987). Their algorithm, which is mathematically more complicated than Eq. 6, holds that the modification threshold changes nonlinearly with the time-averaged postsynaptic activity. Low activity levels cause a decrease in the threshold, effectively enabling weak inputs to increase their strength. High levels drive the threshold up. This prevents saturation and stabilizes the network in a state of maximum selectivity.

Later we consider algorithms that have both pre- and postsynaptic balance points.

ACTIVITY TRACE PRODUCT RULE    Recall that LTP can be induced even with a brief time interval between the end of a presynaptic stimulation and the subsequent onset of a sufficiently strong postsynaptic depolarization (Levy & Burger 1987). Similarly, in certain variations or extensions of a Hebb-like algorithm, the product of earlier presynaptic activity (at time $t-\tau$) and later postsynaptic activity (at time $t$) is what generates the changes (Tesauro 1986, Klopf 1988, Sejnowski & Tesauro 1989). To allow for a "forward-pairing trace period" (cf. Levy & Burger 1987, Brown et al 1989), the presynaptic term in Eq. 6 can be offset by time $\tau$. The differential effects of several values of $\tau$ can be included by modifying Eq. 6 as follows (cf. Klopf 1988, 1989):

$$w_{ij}(t+1) = w_{ij}(t)+g[a_i(t)] \sum_{\tau=0}^{k} \{h[a_j(t-\tau)]c_\tau - w_{ij}(t-\tau)\}, \qquad 7.$$

where $c_\tau$ is a positive coefficient that determines the contribution of each offset.

ACTIVITY COVARIANCE RULE    Sejnowski (1977a,b) proposed a modification rule in which changes in synaptic strength are proportional to the covariance between pre- and postsynaptic spiking (see Eqs. 7–15 of Sejnowski 1981). If we substitute activity for spiking, a discrete-time version of Sejnowski's learning algorithm can be written as follows:

$$w_{ij}(t+1) = w_{ij}(t)+\text{cov}[a_j(t), a_i(t)], \qquad 8.$$

where

$$\text{cov}[a_j(t), a_i(t)] = \gamma[a_j(t)a_i(t)-\bar{a}_j(t)\bar{a}_i(t)], \qquad 9.$$

$\gamma$ is a proportionality constant that determines the learning rate, and $\bar{a}_j(t)$ and $\bar{a}_i(t)$ are the expected values of the activities at time $t$ or the average values over some previous time interval. According to this scheme, $w_{ij}$ increases if pre- and postsynaptic activities are positively correlated and decreases if they are negatively correlated. Uncorrelated activity has no effect on $w_{ij}$. Equation 8 solves the problem of automatic growth owing to spurious coincidences.

Linsker (1986) has used a covariance-type rule to simulate aspects of experience-dependent self-organization of the visual system. Extending this work, Linsker (1988a,b) has proposed an organizing principle for understanding simple layered networks in terms of Shannon information theory. He has shown that Hebb rules can give rise to units with the property of *maximum information preservation* in which each unit maximizes the information that the output signal conveys about the input signal, subject to constraints and noise. By maximizing the Shannon information rate of its output, each unit essentially extracts and summarizes the most relevant features of its input. Linsker has termed this the *informax* principle. What this means is that, within certain limits, simple layered networks comprised of Hebbian connections can learn in an "unsupervised" fashion to perform a principle-components analysis on the incoming information. This ability to self-organize without a teacher seems well-suited to experience-dependent or experience-expectant aspects of perceptual development.

FURTHER GENERALIZATIONS    We have mentioned three types of generalized Hebb algorithms. In Eq. 6 the direction of modification is determined by the level of presynaptic activity. The presynaptic balance point in Eq. 6 is illustrated graphically in Figure 4 (*right curve*). In the type of rule proposed by Cooper's group (Bear et al 1987), the direction of modification is controlled by the level of postsynaptic activity. There is a postsynaptic balance point. In the type of rule that Sejnowski (1977) explored, the balance point is not localized on either side of the synapse. In Sejnowski's covariance rule, the direction of change is a joint function of both pre- and postsynaptic activity (Eq. 9). This is also true of Linsker's (1986, 1988b) rule, a discrete-time version of which can be written as follows:

$$w_{ij}(t+1) = w_{ij}(t) + F[(a_j(t) - A_j)(a_i(t) - A_i)], \qquad 10.$$

where $A_j$ and $A_i$ are constants. If $F$ is a linear functional, Eq. 10 may be expanded to:

$$w_{ij}(t+1) = w_{ij}(t) + \alpha a_j(t)a_i(t) + \beta a_i(t) + \gamma a_j(t) + \delta. \qquad 11.$$

This bilinear equation is identical to that discussed by Palm (1984; see also Linsker 1988b). It contains one interactive term with coefficient $\alpha$, two noninteractive terms with coefficients $\beta$ and $\gamma$, and a constant term $\delta$. If $\alpha$ is positive and $\beta$ and $\gamma$ are negative, then Eq. 11 represents a generalized Hebb rule in which $\beta a_i(t)$ is a noninteractive type of heterosynaptic depression and $\gamma a_j(t)$ is a noninteractive type of homosynaptic depression.
    Equation 11 can be further generalized as follows:

$$w_{ij}(t+1) = w_{ij}(t) + \alpha[a_j(t)a_i(t)] + \beta[a_i(t)] + \gamma[a_j(t)] + \delta, \qquad 12.$$

where $\alpha$, $\beta$, and $\gamma$ are no longer simple coefficients but instead functionals that depend upon pre- and/or postsynaptic activity. The idea is the same, however. There is an interactive form of enhancement and two noninteractive forms of depression. Equation 12 has several attractive features: it appears consistent with both types of ocular dominance plasticity effects mentioned above; it avoids the problem of automatic growth due to spurious coincidences; and it is a plausible implementation in the sense that neurobiological analogs to each of its activity terms exist (Kairiss et al 1988). Equation 12 also raises several interesting questions. Do the required three forms of plasticity co-occur in the same synaptic system? How persistent are they? What is the best mathematical representation of the activity-modification relationships? Is anything gained computationally by replacing the two noninteractive forms of depression with a single interactive form of depression? Do interactive forms of depression exist?

## Global Control of Learning

Several investigators have considered the possibility of "now print" (Livingston 1967a,b, Anderson 1985, von der Malsburg & Bienenstock 1986) or "global teacher" (Brown et al 1988b) signals. Levy & Burger (1987) address this possibility at the algorithmic level by introducing a multiplicative variable $P$, which takes on values between 0 and 1. Adding this variable to Eq. 6 yields:

$$w_{ij}(t+1) = w_{ij}(t) + g[a_i(t)] \{h[a_j(t)]c - w_{ij}(t)\} P. \qquad 13.$$

Equation 13 assumes that three multiplicative terms corresponding to presynaptic activity, postsynaptic activity, and overall plasticity can be isolated meaningfully. If there is a physiological correlate of the $P$ term, it could be a neuromodulatory substance such as norepinephrine (cf. Bear & Singer 1986, Kasamatsu 1987) or acetylcholine (cf. Williams & Johnston 1988, Metherate et al 1987, Sillito 1987), as mentioned above.

An alternative algorithm, suggested by Alspector et al (1988), involves "excess reinforcement." The change in synaptic weight is proportional to the correlation between the activity product and a global reinforcement signal $r$:

$$R = \langle ra_ia_j \rangle - \langle r \rangle \langle a_ia_j \rangle, \qquad 14.$$

where $\langle \ \rangle$ indicates an expected value or a time-averaged quantity. The value of $r$ is set to $+1$ if the output is correct, and $-1$ if incorrect. The＊ learning algorithm increases $w_{ij}$ if $R > 0$, and decreases $w_{ij}$ if $R < 0$. An electronic implementation of a neural network using this algorithm is discussed in Alspector et al (1988).

## SUMMARY AND CONCLUSIONS

We have examined the evolution of the concept of a Hebbian synaptic modification and have suggested a contemporary definition. The biophysical mechanism demonstrated in vitro to control the induction of one type of hippocampal LTP has been shown to satisfy our definition of a Hebbian synaptic modification. Whether this biophysical mechanism is involved in the organization of behavior in the manner that Hebb originally envisioned remains to be seen. We have also summarized several modification algorithms that have been explored in theoretical studies of learning in adaptive networks. These algorithms also satisfied our definition of a Hebbian modification, but their relationships to known neurobiology require further exploration. By reviewing the biophysical mechanisms and formal algorithms together, we have exposed obvious similarities and differences. Such comparisons may help bridge the gap between computational theory and knowledge of the neurobiology of use-dependent synaptic change. Current models of LTP reveal that the activity-modification relationships are extremely sensitive to the biophysical/molecular details. The activity-modification relationships obviously can have a major influence on adaptive neurodynamics at the network level. As more accurate representations of the biological complexity and diversity are introduced into adaptive network simulations, we expect to gain new insights into the classes of computation that particular networks are capable of performing.

ACKNOWLEDGMENTS

We thank James Anderson, Thomas Carew, Nelson Donegan, Harry Klopf, Terrence Sejnowski, and Anthony Zador for useful discussion or comments on this manuscript. Supported by Air Force Office of Scientific Research, Office of Naval Research, and Defence Advanced Research Projects Agency.

*Literature Cited*

Abrams, T. W., Kandel, E. R. 1988. Is contiguity detection in classical conditioning a system or a cellular property? Learning in *Aplysia* suggests a possible molecular site. *Trends Neurosci.* 11: 128–35

Alspector, J., Allen, R. B., Hu, V., Satyanarayana, S. 1988. Stochastic learning networks and their electronic implementation. In *Neural Information Processing Systems*, ed. D. Z. Anderson, pp. 9–21. New York: Am. Inst. Phys.

Anderson, J. A. 1985. What Hebb synapses build. See Levy et al 1985, pp. 153–73

Anderson, J. A., Rosenfeld, E., eds. 1988. *Neurocomputing: Foundations of Research*, pp. 4–14. Cambridge: MIT Press

Arbib, M. A. 1987. *Brains, Machines and Mathematics*. New York: Springer-Verlag. 2nd ed.

Artola, A., Singer, W. 1987. Long-term potentiation and NMDA receptors in rat visual cortex. *Nature* 330: 649–52

Baranyi, A., Fehér, O. 1981. Synaptic facilitation requires paired activation of convergent pathways in the neocortex. *Nature* 290: 413–15

Baranyi, A., Szente, M. B. 1987. Long-lasting potentiation of synaptic transmission requires postsynaptic modifications in the neocortex. *Brain Res.* 423: 378–84

Barrionuevo, G., Brown, T. H. 1983. Associative long-term potentiation in hippocampal slices. *Proc. Natl. Acad. Sci. USA* 80: 7347–51

Baxter, D. A., Bittner, G. D., Brown, T. H. 1985. Quantal mechanism of long-term synaptic potentiation. *Proc. Natl. Acad. Sci. USA* 82: 5978–82

Bear, M. F., Cooper, L. N. 1989. Molecular mechanisms for synaptic modification in the visual cortex: Interaction between theory and experiment. In *Advances in Connectionist Theory: Neurobiological Aspects*, ed. D. Rumelhart, M. Gluck. In press

Bear, M. F., Cooper, L. N., Ebner, F. F. 1987. A physiological basis for a theory of synapse modification. *Science* 237: 42–48

Bear, M. F., Singer, W. 1986. Acetylcholine, norepinephrine and the extrathalamic control of visual cortical plasticity. *Nature* 320: 172–76

Bienenstock, E. L., Cooper, L. N., Munro, P. W. 1982. Theory for the development of neuron selectivity: Orientation specificity and binocular interaction in visual cortex. *J. Neurosci.* 2: 32–48

Bindman, L., Murphy, K. P. S. J., Pockett, S. 1988. Postsynaptic control of the induction of long-term changes in efficacy of transmission at neocortical synapses in slices of rat brain. *J. Neurophysiol.* 60: 1053–65

Bliss, T. V. P., Gardner-Medwin, A. R. 1973. Long-lasting potentiation of synaptic transmission in the dentate area of the unanaesthetized rabbit following stimulation of the perforant path. *J. Physiol.* 232: 357–74

Bliss, T. V. P., Lømo, T. 1973. Long-lasting potentiation of synaptic transmission in the dentate area of the anesthetized rabbit following stimulation of the perforant path. *J. Physiol.* 232: 331–56

Bliss, T. V. P., Lynch, M. A. 1988. Long-term potentiation of synaptic transmission in the hippocampus: Properties and mechanisms. See Landfield & Deadwyler 1988, pp. 3–72

Briggs, C. A., Brown, T. H., McAfee, D. A. 1985. Neurophysiology and pharmacology of long-term potentiation in the rat sympathetic ganglion. *J. Physiol.* 359: 503–21

Brown, T. H., Chang, V. C., Ganong, A. H., Keenan, C. L., Kelso, S. R. 1988a. Biophysical properties of dendrites and spines that may control the induction and expression of long-term synaptic potentiation. See Landfield & Deadwyler 1988, pp. 197–260

Brown, T. H., Chapman, P. F., Kairiss, E. W., Keenan, C. L. 1988b. Long-term synaptic potentiation. *Science* 242: 724–28

Brown, T. H., Ganong, A. H., Kairiss, E. W., Keenan, C. L., Kelso, S. R. 1989. Long-term potentiation in two synaptic systems of the hippocampal brain slice. See Byrne & Berry 1989, pp. 266–306

Brown, T. H., McAfee, D. 1982. Long-term synaptic potentiation in superior cervical ganglion. *Science* 215: 1411–13

Brown, T. H., Zador, A. 1990. The hippocampus. In *The Synaptic Organization of the Brain*, ed. G. M. Shepherd. New York: Oxford. In press

Bruner, J., Kennedy, D. 1970. Habituation: Occurrence at a neuromuscular junction. *Science* 169: 92–94

Byrne, J. H. 1987. Cellular analysis of associative learning. *Physiol. Rev.* 67: 329–439

Byrne, J. H., Berry, W. O., eds. 1989. *Neural Models of Plasticity*. New York: Academic

Cajal, S. Ramón y. 1911. *Histologie du Systeme Nerveux*, Vol. 2. Paris: Maloine

Callaway, E. M., Soha, J. M., Van Essen, D. C. 1987. Competition favors inactive over active motor neurons during synapse elimination. *Nature* 328: 422–26

Callaway, E., Soha, J. M., Van Essen, D. C. 1989. Differential loss of neuromuscular connections according to activity level and spinal position of neonatal rabbit soleus motor neurons. *J. Neurosci.* 9: 1806–24

Carew, T. J., Hawkins, R. D., Abrams, T. W., Kandel, E. R. 1984. A test of Hebb's postulate at identified synapses which mediate classical conditioning in *Aplysia*. *J. Neurosci.* 4: 1217–24

Carew, T. J., Sahley, C. L. 1986. Invertebrate learning and memory: From behavior to molecules. *Annu. Rev. Neurosci.* 9: 435–87

Castellucci, V. F., Kandel, E. R. 1974. A quantal analysis of the synaptic depression underlying habituation of the gill-withdrawal reflex in *Aplysia*. *Proc. Natl. Acad. Sci. USA* 71: 5004–8

Churchland, P. S., Koch, C., Sejnowski, T. J. 1990. What is computational neuroscience? In *Computational Neuroscience*, ed. E. Schwartz. Cambridge: MIT Press. In press

Cline, H. T., Debski, E. A., Constantine-Paton, M. 1987. *N*-methyl-D-aspartate receptor antagonist desegregates eye-specific stripes. *Proc. Natl. Acad. Sci. USA* 84: 4342–45

Collingridge, G. L., Bliss, T. V. P. 1987. NMDA receptors: Their role in long-term potentiation. *Trends Neurosci.* 10: 288–93

Collingridge, G. L., Kehl, S. J., McLennan, H. 1983. Excitatory amino acids in synaptic transmission in the Schaffer collateral-commissural pathway of the rat hippocampus. *J. Physiol.* 334: 33–46

Connors, B. W., Bear, M. F. 1988. Pharmacological modulation of long term potentiation in slices of visual cortex. *Soc. Neurosci. Abstr.* 14: 744

Constantine-Paton, M., Cline, H. T., Debski, E. 1990. Patterned activity, synaptic convergence, and the NMDA receptor in developing visual pathways. *Annu. Rev. Neurosci.* 13: 129–54

Cooper, L. N., Liberman, F., Oja, E. 1979. A theory for the acquisition and loss of neuron specificity in visual cortex. *Biol. Cybern.* 33: 9–28

Cotman, C. W., Monaghan, D. T., Ganong, A. H. 1988. Excitatory amino acid neurotransmission: NMDA receptors and Hebb-type synaptic plasticity. *Annu. Rev. Neurosci.* 11: 61–80

Crepel, F., Krupa, M. 1988. Activation of protein kinase C induces a long-term depression of glutamate sensitivity of cerebellular Purkinje cells. An in vitro study. *Brain Res.* 458: 397–401

Crick, F. H. C., Asanuma, C. 1986. Certain aspects of the anatomy and physiology of the cerebral cortex. In *Parallel Distributed Processing: Explorations in the Microstructure of Cognition*, ed. D. L. McClelland, D. E. Rumelhart, 2: 333–71. Cambridge: MIT Press

deJonge, M., Racine, R. J. 1985. The effects of repeated induction of long-term potentiation in the dentate gyrus. *Brain Res.* 328: 181–85

Denker, J. S. 1986. Neural network models of learning and adaptation. *Physica D* 22: 216–32

Dingledine, R. 1986. NMDA receptors: What do they do? *Trends Neurosci.* 9: 47–49

Donegan, N. H., Gluck, M. A., Thompson, R. F. 1989. Integrating behavioral and biological models of classical conditioning. In *Computational Models of Learning in Simple Neural Systems*, ed. R. D. Hawkins, G. H. Bower. New York: Academic. In press

Donegan, N. H., Wagner, A. R. 1987. Conditioned diminution and facilitaion of the UR: A sometimes opponent-process interpretation. In *Classical Conditioning*, ed. I. Gormezano, W. F. Prokasy, R. F. Thompson, pp. 339–69. Hillsdale, NJ: Erlbaum

Eccles, J. C. 1983. Calcium in long-term potentiation as a model for memory. *Neuroscience* 10: 1071–81

Edelman, G. M., Gall, W. E., Cowan, W. M., eds. 1987. *Synaptic Function.* New York: Wiley

Ekerot, C.-F., Kano, M. 1985. Long-term depression of parallel fibre synapses following stimulation of climbing fibres. *Brain Res.* 342: 357–60

Finkel, L. H., Edelman, G. M. 1987. Population rules for synapses in networks. See Edelman et al 1987, pp. 711–58

Forsythe, I. D., Westbrook, G. L. 1988. Slow excitatory postsynaptic currents mediated by *N*-methyl-D-aspartate receptors on cul-

tured mouse central neurones. *J. Physiol.* 396: 515–33

Frégnac, Y., Schulz, D. 1989. Hebbian synapses in visual cortex. In *Seeing Contour and Colour*, ed. K. J. Kilikowski. Oxford: Pergamon. In press

Frégnac, Y., Schulz, D., Thorpe, S., Bienenstock, E. 1988. A cellular analogue of visual cortical plasticity. *Nature* 333: 367–70

Gamble, E., Koch, C. 1987. The dynamics of free calcium in dendritic spines in response to repetitive synaptic input. *Science* 236: 1311–15

Gold, P. E., McGaugh, J. L. 1975. A single-trace, two process view of memory storage processes. In *Short-Term Memory*, ed. D. Deutsch, J. A. Deutsch, pp. 355–78. New York: Academic

Greenstein, Y. J., Pavlides, C., Winson, J. 1988. Long-term potentiation in the dentate gyrus is preferentially induced at theta rhythm periodicity. *Brain Res.* 438: 331–34

Grossberg, S. 1976. Adaptive pattern classification and universal recoding: I. Parallel development and coding of neural feature detectors. *Biol. Cybern.* 23: 121–34

Gustafsson, B., Wigström, H., Abraham, W. C., Huang, Y.-Y. 1987. Long-term potentiation in the hippocampus using depolarizing current pulses as the conditioning stimulus to single volley synaptic potentials. *J. Neurosci.* 7: 774–80

Hawkins, R. D., Kandel, E. R. 1984. Is there a cell biological alphabet for simple forms of learning? *Psychol. Rev.* 91: 375–91

Hayek, F. A. 1952. *The Sensory Order: An Inquiry into the Foundations of Theoretical Psychology*. Chicago: Univ. Chicago Press

Hebb, D. O. 1949. *The Organization of Behavior*. New York: Wiley

Hopfield, J. J., Feinstein, D. I., Palmer, R. G. 1983. "Unlearning" has a stabilizing effect in collective memories. *Nature* 304: 158–59

Hopfield, J. J., Tank, D. W. 1986. Computing with neural circuits: A model. *Science* 233: 625–33

Hopkins, W. F., Johnston, D. 1988. Noradrenergic enhancement of long-term potentiation at mossy fiber synapses in the hippocampus. *J. Neurophysiol.* 59: 667–87

Ito, M. 1989. Long-term depression. *Annu. Rev. Neurosci.* 12: 85–102

Jahr, C. E., Stevens, C. F. 1987. Glutamate activates multiple single channel conductances in hippocampal neurons. *Nature* 325: 522–25

James, W. 1890. *Psychology: Briefer Course.* Cambridge: Harvard Univ. Press

Johnston, D., Brown, T. H. 1984. Biophysics and microphysiology of synaptic transmission in hippocampus. In *Brain Slices*, ed. R. Dingledine, pp. 51–86. New York: Plenum

Johnston, D., Hopkins, W. F., Gray, R. 1988. Noradrenergic enhancement of long-term synaptic enhancement. See Landfield & Deadwyler 1988, pp. 355–76

Johnston, D., Hopkins, W. F., Gray, R. 1989. The role of norepinephrine in long-term potentiation at mossy fibers synapses in the hippocampus. See Byrne & Berry 1989, pp. 307–28

Kairiss, E. W., Keenan, C. L., Brown, T. H. 1988. A biologically plausible implementation of a Hebbian covariance algorithm. *Soc. Neurosci. Abstr.* 14: 832

Kandel, E. R. 1976. *The Cellular Basis of Behavior*. San Francisco: Freeman

Kandel, E. R., Klein, M., Hochner, B., Shuster, M., Siegelbaum, S. A., et al. 1987. Synaptic modulation and learning: New insights into synaptic transmission from the study of behavior. See Edelman et al 1987, pp. 471–518

Kasamatsu, T. 1987. Norepinephrine hypothesis for visual and cortical plasticity: Thesis, antithesis, and recent development. *Curr. Top. Dev. Biol.* 21: 367–89

Kauer, J., Malenka, R. C., Nicoll, R. A. 1988. A persistent postsynaptic modification mediates long-term potentiation in the hippocampus. *Neuron* 1: 911–17

Kelso, S. R., Brown, T. H. 1986. Differential conditioning of associative synaptic enhancement in hippocampal brain slices. *Science* 232: 85–87

Kelso, S. R., Ganong, A. H., Brown, T. H. 1986. Hebbian synapses in hippocampus. *Proc. Natl. Acad. Sci. USA* 83: 5326–30

Kety, S. 1972. Brain catecholamines, affective states and memory. In *The Chemistry of Mood, Motivation, and Memory*, ed. J. L. McGaugh, pp. 65–80. New York: Raven

Kimura, F., Tsumoto, T., Nishigori, A., Shirokawa, T. 1988. Long-term synaptic potentiation and NMDA receptors in the rat pup visual cortex. *Soc. Neurosci. Abstr.* 14: 188

Kleinschmidt, A., Bear, M. F., Singer, W. 1987. Blockade of "NMDA" receptors disrupts experience-dependent plasticity of kitten striate cortex. *Science* 238: 355–58

Klopf, A. H. 1988. A neuronal model of classical conditioning. *Psychobiology* 16: 85–125

Klopf, A. H. 1989. Classical conditioning phenomena predicted by a drive-reinforcement model of neuronal function. See Byrne & Berry 1989, pp. 104–32

Kohonen, T. 1984. *Self-Organization and Associative Memory*. Berlin/Heidelberg: Springer-Verlag

Kohonen, T. 1987. *Content-Addressable Memories*. New York: Springer-Verlag. 2nd ed.

Kohonen, T. 1988. The "neural" phonetic typewriter. *Computer* 21: 11–22

Kohonen, T., Lehtio, P., Rovano, J. 1974. Modelling of neural associative memory. *Ann. Acad. Sci. Fenn. Ser. A5* 167: 1–18

Kohonen, T., Oja, E. 1976. Fast adaptive formation of orthogonalizing filters and associative memory in recurrent networks of neuron-like elements. *Biol. Cybern.* 21: 85–95

Konorski, J. 1948. *Conditioned Reflexes and Neuron Organization*. London: Cambridge Univ. Press

Krasne, F. B. 1976. Invertebrate systems as a means of gaining insight into the nature of learning and memory. In *Neural Mechanisms of Learning and Memory*, ed. M. R. Rosenzweig, E. L. Bennett, pp. 401–29. Cambridge: MIT Press

Krasne, F. B. 1978. Extrinsic control of intrinsic neuronal plasticity: An hypothesis from work on simple systems. *Brain Res.* 140: 197–216

Landfield, P. W., Deadwyler, S. A., eds. 1988. *Long-Term Potentiation: From Biophysics to Behavior*. New York: Liss

Larson, J., Lynch, G. 1986. Induction of synaptic potentiation in hippocampus by patterned stimulation involves two events. *Science* 232: 985–88

Larson, J., Wong, D., Lynch, G. 1986. Patterned stimulation at the theta frequency is optimal for the induction of hippocampal long-term potentiation. *Brain Res.* 368: 347–50

Lee, K. S. 1983. Cooperativity among afferents for the induction of long-term potentiation in the CA1 region of the hippocampus. *J. Neurosci.* 3: 1369–72

Levy, W. B. 1985. Associative changes at the synapse: LTP in the hippocampus. See Levy et al 1985, pp. 5–33

Levy, W. B., Anderson, J. A., Lehmkuhle, S., eds. 1985. *Synaptic Modification, Neuron Selectivity, and Nervous System Organization*. Hillsdale, NJ: Erlbaum

Levy, W. B., Burger, B. 1987. *IEEE 1st Int. Conf. on Neural Networks, San Diego*, 4: 11–15. San Diego: IEEE

Levy, W. B., Desmond, N. L. 1985. The rules of elemental synaptic plasticity. See Levy et al 1985, pp. 105–21

Levy, W. B., Steward, O. 1979. Synapses as associative memory elements in the hippocampal formation. *Brain Res.* 175: 233–45

Levy, W. B., Steward, O. 1983. Temporal contiguity requirements for long-term associative potentiation/depression in the hippocampus. *Neuroscience* 8: 791–97

Linsker, R. 1986. From basic network principles to neural architecture: Emergence of spatial-opponent cells. *Proc. Natl. Acad. Sci. USA* 83: 7508–12

Linsker, R. 1988a. Towards an organizing principle for a layered perceptual network. In *Neural Information Processing Systems*, ed. D. Andersen, pp. 485–94. New York: Am. Inst. Phys.

Linsker, R. 1988b. Self-organization in a perceptual network. *Computer* 21: 105–17

Livingston, R. B. 1967a. Brain circuitry relating to complex behavior. See Quarton et al 1967, pp. 499–515

Livingston, R. B. 1967b. Reinforcement. See Quarton et al 1967, pp. 568–77

MacDermott, A. B., Mayer, M. L., Westbrook, G. L., Smith, S. J., Barker, J. L. 1986. NMDA-receptor activation increases cytoplasmic calcium concentration in cultured spinal cord neurones. *Nature* 321: 519–22

Malenka, R. C., Kauer, J. A., Zucker, R. S., Nicoll, R. A. 1988. Postsynaptic calcium is sufficient for potentiation of hippocampal synaptic transmission. *Science* 242: 81–84

Malinow, R., Miller, J. P. 1986. Postsynaptic hyperpolarization during conditioning reversibly blocks induction of long-term potentiation. *Nature* 320: 529–30

Mayer, M. L., Westbrook, G. L. 1987. The physiology of excitatory amino acids in the vertebrate central nervous system. *Prog. Neurobiol.* 28: 197–276

Mayer, M. L., MacDermott, A. B., Westbrook, G. L., Smith, S. J., Barker, J. L. 1987. Agonist- and voltage-gated calcium entry in cultured mouse spinal cord neurons under voltage clamp measured using arsenazo III. *J. Neurosci.* 7: 3230–44

Mayer, M. L., Westbrook, G. L., Guthrie, P. B. 1984. Voltage-dependent block by $Mg^{2+}$ of NMDA responses in spinal cord neurones. *Nature* 309: 261–63

McGaugh, J. L. 1988. Modulation of memory storage processes. In *Memory: An Interdisciplinary Approach*, ed. P. R. Solomon, G. R. Goethals, C. M. Kelley, B. R. Stephens. New York: Springer-Verlag. In press

McGaugh, J. L. 1989. Involvement of hormonal and neuromodulatory systems in the regulation of memory storage. *Annu. Rev. Neurosci.* 12: 255–87

McNaughton, B. L., Douglas, R. M., Goddard, G. V. 1978. Synaptic enhancement in fascia dentata: Cooperativity among coactive afferents. *Brain Res.* 157: 277–93

Merzenich, M. M., Nelson, R. J., Stryker, M. P., Cynader, M. S., Schopmann, A., Zook, J. M. 1984. Somatosensory cortical

map changes following digit amputation in adult monkeys. *J. Comp. Neurol.* 224: 591–605

Metherate, R., Tremblay, N., Dykes, R. W. 1987. Acetylcholine permits long-term enhancement of neuronal responsiveness in cat primary somatosensory cortex. *Neuroscience* 22: 75–82

Miller, M. W., Lee, S. C., Krasne, F. B. 1987. Cooperativity-dependent long-lasting potentiation in the crayfish lateral giant escape reaction circuit. *J. Neurosci.* 7: 1081–92

Miller, S. G., Kennedy, M. B. 1986. Regulation of brain type II $Ca^{2+}$/calmodulin-dependent protein kinase by autophosphorylation: A $Ca^{2+}$-triggered molecular switch. *Cell* 44: 861–70

Monaghan, D. T., Cotman, C. W. 1985. Distribution of NMDA-sensitive L-[3H]-glutamate binding sites in rat brain as determined by quantitative autoradiography. *J. Neurosci.* 5: 2909–19

Montorolo, P. G., Kandel, E. R., Schacher, S. 1988. Long-term heterosynaptic inhibition in *Aplysia*. *Nature* 333: 171–74

Muller, D., Joly, M., Lynch, G. 1988. Contributions of quisqualate and NMDA receptors to the induction and expression of LTP. *Science* 242: 1694–97

Muller, D., Lynch, G. 1988. Long-term potentiation differentially affects two components of synaptic responses in hippocampus. *Proc. Natl. Acad. Sci. USA* 85: 9346–50

Nowak, L., Bregestovski, P., Ascher, P., Herbet, A., Prochiantz, A. 1984. Magnesium gates glutamate-activated channels in mouse central neurones. *Nature* 307: 462–66

Pacelli, G. J., Naghdi, F., Kelso, S. R. 1987. Possible role of inhibition in LTP induction. *Soc. Neurosci. Abstr.* 13: 977

Palm, G. 1982a. *Neural Assemblies: An Alternative Approach*. Berlin/Heidelberg/New York: Springer-Verlag

Palm, G. 1982b. Rules for synaptic changes and their relevance for the storage of information in the brain. In *Cybernetics and Systems Research*, ed. R. Trappl, pp. 277–80. Amsterdam: North-Holland

Palm, G. 1984. Associative networks and cell assemblies. In *Brain Theory*, ed. G. Palm, A. Aertsen, pp. 211–28. Berlin/Heidelberg: Springer-Verlag

Pearson, J. C., Finkel, L. H., Edelman, G. M. 1987. Plasticity in the organization of adult cerebral cortical maps: A computer simulation based on neuronal group selection. *J. Neurosci.* 7: 4209–23

Quarton, G. C., Melnechuk, T., Schmitt, F. O., eds. 1967. *The Neurosciences*. New York: Rockefeller Univ. Press

Racine, R. J., deJonge, M. 1988. Short-term and long-term potentiation in projection pathways and local circuits. See Landfield & Deadwyler 1988, pp. 167–200

Racine, R. J., Kairiss, E. W. 1987. Long-term potentiation phenomena: The search for the mechanisms underlying memory storage processes. In *Neuroplasticity, Learning, and Memory*, ed. N. W. Milgram, C. M. MacLeod, T. L. Petit, pp. 173–97. New York: Liss

Racine, R. J., Milgram, N. W., Hafner, S. 1983. Long-term potentiation phenomena in the rat limbic forebrain. *Brain Res.* 260: 217–31

Ranck, J. B. 1964. Synaptic "learning" due to electroosmosis: A theory. *Science* 144: 187–89

Rauschecker, J. P., Hahn, S. 1987. Ketamine-xylazine anaesthesia blocks consolidation of ocular dominance changes in kitten visual cortex. *Nature* 326: 183–85

Reiter, H. O., Stryker, M. P. 1988. Neural plasticity without postsynaptic action potentials: Less-active inputs become dominant when kitten visual cortical cells are pharmacologically inhibited. *Proc. Natl. Acad. Sci. USA* 85: 3623–27

Rescorla, R. A., Wagner, A. R. 1972. A theory of Pavlovian conditioning: Variations in the effectiveness of reinforcement and nonreinforcement. In *Classical Conditioning II*, ed. A. H. Black, W. F. Prokasy, pp. 64–99. New York: Appleton-Century-Crofts

Reymann, K. G., Malisch, R., Schulzeck, K., Brodemann, R., Ott, T., et al. 1985. The duration of long-term potentiation in the CA1 region of the hippocampal slice. *Brain Res. Bull.* 15: 249–55

Rinaldi, P. C., Ganong, A. H., Brown, T. H. 1987. Low-frequency synaptic depression recorded intracellularly from hippocampal granule cells. *Soc. Neurosci. Abstr.* 13: 977

Rose, G. M., Dunwiddie, T. V. 1986. Induction of hippocampal long-term potentiation using physiologically patterned stimulation. *Neurosci. Lett.* 69: 244–48

Rosenblatt, F. 1967. Recent work on theoretical models of biological memory. In *Computer and Information Sciences—II*, ed. J. T. Tou, pp. 33–56. New York: Academic

Sastry, B. R., Goh, J. W., Auyeung, A. 1986. Associative induction of posttetanic and long-term potentiation in CA1 neurons of rat hippocampus. *Science* 232: 988–90

Schmajuk, N. A., Moore, J. W. 1988. The hippocampus and the classically conditioned nictitating membrane response: A real-time attentional-associative model. *Psychobiology* 16: 20–35

Sejnowski, T. J. 1977a. Strong covariance with nonlinearly interacting neurons. *J. Math. Biol.* 4: 303–21

Sejnowski, T. J. 1977b. Statistical constraints on synaptic plasticity. *J. Theor. Biol.* 69: 385–89

Sejnowski, T. J. 1981. Skeleton filters in the brain. In *Parallel Models of Associative Memory*, ed. G. E. Hinton, J. A. Anderson, pp. 49–82. Hillsdale, NJ: Erlbaum

Sejnowski, T. J., Churchland, P. S. 1989. Brain and cognition. In *Foundation of Cognitive Science*, ed. M. I. Posner. Cambridge: MIT Press. In press

Sejnowski, T. J., Koch, K., Churchland, P. S. 1988. Computational neuroscience. *Science* 241: 1299–1306

Sejnowski, T. J., Tesauro, G. 1989. The Hebb rule for synaptic plasticity: Algorithms and implementations. See Byrne & Berry 1989, pp. 94–103

Shatz, C. J., Stryker, M. P. 1978. Ocular dominance in layer IV of the cat's visual cortex and the effects of monocular deprivation. *J. Physiol.* 281: 267–83

Sillito, A. M. 1987. Synaptic processes and neurotransmitters operating in the central visual system: A systems approach. See Edelman et al 1987, pp. 328–72

Singer, W. 1988. Ontogenic self-organization and learning. In *Brain Organization and Memory: Cells, Systems, and Circuits*, ed. J. McGaugh. New York: Oxford Univ. Press. In press

Singer, W., Kleinschmidt, A., Bear, M. F. 1986. Infusion of an NMDA receptor antagonist disrupts ocular dominance plasticity in kitten striate cortex. *Soc. Neurosci. Abstr.* 12: 786

Stanton, P. K., Sejnowski, T. J. 1989. Associative long-term depression in the hippocampus: Induction of synaptic plasticity by Hebbian covariance. *Nature* 339: 215–18

Staubli, U., Lynch, G. 1987. Stable hippocampal long-term potentiation elicited by "theta" pattern stimulation. *Brain Res.* 435: 227–34

Stent, G. S. 1973. A physiological mechanism for Hebb's postulate of learning. *Proc. Natl. Acad. Sci. USA* 70: 997–1001

Sutton, R. S., Barto, A. G. 1981. An adaptive network that constructs and uses an internal model of its world. *Cogn. Brain Theor.* 4: 217–46

Tanzi, E. 1893. I fatti e le induzioni nell 'odierna istologia del sistema nervoso. *Riv. Sper. Freniatr. Med. Leg. Alienazioni Ment. Soc. Ital. Psichiatria* 19: 419–72

Tesauro, G. 1986. Simple neural models of classical conditioning. *Biol. Cybern.* 55: 187–200

Teyler, T. J., DiScenna, P. 1987. Long-term

potentiation. *Annu. Rev. Neurosci.* 10: 131–61

Tsien, R. Y., Zucker, R. S. 1986. Control of cytoplasmic calcium with photolabile tetracarboxylate 2-nitrobenzhydrol chelators. *Biophys. J.* 50: 843–53

Volper, D. J., Hampson, S. E. 1987. Learning and using specific instances. *Biol. Cybern.* 57: 57–71

von den Malsburg, C. 1973. Self-organization of orientation sensitive cells in the striate cortex. *Kybernetik* 14: 85–100

von den Malsburg, C., Bienenstock, E. 1986. Statistical coding and short-term synaptic plasticity: A scheme for knowledge representation in the brain. In *Disordered Systems and Biological Organization*, ed. E. Bienenstock, F. Fogelman Soulié, G. Weisbuch, pp. 247–72. Berlin/Heidelberg: Springer-Verlag

Walters, E. T., Byrne, J. H. 1985. Long-term enhancement produced by activity-dependent modulation of *Aplysia* sensory neurons. *J. Neurophysiol.* 5: 662–72

Wickens, J. 1988. Electrically coupled but chemically isolated synapses: Dendritic spines and calcium in a rule for synaptic modification. *Prog. Neurobiol.* 31: 507–28

Wiesel, T. N., Hubel, D. H. 1963. Single-cell responses in striate cortex of kittens deprived of vision in one eye. *J. Neurophysiol.* 26: 1003–17

Wiesel, T. N., Hubel, D. H. 1965. Comparison of the effects of unilateral and bilateral eye closure on cortical unit responses in kittens. *J. Neurophysiol.* 28: 1029–40

Wigström, H., Gustafsson, B. 1983. Large long-lasting potentiation in the dentate gyrus *in vitro* during blockade of inhibition. *Brain Res.* 275: 153–58

Wigström, H., Gustafsson, B., Huang, Y.-Y., Abraham, W. C. 1986. Hippocampal long-term potentiation is induced by pairing single afferent volleys with intracellularly injected depolarizing pulses. *Acta Physiol. Scand.* 126: 317–19

Williams, S., Johnston, D. 1988. Muscarinic depression of long-term potentiation in CA3 hippocampal neurons. *Science* 242: 84–87

Wood-Jones, F., Porteus, S. D. 1928. *The Matrix of the Mind.* Honolulu: Mercantile

Woodworth, R. S. 1921. *Psychology: A Study of Mental Life.* New York: Holt

Zador, A., Koch, C., Brown, T. H. 1990. Biophysical model of a Hebbian synapse. *Proc. Int. Joint Conf. Neural Networks 1990.* In press

Ziegler, H. E. 1900. Theoretisches zur Tierpsychologie und vergleichender Neurophysiologie. *Biol. Central.* 20: In press

Zucker, R. S. 1972. Crayfish escape behavior and central synapses. I. Physiological mechanisms underlying behavioral habituation. *J. Neurophysiol.* 35: 621–37

Zucker, R. S., Bruner, J. 1977. Long-lasting depression and the depletion hypothesis at crayfish neuromuscular junctions. *J. Comp. Physiol.* 121: 223–40

# SUBJECT INDEX

# CUMULATIVE INDEXES

## CONTRIBUTING AUTHORS, VOLUMES 9–13

# CHAPTER TITLES, VOLUMES 9–13